Contents

5.   STRESS

6.   DEFORMATION

7.   BEHAVIOUR OF ROCKS UNDER STRESS

STRUCTURAL GEOLOGY
Fundamentals and Modern Developments

Related Pergamon Titles of Interest

Books

CONDIE
Plate Tectonics and Crustal Evolution, 3rd edition

GREEN
Exploration with a Computer: Geoscience Data Analysis Applications

HARDAGE
Vertical Seismic Profiling Part A: Principles, 2nd edition

LISLE
Geological Strain Analysis: A Manual for the Rf/Ø Technique
Geological Structures and Maps: A Practical Guide

Journals

Computers & Geosciences

Exploration & Mining Geology

Journal of African Earth Sciences

Journal of Geodynamics

Journal of South American Earth Sciences

Journal of South-East Asian Earth Sciences

Journal of Structural Geology

Full details of all Pergamon publications/
free sample copy of any Pergamon journal available
on request from your nearest Pergamon office.

STRUCTURAL GEOLOGY

Fundamentals and Modern Developments

S. K. GHOSH

*Department of Geological Sciences,
Jadavpur University, Calcutta, India*

PERGAMON PRESS

OXFORD · NEW YORK · SEOUL · TOKYO

U.K.	Pergamon Press Ltd., Headington Hill Hall, Oxford OX3 0BW, England
U.S.A.	Pergamon Press, Inc., 660 White Plains Road, Tarrytown, New York 10591-5153, USA
KOREA	Pergamon Press Korea, KPO Box 315, Seoul 110-603, Korea
JAPAN	Pergamon Press Japan, Tsunashima Building Annex, 3-20-12 Yushima, Bunkyo-ku, Tokyo 113, Japan

First edition 1993

Library of Congress Cataloging in Publication Data
Ghosh, Santi Kumar.
Structural geology: fundamentals and modern
developments/S.K. Ghosh.—1st ed.
p. cm.
1. Geology, Structural.   I. Title.
QE501.G52 1993   551.8—dc20   92-31294

ISBN 0 08 041879 1 Hardcover

ISBN 0 08 041878 3 Flexicover

Printed in Great Britain by B.P.P.C. Wheatons Ltd, Exeter

TO RAJA

CONTENTS

8. FINITE HOMOGENEOUS DEFORMATION

9. PROGRESSIVE DEFORMATION

CONTENTS

CONTENTS

CONTENTS

CONTENTS

21. DUCTILE SHEAR ZONES

22. TIME RELATION BETWEEN
CRYSTALLIZATION AND DEFORMATION

Preface

THERE has been an explosive growth of structural geology from the nineteen fifties and especially in the last two decades. Apart from the formulation of detailed methods of macroscopic geometrical analysis in structurally complex terrains, significant developments have taken place in the fields of strain analysis, buckle folding, formation of ductile shear zones, faulting and recognition of deformation mechanisms in rocks, to name a few. I have attempted in this book to give a comprehensive account of these modern developments within the broad framework of the fundamental concepts and principles. Each chapter of the book begins with introductory and basic concepts, emphasizes the importance of critical field relations and microstructures for reconstructing the kinematic history, discusses the genetic aspects from observations of field work, experiments and theoretical analyses, and, wherever possible, identifies the areas of uncertainties. There is a small variation in the level of treatment in certain chapters. This was dictated partly by the necessity of keeping the book to a reasonable size and partly by the need to minimize an overlapping with recently published books. Thus, while theories of buckling and boudinage are treated in considerable detail, the chapter on strain measurement has been greatly condensed since this topic has been extensively discussed by Ramsay & Huber.

There is nowadays a tendency among students to concentrate solely on current literature and ignore earlier works. I would like to emphasize that many of the early studies are not merely of historical interest; they contain a wealth of detailed observations which still remain pertinent. I hope that the large number of references of both earlier and modern works would encourage the students to go through the earlier classic works from time to time.

The book could not have been written without the constant encouragement of Sheila and Abhik Ghosh and without the unstinting help of the group of structural geologists in the Jadavpur University. My heartfelt thanks go to D. Khan for the large number of line drawings and photographic enlargements and for his help in preparing the index, to S. K. Deb for typing the manuscript through successive stages of revision and to N. Mandal for critically reading the manuscript. I thank

Sudipta Sengupta for allowing me to use certain materials from her unpublished work and for her permission to reproduce photographs and field sketches of mylonites, natural and experimental deformed boudins, syntectonic migmatites and deformed mafic dykes in gneissic terrains. I am indebted to S. K. Chanda for his critical comments on the chapter of primary structures and for allowing me to take photographs of primary structures from his collection of specimens. I am grateful to Carl Jacobson for providing me with a set of photographs of microstructures some of which have been reproduced in Chapter 22. The photograph of Fig. 14.9 was provided by A. B. Roy. I also thank M. S. Paterson, S. M. Schmid, E. H. Rutter and R. B. Smith for their permission to reproduce figures from their published works. Lastly, I thank Peter Henn and David Dickinson of Pergamon Press for taking an active interest during the publication of the book.

A book of this kind owes much to numerous discussions with professional colleagues during field work. I cherish the memory of such shared experience of many years of field work with Kshitin Naha, Ashit Baran Roy and Sudipta Sengupta.

CHAPTER 1

Introductory Concepts

1.1. Geometric, kinematic, dynamic and historical aspects of structural geology

Structural geology deals with the deformation of rocks. After their formation, sedimentary and igneous rocks may remain undisturbed or are deformed to different degrees. A volume of rock may change shape, rotate bodily, fracture or be displaced from one place to another. Such changes may be visible to us, for example, by the tilting of horizontal strata, by development of folds in originally planar beds, by distortion of pebbles, fossils and mineral grains in rocks and by the fragmented character of an originally continuous bed. If these features are in a large scale, their three-dimensional forms are not visible to us; they can be studied only on the eroded surface of the earth. One of the primary objectives of structural geology is to determine three-dimensional forms of these structures mainly from observations on the surface. The first step of *structural analysis* of an area is to study the *geometry of the structures*, i.e. to study their morphology and orientation (or attitude), both by direct observation of small structures in the outcrop and by reconstruction of large structures.

The geometry of the structures may sometimes be used as what has been described by Wilson (1961, p. 425) as "tectonic weathercocks". In other words they may enable us to determine the magnitude, the direction and the sense of movement in the rocks. *Kinematic analysis* consists of a study of such movements. The movement may be a *translation*, a *rotation* or a *strain* (Fig. 1.1). A kinematic analysis can only be done after a thorough study of the geometry of the structures.

Dynamic analysis is concerned with determining the stresses within the mass of rocks or the forces at its boundary. This is usually attempted when an elastic deformation of the rocks is followed by fracturing. The behaviour of rocks during permanent deformation is complex; rocks at depth do not generally behave as a Newtonian liquid in which the stress is proportional to the strain rate. Thus, unless the relation between the stress and the strain or the strain rate is exactly known, a dynamic analysis cannot be made from the geometry and the kinematics. A major field of structural geology is, therefore, devoted to a study of the behaviour of different types of rocks under stress at high temperatures and pressures in the laboratory.

Apart from these three aspects, geometric, kinematic and dynamic, structural geology is also concerned with the *evolution of structures*. By this we do not only mean the history of development of structures from an initial to a mature stage, but also the history of overprinting of structures of an earlier generation by those of a later. The term *tectonics* is used in the same sense as structural geology when it refers not only to the geometry but also to the kinematics or the kinematic evolution.

1.2. Penetrative and non-penetrative structures, fabric of a tectonite

Within a particular domain of rocks the structures may be either *penetrative* or *non-penetrative*. When a structural element is present throughout the rock with an approximately even distribution, it is regarded as penetrative. Evidently, whether a structure is regarded as penetrative or not depends upon the scale of observation. A structure may be penetrative in the scale of a single outcrop but may be non-penetrative in the microscopic scale. Thus, for example, in the scale of an outcrop, the schistosity is a penetrative structure while slickensides (polished and striated surfaces of slip) are non-penetrative.

Within the rock, the small elements, such as mineral grains or groups of grains, may be arranged in a characteristic manner. This spatial arrangement is the *fabric* of the rock. There may be a *shape fabric* when only the external shape of the elements is considered. A *crystallographic fabric* refers to the spatial arrangement of crystal lattices.

Tectonite is a rock in which the fabric has been impressed by a deformation (Sander 1930). The fabric elements of a tectonite often show a *preferred orientation*. The preferred orientation may be either of the shape fabric or of the crystallographic fabric or of both. Thus, the platy minerals in a tectonite may be aligned sub-parallel to one another and give rise to a *cleavage* or *schistosity* (Fig. 1.2a). Similarly, a *mineral lineation* may develop by the preferred orientation of needle-shaped minerals (Fig. 1.2b).

1.3. Introduction to some common geologic structures

The geometry of a structure is described in terms of its *structural elements*. The structural elements may be either *planar* or *linear*.

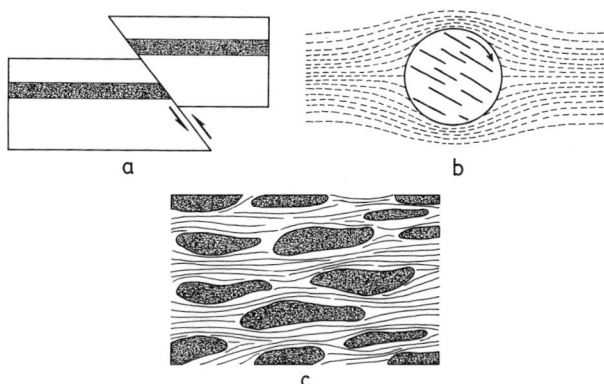

FIG. 1.1. (a) Faulting of a layered sequence. The upper block has undergone a translation along the fault. (b) The circular body represents the section of a spherical crystal of garnet in a matrix of mica schist. The orientation of the cleavage in the matrix is shown by thin dashed lines. The crystal of garnet has overgrown the cleavage and the traces of cleavage occur as trails of inclusions in it. After crystallization, the garnet has undergone a rigid body rotation as indicated by the angle between the trails of inclusions and the external cleavage. (c) Elongated grains of quartz (dotted) in a mica schist. The elongated shapes have resulted from strain.

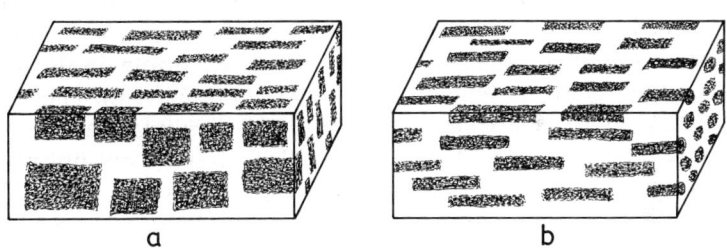

FIG. 1.2. (a) Schistosity, a type of planar structure, and (b) mineral lineation, a type of linear structure, marked by preferred orientation of platy and needle-shaped minerals.

Cleavage is a type of planar structural element and mineral lineation is a type of linear structural element. *Bedding* or *stratification* is another type of planar structure. This is a type of *primary structure*. These are formed by depositional or igneous processes and are present prior to the deformation. Although primary structures are not produced by deformation, their recognition is of great importance in the structural analysis of an area. The structures that are produced by deformation of rocks are *secondary structures*.

Bedding or any other initially planar structure serves as a reference plane whose change in orientation enables us to analyse the deformation of a domain of rock. In most deformed terrains the bedding is distorted to a wavy form. The resulting structures are known as *folds*

3

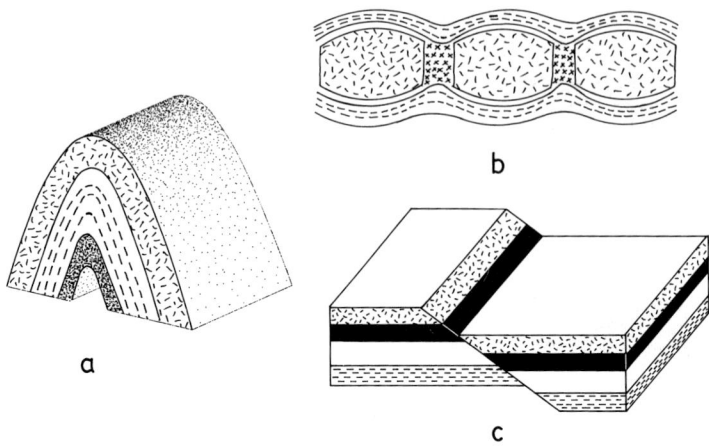

FIG. 1.3. (a) Fold in a layered sequence. (b) Transverse section of a boudinage structure. The central unit was an initially continuous bed which had been broken up into fragments or boudins under layer-parallel extension. (c) Faulting in a layered sequence.

(Fig. 1.3a). Folds, in general, are produced by a shortening parallel to the layering in the rocks. A structure may also develop by an extension or an overall lengthening of a layer. Thus, for example, a brittle layer under extension may break up into a number of fragments aligned in a row and give rise to a *boudinage structure* (Fig. 1.3b). Brittle rocks may also develop *faults*; these are fractures along which the two blocks on either side have moved past each other (Fig. 1.3c). Faulting may give rise to the movement of a block of rock through a distance of several kilometres. Such spectacular structures are known as *overthrusts*.

1.4. Symmetry of structures

The symmetry of a structure or a fabric is described in the same manner as in crystallography. A structure or a fabric has an axial symmetry when there is an infinite number of symmetry planes intersecting along an axis. A rock showing a mineral lineation marked by needle-shaped minerals possesses an axial symmetry (Fig. 1.4a). An orthorhombic symmetry is shown by structures which have two planes of symmetry at a right angle to each other. Thus, for example, a symmetrical fold as shown in Fig. 1.4b has an orthorhombic symmetry. A structure has a monoclinic symmetry when it has a single plane of symmetry. A fold with a monoclinic symmetry is asymmetrical in transverse section and has a single plane of symmetry normal to the fold axis (Fig. 1.4c). A structure with a triclinic symmetry does not have any plane of symmetry (Fig. 1.4d). The symmetry of a structure may be

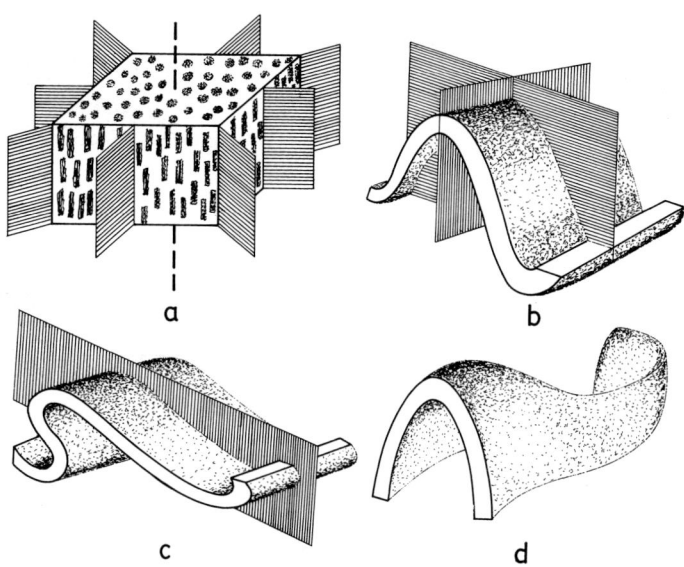

FIG. 1.4. (a) A rock showing a mineral lineation marked by needle-shaped minerals has an axial symmetry; any plane parallel to the lineation is a plane of symmetry. (b) A symmetrical cylindrical fold shows an orthorhombic symmetry with two mutually perpendicular planes of symmetry. (c) An asymmetrical fold shows a monoclinic symmetry with a single plane of symmetry perpendicular to the fold axis. (d) A fold with triclinic symmetry.

different in different scales. Thus, the symmetry of a large fold may be orthorhombic while the smaller folds on its flanks are monoclinic.

The symmetry of a structure or a fabric often reflects the symmetry of movement (Sander 1930, 1948, Knopf & Ingerson 1938, Paterson & Weiss 1961). However, this symmetry argument should be used with some caution. The geometry of the final structure of the rock is controlled by its initial configuration. Thus, depending on the initial orientation of the planar structures undergoing folding, the geometry of the folds may be either orthorhombic or monoclinic in an overall orthorhombic plan of deformation (Fig. 1.5). Similarly, a fabric may result from the superposition of two different events of deformation. The final fabric, containing elements of the earlier fabric, may not then reflect the symmetry of movement of either events of deformation.

1.5. Scale of structures

For all aspects of structural geology the scale of the observed structure should be clearly specified. There are several reasons for this. The methods of study of structures in different scales are not the same. The structures that are studied by a transmission electron microscope may

5

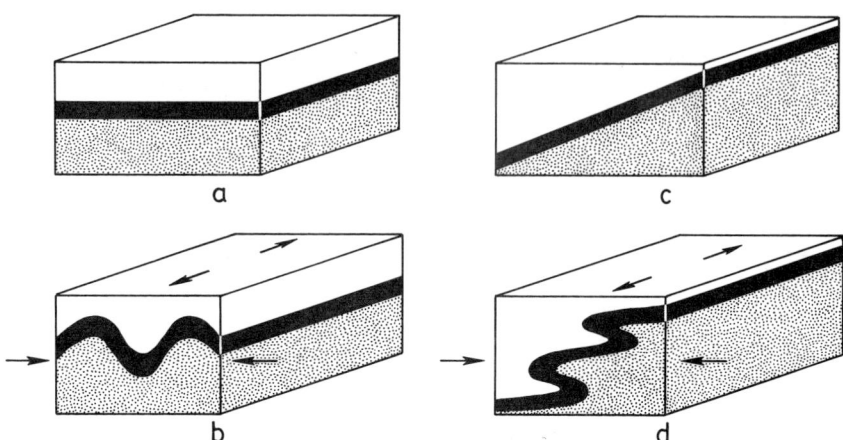

FIG. 1.5. The symmetry of a structure may not reflect the symmetry of movement in all cases. (a) and (b) show the development of a symmetrical fold with orthorhombic symmetry by a layer-parallel compression. (c) and (d) show the development of an asymmetrical fold with monoclinic symmetry from a layer oblique to the principal direction of shortening. The symmetry of bulk movement is orthorhombic in both cases.

be different and may require a different type of interpretation than the somewhat larger features that are seen under an optical microscope. Similarly, the methods of analysis of structures in the scale of a few kilometres are quite different from those that are used to study structures in the scale of a few decimetres. Moreover, the geometry of a structural element generally changes from place to place. If a structural element maintains a uniform orientation within a particular domain, then that domain is regarded as *structurally homogeneous* with respect to that element. The structures become more and more inhomogeneous as the size of the domain increases. Lastly, there is another reason for considering the scale of a geological structure. The gross behaviour of a body depends to a certain extent on its size. One can build a small cube of clay in the scale of a few centimetres, but a large cube of clay, say the size of a building, will collapse under its own weight. In a large structure, in the scale of a few kilometres or tens of kilometres, the force of gravity is an important factor. On the other hand, for rock structures in the scale of a hand specimen the force of gravity is negligible in comparison with the resistive forces (for instance, the viscous forces within the rocks at depth). Hence, while the force of gravity can be neglected in a theory of development of a small structure (such as a small fold), it must be taken into account in a theory about the mode of origin of a large structure (such as a fold spanning a few kilometres).

An earlier system of scales (Bailey 1935) used by alpine geologists consisted of the following divisions:

(1) *Microscopic scale.*
(2) *Small scale* or the scale of a hand specimen.
(3) *Intermediate scale* or the scale of a single outcrop in which the structure can be seen at one glance and there is no necessity of mapping.
(4) *Large scale* in which the structure is so large that it can be studied as a whole only by preparing a map.

Alternatively, we may use the following scale (Weiss 1959a):

(1) *Sub-microscopic scale.* In this scale the size of the structure is too small to be observed under an optical microscope.
(2) *Microscopic scale.*
(3) *Mesoscopic scale.* This is the scale of both the hand specimen and a single outcrop. Indeed, for most cases, the method of study of structures in small and intermediate scales are the same. Hence, most structural geologists use the term mesoscopic instead of distinguishing between the small and intermediate scales.
(4) *Macroscopic scale.* This is the scale of a map and is synonymous with large scale.

1.6. Geological and structural map

The first and the most important step of structural analysis of an area is to prepare a geological and structural map. All subsequent interpretations of the structures of the area will depend upon the observations recorded on this map.

Apart from mapping the lithological contacts, the structural geologist measures the attitudes of all types of planar and linear structures at suitable intervals and plots them directly on the map. The geometry of the mesoscopic structures at individual locations are described and sketched in the field notebook and the best examples are photographed. In comparatively simple areas the three-dimensional form of the structures can be represented by preparing a section along a suitable line in the map.

The preparation of a good geological map needs time and considerable skill. It is also an exciting experience to see the geology of the area slowly unfold as the mapping progresses. The detailed procedure of geological mapping (Lahee 1952) is outside the scope of the present book. The methods of measuring the attitudes of structural elements and of recording them on the map will be described in Chapter 3.

CHAPTER 2

Primary Structures

2.1. Importance of primary structures

In structural geology we are mostly concerned with those structures which are produced by deformation of the rocks. There are, however, certain structures which form by igneous processes or by certain mechanical processes during the time of sediment accumulation and before lithification; these are described as primary structures (Pettijohn 1975). In this chapter we shall be mostly concerned with primary sedimentary structures which are produced by the action of currents or waves or are produced by penecontemporaneous deformation.

An undeformed sedimentary terrain shows a succession of sub-horizontal beds. The individual bed or stratum may be of different thicknesses. A single bed may be massive but, more commonly, each bed shows a fine internal layering or lamination marked by slight differences in colour, composition or texture. The interfaces between these sedimentary layers represent depositional surfaces. The planes parallel to them are called *stratification*. In many terrains the stratification or bedding can be recognized even when the rocks are metamorphosed. Stratification in such rocks can be easily recognized from the presence of certain sedimentary structures such as cross-stratification, from the presence of a fine lamination parallel to interfaces of beds or when the earliest cleavage intersects the layering. On the other hand, in metamorphic terrains it is difficult to decide whether the colour banding parallel to cleavage represents a bedding or metamorphic banding.

The presence of stratification greatly facilitates the structural analysis of an area because the bedding and the interfaces of the beds provide us with a set of marker surfaces which were more or less planar and horizontal before the onset of deformation. In the absence of such markers, most faults would have gone undetected. Even where beds of mappable dimension are absent, the attitudes of bedding can give us significant information regarding the nature of deformation.

The large-scale structure of an area can be extremely complex; the beds may be distorted into complex forms, and large masses of rocks may be transported on a low angle fault surface to a distance of several kilometres or tens of kilometres. The recognition of the large-scale structures and of the structural history of such an area is virtually impossible without a thorough knowledge of the stratigraphic succession of the rocks. Indeed, as the geological investigations in the Alps have shown, the structural and stratigraphic studies in such areas must go hand in hand, each helping the other. Even where the stratigraphic ages of the rocks are not known it may be possible to determine their relative ages from certain sedimentary structures which indicate the *facing* or the *younging* direction of the strata, i.e. the direction in which successively younger rocks occur. The younging directions enable us to find out, for example, whether the rocks occur in the normal order of superposition or are inverted by folding or whether the outcrops of two parallel beds of the same lithology are of different ages or are repetitions of the same bed by isoclinal folding.

The character of the sediments and the occurrence of certain types of sedimentary structures enable us to have a rough idea about the depth of water in which the sediments were deposited. In many regions we find very thick prisms of shallow water sediments. This clearly indicates that the floor of the depositional basin subsided during sedimentation. In coal basins, in particular, where sub-surface data are plentiful because of drilling and mining operations, the sedimentary record may enable us to determine the pattern of variation of the vertical movement of the basin floor and to record the history of vertical movements (Krumbein & Sloss 1963). For instance, in some coal basins the thickness of some of the coal seams is much greater on the hanging wall side of the normal faults than on their foot wall side, indicating thereby that the faulting was broadly contemporaneous with sedimentation. The sedimentary record may also show that the movement on the major faults was not continuous but intermittent.

2.2. Structures that indicate facing

For geometrical analysis of large-scale folds, it is essential to know whether the beds in an outcrop are facing upward or are overturned. In the absence of diagnostic fossils, we have to depend entirely on sedimentary structures to determine the facing direction. In areas where stratigraphic inversion is likely to have occurred, either regionally or locally, the facing should be noted during field work and should be plotted in the map along with other structural data.

One of the commonest indicators of facing in non-cohesive granular sediments is the *cross-stratification* (or *current bedding* or *cross-*

lamination) in which the internal laminations of a bed are inclined to the general bedding or the principal surface of deposition (Fig. 2.1a). The inclined internal layers of the bed are described as foreset layers. A cross-bedded unit, a sandstone or quartzite or grainy limestone, may range in thickness from a few millimetres to a few tens of metres but is commonly less than a metre in thickness. A cross-bedded unit may be tabular or trough shaped. In the tabular type the foreset layers meet the general bedding plane roughly along straight lines while in the trough type of cross-bedding the foreset laminae intersect the bedding plane along curved lines. In vertical sections parallel to the current directions the foreset layers may be planar or sigmoidally curved or are concave upward (Fig. 2.1a). In the last type the cross-laminations are tangential to the general bedding at the bottom of the bed while at the top of the unit the cross-laminations are truncated by the general bedding. The truncation resulted from current-induced erosion before the deposition of the next set of strata. It is because of the presence of this truncation surface that we are able to determine the facing of the bed.

In subaqueous deposits the cross-bedded structure develops due to migration of current-induced sand waves which have very gentle slopes against the current direction and have steeper slope on the lee side. The angle of inclination of the cross-laminations on the lee side cannot significantly exceed 30°, the angle of repose of sand. Larger angles between the foreset laminae and the general bedding therefore invariably indicate subsequent deformation.

Ripple mark is another common sedimentary structure which is frequently used to determine the facing. Ripples are of two types: *current ripple marks* and *wave ripple marks* (Figs 2.1, 2.2). When a current of water flowing over a sandy bottom exceeds a certain velocity, the surface of the sand is thrown into a series of asymmetric waves of more or less uniform wavelength. The waves have gentle slopes against the current direction and have steeper slope on the lee side (Figs 2.1b, 2.2a). Current ripple marks therefore indicate the direction of palaeocurrent; it is at a right angle to the crest line or trough line of a ripple and is towards the steeper slope of the ripple. Unfortunately, the shape of a current ripple does not usually have any feature which characterizes the top or the bottom of a ripple. Hence, a current ripple mark and its inverted casts are alike and cannot be distinguished. Although the trough of the ripple can sometimes be identified by an accumulation of heavy minerals, the current ripple mark is generally useless for determination of younging direction.

The ripples on the sandy bottom usually migrate down the current direction. This causes the development of miniature cross-laminations. While the ripple marks appear as a surface feature, the fine cross-laminations (Fig. 2.3) usually appear as an internal feature (Pettijohn

FIG. 2.1. (a) Cross-lamination in Pakhal limestone of Pranhita-Godavari Valley, India. (b) Current ripple mark and (c) wave ripple mark in Vindhyan sandstone, Maihar, India.

1975). The curved cross-laminations are truncated at the top by the general bedding or by the overlying set of cross-laminated unit (Fig. 2.2b) and can therefore be used to determine the facing direction.

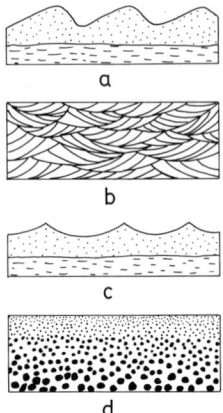

FIG. 2.2. (a) Current ripple mark, (b) ripple cross-lamination, (c) wave ripple mark and (d) graded bedding.

Wave ripple marks are produced on the shallow sandy bottom of a standing body of water agitated by oscillatory motion at the water surface. The wave ripple marks often have sharp crests and rounded troughs. The shapes of wave ripples are therefore good indicators of the younging direction (Figs 2.1c, 2.2c).

Some sandstone beds show a concentration of the coarsest material at the bottom, a gradual reduction of grain size as we go upward and with the finest grains occurring at the top of the bed. The structure resulting from this graded succession of grain size is known as *graded bedding* (Fig. 2.2d). The occurrence of such graded beds indicates the stratigraphic order. Graded beds often occur in succession, one above the other. Although graded bedding has been reported from different types of rocks, it is most common in deep sea sandstones known to have been emplaced by turbidity current (Kuenen & Migliorini 1950). Although inverse grading (Blatt *et al.* 1980), with upward coarsening of grain size within a bed, is known to occur, it is rather rare.

Cross-stratification, ripple mark and graded bedding are the three most important way-up indicators. There are in addition a number of less common indicators. Some of these are described below.

Sole marks are produced as elongate depressions on the firm surface of mud and which is later covered by sand. The sole mark therefore appears as casts on the under surface of the sandy bed. The younging direction is obtained from our knowledge that these structures develop only on the bottom surface of a sandy bed. There are two main types of sole marks, the *flute casts* and the *groove casts* (Fig. 2.4a, b). The flute casts are elongate and somewhat triangular in shape, with a bulging nose pointing up-current and which widen and flatten out in the direction of current flow. Flute casts are therefore often used to determine

FIG. 2.3. Ripple cross-lamination in micaceous quartzite, Ghatsila, India.

the direction of palaeocurrent. Flute casts are produced by the scouring action of eddies on a surface of mud. The groove casts are elongate structures which are parallel to the direction of current flow, although the sense of flow is not indicated by them. The groove cast (Fig. 2.4b) is thought to be one type of *tool mark*, i.e. a mark produced by the down-current movement of a relatively coarse object along a muddy surface.

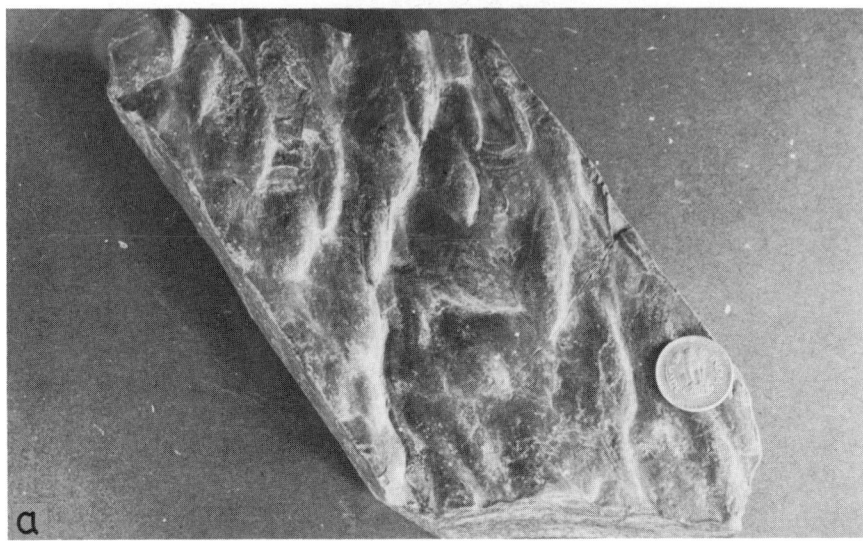

Fig. 2.4. (a) Flute cast in siltstone. (b) Tool marks in fine-grained sandstone. The continuous mark on the right-hand side in (b) is a groove cast. The observed surfaces of both specimens are bottom surfaces. Pakhals, Pranhita-Godavari Valley, India.

Unlike flutes and groove casts, *load casts* are produced by the sinking of a freshly deposited bed of sand as a series of pockets into a soft mud. The downward closures of the pockets are generally broad or rounded, while the upward closures are narrow and sometimes flame-shaped (Fig. 2.5a). Unlike the flutes and grooves which cut through the bedding of the

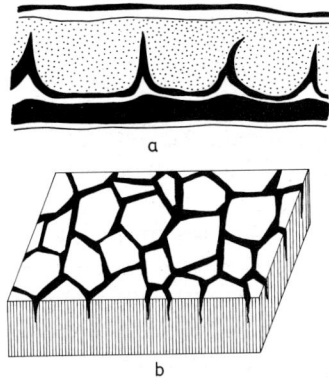

FIG. 2.5. (a) Load cast of sandstone in shale. The load casts have sharp upward-pointed crests and lobate troughs. (b) Mud cracks. In vertical section the cracks taper downward.

underlying shale, the bedding in both the sandy and the shaly units are deformed during the formation of the load structure. The load casts give us the younging direction because they form at the basal part of sandstone beds. In certain cases the shape of the load cast, with its broad or rounded downward projection into the shale and the narrow or sharp upward projections, can also be used to determine the direction of facing.

In comparison with the sedimentary structures described above, *mud cracks* or *sun cracks* (Fig. 2.5b) are rather rare. When a clay-rich surface of deposition is exposed to the air, a polygonal network of cracks appears due to shrinking of the clay during drying. If the surface is once again submerged and is covered by sand, the cracks are filled up by the sand. The shape of the mud cracks is then preserved as casts on the undersurface of the bed of sandstone. In sections normal to the depositional surface, the mud cracks taper downward. The direction of younging can be determined from this feature.

Stromatolite, an organo-sedimentary structure, occurs in certain late Precambrian carbonate rocks. A common variety of stromatolite grows as columns normal to the bedding and has upward convex fine laminations which indicate the direction of younging. In folded terrains the angle between the columns and the general bedding does not remain a right angle. The change in the right angle can then give us both the magnitude and the sense of bedding-parallel shear strain. The stromatolitic structure is often very well preserved even when the rocks are metamorphosed.

Shells of lammelibranchs and brachiopods which roll on the depositional surface come to lie in their stable position with the convex side pointing upward. Burrows and borings are often U-shaped and concave upward. The casts of animal trails are also concave upward. Although such features indicate the facing, they are rarely noticed and are seldom used in the structural study of deformed rocks. Structures produced by penecontemporaneous deformation may also indicate the top direction of the strata. These will be described in some detail in the next section.

Before leaving this section, we should note that the stratigraphic order can be determined from the primary structures of some lava flows. Some lavas are consistently *vesicular* towards the top of the flow, while the pillows of the *pillow lavas* have upward convex surfaces. For a detailed description of primary structures which indicate the facing of a sedimentary sequence, the reader should consult Shrock (1948).

2.3. Structures produced by penecontemporaneous deformation

There is a variety of sedimentary structures which form by deformation of sediments after their deposition but before consolidation and while the sediments are still in their depositional environment. The deformation is generally described as penecontemporaneous. There is no satisfactory classification of this rather large group of structures. Moreover, the modes of development of many of these interesting structures are not well studied, perhaps because the processes involved are in the field intermediate between those of sedimentary geology and structural geology.

Overturned cross-bedding (Fig. 2.6) is a common type of structure formed by penecontemporaneous deformation. As we have seen, the angle between the general bedding and the foreset layers cannot exceed 30–34°, the angle of repose of sand. In one type of penecontemporaneous deformation the upper part of the foreset layers are steepened and even overturned by a bedding-parallel shearing movement the sense of which is invariably in the direction of current flow. That the deformation is penecontemporaneous is further indicated by the fact that it is intrastratal and the bed as a whole is not involved in this deformation. The intensity of the penecontemporaneous deformation is often so large that the foreset layers are deformed to isoclinal folds.

Convolute lamination is the general term to include all types of intrastratal folds of penecontemporaneous origin. The individual layers can generally be traced through the train of folds. In many types of convolute lamination the synclines are broad and rounded while the anticlines are narrow or sharp-hinged (Fig. 2.7). Moreover, convolute laminations often show a sharp truncation of the structures at an

FIG. 2.6. Overturned cross-lamination (in the uppermost unit) in quartzite from Ghatsila, India.

erosion surface at the top. Convolute laminations, therefore, help us to identify the top direction of the beds. In general, the convolute folds are not cylindrical structures and the contortions are visible in all sections perpendicular to the bedding. In sections sub-parallel to the general bedding, they often show closed outcrops of layers and no preferred trend of the axial traces. The intrastratal folds may have diverse shapes in vertical sections, and may range from recumbent to upright; however, the contortions on the outermost or lowermost layers are often roughly symmetrical about the axial surface traces perpendicular to the trace of the general bedding (Ghosh & Lahiri 1983, 1990). This suggests that at least certain types of convolute laminations are produced by vertical movements in unconsolidated sediments.

The development of convolute laminations may, in certain cases, be associated with some lateral movement. On the other hand, the pene-contemporaneous folds which are produced by a significant amount of unidirectional lateral movement are mostly *slump folds*. The slump structures result from downslope sliding of a mass of unconsolidated sediments. If the lateral movement is not very large, the slump folds appear as a train of décollement folds. These may show a consistent sense of overturning in the downslope direction. Where a consistent overturning is absent, the geometry of the folds may be difficult to distinguish from convolute laminations. However, unlike convolute laminations, slumping generally affects a number of beds and the slumped unit is usually much thicker than a unit showing convolute laminations. Indeed, a slump sheet may even be of mappable dimension (e.g. Jones 1937). The slump folds usually have identifiable hinge

FIG. 2.7. Convolute laminations in micaceous quartzite from Ghatsila, India. The shapes of the convolutes have been modified by diastrophic movement.

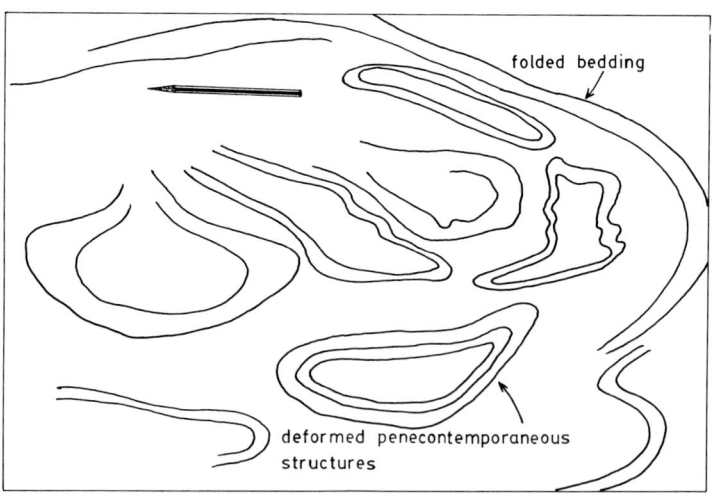

FIG. 2.8. Closed outcrops of penecontemporaneous contortions on a horizontal surface. The contortions have been deformed by diastrophic folds. The trace of axial plane cleavage, the earliest cleavage in this area, is parallel to the pencil. Mica schist in Ghatsila, India.

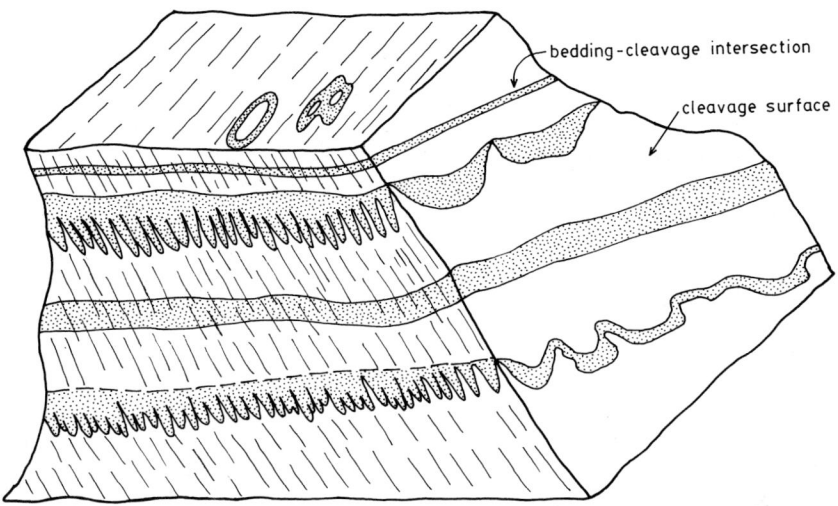

Fig. 2.9. Bedding and cleavage occurring at a high angle in an outcrop of mica schist. The trace of bedding on the cleavage shows a gentle plunge. The penecontemporaneous contortions, as seen on the cleavage surface, are open. The contortions have been greatly tightened on a plane (the front face of the outcrop) at a large angle to the bedding–cleavage intersection. Ferry point, Ghatsila, India.

lines, although their trends may show considerable variation. If the lateral movement of slumping is large, the individual layers may be torn apart and the separate segments are more or less independently folded. Slumping may also cause a partial liquefaction; the resulting slump structure may show chaotic folds of the separate torn fragments or even a conglomerate- or breccia-like feature. It should be noted that both convolute laminations and slump folds often have strongly curved hinge lines or may be dome-and-basin-like structures. An arbitrary section through such structures may show closed outcrops of the bedding laminae. Hence, without a three-dimensional analysis the structures should not be described as "balls".

Where occurring in association with tectonic structures (Fig. 2.8), the penecontemporaneous structures may be difficult to identify. If they are confused with tectonic structures, the structural interpretation may be quite misleading. The penecontemporaneous structures can be distinguished from tectonic structures by the following features:

(1) Convolute laminations are intrastratal and the bed as a whole or the bounding layers do not share the same deformation.

(2) Penecontemporaneous structures are often truncated by an erosion surface.

(3) The morphology of the penecontemporaneous contortions is usually quite distinct from that of the much more abundant tectonic

folds of an area. In areas in which the intensity of later deformation is large, the shapes of the penecontemporaneous folds are greatly modified (Fig. 2.8) and in sections perpendicular to the regional fold axis the axial traces of the deformed contortions may become sub-parallel to the traces of the cleavage (Fig. 2.9). The contortions and their intra-stratal nature can, however, be identified in outcrop faces parallel to the cleavage; in such sections the traces of the contortions are visible, while traces of the general bedding (parallel to the tectonic fold axis) remain uncontorted. In other outcrop faces also, the penecontemporaneous folds may be distinguished by their anomalous styles. Figure 2.9, for instance, is from an area where the diastrophic folds are known to be cylindrical and gently plunging and hence the elliptical closures can be identified as penecontemporaneous folds, the shapes of which can also be clearly distinguished in sections parallel to the axis of the diastrophic folds.

(4) Where the shapes of the penecontemporaneous folds are not so greatly modified, they can be easily identified when their axial surface traces are cut across by the traces of the earliest cleavage of the area.

CHAPTER 3

Structural Elements and Their Attitudes

3.1. Planar and linear structures

Geologic structures are described in terms of a few simple geometrical elements. These structural elements are of two types: *planar structures* and *linear structures*. The most commonly occurring penetrative planar structures are the stratification or bedding and the cleavage. The trace of bedding on the cleavage or the trace of the cleavage on the bedding constitutes an intersection lineation. This is a type of linear structural element. As mentioned in Chapter 1, another common type of linear structure is the mineral lineation marked by preferred orientation of elongate grains. The planar structures are often referred to as S-planes and the linear structures by the symbol L. Thus, rocks which show a mineral lineation lying on a cleavage are sometimes described as L-S tectonites. S-tectonites are those which show only a single dominant cleavage and L-tectonites show a single dominant lineation.

3.2. Attitudes of planar and linear structures

The attitude of a linear structure is described by its *trend* and *plunge*. Take a line parallel to the linear structure and, from any two of its points, drop normals to a plane. The normals will intersect the plane at two points. The line joining them is the projection of the linear structure on the plane. The trend of a linear structure is the direction of its projection on a horizontal plane (Fig. 3.1). Alternatively, imagine a vertical plane containing the linear structure. The geographic direction of a horizontal line on this plane represents the trend of the linear structure. Of the two directions towards either side of this horizontal line, the direction in which the lineation points downward is taken to be its trend. The angle between the linear structure and the same horizontal line is the plunge of the linear structure (Fig. 3.1).

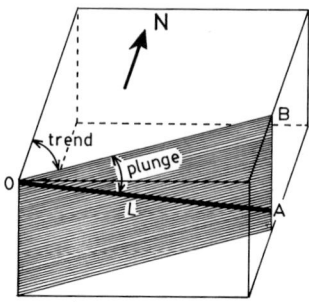

FIG. 3.1. Definition of trend and plunge of a linear structure (*L*). *OBA* is a vertical plane through *L*, with *OB* as a horizontal line lying on the plane.

The attitude of a planar structure is represented by its *strike* and *dip angle* or by its dip and the *dip direction*. Dip is the angle between the planar structure and a horizontal plane. The strike is the geographic direction of a horizontal line occurring on the planar structure. The dip direction refers to the geographic direction of a horizontal line at a right angle to the strike and towards the downward inclination of the planar structure (Fig. 3.2).

3.3. Trend, plunge, dip angle and dip direction as spatial coordinates

The orientation of a line or a plane is represented in coordinate geometry with reference to a coordinate system. The same procedure is followed in geology. Here we take the vertical, the N–S and the E–W directions as the coordinate axes. Let *OX*, *OY* and *OZ* be three mutually perpendicular axes with *OX* parallel to north, *OY* parallel to east and *OZ* as a vertical line (Fig. 3.3). Let *P* be a point in space. *PM* is a normal to the *XY*-plane, with point *M* lying on this plane. θ is the angle between *OM* and the *X*-axis and Φ is the angle between *OP* and the *Z*-axis. The position of the point *P* can be represented by θ, Φ and the distance $r = OP$. These are the spherical coordinates of the point *P*. In representing the attitude of a structural element we do not require to specify its dimension; hence, the coordinate *r* is not required. The attitudes of planar and linear structures are represented by the coordinates θ and Φ.

Let *OP* represent the linear structure, the point *P* lying *below* the *XY*-plane. (It should be noted that a field geologist sees the structures at the ground surface and always keeps in mind that the structures are extending down towards the depth.) The coordinate θ is the trend of the linear structure. The plunge is given by the angle $90° - \Phi$ (Fig. 3.4).

The attitude of a plane is represented in coordinate geometry by the coordinates of its normal. Let *ABCD* represent the planar structure

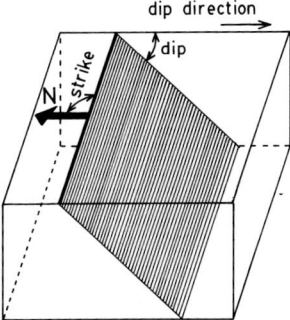

FIG. 3.2. Definition of strike, dip and dip direction of a planar structure.

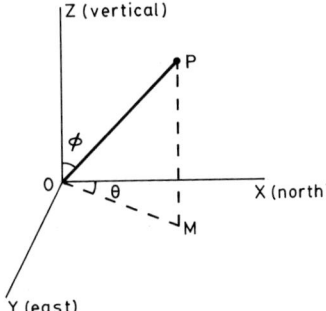

FIG. 3.3. With the north, the east and vertical line as coordinate axes, the orientation of the line OP is represented by the two angles θ and Φ. PM is perpendicular dropped from the point P on the horizontal plane. θ is the angle between X and OM and Φ is the angle between Z and OP.

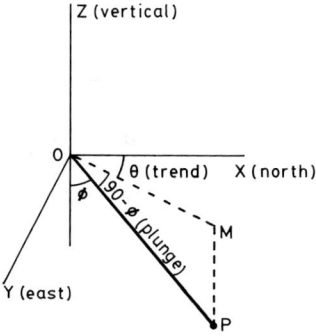

FIG. 3.4. Trend and plunge of a linear structure defined as spatial coordinates with the vertical, the east and the north directions as coordinate axes.

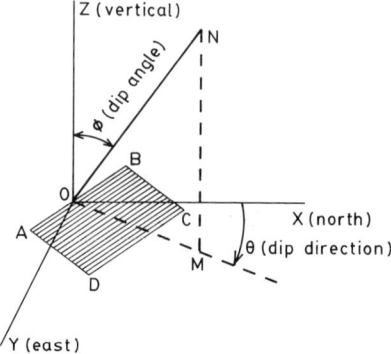

FIG. 3.5. *ON* is the upward directed normal to the planar structure *ABCD*. Φ and θ are its dip and the dip direction.

and *ON* its *upward* directed normal. Its spherical coordinates θ and Φ are the dip direction and the dip angle of the planar structure (Fig. 3.5).

3.4. Measurement of attitudes in the field

The two angles θ and Φ are directly measured in the field with the help of a *clinometer compass*. There are different models of the clinometer compass; the basic principle of construction is similar in all of them. The dial of the compass has an outer circle in which the geographic directions are marked out from 0 to 360°. There is in addition a stylus or clinometer hanging from the centre of the compass dial (Fig. 3.6). When the dial is held in a vertical position, the stylus hangs freely and acts as a plumbline indicating the vertical direction. There is a straight edge in each clinometer compass perpendicular to the plane of its dial. There is also an inner scale in the dial showing 90° divisions on either side of the zero mark. The zero mark points towards the perpendicular to the straight edge.

We defined the dip angle as the angle between the planar structure and a horizontal plane. Since the angle between two planes is defined by the angle between their normals, the dip angle is given by the angle between a vertical line and the normal to the planar structure. The clinometer compass directly measures this angle. To measure the attitude of a planar structure, the straight edge of the compass is placed on it and the compass is rotated till the stylus hangs freely and indicates that the dial is vertical. At this position the zero mark of the inner circle lies along the normal to the planar structure. Since the stylus now points towards the vertical direction, the angle between the vertical and the normal to the plane is directly given by the position of the stylus at the inner scale. This is the dip angle. In this position of the clinometer

24

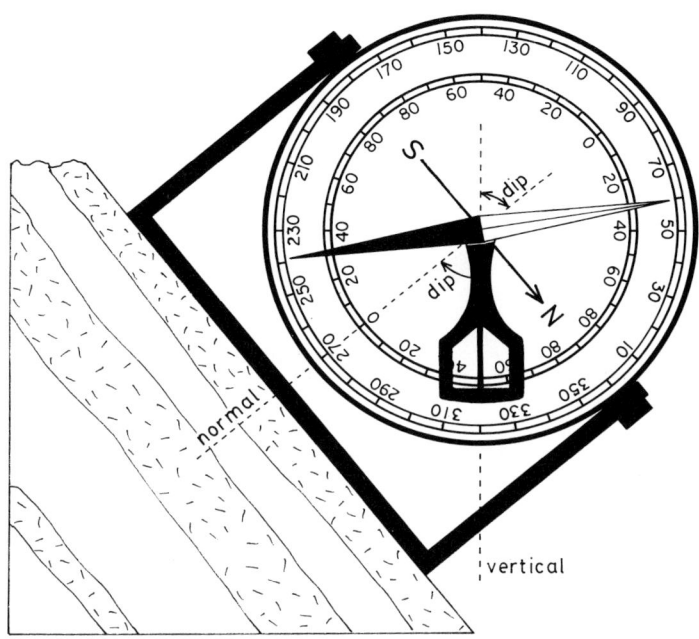

FIG. 3.6. The dip angle Φ, as shown in Fig. 3.5, can be measured directly by means of a clinometer compass. The bridge of the compass is placed on the planar structure aligned parallel to the line of dip (perpendicular to the strike line), and the compass dial is kept in a vertical position. The 0 mark of the inner dial is then parallel to the normal of the planar structure while the freely hanging stylus gives the vertical direction. The angle between these two lines is read off as the dip angle from the position of the stylus at the inner scale.

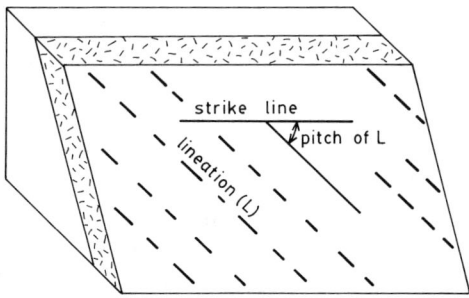

FIG. 3.7. Pitch of a lineation (L) on a planar structure.

compass a line (preferably with a coloured pencil) is drawn on the surface of the planar structure parallel to the straight edge of the compass. This line is down the dip of the planar structure. Another line is

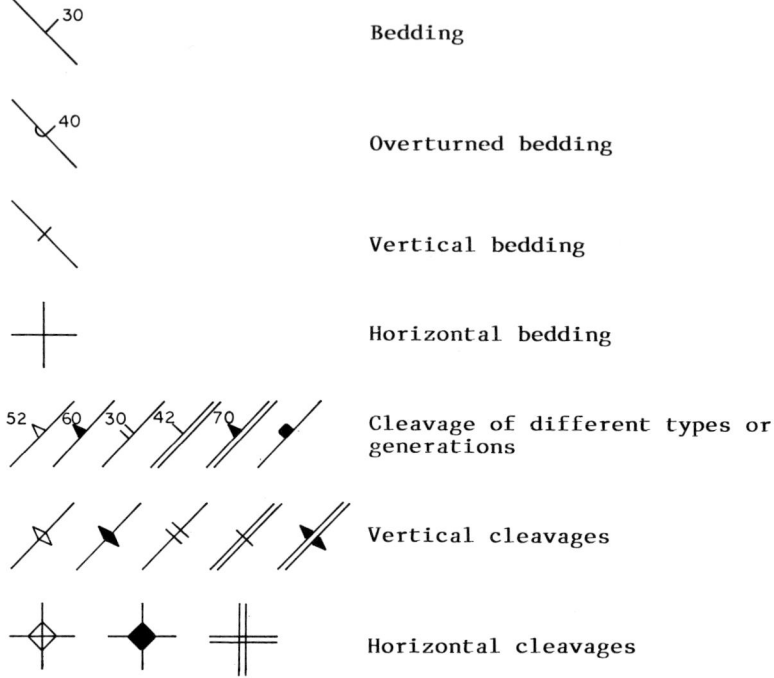

	Bedding
	Overturned bedding
	Vertical bedding
	Horizontal bedding
	Cleavage of different types or generations
	Vertical cleavages
	Horizontal cleavages

FIG. 3.8. Map symbols of planar structures.

then drawn perpendicular to it. This latter line is a horizontal line. The straight edge of the compass is placed parallel to this line and the compass dial is held in a horizontal position so that the magnetic needle swings freely. The reading at the outer scale gives the strike direction. Let us say, we had measured the dip angle as 48° and the strike as 150°. To record the attitude of the planar structure in terms of strike and dip we must now give additional information to indicate whether the dip is towards a north-easterly or south-westerly direction. The attitude is recorded in the following manner: 150/48NE.

To measure the attitude of a linear structure the straight edge of the compass is placed on the outcrop parallel to it and the compass dial is held vertically. The angle of plunge is directly given by the clinometer reading at the inner scale. To measure the trend, the compass is held in a horizontal position and rotated to bring the linear structure in a line with the straight edge while looking vertically downward. While aligning the straight edge one should be careful to keep the N-mark of the dial towards the direction of plunge. In this case, the trend can be read off at the position of the north of the compass needle at the outer scale.

STRUCTURAL ELEMENTS AND THEIR ATTITUDES

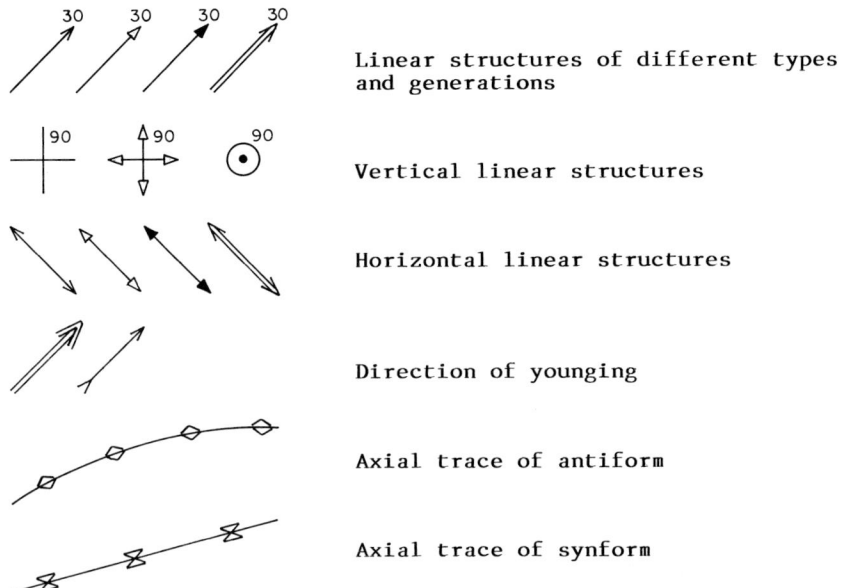

Linear structures of different types and generations

Vertical linear structures

Horizontal linear structures

Direction of younging

Axial trace of antiform

Axial trace of synform

FIG. 3.9. Map symbols of linear structures and of axial surface traces.

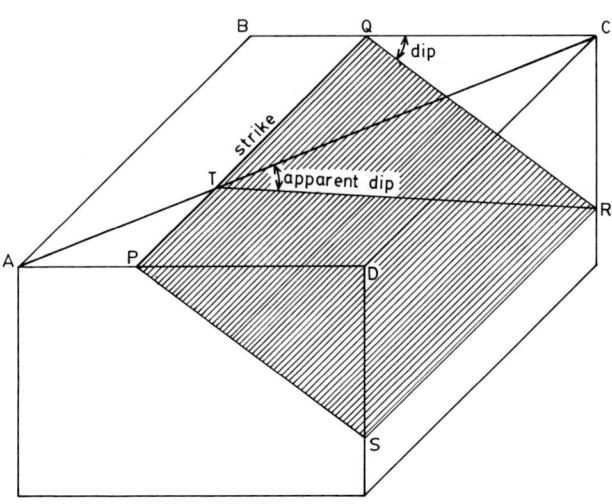

FIG. 3.10. *PQRS* is a planar structure with *PQ* as the strike line. *AC* is a line at an acute angle to the strike. *TCR* is a vertical plane passing through *AC*, with *TR* as its line of intersection with the planar structure. The apparent dip along *AC* is the angle between *AC* and *TR*.

Different symbols are used to represent the attitudes of different kinds of planar and linear structures in the map. These are shown in Figs 3.8 and 3.9.

3.5. The pitch of a line on a plane

It is difficult to measure accurately the trend of a steeply plunging linear structure. Hence, unless the lineation has a gentle plunge it is always recommended to measure its *pitch*. However, the pitch can be measured only when it lies on a plane. The pitch of a line on a plane is defined as the angle between the line and the strike of the plane (Fig. 3.7). The pitch is given as an acute angle; we must indicate from which side of the strike direction the pitch has been measured. As shown in Chapter 4, when the attitude of the plane and the angle of pitch of the line on it are given, the trend and plunge of the lineation can be easily determined from stereographic projection.

3.6. Apparent dip

The surface of a planar structure may not be exposed in all outcrops. We may still be able to determine its attitude if the trace of the planar structure is exposed in differently oriented vertical sections. Evidently, the trace will be horizontal if the section plane is parallel to the strike of the planar structure. The inclination of the trace of the planar structure will increase as the line of section makes a larger angle with the strike. The inclination will be equal to the true dip only if the line of section is at a right angle to the strike line (i.e. along the dip direction). The *apparent dip* is the inclination of the trace of the planar structure on a vertical section that makes an angle with the dip direction (Fig. 3.10). We shall see in Chapter 4 how to determine the attitude of a bed from its apparent dips with the help of stereographic projection.

Stereographic and Equal Area Projections

4.1. Principles of stereographic projection

Stereographic projection is an essential tool of structural geology. It is used to represent the orientations of planar and linear structures of an area and to find out the angular relations among them. Consider a reference sphere with a horizontal equatorial plane and with a vertical diameter intersecting the sphere at its upper and lower poles A and B. A line L passing through the centre of the sphere will meet the lower hemisphere at a point P (Fig. 4.1). A line joining the point P with the upper pole A will meet the equatorial plane at some point P'. P' is then the lower hemisphere stereographic projection of the line L. In a similar way, a plane S passing through the centre of the sphere will intersect the surface of the sphere in a *great circle*. If each point of this great circle in the lower hemisphere is joined to the upper pole or zenith A, the lines will intersect the equatorial plane at points which will lie on a circular arc. This arc is described as a great circle trace. This is the lower hemisphere stereographic projection of the plane S (Fig. 4.2). We could no doubt project in the same way the lines and points of the upper half of the reference sphere on the equatorial plane. However, in structural geology we always use lower hemisphere projections. A circular cone with its apex at the centre of the reference sphere intersects the sphere along a circle. Such circles are known as small circles. If each point of a small circle on the lower hemisphere is joined with the upper pole A, the lines intersect the equatorial plane at points lying along a circle. This too is referred to as a *small circle*. The centre of this small circle is the stereographic projection of the cone axis and the points on the circle are the projections of lines that make a constant angle with the cone axis (Hilbert & Cohn-Vossen 1956, p. 248).

4.2. Stereographic and equal area nets

The equatorial plane of the reference sphere appears as a circle in stereographic projection. It is known as the *primitive circle*. Since this is

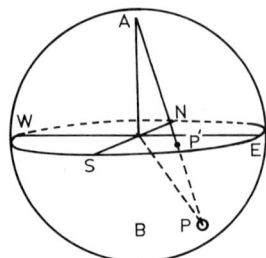

Fɪɢ. 4.1. Principle of lower hemisphere stereographic projection. A line passing through the centre meets the lower hemisphere at *P*. The line joining *P* and *A*, the upper pole of the sphere, meets the equatorial plane at *P'*. This is the stereographic projection of the line.

a horizontal plane, we can mark out the north direction on this plane. Any point occurring at the periphery of this circle represents a horizontal line. As the plunge of the linear structure increases, its point of projection moves closer to the centre. A vertical line projects as a point at the centre of the primitive circle. Figure 4.3a shows a vertical E–W section through the reference sphere. A linear structure plunging 30° towards W meets the lower hemisphere at a point *P*. The line joining *P* and the upper pole *A* intersects the E–W trace of the equatorial plane at the point *Q*. *Q* is the stereographic projection of the linear structure (Fig. 4.3b). A planar structure (*S*) striking N–S and dipping 30°W will appear as a great circle trace in stereographic projection. This is a circular arc which must pass through the N and the S points of the equatorial plane and through the point *Q* on the E–W diameter of the primitive circle (Fig. 4.3b). In the same way we can plot the attitude of any plane in stereographic projection. In actual practice, however, the plotting of planar and linear structures is done with the help of a *stereographic net* or a *Wulff net* (Fig. 4.4a), which shows the great circle traces of a series of planes each striking N–S and dipping either E or W through 0–90° at 2° intervals. In addition, the net also shows a series of small circles about the N and the S poles. The small circles are also at 2° intervals.

It should be noted that equal areas on the surface of the reference sphere may not remain equal in stereographic projection. For this reason the stereographic projection is not suitable when we want to compare the concentration of the plotted points from one part of the projection diagram to another part of it. To enable us to make such comparisons, we generally use the *equal area projection*. The geometrical principles of such a projection will not be discussed here. However, on the basis of such geometrical principles an *equal area net* or *Schmidt net* can be constructed (Fig. 4.4b). The Schmidt net looks quite similar to the stereographic net. It has, like the stereographic net, a series of

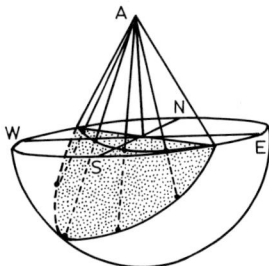

FIG. 4.2. The dotted area is a plane passing through a sphere. It meets the lower hemisphere along a semicircular arc. If each point on this arc is joined to the upper pole *A*, the lines intersect the equatorial plane at points which define a great circle on the stereographic projection.

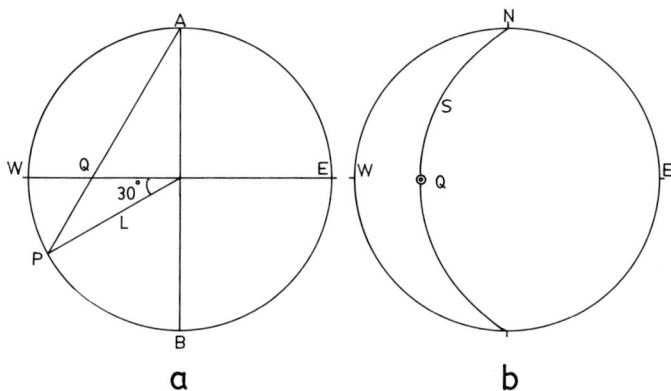

FIG. 4.3. (a) Vertical E–W section through reference sphere. (b) *Q* is the lower hemisphere stereographic projection of a line *L* plunging 30° towards W. The great circle trace is that of a plane (*S*) dipping 30° towards W.

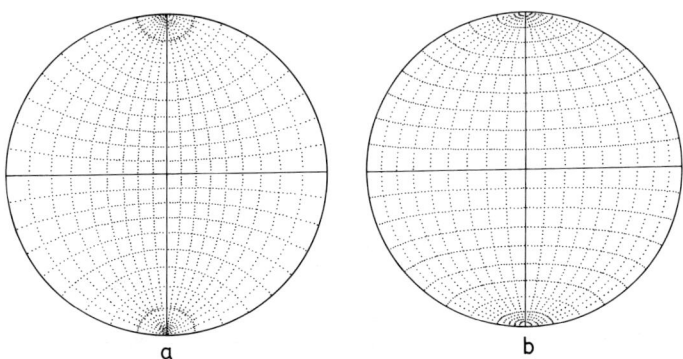

FIG. 4.4. (a) Stereographic net and (b) equal area net showing great circles and small circles at 10° intervals.

great circle traces and a series of small circles at 2° intervals, although, unlike the stereographic net, these arcs are not circular. For the present purpose it is sufficient to know that an equal area net can be used in the same way as a stereographic net. The following procedures of plotting the attitude of a linear or a planar structure are therefore identical for both stereographic and equal area projections.

It is convenient to mount the net, either the Wulff net or the Schmidt net, on a square piece of cardboard or plywood. To use the stereographic net we place a sheet of tracing paper on it and fix it with a board pin at the exact centre of the net. The primitive circle is drawn on it and a N-pointer is drawn on the periphery of the circle.

4.3. Projection of a linear structure

Let us say that a lineation plunges 40° towards 150°. To plot it in stereographic projection we have to go through the following steps:

(1) Rotate the tracing sheet so that its N-pointer coincides with the N-direction of the net.
(2) Mark out the trend or the plunge direction on the periphery (Fig. 4.5a) and rotate the tracing sheet to bring this mark at the N or the S of the net.
(3) Measure from the N (or from the S) of the net the amount of plunge (Fig. 4.5b), from the periphery inward, along the N–S diameter and mark out the position with a point. This is the projection of the linear structure (Fig. 4.5c).

4.4. Projection of a planar structure

The projection of a planar structure as a great circle trace may be called a *cyclographic projection* (Phillips 1971, p.17). The plane may also be represented by its *pole* or π-pole, i.e. the projection of the normal to the plane. The method of polar projection is much more rapid than that of cyclographic projection. Consider a planar structure with strike 30°, dip 60° SE. Alternatively, the attitude may be given as "dip 60° towards 120°". To plot the great circle trace of this planar structure (Fig. 4.6) the following procedure is adopted.

(1) Rotate the tracing sheet to bring its N-pointer to coincide with the N of the net.
(2) Mark out the strike direction, 30° (or the dip direction, 120°) on the periphery (Fig. 4.6a).
(3) Rotate the tracing sheet to bring the strike direction at the N of the net (or, alternatively, to bring the dip direction on the E–W axis of the net).

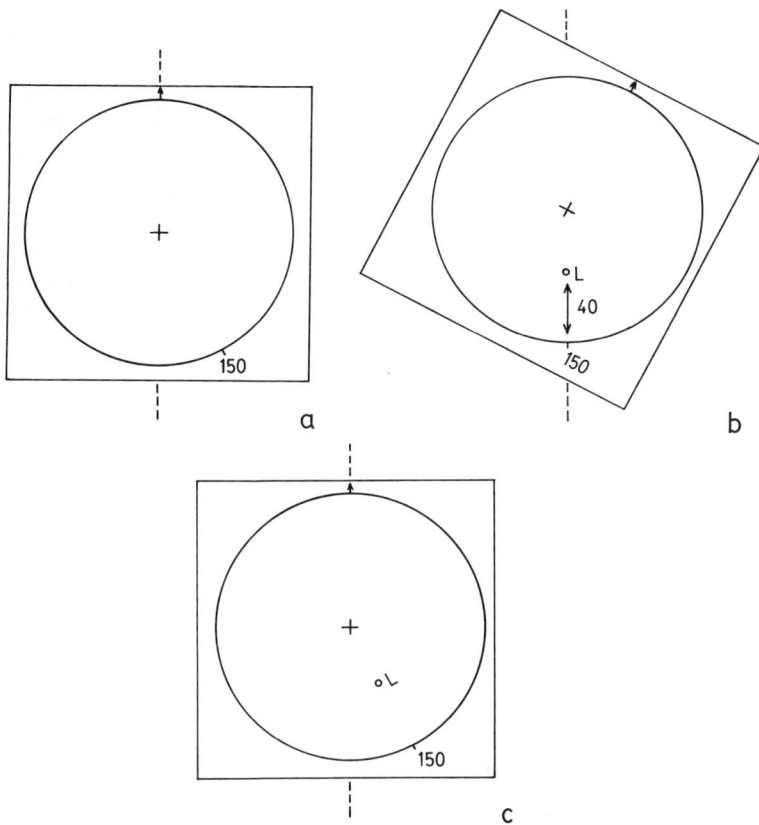

FIG. 4.5. Successive steps of plotting the attitude of a linear structure.

(4a) Draw the great circle for 60° dip, taking care that it is roughly on the eastern side with respect to the N-pointer of the sheet (Fig. 4.6b). To plot the pole to the plane instead of its cyclographic trace, step 4a should be replaced by step 4b.

(4b) Measure the amount of dip *from the centre outward* along the E–W diameter of the net, in a direction opposite to the dip direction, and mark out the position with a dot (Fig. 4.6b). This point is the polar projection of the planar structure (Fig. 4.6c).

4.5. Solution of simple structural problems

The stereographic or equal area net is used to solve a variety of structural problems. The method is much more rapid than any other graphical method. The procedures of solving some simple types of problems are given below.

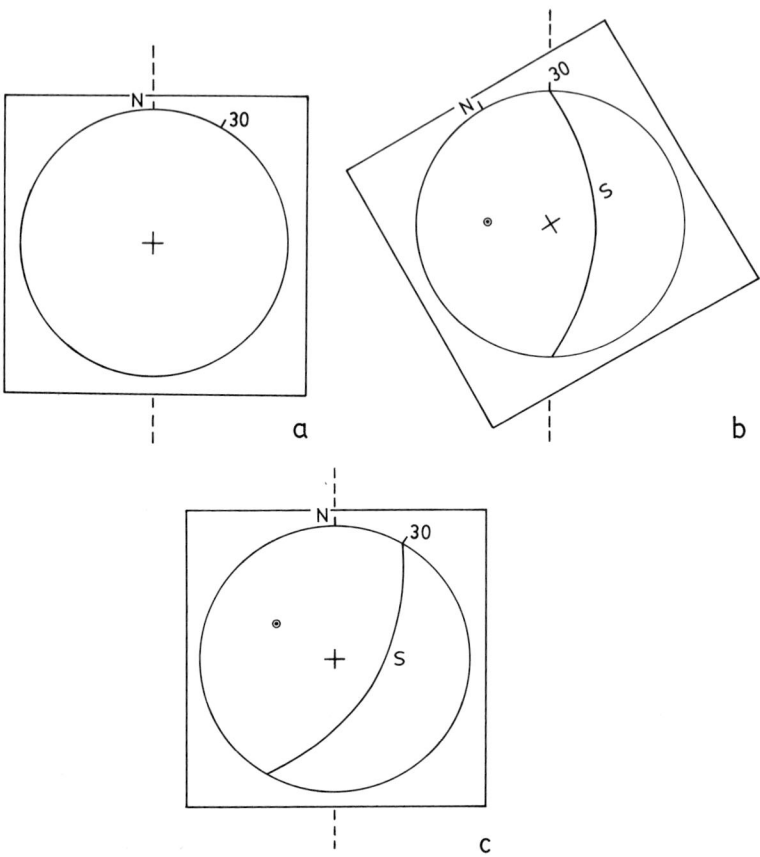

FIG. 4.6. Procedure of plotting a planar structure by its pole (encircled point) and as a great circle trace (*S*).

(a) To determine the *apparent dip* of a planar structure in any direction, draw the great circle trace of the planar structure and then bring the N-pointer of the tracing sheet at the N of the net. Mark out the direction of apparent dip and bring this mark at the N or S of the net. Mark the point of intersection of the great circle trace and the N–S diameter of the net. The apparent dip is the angular distance of this point from the periphery of the net (Fig. 4.7).

(b) To find out the *attitude of a planar structure* from its apparent dips in two directions, mark out the two directions of apparent dip, with the N of the tracing placed at the N of the net. Bring in turn each of these marks at the N point of the net and mark out a point on the N–S diameter so that the point occurs at an angular distance equal to the corresponding apparent dip measured from the periphery. Rotate the

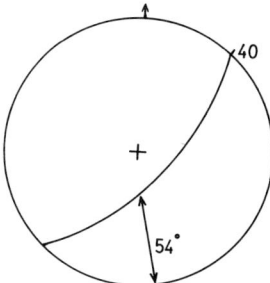

FIG. 4.7. Determination of apparent dip by stereographic projection.

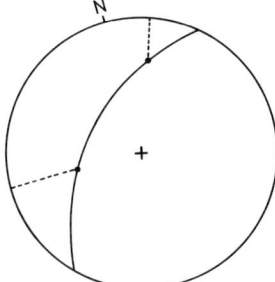

FIG. 4.8. Determination of the attitude of a planar structure from its apparent dips in two directions.

tracing sheet till the two points lie on a great circle of the net. The attitude of the planar structure is given by this great circle (Fig. 4.8).

(c) The direct measurement of the trend of a steeply plunging linear structure may involve some error. Hence, for a linear structure occurring on a S-plane, it is preferable to measure the attitude of the S-plane and the *pitch of the linear structure* and then to determine the trend and plunge of the linear structure from stereographic projection. Let us take the concrete example of a planar structure (*S*) with strike 2°, dip 20°W. A lineation (*L*) lying on it pitches 55° northerly. To determine the trend and plunge of *L*, plot the great circle trace of S-plane, mark out its strike direction (2°) on the primitive and revolve the tracing to bring this mark at the N of the net. Measure from this mark the angular distance of 55° (the amount of pitch) along the arc of the great circle trace of the S-plane and mark out a point. This is the projection of *L*. Rotate the tracing to bring this point on the N-diameter of the net and determine the plunge by measuring the angular distance between N and *L*. With the tracing in this position

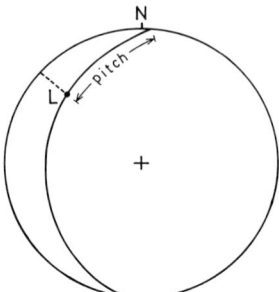

FIG. 4.9. Determination of trend and plunge of a linear structure (L) from its pitch on a plane.

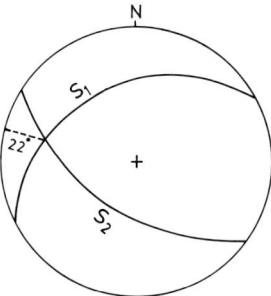

FIG. 4.10. Determination of the attitude of the line of intersection of two planes, S_1 and S_2, in stereographic projection.

mark off the point where the N diameter meets the primitive. This point will give the trend of L. Revolve the tracing to bring the N of the tracing at the N of the net and read off the trend of L (Fig. 4.9).

(d) The attitude of the *line of intersection of two planes* can be determined by plotting the great circle traces of the two planes by following the procedure given in section 4.4. The point at which the great circles intersect represents the line of intersection of the planes. Its trend and plunge can be determined in the same way as in (c) (Fig. 4.10).

(e) To determine the *angle between two linear structures*, L_1 and L_2, their attitudes are plotted by the method given in section 4.3. Then, revolve the tracing till L_1 and L_2 lie on a great circle of the net and measure the acute angle between L_1 and L_2 along the great circle arc (Fig. 4.11).

(f) To find out the *attitude of a plane which bisects the angle between two given planes*, S_1 and S_2, plot the poles to S_1 and S_2 as in section 4.4, bring these poles on a great circle of the net and plot the pole P to this

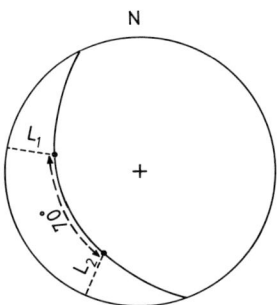

FIG. 4.11. Determination of the angle between two linear structures, L_1 and L_2, from stereographic projection.

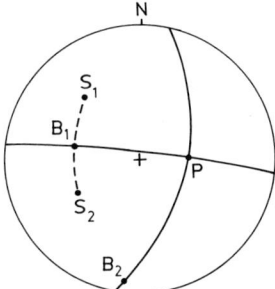

FIG. 4.12. Plotting of the attitude of a plane which bisects the angle between two given planes.

great circle. Measure the angle between the two S-poles along the great circle arc. Say, the acute angle between the poles is 64° and obtuse angle is 116°. Plot the acute bisectrix B_1 on the great circle at an angle of 32° on one side of an S-pole and plot the obtuse bisectrix B_2 at an angle of 58° on the other side. The planes represented by the great circles through B_1 and P and through B_2 and P will bisect, respectively, the obtuse and the acute angles between S_1 and S_2 (Fig. 4.12).

4.6. Rotation of structural elements around a horizontal axis

In certain circumstances we have to find out the orientation of planar or a linear structure after it is rotated through a certain angle around a horizontal axis. Let us consider a concrete example in which the axis of a horizontal isoclinal fold (Fig. 4.13a) trends 80°, with the bedding in each limb dipping at an angle of 60°N. The foreset laminae of cross-laminations on the inverted limb dip 72° on a bearing of 40°. The problem is to determine the orientation of the foreset laminae when the bedding was horizontal and upward facing. The problem can be solved only if we assume that, during folding, the foreset laminae were bodily rotated and did not undergo any internal distortion. To solve this problem under this assumption, the inverted limb has to be rotated around the fold axis through an angle of 120° to make it horizontal and upward facing. Plot the fold axis (F) and the poles to the bedding (P) and the foreset laminae (Q) of the inverted limb (Fig. 4.13b), noting that the strike of the bedding is the same as the trend of the horizontal fold axis, viz. 80°. To make the inverted limb horizontal and upward facing it has to be rotated through the vertical. Since F is the axis of rotation, the locus of the bedding pole will be at a constant angle of 90° with F. In other words, P will move along the diameter at 90° with F. Revolve the tracing to make F coincide with the N or S of the net. Move P along the E–W diameter of the net till, after a rotation of 30° it comes to lie at the periphery. A further rotation in the same sense will make the normal to the inverted limb plunge in the diametrically opposite direction. Hence, with continued rotation P moves inward from the western end of the net till it comes to lie at the centre. Let this new position of P be P'. Thus, P has been rotated through an angle of 120° to its new position P' (Fig. 4.13c). At this position the normal to the limb is vertical and the limb itself has become horizontal. The foreset laminae must undergo the same rotation around the fold axis. Move the point Q eastward along the small circle till it reaches the periphery. Find the diametrically opposite point on the periphery and move the point further eastward along the small circle till the total rotation equals 120°. Let this new position of Q be Q' (Fig. 4.13c). This is the pole to the foreset laminae when the bedding in the inverted limb is made horizontal. The attitude of the foreset laminae can now be read off as 14/51W (Fig. 4.13d).

4.7. Rotation of structural elements around an inclined axis

In certain problems we have to rotate the structural elements around an inclined axis. During the rotation around such an axis, the pole to a plane, or the projection point of a line, will move along a small circle occurring at a constant angle with the inclined axis of rotation. Such small circles are not given in the projection net. They have either to be drawn by locating a series of points at a constant angle from the rotation axis or we have to follow a method in which the existing small

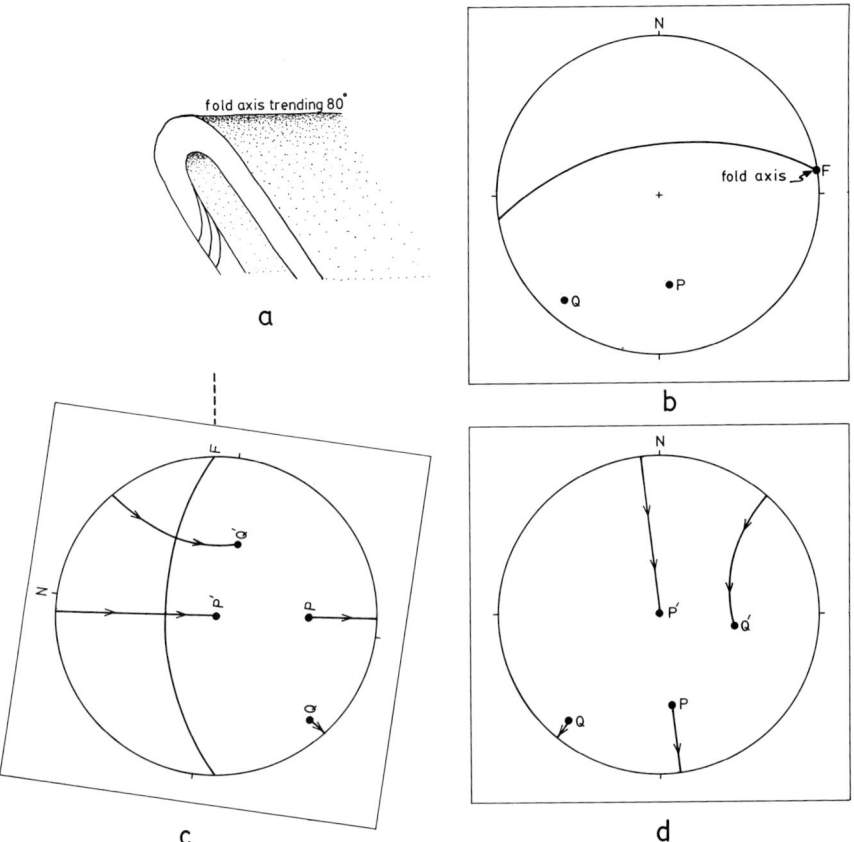

FIG. 4.13. Successive steps of rotating a structural element around a horizontal axis.

circles of the net can be utilized to solve the problem. The latter method is described below.

Consider a bed with an attitude 30/40E. In one place the bed is affected by a rotational fault with the attitude 60/56SE. The rotation of the fault is such that the western side of the hanging wall block has gone down through an angle of 40°. The problem is to determine the orientation of the bed on the hanging wall block. In order to solve this problem we proceed in three steps:

(1) Plot the pole to the bed (P) and the pole to the fault plane Q (Fig. 4.14a). Q is the inclined axis of rotation which plunges at an angle of 30°. In the first step we choose a horizontal rotation axis R_1 at an angle of 90° to Q. Revolve the tracing to bring Q on the E–W diameter of the net. Mark out the point R_1 at the N or the S of the net. Move Q along this diameter through an angle of 30° to bring it at the periphery of the

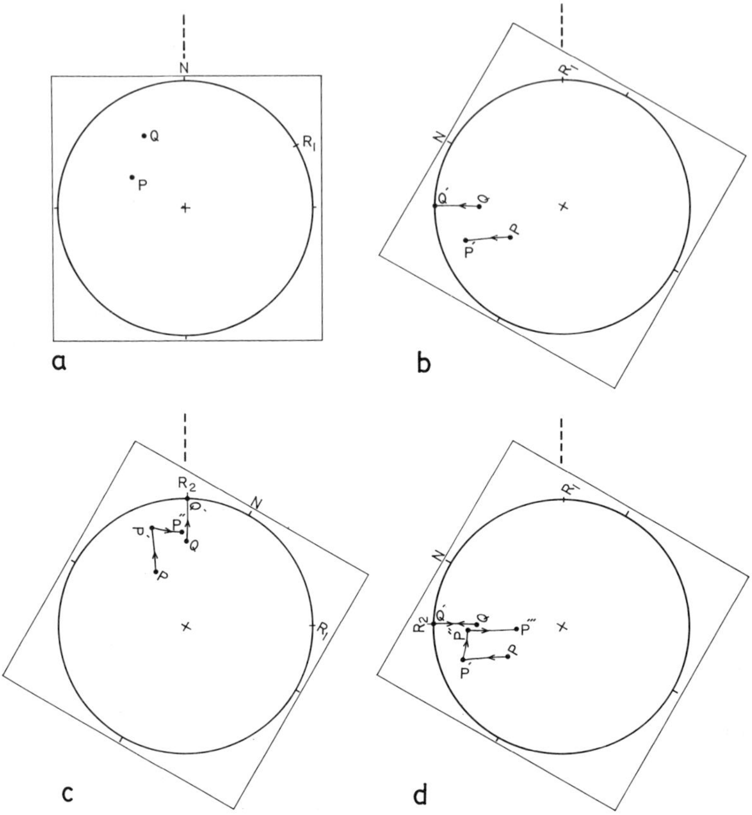

FIG. 4.14. Successive steps of rotating a structural element around an inclined axis.

primitive. Let this new position of Q be Q'. Move P along the small circle on which it lies through the same angle of 30° in the same sense and let this new position be P' (Fig. 4.14b). Note that at the end of the first step the fault plane has become vertical. Its pole $Q' = R_2$ can now be used as a rotation axis for the second step.

(2) Revolve the tracing to bring Q' at the N of the net (Fig. 4.14c). Locate the small circle passing through P'. The sense of rotation on the fault is such that all points on the lower hemisphere projection moves from left to right. Move P' along the small circle from left to right through an angle of 40°. Let this new position of P' be P''.

(3) In the third step (Fig. 4.14d) the fault plane is rotated back to its original attitude by rotating around R_1. Bring R_1 at the N or S of the net. Rotate back Q' to its original position Q. Evidently this involves a rotation through an angle of 30°. Move P'' in a small circle through an angle of 30° in the same sense. Let this position be P'''. This is the pole to the rotated bed in the hanging wall.

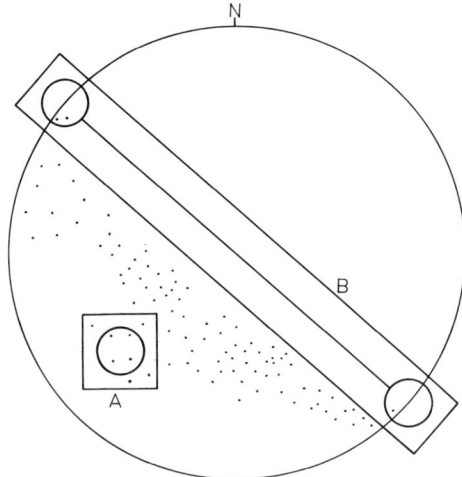

Fɪɢ. 4.15. Use of counters for contouring a pole diagram. The area of the circle in the counter represents 1 per cent of the area of the equal area net.

4.8. Contour diagrams

A contour diagram is prepared to bring out the pattern of density distribution of a large number of poles plotted in a single diagram. To prepare a contour diagram the data are plotted on an equal area projection. The contouring is done on a separate piece of tracing sheet placed over the point diagram. We must know the total number of data points in the diagram. Let us consider an example in which there are 160 data points in the diagram. The concentration of the data points is determined by counting out the number of data points in each 1 per cent area of the diagram. Let us assume, for example, that the equal area net is of 10 cm radius. A circle of 1 cm radius will then represent 1 per cent area of the diagram. We can use a square piece of cardboard or stiff plastic sheet with a circular hole of 1 cm radius (Fig. 4.15). A single point occurring within this circle represents 0.6 per cent of data points. Similarly, two points represent 1.3 per cent, three points represent 1.9 per cent and so on. The counter is moved from place to place all over the diagram. At each position the percentage of poles is written at the centre of the counting circle. For percentage calculation at the periphery of the primitive circle, a special type of counter is used. For the present example with a primitive circle of 10 cm radius, we require a strip of cardboard or plastic with two circular holes of 1 cm radius, the distance between the centres of the circles being 20 cm. The centre of the strip is placed at the centre of the net, the total number of points within both the peripheral circles are counted and the percentage is written out at both their centres lying on the primitive circle. After the percentages are written out all over the diagram wherever there are data points, we have to select a small number of contours which will best bring out the pattern of density distribution of the data points. Each of these contours is drawn as a smooth line through the corresponding percentage points. While drawing contours near the periphery, it should be ensured that a contour line ending at the periphery must also be present at the diametrically opposite point of the diagram. Further details of the contouring technique are given by Turner & Weiss (1963, pp. 58–64).

CHAPTER 5

Stress

5.1. Force

The state of motion or equilibrium of a body depends on its mechanical interaction with other bodies. *Force* is a quantitative measure of this mechanical interaction. It is a vector quantity and hence is characterized by its magnitude, direction and point of application. The SI unit of force is newton (N), the force necessary to give acceleration of 1 metre per second per second to 1 kilogram of mass. Dyne is the force which will produce an acceleration of 1 centimetre per second per second in a gram mass. The kilogram of force (kgf) is often used in engineering. 1 kgf $= 9.81$ N and 1 dyne $= 10^{-5}$ N.

The projection of a force on an axis is equal to the product of the magnitude of the force and the cosine of the angle between the direction of the force and the positive direction of the axis. Thus, if F is the magnitude of a force $\mathbf{F}$ and α, β and γ are the angles it makes with the axes of a cartesian coordinate system, the components of the force are

$$F_x = F \cos \alpha, F_y = F \cos \beta, F_z = F \cos \gamma \text{ and}$$
$$F^2 = F_x^2 + F_y^2 + F_z^2, \text{ since } \cos^2 \alpha + \cos^2 \beta + \cos^2 \gamma = 1.$$

The *resultant* of two forces acting on one point of a body is a force at the same point and is represented by the diagonal of a parallelogram obtained with the two given forces as its sides (Fig. 5.1). By constructing successive parallelograms we can obtain the resultant of any number of concurrent forces (forces whose lines of action meet at a point). The *principal vector* of a system of forces is the geometric sum of all the forces in the system. Thus, the principal vector $\mathbf{R}$ of a system of forces $\mathbf{F}_1$, $\mathbf{F}_2$, $\mathbf{F}_3$ and $\mathbf{F}_4$ is $\mathbf{R} = \mathbf{F}_1 + \mathbf{F}_2 + \mathbf{F}_3 + \mathbf{F}_4$. Since vector addition obeys the parallelogram law, the resultant of a system of concurrent forces can also be obtained by vector addition of the separate forces. In other words, for concurrent forces the resultant and the principal vector are the same. For non-concurrent forces, however, they have to be distinguished (Fig. 5.2).

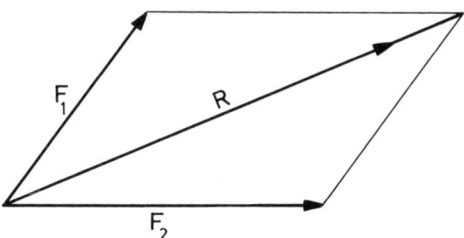

FIG. 5.1. Resultant **R** of two concurrent forces **F**$_1$ and **F**$_2$.

The forces acting on a body may be divided into two groups, *external forces* and *internal forces*. The external forces refer to the action of other bodies on the particles of a given body. We may distinguish between two types of external forces, *surface forces* acting on the surface of a body and *body forces* acting on unit mass or unit volume of the body. Internal forces represent the interaction between the particles in the body.

5.2. Stress vector

Under the action of the external forces the positions of the particles in the body change, i.e. the body deforms. As a result internal forces arise in the body. To determine them we need to apply the principles of statics which deal with external forces only. It is, however, possible to convert the internal forces to external forces by the well-known method of sections. Let us imagine that the body is cut into two parts by a plane and one of the parts is removed (Fig. 5.3). Since the remaining part must be in equilibrium, the action of the removed part should be replaced by forces acting on the surface of the section of the remaining part. These forces, which can now be classified as surface forces, should be so chosen that together with the forces within the remaining part of the body, they will constitute a balanced system of forces.

Let us now isolate a small area ΔA of the section. The surface forces acting on ΔA can be replaced by a single force, say ΔF, acting at its centre. Let us now reduce the area ΔA around its central point indefinitely. The limit of the ratio

$$\lim_{\Delta A \to 0} (\Delta F / \Delta A) = T_n \qquad (5.1)$$

is called the *stress vector* or *traction* at a point on the surface element with normal **n**. It follows from the definition that the dimension of the stress vector is force/(length)^2. The stress vector can be resolved into

44

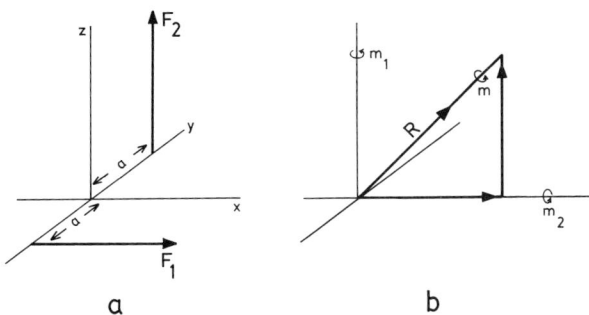

FIG. 5.2. Principal vector **R** of two non-concurrent forces **F**$_1$ and **F**$_2$. The system of forces F_1 and F_2 is reduced to a single force **R** along with the principal moment **m**.

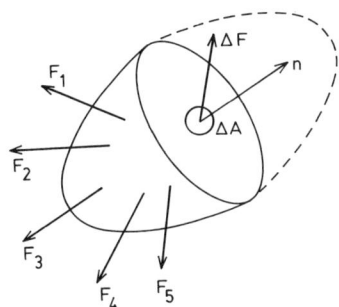

FIG. 5.3. Stress vector at a point on a surface element with normal **n**. The stress vector is the limit of the ratio $\Delta F/\Delta A$ as ΔA, the surface element, tends to 0.

two components: (1) σ_n, the *normal stress* directed along the normal to the surface, and (2) τ_n, the *shearing stress* directed along the line of intersection of the surface and the plane passing through $\mathbf{T}_n$ and **n**.

5.3. State of stress at a point

It is evident from the foregoing discussion that the magnitude and orientation of the stress vector at a point depends on the orientation of the surface on which the stress vector acts. To represent completely the state of stress at a given point it is sufficient to give the stress vector on each of three mutually perpendicular planes. Let us choose a system of rectangular cartesian axes, x, y and z with the origin at point O and consider the stress vector on (1) a plane perpendicular to the x-axis, (2) a plane perpendicular to the y-axis and (3) a plane perpendicular to the

45

z-axis. The stress vector on any one of these planes can be resolved into three components parallel to the coordinate axes. Consider the stress vector on a plane perpendicular to the x-axis. Its component parallel to the x-axis is a normal stress. Let us call it σ_x where the symbol σ indicates that it is a normal stress and the subscript x indicates that it acts on a plane perpendicular to the x-axis. The other two components (Fig. 5.4a) acting parallel to the y- and z-axes, respectively, lie on the chosen plane (i.e. the plane perpendicular to the x-axis) and are therefore shear stresses. Let us designate them as τ_{xy} and τ_{xz}. The first suffix x indicates that the plane on which the stress component acts is perpendicular to the x-axis. The second suffix indicates the direction in which the stress component acts (Fig. 5.4a). σ_x, τ_{xy} and τ_{xz} are thus the components of the traction or stress vector at the point O on a plane whose outer normal is along the x coordinate axis. Similarly, at the same point O we can choose a plane perpendicular to the y coordinate axis. The components of the stress vector on this plane are the normal stress σ_y and shear stresses τ_{yx} and τ_{yz} (Fig. 5.4b). Lastly, on a plane perpendicular to the z-axis the components of the stress vector are σ_z, τ_{zx} and τ_{zy} (Fig. 5.4c). The state of stress at the point O can thus be completely represented by the *elements of the following* matrix:

$$\begin{bmatrix} \sigma_x & \tau_{xy} & \tau_{xz} \\ \tau_{yx} & \sigma_y & \tau_{yz} \\ \tau_{zx} & \tau_{zy} & \sigma_z \end{bmatrix}. \tag{5.2}$$

The matrix (5.2) may be called the *stress matrix*. Figure 5.5a emphasizes that the single subscript of the normal stress σ and the first subscript of the shear stress τ denote the direction of normal to the plane, while the second subscript of the shear stress denotes the direction of the stress component.

A normal stress is *tensile* if it tends to pull the material on one side of the plane from that on the other side. It is *compressive* if it tends to push the material from one side towards the other side. We shall follow the convention that a tensile stress is positive and a compressive stress is negative. It should be noted that the reverse convention is also followed by many authors.

The nine elements of the stress matrix are the components of a single entity, the stress at the point O. It is not a vector since a vector has only three components. It will be shown later that the nine elements of the matrix (5.2) are the components of a second order cartesian tensor known as the stress tensor.

The notation used in (5.2) is not unique. Instead of using separate symbols for the normal and the shear stress, we may use the same symbol, say σ or τ. The normal stresses can then be designated as σ_{xx}, σ_{yy} and σ_{zz} or τ_{xx}, τ_{yy} and τ_{zz}. The notation can also be made simpler in the

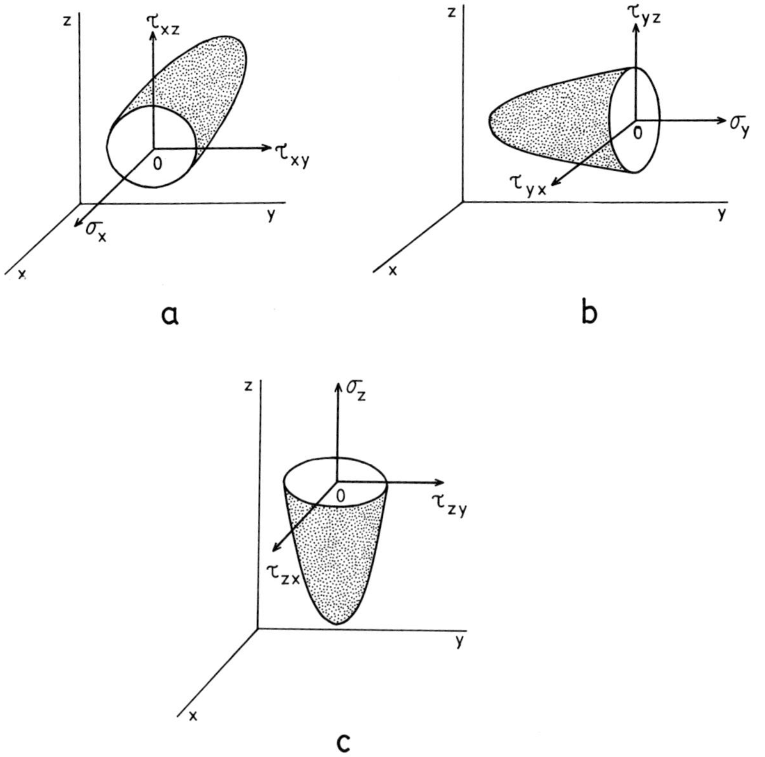

FIG. 5.4. Normal and shear stress components of stress vector at a point O, when the plane through O is (a) perpendicular to the x-axis, (b) perpendicular to the y-axis and (c) perpendicular to the z-axis.

following way. Let us designate the x, y and z coordinate axes as x_1, x_2 and x_3 axes. The components of the stress (Fig. 5.5b) can then be represented by the following matrix:

$$\begin{bmatrix} \sigma_{11} & \sigma_{12} & \sigma_{13} \\ \sigma_{21} & \sigma_{22} & \sigma_{23} \\ \sigma_{31} & \sigma_{32} & \sigma_{33} \end{bmatrix}. \tag{5.3}$$

5.4. Cartesian tensors

The components of stress, as given by (5.2), depend upon our choice of the coordinate axes x, y and z or x_1, x_2 and x_3. If we choose a different set of rectangular cartesian axes, say x', y', z' or x'_1, x'_2, x'_3, the physical entity of stress at the point will remain unchanged; its components, however, with reference to the new coordinate axes, will change. The

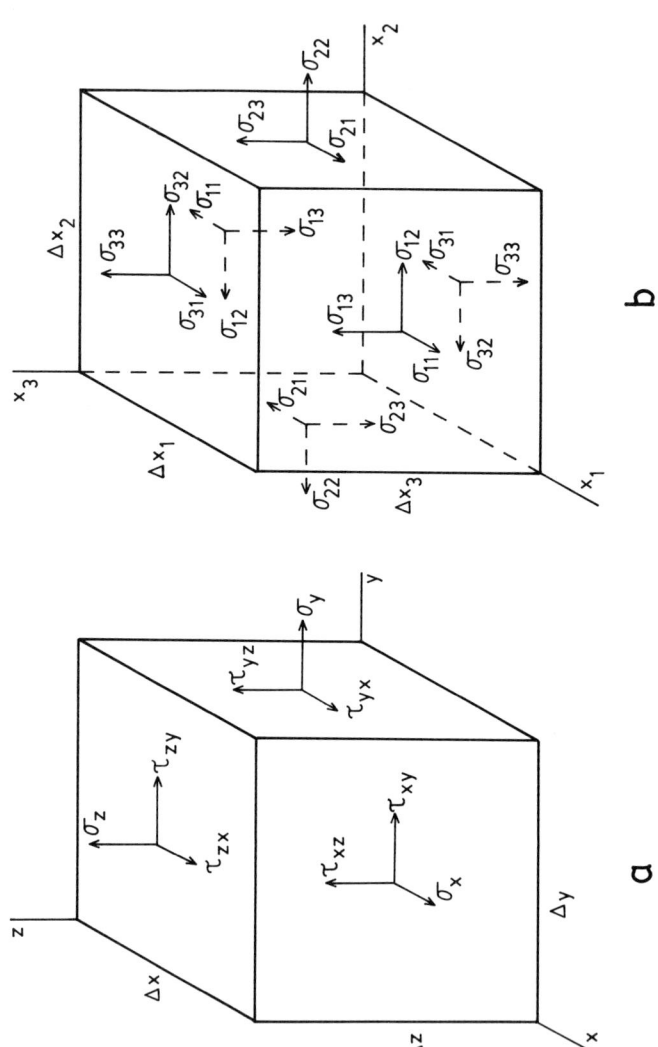

FIG. 5.5. The nine elements of the stress matrix at a point. The stresses are shown on an infinitesimally small cube, the faces of which are perpendicular to the coordinate axes. A normal stress is sometimes indicated by a single subscript (such as σ_y) indicating the direction of the normal. A shear stress is represented by two subscripts; the first indicates the direction of the normal and the second denotes the direction of the stress component. A normal stress may also be represented by two suffixes (such as σ_{22} or as σ_{yy}).

arbitrariness of the choice of the coordinate axes can be removed if we can show how the components are transformed when we pass from one system of rectangular coordinates to another.

To clarify this problem let us first consider how the coordinates of a point P change when the coordinate axes are rotated (Fig. 5.6). Let OX_1, OX_2, OX_3 be the old coordinate axes and OX'_1, OX'_2, OX'_3 be the new coordinate axes. The direction cosines of OX'_1, OX'_2, OX'_3 relative to the system OX_1, OX_2, OX_3 are given in the following table:

	x'_1	x'_2	x'_3
x_1	l_{11}	l_{12}	l_{13}
x_2	l_{21}	l_{22}	l_{23}
x_3	l_{31}	l_{32}	l_{33}

(5.4)

For instance $l_{12}=\cos(OX_1, OX'_2)$. The equations of coordinate transformation are then

$$x'_1 = l_{11}x_1 + l_{21}x_2 + l_{31}x_3$$
$$x'_2 = l_{12}x_1 + l_{22}x_2 + l_{32}x_3 \qquad (5.5)$$
$$x'_3 = l_{13}x_1 + l_{23}x_2 + l_{33}x_3$$

and

$$x_1 = l_{11}x'_1 + l_{12}x'_2 + l_{13}x'_3$$
$$x_2 = l_{21}x'_1 + l_{22}x'_2 + l_{23}x'_3 \qquad (5.6)$$
$$x_3 = l_{31}x'_1 + l_{32}x'_2 + l_{33}x'_3.$$

The reader may verify that, for the two-dimensional case, eqns. (5.5) and (5.6) reduce to the following familiar expressions:

$$x' = x\cos\theta + y\sin\theta$$
$$y' = -x\sin\theta + y\cos\theta \qquad (5.7)$$

and

$$x = x'\cos\theta - y'\sin\theta$$
$$y = x'\sin\theta + y'\cos\theta \qquad (5.8)$$

where x and y replace x_1 and x_2 while x' and y' replace x'_1 and x'_2. θ is here the angle of rotation from OX to OX' axis.

Let us now follow the convention that when a suffix is repeated it indicates a summation obtained by assigning all possible values to the identical suffixes. If there are remaining suffixes they are unaltered during the summation. Thus,

$$a_i a_i = a_1 a_1 + a_2 a_2 + a_3 a_3$$
$$= a_1^2 + a_2^2 + a_3^2$$

and $a_{ij} b_j$ represents three entities:

$$a_{1j} b_j = a_{11} b_1 + a_{12} b_2 + a_{13} b_3$$
$$a_{2j} b_j = a_{21} b_1 + a_{22} b_2 + a_{23} b_3$$
$$a_{3j} b_j = a_{31} b_1 + a_{32} b_2 + a_{33} b_3.$$

With such a *summation convention*, eqns. (5.5) and (5.6) can be written in a very compact form:

$$x'_j = l_{ij} x_i$$
$$x_i = l_{ij} x'_j. \tag{5.9}$$

We have been considering so far the coordinates of a point P (Fig. 5.6). Let the directed line segment OP, from the origin O to the point P, represent a vector, say a force $\mathbf{F}$. It is obvious that its components in the old and the new coordinate systems will be related by equations similar to (5.9), i.e.

$$F'_j = l_{ij} F_i$$
$$F_i = l_{ij} F'_j. \tag{5.10}$$

We are now in a position to define a *cartesian tensor of first order*. It is an entity which can be represented by a set of three numbers or components and, with respect to any two systems of rectangular axes, the components are connected by the relation

$$a'_j = l_{ij} a_i$$
$$a_i = l_{ij} a'_j. \tag{5.11}$$

l_{ij} is the direction cosine as defined above. Comparing (5.10) with (5.11) we find that a vector is a tensor of first order.

In a similar way, a *tensor of second order* is defined as an entity which can be represented as a two-suffix set whose components a_{ij} and a'_{pq}, with reference to any two systems of rectangular axes, are connected by the relation

$$a'_{pq} = l_{ip} l_{jq} a_{ij}. \tag{5.12}$$

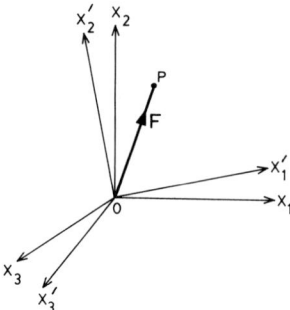

FIG. 5.6. The coordinates of a point P change when the coordinate axes OX_1, OX_2, OX_3 are rotated to OX'_1, OX'_2, OX'_3.

The equations of transformation from the primed to the unprimed coordinates have the form

$$a_{pq} = l_{pi} l_{qj} a'_{ij}. \tag{5.13}$$

The so-called quotient law of tensors is often found to be useful in establishing the tensorial character of an entity which can be represented by a two-suffix set. Consider the equation

$$a_{ij} b_i = c_j \tag{5.14}$$

where b_i and c_j are vectors. According to the quotient law of tensor, if eqn. (5.14) holds good, then a_{ij} is a second order tensor.

There are two special tensors which are often useful in writing out a tensor equation. The *Kronecker tensor* δ_{ij} is defined as

$$\delta_{ij} = 0, \text{ if } i \neq j,$$
$$\delta_{ij} = 1, \text{ if } i = j. \tag{5.15}$$

The *alternate tensor* ε_{ijk} is defined as

$$\varepsilon_{ijk} = 0, \text{ if any two of } i, j, k \text{ are equal,}$$
$$\varepsilon_{ijk} = 1, \text{ if } ijk \text{ is a cyclic permutation of } 1, 2, 3, \tag{5.16}$$
$$\varepsilon_{ijk} = -1, \text{ if } ijk \text{ is an anticyclic permutation of } 1, 2, 3.$$

For unequal values of the suffixes

$$\varepsilon_{123} = \varepsilon_{231} = \varepsilon_{312} = \quad 1$$
$$\varepsilon_{132} = \varepsilon_{213} = \varepsilon_{321} = -1.$$

The reader should notice that the symbol l_{ij} as used above was defined as the cosine of the angle between OX_i and OX'_j axes. Some authors use this symbol as the cosine of the angle between OX'_i and OX_j. If the direction cosine of the latter convention is designated as l'_{ij}, then

$$l'_{ij} = l_{ji}.$$

In this convention, eqn. (5.11) will read as

$$a'_j = l'_{ji}a_i$$
$$a_i = l'_{ji}a_j \qquad (5.17)$$

and eqns. (5.12) and (5.13) will read as

$$a'_{pq} = l'_{pi}l'_{qj}a_{ij}$$
$$a_{pq} = l'_{ip}l'_{jq}a'_{ij}. \qquad (5.18)$$

The meaning of the direction cosines l'_{ij} are shown in the following table.

	x_1	x_2	x_3
x'_1	$l'_{11} = l_{11}$	$l'_{12} = l_{21}$	$l'_{13} = l_{31}$
x'_2	$l'_{21} = l_{12}$	$l'_{22} = l_{22}$	$l'_{23} = l_{32}$
x'_3	$l'_{31} = l_{13}$	$l'_{32} = l_{23}$	$l'_{33} = l_{33}$

$$(5.19)$$

5.5. Stress tensor

Consider an element in the form of a tetrahedron bounded by three planes parallel to the coordinate planes and the fourth plane intersecting the x_1, x_2 and x_3 coordinate axes at points a, b and c, respectively (Fig. 5.7a). h is the perpendicular distance of the origin O from the plane abc. Let $\mathbf{n}$ be the outward normal to the plane abc and let its direction cosines be l_1, l_2 and l_3. If A is the area of the face abc, then the areas of the other faces of the tetrahedron can be found as the projections of A on the x_1x_2-, x_1x_3- and x_2x_3-planes:

$$\text{Area of triangle } Obc = l_1A,$$
$$\text{Area of triangle } Oac = l_2A,$$
$$\text{Area of triangle } Oab = l_3A.$$

Consider a stress vector acting on the plane abc. Its components along the coordinate axes are p_1, p_2 and p_3. Let the components of the stress at the point O with reference to the coordinate axes be as given by the

matrix (5.3). Let us now determine the forces acting on each face of the tetrahedron. In terms of the components of the stress vector, the forces acting on the face abc are p_1A, p_2A and p_3A along the coordinate axes x_1, x_2, x_3, respectively. Similarly, the forces acting on the face Obc are

$$-\sigma_{11}l_1A, \quad -\sigma_{12}l_1A \text{ and } -\sigma_{13}l_1A;$$

the forces on the face Oac are

$$-\sigma_{21}l_2A, \quad -\sigma_{22}l_2A \text{ and } -\sigma_{23}l_2A;$$

lastly, the forces acting on the face Oab are

$$-\sigma_{31}l_3A, \quad -\sigma_{32}l_3A \text{ and } -\sigma_{33}l_3A.$$

The negative signs indicate that the forces on the three faces Obc, Oac and Oab are directed towards the negative directions of the coordinate axes (compare Fig. 5.5b). For equilibrium of the tetrahedron, the sum of the forces along each of the coordinate axes must vanish. Let us now suppose that the tetrahedron shrinks to the point O as $h \rightarrow 0$. Then,

$$p_1A - \sigma_{11}l_1A - \sigma_{21}l_2A - \sigma_{31}l_3A = 0,$$
$$p_2A - \sigma_{12}l_1A - \sigma_{22}l_2A - \sigma_{32}l_3A = 0,$$
$$p_3A - \sigma_{13}l_1A - \sigma_{23}l_2A - \sigma_{33}l_3A = 0,$$

or,

$$p_1 = \sigma_{11}l_1 + \sigma_{21}l_2 + \sigma_{31}l_3,$$
$$p_2 = \sigma_{12}l_1 + \sigma_{22}l_2 + \sigma_{32}l_3, \qquad (5.20)$$
$$p_3 = \sigma_{13}l_1 + \sigma_{23}l_2 + \sigma_{33}l_3.$$

By using the convention of summation over a repeated suffix, this set of three equations can be written as

$$p_i = \sigma_{ji}l_j. \qquad (5.21)$$

Since l_j, the unit normal to the plane abc, is a vector, we can apply the quotient law of tensors [eqn. (5.14)] and conclude that the stress σ_{ji} at the point O is a second order tensor, the *stress tensor*.

The coordinate axes x_1, x_2 and x_3 were chosen arbitrarily. However, by the use of eqn.(5.12), the components of stress at the point O for any other set of rectangular coordinate axes can be easily determined:

$$\sigma'_{pq} = l_{ip}l_{jq}\sigma_{ij}. \qquad (5.22)$$

Equation (5.22) represents a set of nine equations. Thus, for example,

$$\sigma'_{11} = l_{i1}l_{j1}\sigma_{ij}$$
$$= l_{11}l_{j1}\sigma_{1j} + l_{21}l_{j1}\sigma_{2j} + l_{31}l_{j1}\sigma_{3j}$$

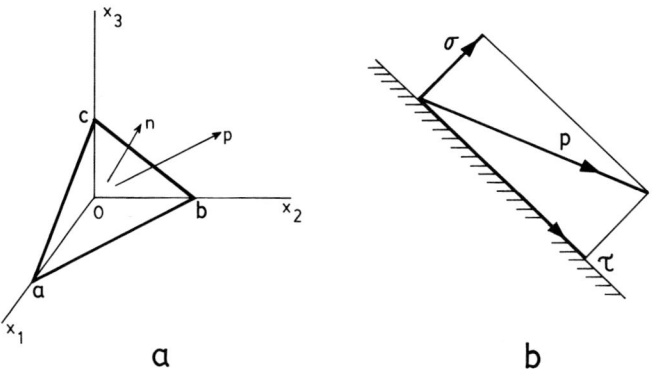

FIG. 5.7. (a) Tetrahedron $Oabc$ with the inclined face abc having an area A and outward normal $\mathbf{n}$ with direction cosines l_1, l_2, l_3. $\mathbf{p}$ is a stress vector acting on the plane abc. (b) The stress vector $\mathbf{p}$, the normal stress (σ) and the shear stress (τ) on an inclined plane.

$$= l_{11}l_{11}\sigma_{11} + l_{11}l_{21}\sigma_{12} + l_{11}l_{31}\sigma_{13}$$
$$+ l_{21}l_{11}\sigma_{21} + l_{21}l_{21}\sigma_{22} + l_{21}l_{31}\sigma_{23}$$
$$+ l_{31}l_{11}\sigma_{31} + l_{31}l_{21}\sigma_{32} + l_{31}l_{31}\sigma_{33},$$

or,

$$\sigma'_{11} = l^2_{11}\sigma_{11} + l^2_{21}\sigma_{22} + l^2_{31}\sigma_{33} + l_{11}l_{21}(\sigma_{12} + \sigma_{21})$$
$$+ l_{11}l_{31}(\sigma_{13} + \sigma_{31}) + l_{21}l_{31}(\sigma_{23} + \sigma_{32}).$$

The other elements of σ_{ij} can be found in the same way.

It should be noted that in eqn.(5.22) we have used the direction cosines of the new system of coordinates relative to the old system of coordinates. If we use the direction cosines of the old system relative to the new system of coordinates, then the transformation is:

$$\sigma'_{pq} = l'_{pi} l'_{qj}\sigma_{ij}, \tag{5.23}$$

where $l'_{ij} = l_{ji}$.

For a body in equilibrium under the action of forces, the rules of statics demand that:

(1) the sum of all the moments along each coordinate axis must vanish, and

(2) the sum of all the forces along each coordinate axis must vanish.

To satisfy the first of these conditions, let us assume that the areas of the faces of the unit cube shown in Fig. 5.5b are so small that the change in stress over the face is negligible. For the equilibrium of the cube

$$(\sigma_{12} \triangle x_2 \triangle x_3) \triangle x_1 = (\sigma_{21} \triangle x_1 \triangle x_3) \triangle x_2.$$

Therefore,

$$\sigma_{12} = \sigma_{21}. \qquad\qquad (5.24a)$$

In the same way it can be shown that

$$\sigma_{32} = \sigma_{23} \qquad\qquad (5.24b)$$

$$\sigma_{13} = \sigma_{31}. \qquad\qquad (5.24c)$$

Equations (5.24) show that, of the nine components of stress, only six are independent. The stress matrix, as shown in (5.2) or (5.3), is thus a symmetric matrix and the stress tensor is a symmetric tensor with

$$\sigma_{ij} = \sigma_{ji}. \qquad\qquad (5.25)$$

5.6. Stress quadric, principal axes and invariants of stress

Stress quadric

Equations (5.20) give the components, p_1, p_2 and p_3 of the stress vector across a plane inclined to the coordinate axes. The projections of these components on the normal to the plane are $l_1 p_1$, $l_2 p_2$ and $l_3 p_3$, where l_1, l_2, l_3 are the direction cosines of the normal $\mathbf{n}$. The sum of these projections gives the normal stress across the plane:

$$\sigma = l_1 p_1 + l_2 p_2 + l_3 p_3. \qquad\qquad (5.26)$$

Substituting the expressions for p_1, p_2 and p_3 from eqns. (5.20) into eqn. (5.26), and taking into account that $\sigma_{ij} = \sigma_{ji}$, we have

$$\sigma = l_1^2 \sigma_{11} + l_2^2 \sigma_{22} + l_3^2 \sigma_{33} + 2 l_1 l_2 \sigma_{12} + 2 l_2 l_3 \sigma_{23} + 2 l_3 l_1 \sigma_{31}. \quad (5.27)$$

This equation gives the normal stress at a point O on a plane at any orientation with respect to the coordinate axes.

Let us now choose a length h measured from the point O on the normal to the plane. Let x_1, x_2, x_3 be the coordinates of the end point of h. The direction cosines of the normal can be expressed in the following way:

$$l_1 = x_1/h, \, l_2 = x_2/h, \, l_3 = x_3/h.$$

Substituting these expressions in eqn. (5.27), we find

$$\sigma_{11}(x_1/h)^2 + \sigma_{22}(x_2/h)^2 + \sigma_{33}(x_3/h)^2 + 2\sigma_{12}x_1 x_2/h^2$$
$$+ 2\sigma_{23}x_2 x_3/h^2 + 2\sigma_{31}x_3 x_1/h^2 = \sigma$$

or,

$$\sigma_{11}x_1{}^2 + \sigma_{22}x_2{}^2 + \sigma_{33}x_3{}^2 + 2\sigma_{12}x_1x_2 + 2\sigma_{23}x_2x_3 + 2\sigma_{31}x_3x_1 = c, \quad (5.28)$$

where

$$c = h^2\sigma. \qquad (5.29)$$

We now prescribe the condition that the length h can be adjusted in such a manner that $h^2\sigma$ remains a constant. Whether the constant c is positive or negative depends on the sign of σ, i.e. on whether σ is tensile or compressive.

Equation (5.28) is in the form of the general equation of quadric surfaces in cartesian coordinates. The surface represented by eqn.(5.28) is known as the *stress quadric of Cauchy*. Equation (5.29) shows that the normal stress σ in a particular direction is inversely proportional to the square of the radius vector in that direction.

Principal axes of stress

The quadric surface represented by (5.28) may be an ellipsoid or a hyperboloid of one sheet or two sheets. It is known from coordinate geometry that such a surface, in general, has three mutually perpendicular principal axes. If the coordinate axes are chosen along these principal axes, the terms containing x_1x_2, x_2x_3 and x_3x_1 in eqn.(5.28) vanish. These axes are known as the *principal axes of stress* and the normal stresses along them are known as the principal stresses. Along the principal axes the shear stresses vanish and the normal stress has a maximum, a minimum and an intermediate value.

Values of principal stresses

If a given plane is perpendicular to a principal axis, the stress vector or the total stress acting on it is directed along its normal and is a principal stress. Let this principal stress be represented by σ. Its components along the x_1, x_2 and x_3 coordinate axes are

$$p_1 = \sigma l_1, p_2 = \sigma l_2, p_3 = \sigma l_3$$

where l_1, l_2, l_3 are the direction cosines of the normal to the plane. Substituting these expressions in eqns. (5.20) we have

$$(\sigma_{11} - \sigma)l_1 + \sigma_{21}l_2 + \sigma_{31}l_3 = 0,$$
$$\sigma_{12}l_1 + (\sigma_{22} - \sigma)l_2 + \sigma_{32}l_3 = 0, \qquad (5.30)$$
$$\sigma_{13}l_1 + \sigma_{23}l_2 + (\sigma_{33} - \sigma)l_3 = 0.$$

The determinant D of this system of equations must vanish if l_1, l_2 and l_3 are not all zero.

$$D = \begin{vmatrix} (\sigma_{11} - \sigma) & \sigma_{21} & \sigma_{31} \\ \sigma_{12} & (\sigma_{22} - \sigma) & \sigma_{32} \\ \sigma_{13} & \sigma_{23} & (\sigma_{33} - \sigma) \end{vmatrix} = 0. \tag{5.31}$$

Equation (5.31) is a cubic equation in σ,

$$\sigma^3 - I_1\sigma^2 + I_2\sigma - I_3 = 0 \tag{5.32}$$

where

$$I_1 = \sigma_{11} + \sigma_{22} + \sigma_{33} \tag{5.33a}$$

$$I_2 = \begin{vmatrix} \sigma_{11} & \sigma_{12} \\ \sigma_{12} & \sigma_{22} \end{vmatrix} + \begin{vmatrix} \sigma_{11} & \sigma_{31} \\ \sigma_{31} & \sigma_{33} \end{vmatrix} + \begin{vmatrix} \sigma_{22} & \sigma_{23} \\ \sigma_{23} & \sigma_{33} \end{vmatrix}$$

$$= (\sigma_{11}\sigma_{22} + \sigma_{22}\sigma_{33} + \sigma_{33}\sigma_{11}) - (\sigma_{12}^2 + \sigma_{23}^2 + \sigma_{31}^2) \tag{5.33b}$$

$$I_3 = \begin{vmatrix} \sigma_{11} & \sigma_{12} & \sigma_{31} \\ \sigma_{12} & \sigma_{22} & \sigma_{23} \\ \sigma_{31} & \sigma_{23} & \sigma_{33} \end{vmatrix}$$

$$= \sigma_{11}\sigma_{22}\sigma_{33} + 2\sigma_{12}\sigma_{23}\sigma_{31} - \sigma_{11}\sigma_{23}^2 - \sigma_{22}\sigma_{31}^2 - \sigma_{33}\sigma_{12}^2. \tag{5.33c}$$

The cubic equation (5.32) has three real roots σ_1, σ_2 and σ_3 which are the values of the three principal stresses. If we choose the coordinate axes along the directions of principal stresses the stress matrix takes the following form:

$$\begin{bmatrix} \sigma_1 & 0 & 0 \\ 0 & \sigma_2 & 0 \\ 0 & 0 & \sigma_3 \end{bmatrix}. \tag{5.34}$$

Directions of principal axes

Once the values of principal stresses are determined, the direction cosines of the principal axes can be found in the following manner. Substitute the value of σ_1 for σ in any two of the equations (5.30). By eliminating, say, l_1, we can get a single equation with l_2 and l_3. This equation will give the ratio $l_2:l_3$. In a similar way if we eliminate l_3, we can get the ratio $l_1:l_2$. We therefore get three numbers a_1, a_2, a_3 proportional to l_1, l_2, l_3. The direction cosines are obtained by dividing each of a_1, a_2, and a_3 by $(a_1^2 + a_2^2 + a_3^2)^{\frac{1}{2}}$. In the same way, by substituting the value of σ_2 or σ_3 in two equations of (5.30) we can get the direction cosines of the σ_2-axis or σ_3-axis.

Invariants of stress

I_1, I_2 and I_3, as given by eqns. (5.33), are known as the *invariants of stress*, since their values do not change when the coordinate axes are rotated. Since the shear stresses vanish along the principal directions, the expressions for I_1, I_2 and I_3 become simpler when the principal directions are chosen as the coordinate axes:

$$I_1 = \sigma_{11} + \sigma_{22} + \sigma_{33} = \sigma_1 + \sigma_2 + \sigma_3$$

$$I_2 = (\sigma_{11}\sigma_{22} + \sigma_{22}\sigma_{33} + \sigma_{33}\sigma_{11}) - (\sigma_{12}^2 + \sigma_{23}^2 + \sigma_{31}^2)$$
$$= \sigma_1\sigma_2 + \sigma_2\sigma_3 + \sigma_3\sigma_1$$

$$I_3 = \sigma_{11}\sigma_{22}\sigma_{33} + 2\sigma_{12}\sigma_{23}\sigma_{31} - \sigma_{11}\sigma_{23}^2 - \sigma_{22}\sigma_{31}^2 - \sigma_{33}\sigma_{12}^2$$
$$= \sigma_1\sigma_2\sigma_3 . \tag{5.35}$$

A numerical example

Let the components of stress at a point be represented by the following matrix

$$\begin{bmatrix} 0.3919 & 0.4816 & 0.1202 \\ 0.4816 & 0.5916 & 0.1343 \\ 0.1202 & 0.1343 & -0.9836 \end{bmatrix}$$

with reference to some coordinate system x_1, x_2, x_3. We shall first calculate the first, second and the third invariants of stress. We find that for the stress matrix, as given above

$$I_1 \approx 0, I_2 \approx -1, I_3 \approx 0.$$

Substituting these values in eqn. (5.32) we have

$$\sigma^3 - \sigma = 0.$$

The three roots of this equation are

$$\sigma_1 = 1, \sigma_2 = 0, \sigma_3 = -1.$$

Substituting the value of σ_1 in the first two equations of (5.30) we get

$$-0.6081 l_1 + 0.4816 l_2 + 0.1202 l_3 = 0,$$
$$0.4816 l_1 - 0.4084 l_2 + 0.1343 l_3 = 0.$$

Eliminating l_1, we find

$$-0.0164 l_2 + 0.1396 l_3 = 0.$$

Or, $l_3/l_2 = 0.1175$.

In a similar way, by eliminating l_3 we find

$$l_1/l_2 = 0.7848.$$

Thus, the direction ratios are

$$a_1 = 0.7848, a_2 = 1, a_3 = 0.1175.$$

Dividing each of these by $(a_1^2 + a_2^2 + a_3^2)^{\frac{1}{2}} = 1.2766$, we have the direction cosines of the σ_1-axis:

$$l_1 = 0.6148, l_2 = 0.7833, l_3 = 0.0920.$$

The direction cosines of σ_2- and σ_3-axes are found in the same way.

5.7. Maximum shearing stresses

Equation (5.27) shows the manner in which the value of normal stress changes in different directions. We shall next find out the shear stress along a plane. For this purpose it is convenient to choose the coordinate axes along the principal axes of stress. Let l_1, l_2, l_3 be the direction cosines of the normal to a plane which is oblique to the coordinate axes. Let $\mathbf{p}$ be the stress vector across this plane and σ the normal stress (Fig. 5.7b). Since there are no shear stresses along the principal axes of stress, the components of $\mathbf{p}$ along the coordinate axes are, by eqn. (5.20):

$$p_1 = \sigma_1 l_1, p_2 = \sigma_2 l_2, p_3 = \sigma_3 l_3. \tag{5.36}$$

The magnitude of p is given by the equation

$$p^2 = \sigma_1^2 l_1^2 + \sigma_2^2 l_2^2 + \sigma_3^2 l_3^2. \tag{5.37}$$

Further, by eqn. (5.26), we have

$$\sigma = l_1^2 \sigma_1 + l_2^2 \sigma_2 + l_3^2 \sigma_3. \tag{5.38}$$

From Fig. 5.7b we can see that the relation among p, σ and τ is

$$p^2 = \sigma^2 + \tau^2, \tag{5.39}$$

where τ is the total shearing stress on the inclined plane. From eqns. (5.37), (5.38) and (5.39) we find

$$\tau^2 = p^2 - \sigma^2$$
$$= (\sigma_1^2 l_1^2 + \sigma_2^2 l_2^2 + \sigma_3^2 l_3^2) - (l_1^2 \sigma_1 + l_2^2 \sigma_2 + l_3^2 \sigma_3)^2,$$

or,

$$\tau^2 = (\sigma_1 - \sigma_2)^2 l_1^2 l_2^2 + (\sigma_2 - \sigma_3)^2 l_2^2 l_3^2 + (\sigma_3 - \sigma_1)^2 l_3^2 l_1^2. \tag{5.40}$$

Since
$$l_3^2 = 1 - l_1^2 - l_2^2,$$
$$\tau^2 = [(\sigma_1^2 - \sigma_3^2)l_1^2 + (\sigma_2^2 - \sigma_3^2)l_2^2 + \sigma_3^2]$$
$$- [(\sigma_1 - \sigma_3)l_1^2 + (\sigma_2 - \sigma_3)l_2^2 + \sigma_3]^2. \tag{5.41}$$

Equations (5.40) or (5.41) give the magnitude of the total shear stress along any plane the normal to which has direction cosines l_1, l_2 and l_3. The stationary values of τ are obtained from the conditions

$$\frac{\delta\tau}{\delta l_1} = 0, \quad \frac{\delta\tau}{\delta l_2} = 0. \tag{5.42}$$

Differentiating the left- and right-hand terms of eqn. (5.41) with respect to l_1 and l_2 we find the two equations

$$2\tau\frac{\delta\tau}{\delta l_1} = 2l_1(\sigma_1^2 - \sigma_3^2) - 4l_1(\sigma_1 - \sigma_3)[(\sigma_1 - \sigma_3)l_1^2 + (\sigma_2 - \sigma_3)l_2^2 + \sigma_3],$$

$$2\tau\frac{\delta\tau}{\delta l_2} = 2l_2(\sigma_2^2 - \sigma_3^2) - 4l_2(\sigma_2 - \sigma_3)[(\sigma_1 - \sigma_3)l_1^2 + (\sigma_2 - \sigma_3)l_2^2 + \sigma_3].$$

Then, from the conditions (5.42) we have:

$$l_1(\sigma_1^2 - \sigma_3^2) - 2l_1(\sigma_1 - \sigma_3)[(\sigma_1 - \sigma_3)l_1^2 + (\sigma_2 - \sigma_3)l_2^2 + \sigma_3] = 0,$$

$$l_2(\sigma_2^2 - \sigma_3^2) - 2l_2(\sigma_2 - \sigma_3)[(\sigma_1 - \sigma_3)l_1^2 + (\sigma_2 - \sigma_3)l_2^2 + \sigma_3] = 0.$$

If $\sigma_1, \sigma_2, \sigma_3$ are all unequal, we can divide the terms of the first equation by $(\sigma_1 - \sigma_3)$ and the terms of the second equation by $(\sigma_2 - \sigma_3)$ and obtain the following two equations, after simplification:

$$l_1[(\sigma_1 - \sigma_3) - 2\{(\sigma_1 - \sigma_3)l_1^2 + (\sigma_2 - \sigma_3)l_2^2\}] = 0,$$

$$l_2[(\sigma_2 - \sigma_3) - 2\{(\sigma_1 - \sigma_3)l_1^2 + (\sigma_2 - \sigma_3)l_2^2\}] = 0. \tag{5.43}$$

There are four possible solutions to these equations, viz.

$$(1)\ l_1 = l_2 = 0, \qquad (2)\ l_1 \neq 0, l_2 \neq 0,$$
$$(3)\ l_1 \neq 0, l_2 = 0, \qquad (4)\ l_1 = 0, l_2 \neq 0. \tag{5.44}$$

The first solution gives a plane normal to the coordinate axis x_3. We reject this solution since we are considering a plane oblique to the coordinate axes. For the second solution, we subtract the second from the first equation of (5.43) and obtain the relation $(\sigma_1 - \sigma_2) = 0$. This solution is also unacceptable, since our initial assumption was that the principal stresses are unequal. From the third solution $(l_1 \neq 0, l_2 = 0)$ we find

$$(\sigma_1 - \sigma_3) - 2(\sigma_1 - \sigma_3)l_1^2 = 0.$$

Or, since $(\sigma_1 - \sigma_3) \neq 0$,

$$l_1 = \pm\sqrt{\tfrac{1}{2}}.$$

Thus the third solution is

$$l_1 = \pm\sqrt{\tfrac{1}{2}}, l_2 = 0, \qquad l_3 = \pm\sqrt{\tfrac{1}{2}}. \tag{5.45a}$$

When the fourth possible solution of (5.44) is substituted in eqns. (5.43) we find

$$l_1 = 0, l_2 = \pm \sqrt{\tfrac{1}{2}}, l_3 = \pm \sqrt{\tfrac{1}{2}}. \qquad (5.45b)$$

If in deriving eqn. (5.41) we had retained l_3 and eliminated either l_1 or l_2, we would have obtained another solution:

$$l_1 = \pm \frac{1}{\sqrt{2}}, l_2 = \pm \frac{1}{\sqrt{2}}, l_3 = 0. \qquad (5.45c)$$

By substituting the values of direction cosines from (5.45a), (5.45b) and (5.45c) in eqn. (5.41) we get the three stationary values of stress:

$$\tau_2 = \pm \tfrac{1}{2}(\sigma_1 - \sigma_3),$$
$$\tau_1 = \pm \tfrac{1}{2}(\sigma_2 - \sigma_3),$$
$$\tau_3 = \pm \tfrac{1}{2}(\sigma_1 - \sigma_2). \qquad (5.46)$$

τ_1, τ_2 and τ_3 are called the principal shearing stresses. The eqns. (5.45) show that the principal shearing stresses act on the following planes:

(1) A pair of planes intersecting along the σ_2-axis and inclined at $\pm 45°$ with the σ_1- or the σ_3-axis (Fig. 5.8a). The absolute value of the shear stress on these planes is the greatest and has the value $\tau_2 = \tfrac{1}{2}(\sigma_1 - \sigma_3)$.

(2) A pair of planes intersecting along the σ_1-axis and inclined at $\pm 45°$ with the σ_2- or σ_3-axis (Fig. 5.8b); the shear stress on these planes is $\tau_1 = \tfrac{1}{2}(\sigma_2 - \sigma_3)$.

(3) A pair of planes intersecting along the σ_3-axis and inclined at $\pm 45°$ with the σ_1- or σ_2-axis (Fig. 5.8c); the shear stress on these planes is $\tau_3 = \pm \tfrac{1}{2}(\sigma_1 - \sigma_2)$.

5.8. Sign convention for shear stress

A shear stress is positive if it is on the positive face of a cube and is directed towards the positive direction. It is also positive if it is on the negative face of a cube and is directed towards the negative direction (Fig. 5.9a). It is negative if it occurs on the negative face and is directed towards the positive direction or occurs on a positive face and is directed towards the negative direction (Fig. 5.9b). This sign convention (Dieter 1988, p. 18) is in agreement with our earlier finding (eqn. 5.24) that $\tau_{ij} = \tau_{ji}$. Thus, on an infinitesimal cubic element, τ_{12} and τ_{21} both have equal absolute values and have the same sign, but are so disposed that they keep the cubic element in equilibrium (Fig. 5.9c).

Consider a situation where a new set of coordinate axes, x_1', x_2', is obtained by rotating the older set through an angle θ (Fig. 5.9d). PB is a face normal to the x_1'-axis. Since this is a positive face, the shear stress

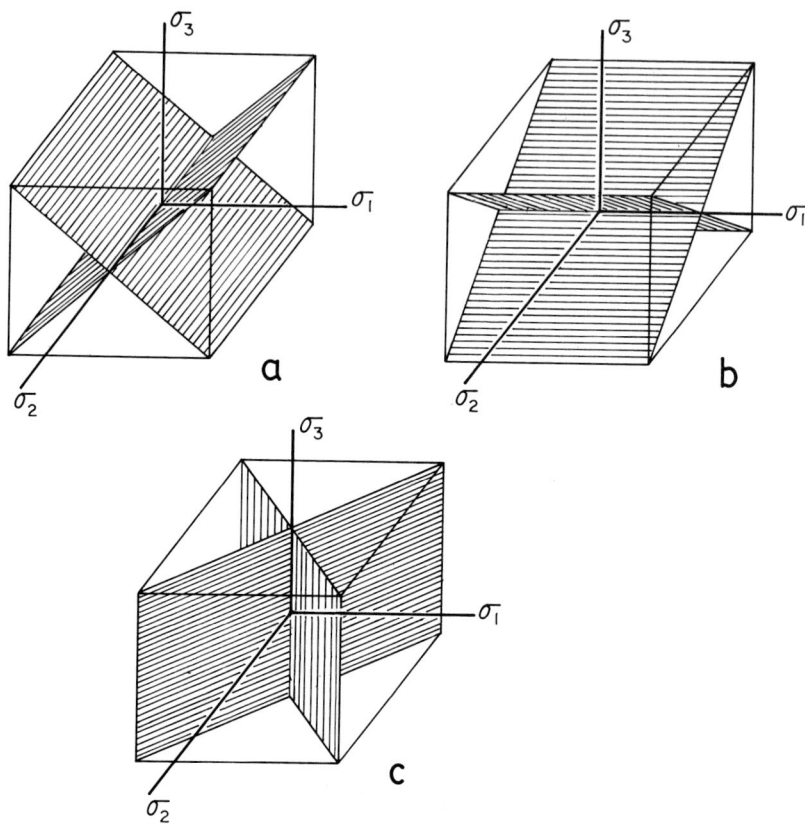

FIG. 5.8. Planes on which the principal shearing stresses act: (a) a pair of planes intersecting along the σ_2-axis, (b) a pair of planes intersecting along the σ_1-axis and (c) a pair of planes intersecting along the σ_3-axis. Each plane is parallel to one principal axis and is at an angle of 45° with the other two.

(τ) on it is positive if it is directed towards the positive direction of the x_2'-axis (Fig. 5.9d).

5.9. Direction of shear stress on an obliquely inclined plane

It is often necessary in rock mechanics, fault-slip analysis and in petrofabrics to determine the magnitude and direction of shear stress on a plane inclined to the principal axes. The magnitude is given by eqn. (5.41). The direction of shear stress is most conveniently obtained by stereographic projection. Two simple methods have been proposed by Lisle (1989) and Means (1989b). Both the methods are described below without giving the explanatory reasons for following the successive

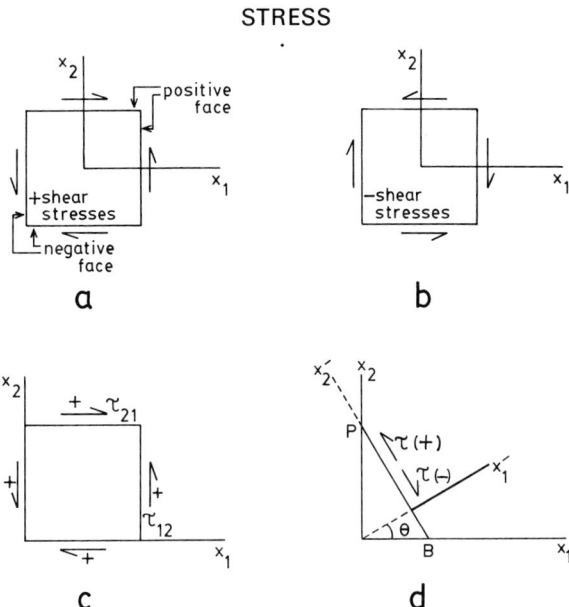

FIG. 5.9. Sign convention for shear stress.

steps. For the explanations the reader should consult the original papers.

First method (Lisle 1989)

(1) Plot the σ_1-, σ_2- and σ_3- axes in stereographic projection. Plot the pole (P) to the given plane (Fig. 5.10).

(2) Determine the value of V from the relation

$$\tan^2 V = (\sigma_2 - \sigma_3)/(\sigma_1 - \sigma_2).$$

(3) Draw a great circle through σ_1 and σ_3 and mark out on this great circle two points A and B on either side of σ_3 at the same angle V with it. Draw a great circle through σ_2 and A and another one through σ_2 and B. These are the two circular sections of the Cauchy stress ellipsoid.

(4) Rotate the tracing sheet to bring the point P in such a position that a great circle through it intersects the circular sections at two points at an equal angle with P. Draw this great circle. Plot the pole to this plane. This is the required orientation of the shear stress τ.

Second method (Means 1989b)

(1) Plot in an *upper hemisphere* stereographic projection the obliquely oriented inclined plane, its pole P, and the σ_1- and σ_3-axes (Fig. 5.11).

(2) Rotate the inclined plane around its strike to make it horizontal. The pole P now occupies the centre of the projection. The positions of σ_1 and σ_3 are shifted accordingly (Fig. 5.11b).

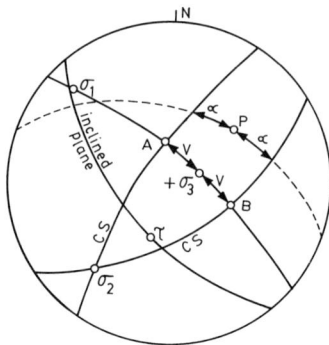

FIG. 5.10. Determination of direction of shear stress on a plane inclined to the principal axes by the method of stereographic projection. After Lisle 1989.

(3) Measure the angle (*a*) between *P* and σ_3 and also measure the angle (*b*) between *P* and σ_1.

(4) Calculate the entities τ_1 and τ_3 from the following formulae:

$$\tau_1 = (\sigma_3 - \sigma_2) \cos a \sin a,$$
$$\tau_3 = (\sigma_2 - \sigma_1) \cos b \sin b.$$

According to the convention followed in this book, tensile stress has been taken as positive and Means's expressions for τ_1 and τ_3 have been changed accordingly.

(5) Draw a diameter along σ_3 and *P*, and from *P* mark out on it a directed line segment equal to the magnitude of τ_1 on a suitable scale. The line segment should be drawn from *P* away from σ_3. Similarly, draw a diameter through σ_1 and *P* and mark out on it a directed line segment equal to the magnitude of τ_3. This line should extend from *P* towards the σ_1 point. Draw the resultant (τ) of the two components τ_1 and τ_3. Its length gives us the magnitude of τ. Extend this line to meet at a point on the periphery of the primitive circle. This is the position of the shear stress τ.

(6) Rotate back the pole *P* to its original position (Fig. 5.11c). The τ point is shifted accordingly. This gives the orientation of the shear stress on the obliquely inclined plane (Fig. 5.11d).

5.10. Splitting up of stress tensor into isotropic and deviatoric components

The sum of two tensors is defined as the tensor whose components are equal to the sums of the respective components of these tensors.

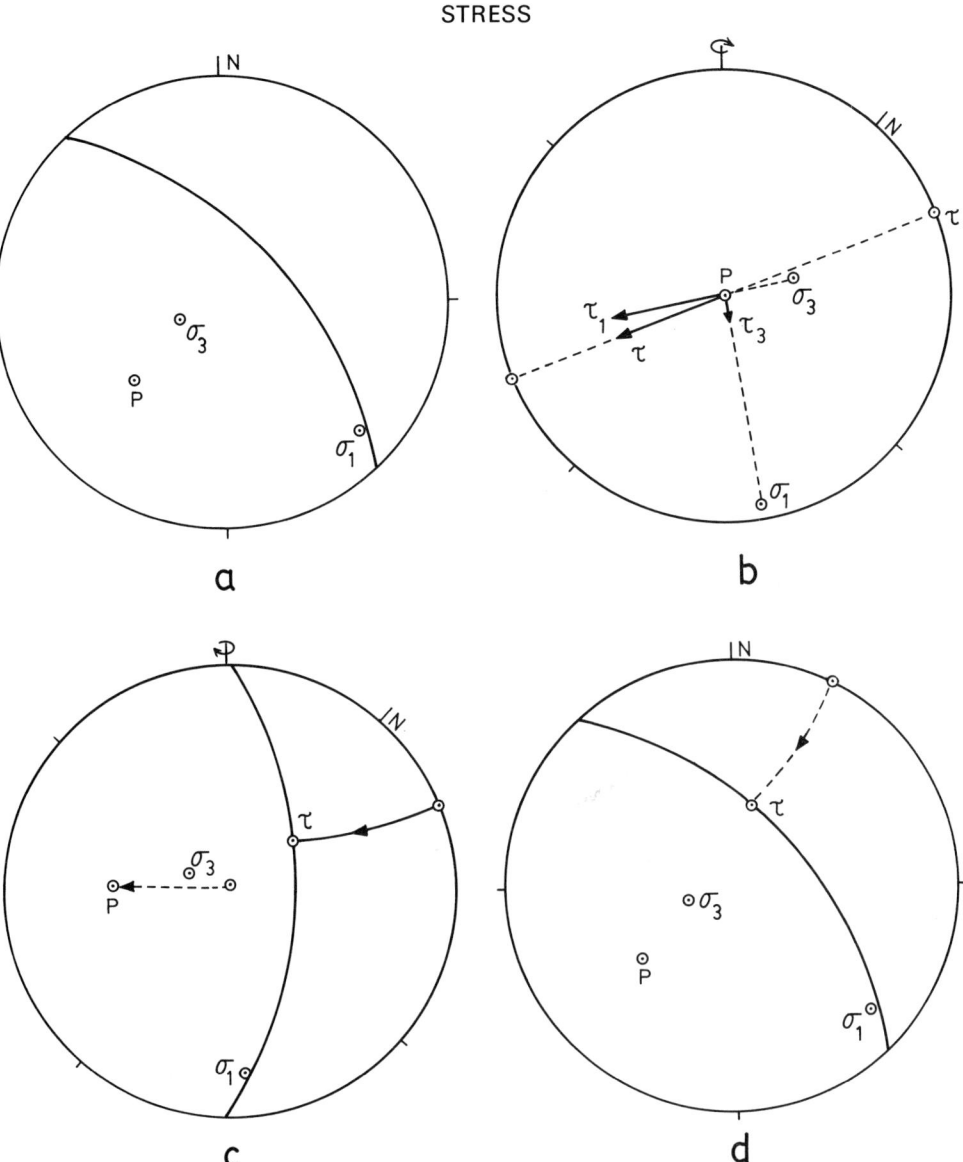

FIG. 5.11. Alternative method of determination of shear stress on an inclined plane in the upper hemisphere of stereographic projection. After Means 1989.

Thus, the stress tensor can be split up in the following manner:

$$\begin{bmatrix} \sigma_{11} & \sigma_{12} & \sigma_{13} \\ \sigma_{21} & \sigma_{22} & \sigma_{23} \\ \sigma_{31} & \sigma_{32} & \sigma_{33} \end{bmatrix} = \begin{bmatrix} (\sigma_{11} - s) & \sigma_{12} & \sigma_{13} \\ \sigma_{21} & (\sigma_{22} - s) & \sigma_{23} \\ \sigma_{31} & \sigma_{32} & (\sigma_{33} - s) \end{bmatrix} + \begin{bmatrix} s & 0 & 0 \\ 0 & s & 0 \\ 0 & 0 & s \end{bmatrix} \quad (5.47)$$

where $s = \frac{1}{3}(\sigma_{11} + \sigma_{22} + \sigma_{33})$.

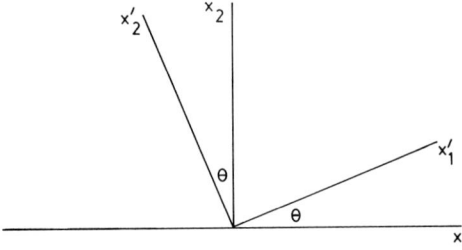

FIG. 5.12. Anticlockwise rotation of coordinate axes through an angle θ.

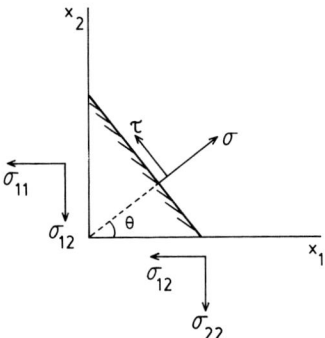

FIG. 5.13. Normal stress (σ) and shear stress (τ) on a plane whose normal makes an angle θ with the direction of σ_{11}.

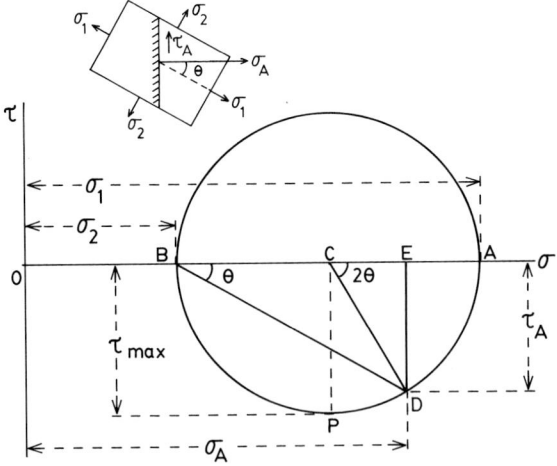

FIG. 5.14. Construction of Mohr circle diagram to determine normal stress (σ_A) and shear stress (τ_A) on a plane whose normal is at an angle of θ with the σ_1-axis. BD is the direction of σ_1.

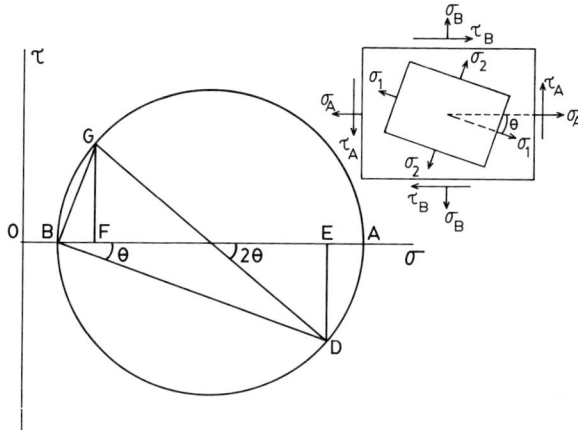

FIG. 5.15. Determination of principal stresses from a Mohr diagram when the stresses on two mutually perpendicular planes are given.

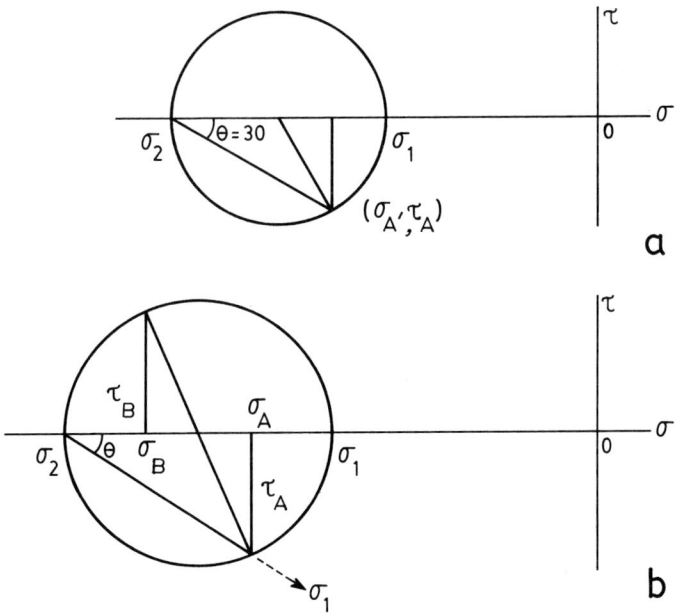

FIG. 5.16. (a) Numerical example showing method of construction of a Mohr diagram to determine stresses on an inclined plane. (b) Numerical example showing method of determining principal stresses when the stresses on two mutually perpendicular planes are given.

67

The first tensor (or matrix) on the right-hand side is called *stress deviator* or *deviatoric stress*, while the second is a spherical tensor in which the three principal stresses are the same. The reason for splitting up the stress tensor in this way is to isolate an all-round tension or compression, equal in all directions, which causes only a change in volume in the body and no change in shape. The remaining part, the deviatoric stress, represents such a state in which the shape of the element changes without any change in volume. In terms of the principal stresses, eqn. (5.47) assumes a simpler form:

$$\begin{bmatrix} \sigma_1 & 0 & 0 \\ 0 & \sigma_2 & 0 \\ 0 & 0 & \sigma_3 \end{bmatrix} = \begin{bmatrix} (\sigma_1 - s) & 0 & 0 \\ 0 & (\sigma_2 - s) & 0 \\ 0 & 0 & (\sigma_3 - s) \end{bmatrix} + \begin{bmatrix} s & 0 & 0 \\ 0 & s & 0 \\ 0 & 0 & s \end{bmatrix} \quad (5.48)$$

where $s = \frac{1}{3}(\sigma_1 + \sigma_2 + \sigma_3)$.

5.11. Plane stress

The analysis of stress becomes simpler in two-dimensional problems where the quantities are independent of the x_3-axis. The stress matrix in this case has only four elements:

$$\begin{bmatrix} \sigma_{11} & \sigma_{12} \\ \sigma_{21} & \sigma_{22} \end{bmatrix} \quad (5.49)$$

of which only three are independent, since

$$\sigma_{21} = \sigma_{12}.$$

If the coordinate axes are rotated through an angle θ (Fig. 5.12), the components of stress will change by eqn. (5.22) in the following manner:

$$\sigma'_{11} = l_{11}^2 \sigma_{11} + l_{21}^2 \sigma_{22} + 2l_{11} l_{21} \sigma_{12},$$
$$\sigma'_{22} = l_{12}^2 \sigma_{11} + l_{22}^2 \sigma_{22} + 2l_{12} l_{22} \sigma_{21},$$
$$\sigma'_{12} = l_{11} l_{12} \sigma_{11} + l_{21} l_{22} \sigma_{22} + (l_{11} l_{22} + l_{21} l_{12})\sigma_{12}. \quad (5.50)$$

Figure 5.12 and the following table show that the direction cosines can all be expressed in terms of the angle θ between the x_1- and x'_1-axes.

	x'_1	x'_2
x_1	$l_{11} = \cos\theta$	$l_{12} = -\sin\theta$
x_2	$l_{21} = \sin\theta$	$l_{22} = \cos\theta$

$$(5.51)$$

Equations (5.50) can therefore be written as

$$\sigma'_{11} = \sigma_{11} \cos^2\theta + \sigma_{22} \sin^2\theta + 2\sigma_{12} \sin\theta \cos\theta,$$

$$\sigma'_{22} = \sigma_{11} \sin^2\theta + \sigma_{22} \cos^2\theta - 2\sigma_{12} \sin\theta \cos\theta, \qquad (5.52)$$

$$\sigma'_{12} = (\sigma_{22} - \sigma_{11}) \sin\theta \cos\theta + \sigma_{12}(\cos^2\theta - \sin^2\theta).$$

Equations (5.52) also give the normal and shear stress on any plane whose normal makes an angle θ with the x_1-axis (Fig. 5.13):

$$\sigma = \sigma_{11} \cos^2\theta + \sigma_{22} \sin^2\theta + 2\sigma_{12} \sin\theta \cos\theta,$$

$$\tau = (\sigma_{22} - \sigma_{11}) \sin\theta \cos\theta + \sigma_{12}(\cos^2\theta - \sin^2\theta). \qquad (5.53)$$

The directions of the principal axes are found from the condition $d\sigma/d\theta = 0$, i.e.

$$2(\sigma_{22} - \sigma_{11}) \sin\theta \cos\theta - 2\sigma_{12} \cos 2\theta = 0$$

or,

$$\tan 2\theta = \frac{2\sigma_{12}}{\sigma_{22} - \sigma_{11}}. \qquad (5.54)$$

Equation (5.54) gives the directions of the σ_1- and σ_2-axes.

For the two-dimensional case, the first two equations of (5.30) become

$$(\sigma_{11} - \sigma)l_1 + \sigma_{21}l_2 = 0,$$
$$\sigma_{12}l_1 + (\sigma_{22} - \sigma)l_2 = 0.$$

If both l_1 and l_2 are not zero, then the determinant of the equations must vanish:

$$(\sigma_{11} - \sigma)(\sigma_{22} - \sigma) - \sigma_{12}^2 = 0,$$

or,

$$\sigma^2 - (\sigma_{11} + \sigma_{22})\sigma + \sigma_{11}\sigma_{22} - \sigma_{12}^2 = 0.$$

Therefore,

$$\sigma = \tfrac{1}{2}(\sigma_{11} + \sigma_{22}) \pm \tfrac{1}{2}[(\sigma_{11} + \sigma_{22})^2 - 4(\sigma_{11}\sigma_{22} - \sigma_{12}^2)]^{\frac{1}{2}},$$

or,

$$\sigma_1 = \tfrac{1}{2}(\sigma_{11} + \sigma_{22}) + \tfrac{1}{2}[(\sigma_{11} - \sigma_{22})^2 + 4\sigma_{12}^2]^{\frac{1}{2}},$$
$$\sigma_2 = \tfrac{1}{2}(\sigma_{11} + \sigma_{22}) - \tfrac{1}{2}[(\sigma_{11} - \sigma_{22})^2 + 4\sigma_{12}^2]^{\frac{1}{2}}. \qquad (5.55)$$

For plane stress the *invariants* are

$$I_1 = \sigma_{11} + \sigma_{22} = \sigma_1 + \sigma_2,$$
$$I_2 = \sigma_{11}\sigma_{22} - \sigma_{12}^2 = \sigma_1\sigma_2. \tag{5.56}$$

5.12. Mohr's circle diagram

If principal axes of stress are chosen as the coordinate axes, eqns. (5.53) have simpler forms

$$\sigma = \sigma_1 \cos^2\theta + \sigma_2 \sin^2\theta,$$
$$\tau = (\sigma_2 - \sigma_1) \sin\theta \cos\theta, \tag{5.57}$$

or, since

$$\cos^2\theta = \tfrac{1}{2}(1 + \cos 2\theta),$$
$$\sin^2\theta = \tfrac{1}{2}(1 - \cos 2\theta),$$
$$\sin\theta \cos\theta = \tfrac{1}{2}\sin 2\theta,$$

$$\sigma = \tfrac{1}{2}(\sigma_1 + \sigma_2) + \tfrac{1}{2}(\sigma_1 - \sigma_2) \cos 2\theta,$$
$$\tau = \tfrac{1}{2}(\sigma_2 - \sigma_1) \sin 2\theta. \tag{5.58}$$

Based on eqns. (5.58), Mohr devised a graphical method of representing the state of stress at a point on an oblique plane. Let us take a rectangular coordinate system with the σ-axis as the abscissa and the τ-axis as the ordinate. Let the origin be at O. We shall consider two types of problem. (1) For certain problems the principal stresses and their directions are known, while the normal and shear stresses on an oblique plane are to be determined. (2) In some other cases, the normal and shear stresses on two mutually perpendicular planes are known. We have to find out the principal stresses and their directions. For both cases the positive angle θ will be measured counterclockwise from the σ_1-axis.

It should be noted that the convention for representing shear stresses in a Mohr diagram is different (e.g. Dieter 1988, p. 25) from the general sign convention for shear stress as given in section 5.8. For the Mohr diagram the shear stress causing a clockwise rotation is plotted on the upper or positive side of the τ-axis and a shear stress causing a counterclockwise rotation is plotted on the lower side of the τ-axis.

(1) When the principal stresses are given, we plot on the σ-axis the segments OA and OB, on a certain scale, equal to the numerical values of σ_1 and σ_2. Tensile stresses are laid off on the positive direction and compressive stress on the negative direction of the σ-axis. A circle is then drawn (Fig. 5.14) with AB as diameter. Let C be the centre of this circle. Then the segment $OC = \tfrac{1}{2}(\sigma_1 + \sigma_2)$ and the radius $CA = \tfrac{1}{2}(\sigma_1 - \sigma_2)$. To determine the normal and shear stresses in a plane the normal to which makes an angle θ with the σ_1-axis, we draw an angle 2θ at point

C, taking its positive value counterclockwise from the σ-axis. Let D be a point on the circle with angle $ACD = 2\theta$. Let DE be the perpendicular on the σ-axis. Then,

$$OE = OC + CE = \tfrac{1}{2}(\sigma_1 + \sigma_2) + CD \cos 2\theta$$
$$= \tfrac{1}{2}(\sigma_1 + \sigma_2) + \tfrac{1}{2}(\sigma_1 - \sigma_2) \cos 2\theta$$

and $ED = -CD \sin 2\theta = -\tfrac{1}{2}(\sigma_1 - \sigma_2) \sin 2\theta$.

Comparing these expressions with eqns. (5.58) we find that the normal and shear stresses are given by the coordinates of the point D. It is evident from Fig. 5.14 that the maximum absolute value of τ is equal to the segment PC or the radius of the circle:

$$\tau_{max} = \tfrac{1}{2}(\sigma_1 - \sigma_2).$$

This maximum shear stress acts on a plane whose normal is oriented at an angle of $45°$ to the σ_1-axis.

(2) Sometimes we are required to solve the opposite problem, i.e. to determine the principal stresses and the principal directions when the normal and shear stresses on any two mutually perpendicular planes are known. Let σ_A, τ_A and σ_B, τ_B be the stresses on two mutually perpendicular planes with $\sigma_A > \sigma_B$.

To find the principal stresses, let us take the rectangular coordinate axes σ and τ (Fig. 5.15) as in the previous case and plot OE and OF corresponding to σ_A and σ_B respectively. A perpendicular ED is then drawn, with $ED = \tau_A$. Another perpendicular FG is drawn equal to ED but directed in the opposite sense. Since the two planes on which the stresses are acting are at a right angle to each other, DG must be a diameter of the stress circle and the point C, the intersection of DG with the σ-axis, must be the centre of the stress circle. Let this circle intersect the σ-axis at points A and B, with $OA > OB$. Then OA and OB give the values of σ_1 and σ_2. BD is the direction of the σ_1-axis and BG is the direction of the σ_2-axis.

We shall now consider two numerical examples:

(1) The principal stresses at some point are $\sigma_1 = -40$ MPa and $\sigma_2 = -80$ MPa. We have to determine from the Mohr circle the normal and shear stresses on a plane whose normal is at angle of $30°$ with the σ_1-axis. The method of construction of the Mohr diagram is shown in Fig. 5.16a. We find from this diagram that $\sigma = -50$ MPa and $\tau = -17.3$ MPa.

(2) In the second example the normal and shear stresses are given on two mutually perpendicular planes, A and B. $\sigma_A = -65$ MPa, $\tau_A = -22.8$ MPa and $\sigma_B = -85$ MPa, $\tau_B = 22.8$ MPa. The construction of the Mohr circle is shown in Fig. 5.16b. We find from the diagram that $\sigma_1 = -50$ MPa and $\sigma_2 = -100$ MPa. The diagram further shows that $\theta = 33°$. This means that σ_1 is oriented at an angle of $33°$ measured

clockwise from the σ_A-axis. θ is positive because, according to our convention, it has been measured anticlockwise from the direction of σ_1.

5.13. Differential equations of equilibrium

When a body is subjected to forces at its boundary, the stresses within it vary from point to point. These variations must be such that an infinitesimal element in the form of a parallelopiped remains in equilibrium. We must now take into account the body force acting on this element. For a static equilibrium of the element the following set of equations holds good.

$$\frac{\delta\sigma_{11}}{\delta x_1} + \frac{\delta\sigma_{21}}{\delta x_2} + \frac{\delta\sigma_{31}}{\delta x_3} + R_1 = 0$$

$$\frac{\delta\sigma_{12}}{\delta x_1} + \frac{\delta\sigma_{22}}{\delta x_2} + \frac{\delta\sigma_{32}}{\delta x_3} + R_2 = 0 \qquad (5.59)$$

$$\frac{\delta\sigma_{13}}{\delta x_1} + \frac{\delta\sigma_{23}}{\delta x_2} + \frac{\delta\sigma_{33}}{\delta x_3} + R_3 = 0$$

where R_1, R_2, R_3 are the components of the body force per unit volume of the element. The three equations may be written in a compact form by using the double suffix notation with summation over the repeated suffix:

$$\sigma_{ji,j} + R_i = 0$$

where $\sigma_{ji,j}$ stands for $\delta\sigma_{ji}/\delta x_j$.

CHAPTER 6

Deformation

6.1. Translation, rotation and deformation

A large part of structural geology is concerned with the geometry of motion of rock bodies or of particles in a rock, without taking into account the forces acting on them. The description of such motion is known as *kinematics*. If, during the period of motion, the distance between any two points in a body remains the same, the body is regarded as a rigid body. A rigid body movement may involve a translation or a rotation or a combination of translation and rotation. The movement of a translational fault, a rotational fault or the rotation of a rigid porphyroblast are examples of such rigid body motion (Fig. 1.1). *Translation* is a motion in which any straight line through the body remains parallel to itself at all stages of the motion. As shown in Fig. 6.1a, translation is not necessarily rectilinear motion. A rigid body *rotation* is a motion in which there are always two points, within the body or in the extended part of a body, which remain motionless. The line which joins the two points is known as the *axis of rotation* (Fig. 6.1b).

In rigid body motion the position of a particle relative to other particles does not change. When the *relative* configuration of particles in a body changes, the body undergoes a *deformation* (Fig. 6.1c). Deformation involves a change in volume or shape or both. In addition, some rigid body motion may be involved in deformation. Deformation is the transformation from the undeformed to the deformed state without specifying the sequence of configurations of the particles through which the total transformation takes place. However, structural geologists are often interested in reconstructing the history of deformation. The reconstruction of such intermediate configurations involves the analysis of *progressive deformation*.

6.2. Measure of strain

Strain is a change in the relative configuration of the particles in an element of body. Strain is measured by change in the length of a line and by the change in the right angle between two lines or between a line and a plane. The strain which is measured by a change in the length of a line is known as *longitudinal strain*. The strain which is measured by a change in the angle is a *shear strain*.

There are several ways in which longitudinal strain is measured. Let l be the distance between two neighbouring particles of a continuous body before deformation and l' be their distance after deformation (Fig. 6.2a). The ratio of change in length to the original length is known as *elongation*:

$$\varepsilon = \frac{l' - l}{l}. \tag{6.1}$$

The ratio l'/l is called the *stretch*. *Quadratic elongation* is the square of stretch:

$$\lambda = \left(\frac{l'}{l}\right)^2. \tag{6.2}$$

Natural strain or *logarithmic strain* is defined as

$$\bar{\varepsilon} = \ln(l'/l). \tag{6.3}$$

Since strain is measured by a ratio of lengths, it is a non-dimensional entity. The interrelations among ε, λ and $\bar{\varepsilon}$ are

$$\lambda = (1 + \varepsilon)^2,$$
$$\lambda = \exp(2\,\bar{\varepsilon}), \tag{6.4}$$
$$\bar{\varepsilon} = \ln(1 + \varepsilon).$$

Shear strain is measured by the change in the right angle between two lines. In Fig. 6.2b, OA and OB are two lines which are perpendicular to each other. After deformation, the angle has changed to AOB'. The change in the right angle is $\psi = $ angle BOB'. The shear strain is defined as

$$\gamma = \tan \psi. \tag{6.5}$$

6.3. Displacement

Consider a particle P in a body in the coordinate system x_1, x_2, x_3. Let the particle P be shifted to the position P' by deformation, the coordinates of P' being $x_1 + u_1$, $x_2 + u_2$ and $x_3 + u_3$. The components of the displacement are then u_1, u_2 and u_3 (Fig. 6.3). If the displacement vector

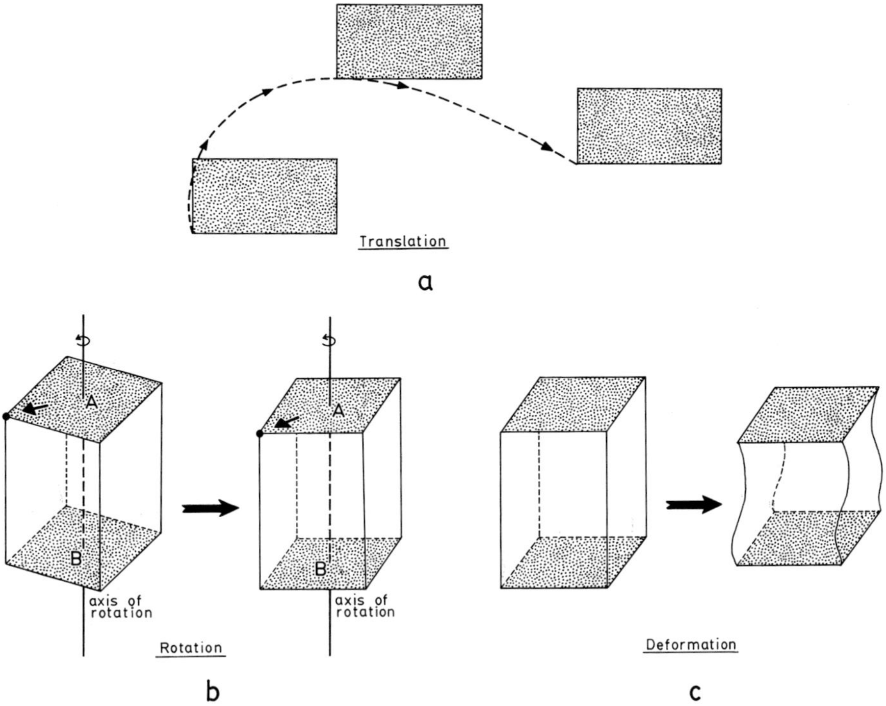

DEFORMATION

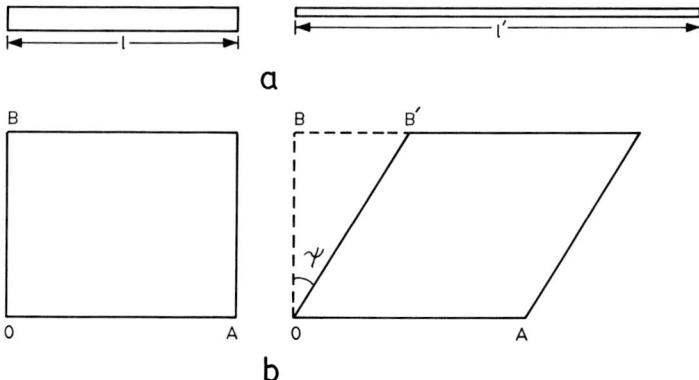

Translation

a

Rotation

b

Deformation

c

FIG. 6.1. Translation, rigid body rotation and deformation.

a

b

FIG. 6.2. Longitudinal strain and shear strain.

is the same for all particles of a body, then the body does not deform. When the body deforms, then in the general case the displacement of a particle is a function of its position:

75

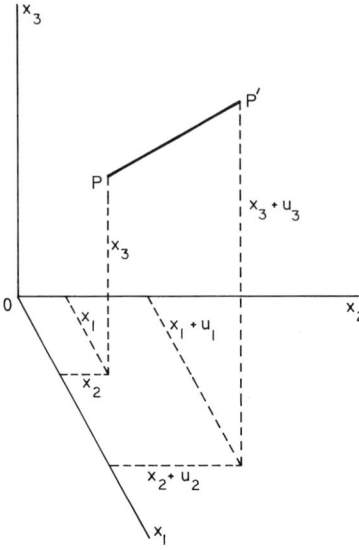

FIG. 6.3. The particle P with coordinates (x_1, x_2, x_3) is displaced to the position P' with coordinates $(x_1 + u_1, x_2 + u_2, x_3 + u_3)$. u_1, u_2, u_3 are the components of displacement.

$$u_1 = f_1(x_1, x_2, x_3)$$
$$u_2 = f_2(x_1, x_2, x_3) \qquad (6.6)$$
$$u_3 = f_3(x_1, x_2, x_3).$$

6.4. Infinitesimal strain tensor and rotation tensor

Deformation tensor

The general theory of strain is so complex that for practical applications it is usually simplified with certain assumptions. If it is assumed that the strains are so small that their squares and products are negligible, the analysis is considerably simplified. This is the case of infinitesimal strain.

Consider in a body any point P (Fig. 6.4) with coordinates (x_1, x_2, x_3). By deformation let the position of the point be shifted to P' with coordinates $(x_1 + u_1, x_2 + u_2, x_3 + u_3)$. The components of the displacement vector PP' are u_1, u_2, u_3. As indicated by eqn. (6.6), the components of the displacement vector are functions of the coordinates. Let Q be a point in the neighbourhood of P. The directed line segment PQ is the vector A_i with components A_1, A_2, A_3. Let us find out an expression for the change in A_i caused by the deformation. Assume

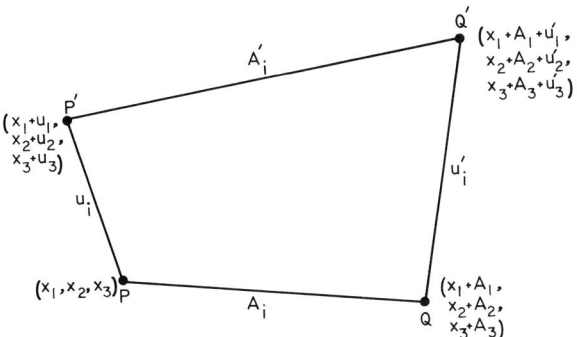

FIG. 6.4. Change in the vector PQ (or A_i) as a result of deformation.

that, by deformation, the vector A_i has changed to the vector A_i' represented in Fig. 6.4 by the directed line segment $P'Q'$. Let us recall that the coordinates of the points P, P' and Q are, respectively, (x_1, x_2, x_3), $(x_1 + u_1, x_2 + u_2, x_3 + u_3)$ and $(x_1 + A_1, x_2 + A_2, x_3 + A_3)$.

Now, the displacement of the point Q to Q' is a function of the original position of Q. Thus, the components of the vector QQ' are

$$u_1' = f_1(x_1 + A_1, x_2 + A_2, x_3 + A_3)$$
$$u_2' = f_2(x_1 + A_1, x_2 + A_2, x_3 + A_3) \tag{6.7}$$
$$u_3' = f_3(x_1 + A_1, x_2 + A_2, x_3 + A_3).$$

Since Q is in the neighbourhood of P, we may neglect the squares and higher powers of A_1, A_2 and A_3. Thus, by expanding the functions (6.7) by Taylor's theorem and neglecting small quantities we find

$$u_1' = f_1(x_1, x_2, x_3) + \frac{\delta f_1}{\delta x_1} A_1 + \frac{\delta f_1}{\delta x_2} A_2 + \frac{\delta f_1}{\delta x_3} A_3$$

$$u_2' = f_2(x_1, x_2, x_3) + \frac{\delta f_2}{\delta x_1} A_1 + \frac{\delta f_2}{\delta x_2} A_2 + \frac{\delta f_2}{\delta x_3} A_3$$

$$u_3' = f_3(x_1, x_2, x_3) + \frac{\delta f_3}{\delta x_1} A_1 + \frac{\delta f_3}{\delta x_2} A_2 + \frac{\delta f_3}{\delta x_3} A_3.$$

where f_1, f_2 and f_3 are given by eqn. (6.6).

The coordinates of the point Q' are $(x_1 + A_1 + u_1')$, $(x_2 + A_2 + u_2')$, $(x_3 + A_3 + u_3')$.

The components of the vector A_i' of Fig. 6.4 are therefore

$$A_1' = (x_1 + A_1 + u_1') - (x_1 + u_1)$$

$$= A_1 + \frac{\delta u_1}{\delta x_1} A_1 + \frac{\delta u_1}{\delta x_2} A_2 + \frac{\delta u_1}{\delta x_3} A_3$$

$$A_2' = (x_2 + A_2 + u_2') - (x_2 + u_2)$$

$$= A_2 + \frac{\delta u_2}{\delta x_1} A_1 + \frac{\delta u_2}{\delta x_2} A_2 + \frac{\delta u_2}{\delta x_3} A_3 \qquad (6.8)$$

$$A_3' = (x_3 + A_3 + u_3') - (x_3 + u_3)$$

$$= A_3 + \frac{\delta u_3}{\delta x_1} A_1 + \frac{\delta u_3}{\delta x_2} A_2 + \frac{\delta u_3}{\delta x_3} A_3,$$

or,

$$A_1' - A_1 = \frac{\delta u_1}{\delta x_1} A_1 + \frac{\delta u_1}{\delta x_2} A_2 + \frac{\delta u_1}{\delta x_3} A_3$$

$$A_2' - A_2 = \frac{\delta u_2}{\delta x_1} A_1 + \frac{\delta u_2}{\delta x_2} A_2 + \frac{\delta u_2}{\delta x_3} A_3 \qquad (6.9)$$

$$A_3' - A_3 = \frac{\delta u_3}{\delta x_1} A_1 + \frac{\delta u_3}{\delta x_2} A_2 + \frac{\delta u_3}{\delta x_3} A_3.$$

These equations represent the change in the vector A_i caused by the deformation and can be expressed in a compact form as

$$\delta A_i = \frac{\delta u_i}{\delta x_j} A_j \qquad (6.10)$$

or, replacing $\delta u_i / \delta x_j$ by the symbol a_{ij},

$$\delta A_i = a_{ij} A_j. \qquad (6.11)$$

By the quotient law, as stated in connection with eqn. (5.14), we find that a_{ij} is a tensor of second order. It is the *deformation tensor* and its components are shown by the elements of the matrix

$$\begin{bmatrix} \dfrac{\delta u_1}{\delta x_1} & \dfrac{\delta u_1}{\delta x_2} & \dfrac{\delta u_1}{\delta x_3} \\[2mm] \dfrac{\delta u_2}{\delta x_1} & \dfrac{\delta u_2}{\delta x_2} & \dfrac{\delta u_2}{\delta x_3} \\[2mm] \dfrac{\delta u_3}{\delta x_1} & \dfrac{\delta u_3}{\delta x_2} & \dfrac{\delta u_3}{\delta x_3} \end{bmatrix} \equiv \begin{bmatrix} \dfrac{\delta u}{\delta x} & \dfrac{\delta u}{\delta y} & \dfrac{\delta u}{\delta z} \\[2mm] \dfrac{\delta v}{\delta x} & \dfrac{\delta v}{\delta y} & \dfrac{\delta v}{\delta z} \\[2mm] \dfrac{\delta w}{\delta x} & \dfrac{\delta w}{\delta y} & \dfrac{\delta w}{\delta z} \end{bmatrix}. \qquad (6.12)$$

In the matrix on the right-hand side the coordinates x_1, x_2, x_3 have been replaced by x, y, z and the displacement components u_1, u_2, u_3 have been replaced by u, v, w.

Splitting up of deformation tensor into strain tensor and rotation tensor

The deformation tensor is asymmetric and can be split up into a symmetric tensor and a skew-symmetric tensor

$$a_{ij} = \tfrac{1}{2}(a_{ij} + a_{ji}) + \tfrac{1}{2}(a_{ij} - a_{ji})$$

or,

$$a_{ij} = \varepsilon_{ij} + \omega_{ij}. \tag{6.13}$$

It will be shown below that the tensor ε_{ij} causes a pure strain, i.e. a change in shape and size while the tensor ω_{ij} causes a rigid body rotation. ε_{ij} is called the *strain tensor* and ω_{ij} is called the *rotation tensor*. The components of the strain tensor can be represented by the matrix

$$\begin{bmatrix} \varepsilon_{11} & \varepsilon_{12} & \varepsilon_{13} \\ \varepsilon_{21} & \varepsilon_{22} & \varepsilon_{23} \\ \varepsilon_{31} & \varepsilon_{32} & \varepsilon_{33} \end{bmatrix} \equiv \begin{bmatrix} \varepsilon_{xx} & \varepsilon_{xy} & \varepsilon_{xz} \\ \varepsilon_{yx} & \varepsilon_{yy} & \varepsilon_{yz} \\ \varepsilon_{zx} & \varepsilon_{zy} & \varepsilon_{zz} \end{bmatrix}. \tag{6.14}$$

Since the strain tensor is symmetric, $\varepsilon_{ij} = \varepsilon_{ji}$. Hence the strain tensor can be expressed through six components:

$$\varepsilon_{11} = \varepsilon_{xx} = \frac{\delta u_1}{\delta x_1} = \frac{\delta u}{\delta x}$$

$$\varepsilon_{22} = \varepsilon_{yy} = \frac{\delta u_2}{\delta x_2} = \frac{\delta v}{\delta y}$$

$$\varepsilon_{33} = \varepsilon_{zz} = \frac{\delta u_3}{\delta x_3} = \frac{\delta w}{\delta z} \tag{6.15}$$

$$\varepsilon_{12} = \varepsilon_{21} = \varepsilon_{xy} = \varepsilon_{yx} = \frac{1}{2}\left(\frac{\delta u_1}{\delta x_2} + \frac{\delta u_2}{\delta x_1}\right) = \frac{1}{2}\left(\frac{\delta u}{\delta y} + \frac{\delta v}{\delta x}\right)$$

$$\varepsilon_{23} = \varepsilon_{32} = \varepsilon_{yz} = \varepsilon_{zy} = \frac{1}{2}\left(\frac{\delta u_2}{\delta x_3} + \frac{\delta u_3}{\delta x_2}\right) = \frac{1}{2}\left(\frac{\delta v}{\delta z} + \frac{\delta w}{\delta y}\right)$$

$$\varepsilon_{31} = \varepsilon_{13} = \varepsilon_{zx} = \varepsilon_{xz} = \frac{1}{2}\left(\frac{\delta u_1}{\delta x_3} + \frac{\delta u_3}{\delta x_1}\right) = \frac{1}{2}\left(\frac{\delta u}{\delta z} + \frac{\delta w}{\delta x}\right).$$

It will now be shown that the first three of these components represent elongations while the last three represent shearing strains. Consider a vector A_i directed parallel to the x_1 axis. Its length parallel to the x_1 axis is A_1 while $A_2 = A_3 = 0$. From eqn. (6.9) we find that

$$\frac{A_1' - A_1}{A_1} = \frac{\delta u_1}{\delta x_1}.$$

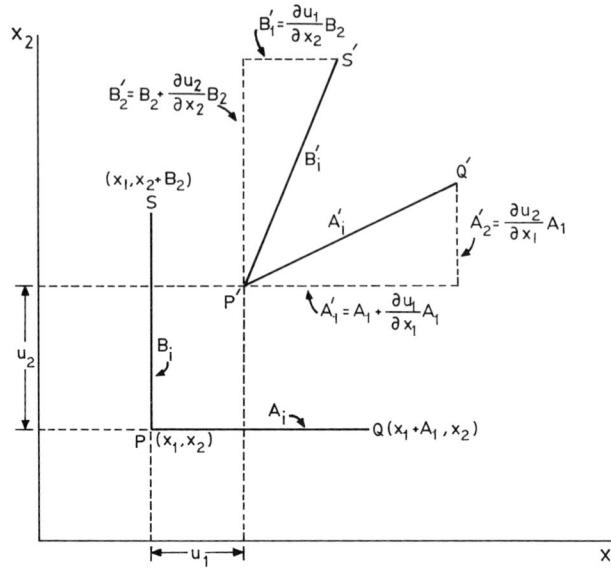

FIG. 6.5. Change in the right angle between two lines, PQ and PS, parallel to the coordinate axes, as a result of deformation.

Thus, $\delta u_1/\delta x_1$ is the longitudinal strain parallel to the x_1 axis. In a similar way it can be shown that $\delta u_2/\delta x_2$ and $\delta u_3/\delta x_3$ are the longitudinal strains along the x_2 and x_3 axes, respectively.

Next, consider two vectors A_i and B_i represented by the directed line segments PQ and PS parallel to the x_1 and x_2 axes respectively. By deformation let these vectors change to A_i' and B_i' represented by the line segments $P'Q'$ and $P'S'$ (Fig. 6.5). The angle between $P'Q'$ and $P'S'$ will not in general be 90°. Let this angle be $(90° + \delta\theta)$, where $\delta\theta$ is the change in the right angle caused by the deformation. The angle can be expressed in terms of the sides of the triangle $P'Q'S'$. In a triangle ABC with sides a, b, c, the angle C included between the sides a and b is given by the formula

$$\cos C = (a^2 + b^2 - c^2)/2ab.$$

Hence, in the triangle $P'Q'S'$,

$$\cos(90° + \delta\theta) = \frac{P'Q'^2 + P'S'^2 - Q'S'^2}{2\,P'Q' \cdot P'S'}. \tag{6.16}$$

Now,
$$P'Q'^2 = A_1'^2 + A_2'^2$$
$$P'S'^2 = B_1'^2 + B_2'^2$$
$$Q'S'^2 = (A_1' - B_1')^2 + (A_2' - B_2')^2.$$

Substituting these expressions for $P'Q'$, $P'S'$ and $Q'S'$ in eqn. (6.16) and then expressing A_1', A_2', B_1', B_2' in terms of A_1, A_2, B_1, B_2 through eqn. (6.8) and noting that (Fig. 6.5) $A_2 = B_1 = 0$, we have

$$\cos(90° + \delta\theta) = \frac{\dfrac{\delta u_1}{\delta x_2} + \dfrac{\delta u_2}{\delta x_1}}{\sqrt{\left\{\left(1 + \dfrac{\delta u_1}{\delta x_1}\right)^2 + \left(\dfrac{\delta u_2}{\delta x_1}\right)^2\right\}\left\{\left(\dfrac{\delta u_1}{\delta x_2}\right)^2 + \left(1 + \dfrac{\delta u_2}{\delta x_2}\right)\right\}}}.$$

The quantities $\delta u_1/\delta x_1$, etc., in the denominator are negligibly small in comparison with unity. Therefore, for a small deformation, this equation can be written as

$$- \sin\delta\theta \approx - \delta\theta = \frac{\delta u_1}{\delta x_2} + \frac{\delta u_2}{\delta x_1}$$

Thus, the angle between the two perpendicular lines PQ and PS is reduced by $[(\delta u_1/\delta x_2) + (\delta u_2/\delta x_1)]$. Hence, from (6.15) we find that

$$2\,\varepsilon_{12} = \frac{\delta u_1}{\delta x_2} + \frac{\delta u_2}{\delta x_1} = \text{say, } \gamma_{12}. \tag{6.17}$$

Since $\delta\theta \approx \tan \delta\theta$, $2\varepsilon_{12}$ can be regarded as a shear strain on the $x_1 x_2$-plane. In the same way it can be shown for other elements of the strain tensor that when $i \neq j$, $2\varepsilon_{ij}$ is the shear strain in the plane normal to the k axis where $k \neq i, k \neq j$. Thus, the three diagonal elements of the matrix (6.14) are the longitudinal strains while the other six elements represent shear strains. In the latter group, only three are independent. The strain tensor, thus, has six independent components:

$$\varepsilon_{xx} = \varepsilon_{11}, \varepsilon_{yy} = \varepsilon_{22}, \varepsilon_{zz} = \varepsilon_{33},$$

$$\varepsilon_{xy} = \varepsilon_{12} = \tfrac{1}{2}\gamma_{12}, \varepsilon_{yz} = \varepsilon_{23} = \tfrac{1}{2}\gamma_{yz} = \tfrac{1}{2}\gamma_{23},$$

$$\varepsilon_{zx} = \varepsilon_{31} = \tfrac{1}{2}\gamma_{zx} = \tfrac{1}{2}\gamma_{31}. \tag{6.18}$$

It should be noted that in (6.18) we have used two symbols for shear strain, ε_{ij} and γ_{ij}, $(i \neq j)$ and with $\gamma_{ij} = 2\varepsilon_{ij}$. γ_{ij}, known as the engineering shear strain, is often used in engineering as well as in geology. It is not a tensor quantity.

ω_{ij}, the skew-symmetric tensor of (6.13), causes a rigid body rotation, with three independent components ω_{23}, ω_{31} and ω_{12}, with $\omega_{ij} = -\omega_{ji}$:

$$\omega_{23} = \frac{1}{2}\left(\frac{\delta u_2}{\delta x_3} - \frac{\delta u_3}{\delta x_2}\right)$$

$$\omega_{31} = \frac{1}{2}\left(\frac{\delta u_3}{\delta x_1} - \frac{\delta u_1}{\delta x_3}\right) \tag{6.19}$$

$$\omega_{12} = \frac{1}{2}\left(\frac{\delta u_1}{\delta x_2} - \frac{\delta u_2}{\delta x_1}\right).$$

ω_{23}, ω_{31} and ω_{12} are the *components of rotation* around the x_1, x_2 and the x_3 axes, respectively. Hence they are also sometimes represented by the symbols ω_1, ω_2 and ω_3. The total *angle of rotation* is

$$\omega = (\omega_1^2 + \omega_2^2 + \omega_3^2)^{\frac{1}{2}}. \tag{6.20}$$

The *orientation of the axis of rotation* is given by the direction cosines

$$(l_1, l_2, l_3) = \left(\frac{\omega_1}{\omega}, \frac{\omega_2}{\omega}, \frac{\omega_3}{\omega}\right). \tag{6.21}$$

The rotation tensor causes a rotation of the vector A_i (the directed line segment PQ in Fig. 6.4) by an angle about a line through the point P and having direction cosines as given by eqn. (6.21).

Strain components in new coordinate axes

The derivation of the equations of infinitesimal strain is analogous to those of stresses. The components of strain will change if we rotate the coordinate axes. If the direction cosines of the new coordinate axes with reference to the old axes are as given according to eqn. (5.4), the components of strain in the new system can be found by an equation similar to that of eqn. (5.12) or eqn. (5.18):

$$\varepsilon'_{pq} = l_{ip} l_{jq} \varepsilon_{ij}. \tag{6.22}$$

This represents a set of nine equations giving the components of the strain tensor. Thus,

$$\varepsilon'_{11} = l_{11}^2 \varepsilon_{11} + l_{21}^2 \varepsilon_{22} + l_{31}^2 \varepsilon_{33} + l_{11}l_{21}(\varepsilon_{12} + \varepsilon_{21})$$
$$+ l_{11}l_{31}(\varepsilon_{13} + \varepsilon_{31}) + l_{21}l_{31}(\varepsilon_{23} + \varepsilon_{32}).$$

Principal strains

The principal extensions ε_1, ε_2 and ε_3 are determined in the same way as the principal stresses [eqns. (5.27) and (5.28)]. We start with equation:

$$\begin{vmatrix} \varepsilon_{11} - \varepsilon & \varepsilon_{21} & \varepsilon_{31} \\ \varepsilon_{12} & \varepsilon_{22} - \varepsilon & \varepsilon_{32} \\ \varepsilon_{13} & \varepsilon_{23} & \varepsilon_{33} - \varepsilon \end{vmatrix} = 0 \tag{6.23}$$

where the left-hand side represents the determinant of the matrix. Or,

$$\varepsilon^3 - I_1' \varepsilon^2 + I_2' \varepsilon - I_3' = 0 \qquad (6.24)$$

where I_1', I_2' and I_3' are the three *invariants* of the infinitesimal strain:

$$
\begin{aligned}
I_1' &= \varepsilon_{11} + \varepsilon_{22} + \varepsilon_{33} \\
I_2' &= \varepsilon_{11} \varepsilon_{22} + \varepsilon_{22} \varepsilon_{33} + \varepsilon_{33} \varepsilon_{11} - (\varepsilon_{12}^2 + \varepsilon_{31}^2 + \varepsilon_{23}^2) \\
I_3' &= \varepsilon_{11} \varepsilon_{22} \varepsilon_{33} + 2\varepsilon_{12} \varepsilon_{31} \varepsilon_{23}
\end{aligned}
$$

$$- \varepsilon_{11} \varepsilon_{23}^2 - \varepsilon_{22} \varepsilon_{31}^2 - \varepsilon_{33} \varepsilon_{12}^2. \qquad (6.25)$$

The principal extensions ε_1, ε_2 and ε_3 are given by the three roots of the cubic equation (6.24).

To determine the direction cosines of the principal axes of infinitesimal strain, substitute the value of ε_1 for ε in any two of the following equations:

$$
\begin{aligned}
(\varepsilon_{11} - \varepsilon)l_1 + \varepsilon_{12} l_2 + \varepsilon_{31} l_3 &= 0 \\
\varepsilon_{12} l_1 + (\varepsilon_{22} - \varepsilon)l_2 + \varepsilon_{23} l_3 &= 0 \\
\varepsilon_{31} l_1 + \varepsilon_{23} l_2 + (\varepsilon_{33} - \varepsilon)l_3 &= 0.
\end{aligned}
\qquad (6.26)
$$

The ratios l_1/l_2 and l_3/l_2 can be obtained from these equations. The direction cosines l_1, l_2, l_3 can then be found from standard procedures of coordinate geometry. In a similar way the direction cosines of ε_2- and ε_3-axes can be determined.

Deviatoric strain and dilatation

Just as in the case of the stress tensor, the infinitesimal strain tensor can be split up into two tensors, a *strain deviator* and a *spherical tensor*:

$$
\begin{bmatrix}
\varepsilon_{11} - e & \varepsilon_{12} & \varepsilon_{13} \\
\varepsilon_{21} & \varepsilon_{22} - e & \varepsilon_{23} \\
\varepsilon_{31} & \varepsilon_{32} & \varepsilon_{33} - e
\end{bmatrix}
+
\begin{bmatrix}
e & 0 & 0 \\
0 & e & 0 \\
0 & 0 & e
\end{bmatrix}
\qquad (6.27)
$$

where $e = \frac{1}{3}(\varepsilon_{11} + \varepsilon_{22} + \varepsilon_{33})$.

The elements of the first matrix of (6.27) represent deviatoric strain which causes a change in shape only. The second term of (6.27) causes equal elongation in all directions. Thus, the shape of an element in the neighbourhood of a point remains similar; only its volume is changed. The volumetric strain or *dilatation* is

$$\triangle = \delta V/V = \varepsilon_{11} + \varepsilon_{22} + \varepsilon_{33} = \varepsilon_1 + \varepsilon_2 + \varepsilon_3. \qquad (6.28)$$

Strains on an oblique plane

In analogy with eqns. (5.27) and (5.40), the *normal* and *shear strains across a plane* are:

$$\varepsilon = l_2^2 \varepsilon_1 + l_2^2 \varepsilon_2 + l_3^2 \varepsilon_3$$

$$\left(\frac{\gamma}{2}\right)^2 = (\varepsilon_1 - \varepsilon_2)^2 l_1^2 l_2^2 + (\varepsilon_2 - \varepsilon_3)^2 l_2^2 l_3^2 + (\varepsilon_3 - \varepsilon_1)^2 l_3^2 l_1^2 \qquad (6.29)$$

where l_1, l_2, l_3 are the direction cosines of the normal to the plane with reference to the principal axes.

For infinitesimal deformation, the absolute values of the *principal shearing strains* can be obtained, in analogy with the stress equations (5.46), as

$$\begin{aligned}
\gamma_2 &= \gamma_{max} = \varepsilon_1 - \varepsilon_3 \\
\gamma_1 &= \varepsilon_2 - \varepsilon_3 \\
\gamma_3 &= \varepsilon_1 - \varepsilon_2.
\end{aligned} \qquad (6.30)$$

The direction cosines of the line along which γ_2 acts are

$$l_1 = \pm \sqrt{\tfrac{1}{2}}, l_2 = 0, l_3 = \pm \sqrt{\tfrac{1}{2}}.$$

In other words, this direction lies on the $\varepsilon_1\varepsilon_3$-plane and occurs at an angle of 45° to the ε_1- or ε_3-axis. The direction of γ_1 lies on the $\varepsilon_2\varepsilon_3$-plane and bisects the angle between the ε_2- and ε_3-axis. The direction of γ_3 lies on the $\varepsilon_1\varepsilon_2$-plane and bisects the angle between the ε_1- and ε_2-axis.

Numerical example

Let us consider a numerical example. Let the strain matrix ε_{ij} with reference to the coordinate system x_i be

$$\begin{bmatrix}
0.1960 & 0.2408 & 0.0601 \\
0.2408 & 0.2958 & 0.0672 \\
0.0601 & 0.0672 & -0.4918
\end{bmatrix}$$

To determine the principal strains, we first find out values of the strain invariants from eqn. (6.25):

$$I_1' = 0, I_2' = -0.25, I_3' = 0.$$

Substituting these values in eqn. (6.24) we have

$$\varepsilon^3 - \tfrac{1}{4}\varepsilon = 0.$$

The three roots of this equation are

$$\varepsilon_1 = \tfrac{1}{2}, \varepsilon_2 = 0, \varepsilon_3 = -\tfrac{1}{2}$$

with $\varepsilon_1 > \varepsilon_2 > \varepsilon_3$. Substituting the value of ε_1 in the first two equations of (6.26) we find

$$l_1/l_2 = 0.8153 \text{ and } l_3/l_2 = 0.1171.$$

Therefore, the direction ratios of the ε_1-axis are $0.8151, 1, 0.1171$. The direction cosines are

$$l_1 = \frac{0.8151}{\sqrt{(0.8151)^2 + (1)^2 + (0.1171)^2}} = \frac{0.8151}{1.2954} = 0.6292$$

$$l_2 = \frac{1}{1.2954} = 0.7720$$

$$l_3 = \frac{0.1171}{1.2954} = 0.0904.$$

Since $\varepsilon_1 + \varepsilon_2 + \varepsilon_3 = 0$, there is no volume change.

Compatibility conditions

The six components of strain ε_{11}, ε_{22}, ε_{33}, γ_{12}, γ_{23}, γ_{31} cannot be prescribed arbitrarily. They are related by the following six compatibility equations:

$$\frac{\delta^2 \varepsilon_{11}}{\delta x_2^2} + \frac{\delta^2 \varepsilon_{22}}{\delta x_1^2} = \frac{\delta^2 \gamma_{12}}{\delta x_1 \, \delta x_2}$$

$$\frac{\delta^2 \varepsilon_{22}}{\delta x_3^2} + \frac{\delta^2 \varepsilon_{33}}{\delta x_2^2} = \frac{\delta^2 \gamma_{23}}{\delta x_2 \, \delta x_3} \qquad (6.31a)$$

$$\frac{\delta^2 \varepsilon_{33}}{\delta x_1^2} + \frac{\delta^2 \varepsilon_{11}}{\delta x_3^2} = \frac{\delta^2 \gamma_{31}}{\delta x_3 \, \delta x_1}$$

$$\frac{\delta}{\delta x_3}\left(\frac{\delta\gamma_{23}}{\delta x_1} + \frac{\delta\gamma_{31}}{\delta x_2} - \frac{\delta\gamma_{12}}{\delta x_3}\right) = 2\frac{\delta^2 \varepsilon_{33}}{\delta x_1 \, \delta x_2}$$

$$\frac{\delta}{\delta x_1}\left(\frac{\delta\gamma_{31}}{\delta x_2} + \frac{\delta\gamma_{12}}{\delta x_3} - \frac{\delta\gamma_{23}}{\delta x_1}\right) = 2\frac{\delta^2 \varepsilon_{11}}{\delta x_2 \, \delta x_3} \qquad (6.31b)$$

$$\frac{\delta}{\delta x_2}\left(\frac{\delta\gamma_{12}}{\delta x_3} + \frac{\delta\gamma_{23}}{\delta x_1} - \frac{\delta\gamma_{31}}{\delta x_2}\right) = 2\frac{\delta^2 \varepsilon_{22}}{\delta x_3 \, \delta x_1}.$$

The compatibility conditions are satisfied automatically when the displacements u_1, u_2, u_3 are already known. If, on the other hand, we first find out the stresses from the given load on the body and then the strains, the procedure must involve the fulfilling of the compatibility conditions.

Infinitesimal strain in two dimensions

The equations of infinitesimal strain for the *two-dimensional case* are easily derived from the foregoing analysis. Thus, from eqn. (6.22) we find that, if OX_1' and OX_2' are the new coordinate axes rotated through an angle θ from OX_1 to OX_1', then the components of strain in the new coordinate system are

$$\varepsilon_{11}' = \varepsilon_{11}\cos^2\theta + \gamma_{12}\sin\theta\cos\theta + \varepsilon_{22}\sin^2\theta$$
$$\varepsilon_{22}' = \varepsilon_{11}\sin^2\theta - \gamma_{12}\sin\theta\cos\theta + \varepsilon_{22}\cos^2\theta \qquad (6.32)$$
$$\gamma_{12}' = (\varepsilon_{22} - \varepsilon_{11})\sin2\theta + \gamma_{12}\cos2\theta.$$

From eqn. (6.32) we can verify that

$$\triangle = \varepsilon_{11}' + \varepsilon_{22}' = \varepsilon_{11} + \varepsilon_{22}$$

and is an invariant quantity. $\triangle$ is the dilatation and, for the two-dimensional case, measures the change in area.

The normal and shear strains along a direction at an angle θ to the OX_1-axis are

$$\varepsilon = \varepsilon_{11}\cos^2\theta + \gamma_{12}\sin\theta\cos\theta + \varepsilon_{22}\sin^2\theta$$
$$\gamma = (\varepsilon_{22} - \varepsilon_{11})\sin2\theta + \gamma_{12}\cos2\theta. \qquad (6.33)$$

To obtain the directions of principal infinitesimal strains we put $\delta\varepsilon/\delta\theta = 0$, and obtain the equation

$$\tan2\theta = \frac{\gamma_{12}}{\varepsilon_{11} - \varepsilon_{22}}. \qquad (6.34)$$

The two values of θ obtained from eqn. (6.34) give the orientations of the ε_1- and ε_2-axis. Substituting each of these values in the first equation of (6.33) we obtain the principal infinitesimal strains ε_1 and ε_2.

Sign convention for longitudinal strain and shear strain

Equation (6.1) shows that elongation in a direction is positive if there is an extension of a line in that direction. It is negative if there is a contraction of the line. This is in conformity with the sign convention for normal stress as followed in this book: a tensile stress is positive and a compressive stress is negative. Again, in conformity with the sign convention for shear stress in Chapter 5, the shear strain is considered as positive (cf. Dieter 1988, p.39) when it rotates a line from one positive axis towards another positive axis (or from one negative axis towards another negative axis). It is negative if it rotates a line from one positive axis towards another negative axis. It should be noted that there are other sign conventions for stress and strain in the geological literature.

CHAPTER 7

Behaviour of Rocks Under Stress

7.1. Elastic, viscous and plastic substances

The theory of stress and the theory of strain, as described in the preceding chapters, do not depend upon the physical nature of the materials. The theory of stress is based essentially on statics and the theory of strain is derived from geometrical considerations alone. To describe the physical behaviour of a material we must describe the relation between stress (or the rate of application of stress) and strain (or the rate of strain). The relation between stress and strain in actual materials is rather complex. Moreover, the exact physical behaviour of rocks at depth is not yet well understood. On the other hand, experiments indicate that the behaviour of certain substances under certain physical conditions (e.g. pressure and temperature) can be idealized into simple mathematical relations between stress and strain. Thus, Robert Hooke found that in many solids under a small tension the stress is proportional to strain

$$\sigma = E \varepsilon$$

or,

$$\sigma = 2\mu \, \varepsilon. \tag{7.1}$$

This is known as Hooke's law. A substance which follows this law is called a *Hooke solid* or a *perfectly elastic solid*. The constant E is called the *modulus of elasticity* or *Young's modulus*. μ is the *shear modulus*. The stress–strain curve of a perfectly elastic solid is shown in Fig. 7.1a. Similarly, it was found by Newton that in certain liquids in laminar motion the shear stress is proportional to the rate of shear strain (Fig. 7.1b)

$$\tau = \eta \, \dot{\gamma}, \tag{7.2}$$

where the constant η is known as the coefficient of viscosity. A substance which follows this law is known as a *Newtonian liquid* or *perfectly viscous*

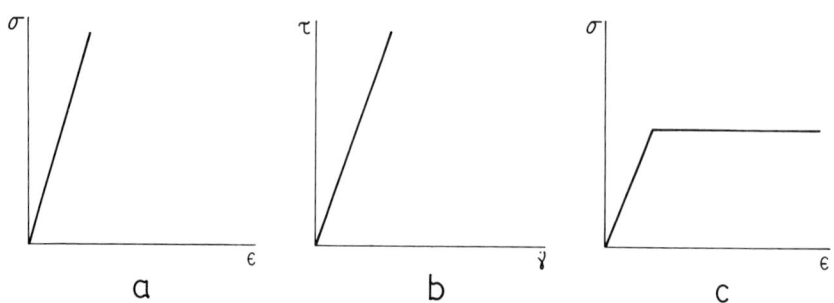

F<small>IG</small>. 7.1. (a) Stress–strain curve for perfectly elastic solid. (b) Relation between shear stress and rate of shear strain in a perfectly viscous liquid. (c) Stress–strain curve for a plastic substance.

liquid. Equation (7.1) or eqn. (7.2), giving the relation between stress and strain (or strain rate), is a *constitutive equation* or *rheological equation*. The behaviour of a substance is completely defined by its constitutive equation.

It was found by Saint Venant that when the stress is increased in certain solids, a stage is reached at which the substance no longer remains elastic; it continuously deforms at a constant stress (Fig. 7.1c). The behaviour of the material beyond the elastic range is known as plastic behaviour. An idealized substance in which the stress does not rise above the yield stress and in which the deformation increases indefinitely at the yield stress is known as a *Saint Venant substance* or a *perfectly plastic substance*. At the *yield stress* the perfectly plastic substance flows at a constant rate.

7.2. Rheological models

The elastic solid, the Newtonian liquid and the perfectly plastic substance are the three classical bodies in terms of which the general behaviour of physical bodies were expressed in earlier times. The behaviours of more complex substances are generally expressed by different manners of combination of elastic, viscous and plastic elements.

It is often convenient to visualize the physical behaviour of a substance in terms of mechanical models. Thus, the behaviour of a perfectly elastic body may be visualized by a helical spring rigidly supported at one end (Fig. 7.2a). The spring extends when a force is applied at the free end, the extension being proportional to the applied force. As in an elastic substance, the extension vanishes when the force is removed. A Newtonian liquid may be represented by a "dashpot", a perforated piston moving in a cylinder containing a viscous liquid such as oil

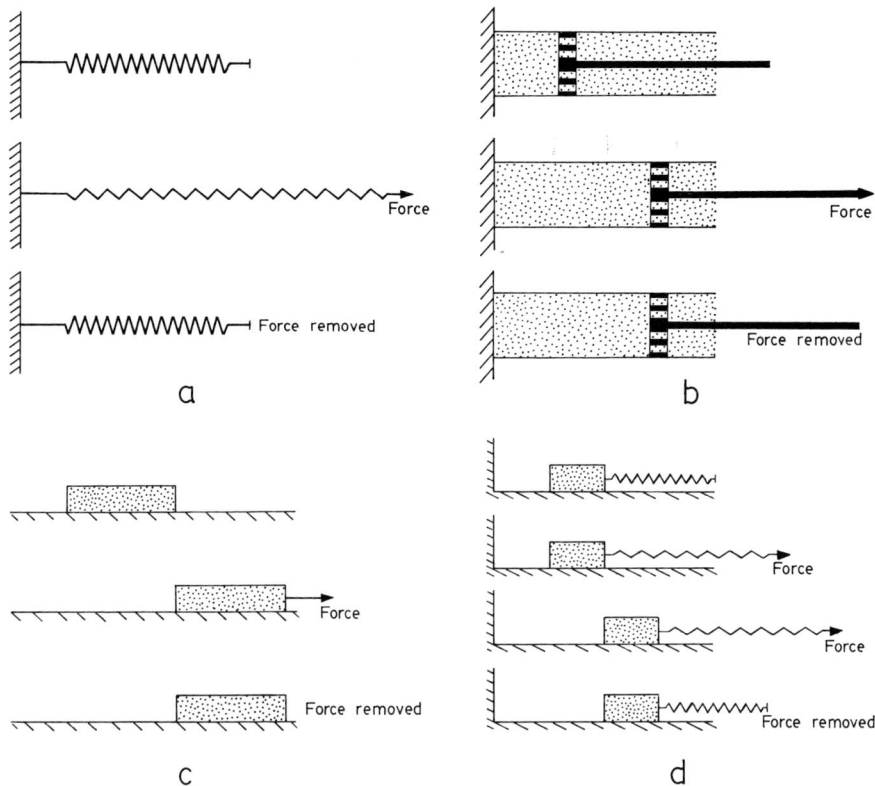

FIG. 7.2. Mechanical models for the behaviour of (a) an elastic solid, (b) a viscous liquid, (c) a perfectly plastic substance and (d) a plastic substance with an elastic deformation before yield stress.

(Fig. 7.2b). When a force, however small, is applied to the piston, it moves at a steady rate, the rate of movement being proportional to the applied force. When the force is removed, the piston does not go back. Lastly, the yield stress of a perfectly plastic substance may be represented by a weight resting on a horizontal surface (Fig. 7.2c). When the weight is pulled by a force it is not displaced as long as the pulling force is less than the frictional resistance. The weight moves continuously when the pulling force just overcomes the frictional resistance. When the force is removed, the weight comes to rest in its new position. This model evidently does not completely represent the behaviour of a plastic substance which should show an elastic deformation below the yield stress. The mechanical model for the substance (e.g. Reiner 1959) can be better represented by a spring attached in series to a weight resting on a surface (Fig. 7.2d).

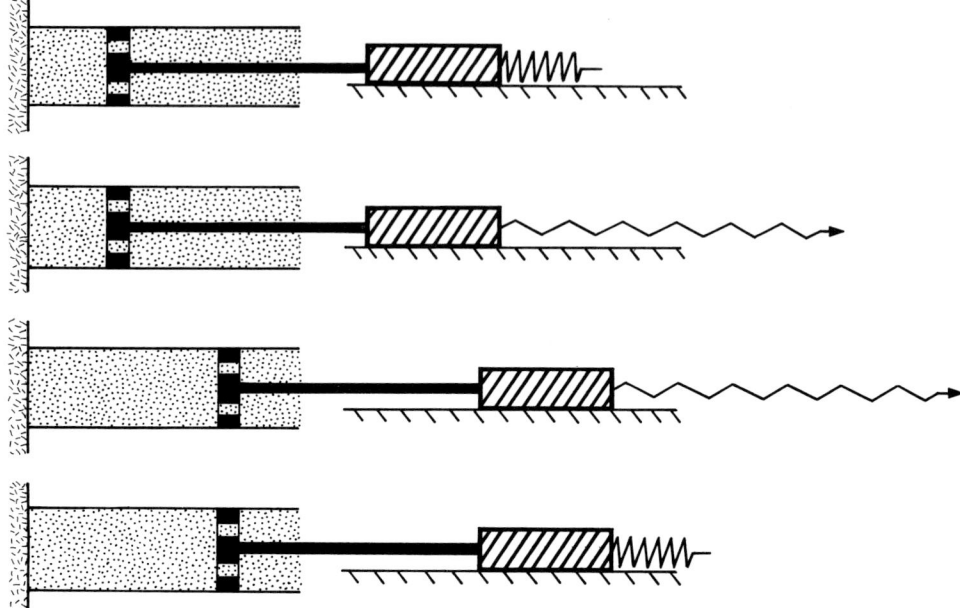

FIG. 7.3. Mechanical model for a Bingham substance.

7.3. More complex substances

The majority of natural substances, including rocks, behave in a complex manner. Their behaviours may be represented by different combinations of three basic elements of our mechanical models — the spring, the dashpot and the weight. Thus, many substances show a viscous deformation in addition to an initial elastic deformation and the presence of a yield stress. The behaviour of such a substance can be represented by the mechanical model of a spring, a friction block and a dashpot arranged in series (Fig. 7.3). Such a substance is called a *Bingham substance*. For a Bingham substance the plot of strain rate against stress shows a straight line as in a Newtonian liquid, but it differs from a Newtonian liquid in having a yield stress (Fig. 7.4).

If a spring and a dashpot are arranged in series, the model would represent the behaviour of a *Maxwell substance* or an *elastico-viscous substance*. At a small constant stress (Fig. 7.5), a Maxwell body shows an instantaneous elastic deformation (represented by the stretching of a spring element) followed by slow viscous deformation at a steady rate (represented by movement in the dashpot). If the force is removed, the elastic deformation is restored but the viscous deformation remains unaltered. If, on the other hand, the total strain is kept constant, the stress has to be reduced gradually. This is called *stress relaxation*. The

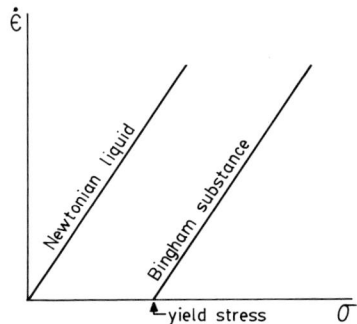

FIG. 7.4. Strain rate–stress curves for a Newtonian liquid and a Bingham substance.

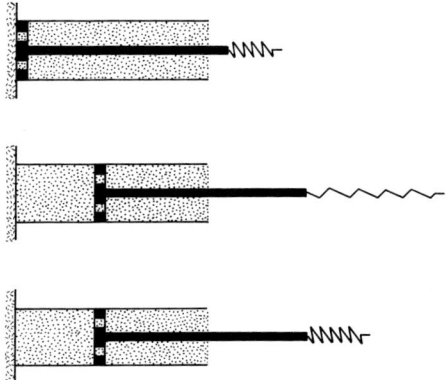

FIG. 7.5. Mechanical model for a Maxwell substance.

elastic strain also disappears gradually. In terms of the model the front end of the spring remains in place while the rear end slowly moves forward, and pulls the piston with it, till the spring regains its original length (Fig. 7.6). If a Maxwell body is suddenly stressed for a very brief interval, the behaviour is essentially like that of an elastic solid. The mechanical model should then be imagined to have a very viscous liquid in the dashpot and a weak spring. The applied force is sufficient to stretch the spring but negligibly small to move the piston through a significant distance. The behaviour similar to that of a Maxwell body, or a Maxwell liquid as it is often called, is shown by many substances. Thus, when a ball of pitch is thrown on the floor, it does not permanently deform on impact. When it is slowly pulled by hand it deforms continuously like a viscous substance. Similarly, the earth's mantle behaves as an elastic body during an earthquake but slowly deforms under small forces of long duration.

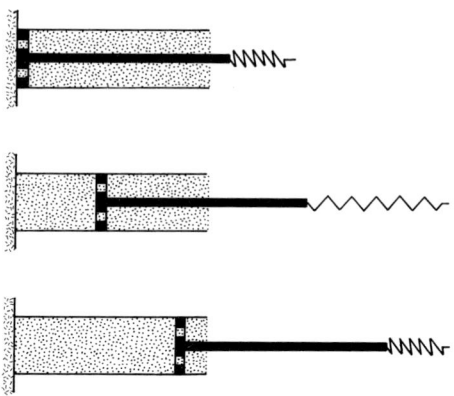

FIG. 7.6. Mechanical model for stress relaxation in a Maxwell body.

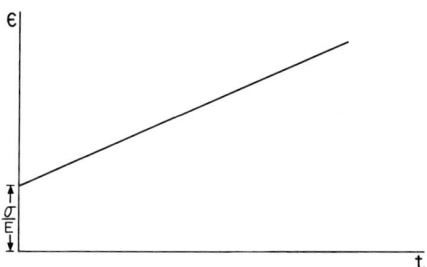

FIG. 7.7. Strain–time curve for Maxwell body.

The total stress in a Maxwell body causes different strains in the elastic and the viscous elements:

$$\varepsilon_1 = \sigma/E, \dot{\varepsilon}_2 = \sigma/\eta. \tag{7.3}$$

If the applied stress is constant, the total strain is:

$$\varepsilon = \varepsilon_1 + \varepsilon_2 = \sigma/E + \sigma t/\eta. \tag{7.4}$$

The strain–time curve for the Maxwell body is shown in Fig. 7.7. The total strain rate is, from eqn. (7.3),

$$\dot{\varepsilon} = \dot{\varepsilon}_1 + \dot{\varepsilon}_2 = \dot{\sigma}/E + \sigma/\eta. \tag{7.5}$$

If the total strain is kept constant, $\dot{\varepsilon} = 0$ and hence

$$\frac{\dot{\sigma}}{E} + \frac{\sigma}{\eta} = 0$$

or,

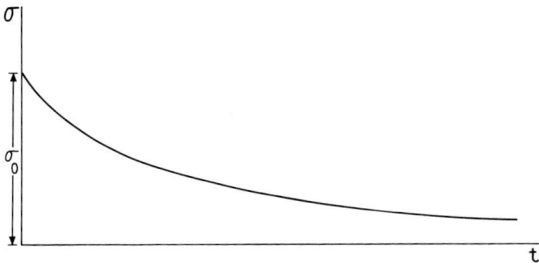

FIG. 7.8. σ-*t* curve for stress relaxation in a Maxwell body. A certain strain ε_0 is kept constant. The stress is then reduced exponentially. The figure shows that a complete relaxation of the stress may take a long time.

$$\int \frac{d\sigma}{\sigma} = - \int \frac{E}{\eta} dt$$

which gives

$$\sigma = \sigma_0 e^{-\frac{Et}{\eta}}.$$ (7.6)

The equation shows that if a certain strain ε_0 is produced by a stress σ_0 and if the strain is kept constant, the stress (σ) has to be reduced exponentially (Fig. 7.8). Since the time of relaxation of stress increases with increasing η, a highly viscous substance, such as rocks at depth, may take a long time for relaxation. Hence small residual stresses may remain in the rocks long after their tectonic deformation. This may be one of the reasons for the development of joints with regular geometrical relations with earlier deformation structures (Price 1959).

In terms of model elements a *Kelvin* (or Voigt) *body* is represented by a spring in parallel with a dashpot (Fig. 7.9). When a Kelvin body is subjected to a force, the elastic deformation does not take place instantaneously but is damped by the viscous element of the body. This feature is known as elastic after effect. Similarly, when the force is removed, the strain does not immediately disappear as in a perfectly elastic body (Fig. 7.10). The constitutive equation of a Kelvin–Voigt body is

$$\sigma = \mu \varepsilon + \eta \dot{\varepsilon}.$$ (7.7)

The model was first introduced by Lord Kelvin to explain the damping of torsional oscillations of metal wires. The model has been used to explain the behaviour of a certain type of soil containing large grains of silt embedded in fine clay and to explain the damping of earthquake waves.

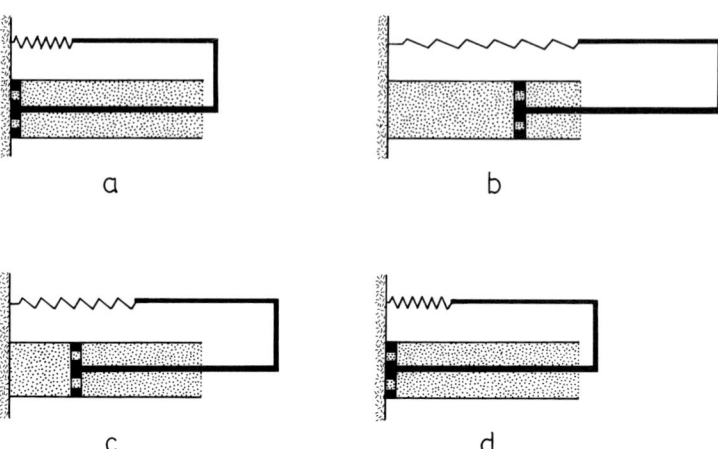

FIG. 7.9. Mechanical model for a Kelvin body.

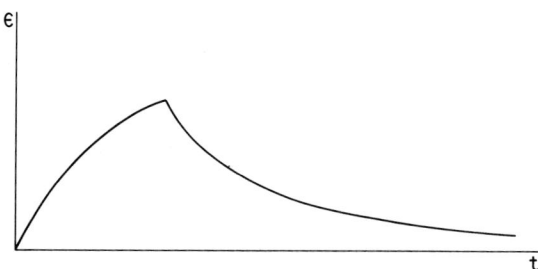

FIG. 7.10. Gradual decrease of strain in a Kelvin body when the force is removed.

The behaviour of rocks at depth and under a time span of millions of years is not yet well understood. The creep experiments give only a rough idea of the behaviour of rocks at depth. A model which explains some of the creep experiments consists of a dashpot, a friction block, a Kelvin element and a spring linked in series (Fig. 7.11). Such a substance shows an instantaneous elastic strain (represented by stretching of the spring), followed by time-dependent elastic deformation (represented by the Kelvin element) and then a viscous deformation (represented by the dashpot) with constant strain rate. The friction block indicates that the rock has a strength which has to be overcome before a permanent deformation can occur.

In a Newtonian liquid the stress is proportional to the strain rate. In many geological situations where the rocks show a viscous behaviour the relation between stress and strain rate is not linear and, generally,

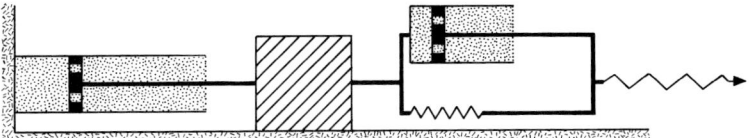

FIG. 7.11. Mechanical model of a complex substance. The model consists of an elastic element, a Kelvin element, a friction block and a viscous element.

the strain rate increases with the applied stress. The material is then said to be *non-linear*. The strain-rate dependence of the stress may be of different types. A commonly used empirical relation is that of power law creep:

$$\dot{\varepsilon} = A\,\sigma^n \qquad (7.8)$$

where A, the constant of proportionality, depends upon pressure, temperature and other factors. If n, the stress exponent, is equal to unity, the relation given by eqn.(7.8) is that of a Newtonian liquid. For $n > 1$ the tangent to the strain rate–stress curve gives us the *effective viscosity*. The effective viscosity generally decreases with increasing stress.

7.4. Brittle, transitional and ductile behaviours of rocks

Rocks do not undergo any permanent deformation at the earth's surface or at shallow depths. When subjected to a progressively increasing load, they undergo an elastic deformation followed by fracture. On the other hand, from observation of geological structures we know that rocks at depth can undergo very large permanent deformations. These observations led geologists to conclude that at high confining pressures and at elevated temperatures the rocks behave in a ductile manner. Later studies by a number of workers (e.g. Von Karman 1911, Griggs 1936, Balsley 1941, Robertson 1955, Handin & Hager 1957, Paterson 1958) quantitatively demonstrated from experiments with rock samples that the behaviour of a rock changes from brittle to ductile with increasing confining pressure. In a similar way experimental studies by Griggs *et al.* (1951, 1953, 1960), Handin & Fairbairn (1955), Turner *et al.* (1956) and Handin & Hager (1958) indicated that ductility of rocks increases with increasing temperature.

The brittle-to-ductile transition is not very sharp. There is generally a spectrum of behaviour ranging from brittle through transitional or semi-brittle to ductile. The term ductility is used here to indicate the capacity of the rock to undergo permanent deformation without the development of macroscopic fractures. Most of the detailed investigations on the brittle–ductile transition have been done with samples of marbles and limestones. At low confining pressures these rocks show a brittle behaviour. The rocks can be rendered ductile at high confining pressures at room temperatures. Heard (1960) has shown that fine-grained Solenhofen limestone can undergo permanent deformation in excess of 10 per cent strain at 1000 atm confining pressure and at 25°C. However, for most silicate rocks we need both a confining pressure and elevated temperature to make them ductile.

The experiments on rock deformation are generally carried out with cylindrical rock specimens surrounded by a fluid which applies a confining pressure. The cylinder and the fluid are separated by a metal jacket. In a tensile test the pressure across the ends of the cylinder is reduced, while in a compression test the pressure at the ends is increased. Although these tests are called triaxial tests, the principal stresses perpendicular to the cylinder axis are equal and have the same value as the confining pressure. The maximum principal stress difference, i.e. the difference between the axial stress and the confining pressure, is called the *differential stress*. It should be noted that the cross-section of the cylinder changes during deformation. Let l_0 and l be the initial and current lengths of a cylinder and A_0 and A be the initial and current cross-sections. If the volume remains constant

$$A_0 l_0 = Al$$

or,

$$A = \frac{A_0 l_0}{l} = \frac{A_0}{1 + \varepsilon}.$$

The stress is obtained by dividing the axial load by the current cross-section A. Similarly, the longitudinal strain, measured by the ratio of change in length to the original length, differs from the instantaneous strain measured with reference to the actual length at a particular time. To have the true strain at a particular time we need to calculate the natural strain or logarithmic strain.

Most modern studies express the differential stress and the confining pressure in the SI system of units. The most commonly used unit is the megapascal (MPa), i.e. meganewton per square metre.

$$1 \text{ MPa} = 10 \text{ bars} = 145 \text{ lb in}^{-2} = 10.197 \text{ kg cm}^{-2},$$
$$100 \text{ MPa} = 1 \text{ kbar}.$$

Rock deformation experiments show the following categories of structures with increasing confining pressures. Structures in the *brittle field* are represented by fractures. These are of two types, *extension fractures* and *shear fractures*. The extension fractures generally develop at little or no confining pressure. In tensile tests they develop normal to the cylinder axis. In compression tests they appear as fractures parallel to the compressive stress (Fig. 7.12a, b). The most common type of

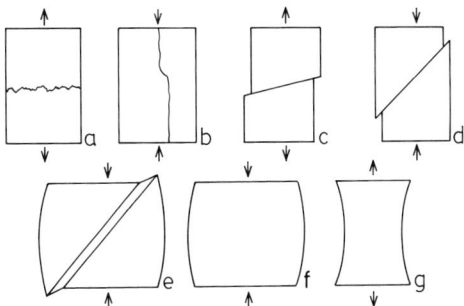

FIG. 7.12. Characteristic structures which appear in rock samples under triaxial tests. (a) and (b) Extension fracture in tensile and compressive tests. (c) and (d) Shear fractures in tensile and compressive tests. (e) Shear zone in compressive tests in the brittle–ductile transitional field. (f) and (g) Ductile deformation in compressive and tensile stress. After Griggs & Handin 1960.

structure in the brittle field at low to moderate confining pressure is shear fracture. In both triaxial extension and compression tests the shear fractures develop at an angle of 20–30° to the principal compressive stress (Fig. 7.12c, d). In certain experimental situations, in the upper part of the brittle field, we sometimes get a conjugate pair of shear fractures (Paterson 1978, fig. 48).

The most conspicuous effect in the semi-brittle or *transitional field* is a widening of the zone of shear failure. There is an intense deformation along this shear zone. The development of such a shear zone (Donath & Faill 1963, Donath *et al.* 1971) is preceded by a bulging or barrelling of the specimen (Fig. 7.12e), resulting from distributed deformation.

In the *ductile field* the deformation becomes more uniformly distributed. In compression tests the specimen shows a bulging or barrelling, while in a tensile test the specimen shows necking (Fig. 7.12f, g), and the development of clean-cut faults and cataclastic shear zones is inhibited.

The brittle–ductile transition has been most thoroughly studied in the fine-grained Solenhofen limestone (Heard 1960). These tests show that the transition takes place at much lower confining pressures in compression tests than in extension tests (Fig. 7.13). Thus, in dry extension experiments at 25°C the brittle–ductile transition takes place at more than 700 MPa, while in dry compression tests at 25°C the transition takes place at 100 MPa. The transition takes place at lower confining pressures when the temperature is increased.

The brittle–ductile transition is associated with characteristic changes in the shape of the stress–strain curve (Fig. 7.14). With increasing confining pressure, the following changes are noticeable (Heard 1960, Paterson 1978, p.162) in triaxial compression tests of marble. At

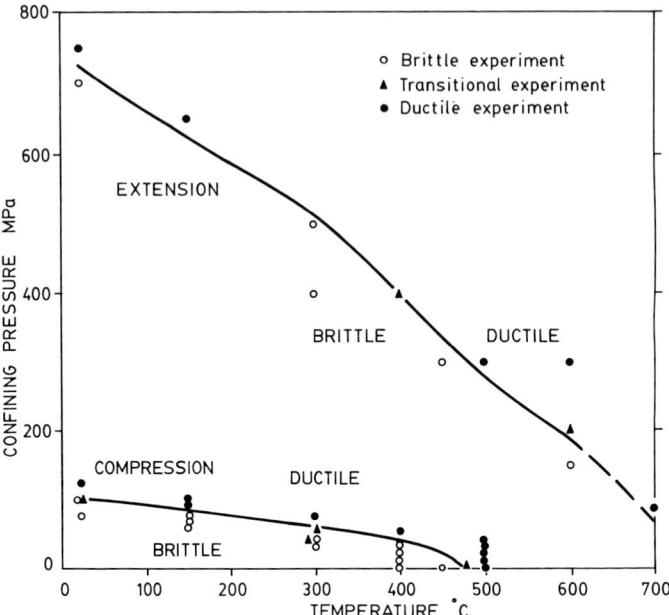

FIG. 7.13. Influence of confining pressure and temperature on the brittle–ductile transition in compression (lower curve) and extension (upper curve). After Heard 1960.

low confining pressures, in the lower part of the brittle field, there is a small amount of strain before failure takes place. The stress–strain curve, after rising steeply to a small (less than 2–3 per cent) value of strain and a relatively small value of differential stress, takes a sudden downward plunge up to the stage of final failure. As the confining pressure is increased, the stress–strain curve rises steeply to a higher value of differential stress, then assumes a gentle slope and reaches a significant amount of strain (about 3–5 per cent strain for brittle–ductile transition) before failure. When the ductile field is reached, the stress–strain curve continues to rise to larger strain and differential stress.

The stress at which the rock fails by brittle fracture is known as the *breaking strength* of the rock (Fig. 7.15a). In certain cases the stress–strain curve, after rising steeply, makes a sudden downward plunge. The stress at which this change occurs is the *yield stress*. The corresponding point on the curve is the *yield point* (Fig. 7.15b). Beyond the yield point the stress–strain curve may continue to move down till rupture, may trend parallel to the strain axis as in an ideal plastic substance (Fig. 7.16e), or may rise upward (Fig. 7.16a). For the last case, a greater stress is required to cause further permanent deformation; this phenomenon is known as *strain hardening* or *work hardening*. On the

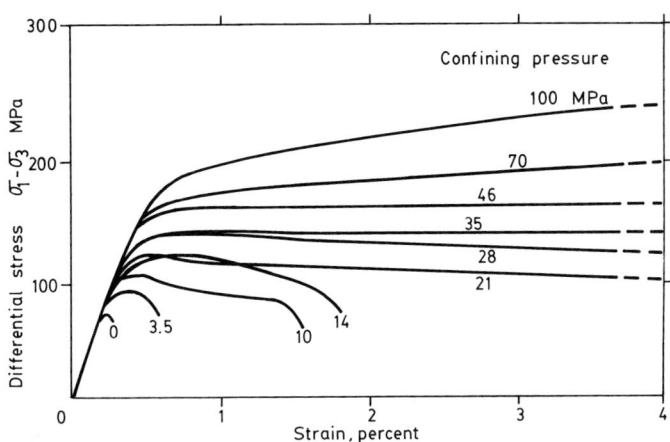

FIG. 7.14. Change in stress–strain curve with confining pressure in marble. After Paterson 1958.

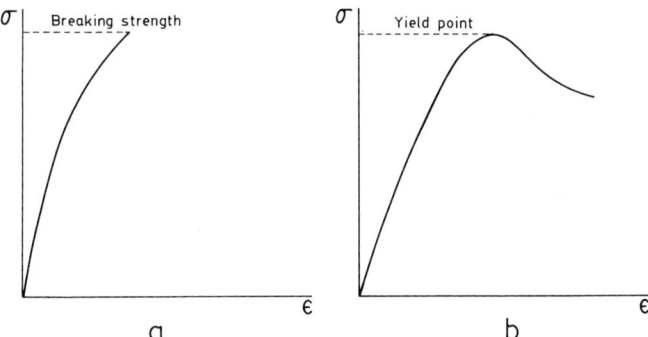

FIG. 7.15. Stress–strain curves showing breaking strength and yield point.

other hand, the stress–strain curve beyond the yield point may continue to move down (Fig. 7.16c). This phenomenon is known as *strain soften-ing*. In many cases the slope of the stress–strain curve changes rather slowly; in such cases there is no sharply defined yield point.

With increase in confining pressure, there is a marked increase in the maximum differential stress preceding brittle shear fracture. This dependence of the differential stress at shear failure on the confining pressure may be approximated by a straight line or more commonly by a curved line (Fig. 7.17) concave towards the pressure axis. For ductile behaviour we have to consider two cases. If the permanent deformation

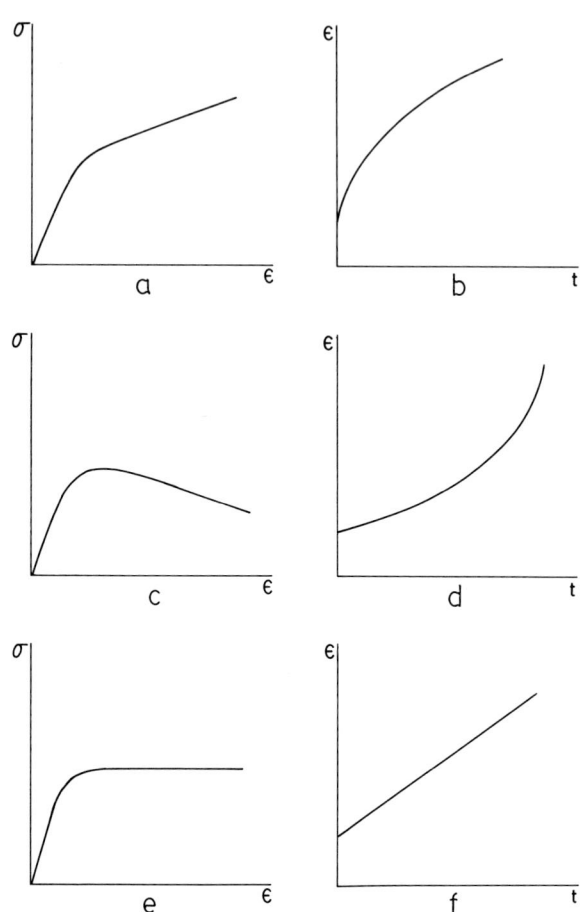

FIG. 7.16. (a) and (b) Stress–strain curve and strain–time (at constant stress) curve for a work-hardening material. (c) and (d) Stress–strain curve and strain–time curve for a material showing strain softening. (e) and (f) Stress–strain curve and ε-*t* curve for a plastic substance.

is entirely by crystal plastic processes, the confining pressure has very little influence on the yield stress (Griggs *et al.* 1951, Handin & Hager 1958, Paterson 1978). If, on the other hand, ductile deformation takes place by cataclastic flow, the stress at which the rock begins to show permanent deformation is fairly sensitive to pressure. However, this pressure sensitivity is much less than that of the fracture stress of brittle rocks. The stress at brittle shear fracture is not very sensitive to temperature. On the other hand, when ductile deformation takes place essentially by crystal plastic processes, the yield stress is markedly sensitive to temperature.

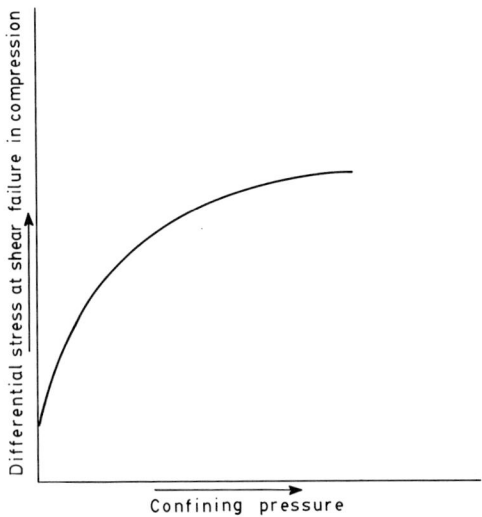

FIG. 7.17. Influence of confining pressure on differential stress at shear failure in compression.

7.5. Influence of time

Geological deformation generally takes place over a long time interval and at a very small strain rate. Hence, experiments on the influence of time on the mechanical behaviour of rocks are of considerable importance. The time-dependent aspects of deformation are commonly studied from three different types of tests: *constant strain-rate tests*, *creep tests* and *stress relaxation tests* at some confining pressure and temperature. In the constant strain-rate tests the specimen is held between pistons which are driven at a constant motor speed. The length of the specimen changes at a constant rate. A constant strain rate is not necessarily maintained at all stages. However, in many tests a constant flow stress is soon established and the main part of the test is at a constant strain rate. In creep tests the sample is subjected to a constant load under which the sample slowly deforms or creeps. For stress relaxation the sample is first permanently deformed up to a certain value of stress and strain. The machine is then stopped. The total strain remains the same, but the stress gradually decreases. The total strain is due to both the elastic and non-elastic parts. Although the total strain remains the same, the elastic strain of the sample gradually turns into non-elastic strain. The strain rate of the non-elastic part is proportional to $\delta\sigma/\delta t$, the rate of stress relaxation.

The experimental results can be represented in various ways (Fig. 7.16), e.g. by stress–strain or stress–time curves at constant strain rates, strain–time curves at constant stresses and/or by plots of log σ against

$-\log \dot{\varepsilon}$. In creep tests, in which the differential stress is kept constant, we generally, but not necessarily, find three successive stages (Fig. 7.18). In the first stage the strain rate (the slope of the strain–time curve) decreases. This stage is known as *primary* or *transient creep*. If the temperature is sufficiently high, this stage is often followed by *secondary* or *steady state creep* during which the strain rate remains more or less constant and the ε-t curve approaches a straight line. If the test is continued further, we may reach the *tertiary stage* when, in compression, the strain rate again starts to fall (Fig. 7.18a). In tensile tests the specimen usually fails by necking and the strain rate rapidly increases (Fig. 7.18b). If we prepare a stress–strain curve from the measurements of a constant strain-rate test, the steady state creep regime may be identified if the curve becomes roughly parallel to the strain axis (Fig. 7.18c). A steady state condition is generally approached only if the temperature is sufficiently high or if the strain rate is sufficiently low. The experiments indicate that an increase in temperature has more or less similar effects as a decrease in strain rate. At comparatively low temperatures or at large strain rates, rocks generally show strain hardening, i.e. a continuous rise of the stress–strain curve.

Rocks behaving in a ductile manner may not have a sharply defined yield point. For this reason and for convenience of comparison of experimental data, the strength of a rock is often measured by the differential stress (*flow stress*) corresponding to 10 per cent strain of the test specimen. In many of the experiments a steady state is reached after a few per cent of strain. For such experiments the strength is the same as the differential stress at which a steady state is reached.

7.6. Flow law for steady state creep

In the foregoing sections we have seen that the behaviour of a rock under a differential stress is influenced by confining pressure, temperature and strain rate. The influence of confining pressure is important mostly in the brittle or semi-brittle field. The flow in the ductile range is comparatively insensitive to pressure. As a first approximation the flow behaviour is therefore a function of three main variables, with the general form of flow law

$$f(\sigma, \dot{\varepsilon}, T) = 0. \tag{7.9}$$

The general form of the constitutive equation for ductile behaviour of rocks under geological situation is still being investigated; however, the experimental results for steady state creep at low differential stresses and at high temperatures and pressures can often be fitted to a *power law*:

$$\dot{\varepsilon} = A\sigma^{n} \exp(-Q/RT) \tag{7.10}$$

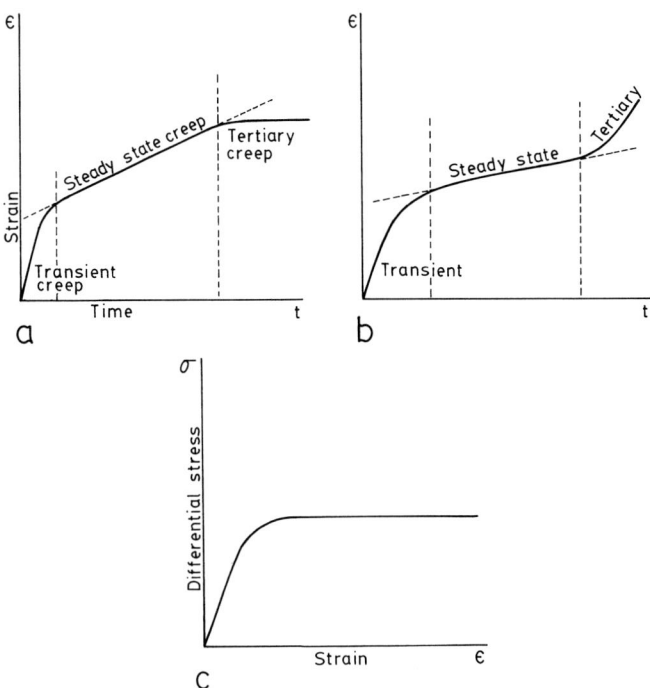

FIG. 7.18. (a) and (b) Transient creep, steady-state creep and tertiary creep in compressive and tensile tests. (c) Stress–strain curve for steady-state creep.

where $\dot{\varepsilon}$ is the strain rate, σ is the maximum differential stress, R is the gas constant, T the absolute temperature and A, Q, and n are empirical parameters. When the temperature remains constant, the strain rate for steady state flow will be proportional to σ^n and consequently the log $\dot{\varepsilon}$–log σ plot will be a straight line. n in eqn.(7.10) is called the *stress exponent*. If $n = 1$, the stress is proportional to the strain rate and the material shows Newtonian flow. For most experimental results for which eqn.(7.10) is a good fit, the values of n lies between 3 to 5 or more (Paterson 1987, p.34). Very low values of n, with $n < 2$, are rather uncommon for rocks, but have been reported from certain deformation regimes in fine-grained rocks (for example, $n = 1.7$ for the fine-grained Solenhofen limestone deformed at high temperature by Schmid *et al.* 1977, or $n = 1.5$ for a fine-grained anhydrite rock as reported by Müller *et al.* 1981).

Flow laws other than the power law have also been found to fit some results of experimental deformation. Thus, for example, a *hyperbolic sine law* (Garofalo 1965)

$$\dot{\varepsilon} = A \exp(- Q/RT) \, [\sinh (\sigma/\sigma_0)]^n \qquad (7.11)$$

or an *exponential law*

$$\dot{\varepsilon} = A \exp(- Q/RT) \exp(\sigma/\sigma_0) \qquad (7.12)$$

with an additional constant σ_0, has been found to fit certain experimental data (e.g. Schmid *et al.* 1980, Müller *et al.* 1981). For $\sigma > \sigma_0$, the hyperbolic sine law corresponds to a power law with a continuously changing value of *n*. For $\sigma < \sigma_0$, sinh $(\sigma/\sigma_0) \approx \sigma/\sigma_0$ and a hyperbolic sine law corresponds closely to the power law (Müller *et al.* 1981), especially when *n* is rather small.

7.7. Other factors influencing flow of rocks

Water weakening of quartz

Quartz is one of the most common minerals in the earth's crust and its plastic deformation is of great importance in structural geology. It is, however, extremely difficult to deform natural quartz in a dry environment in the laboratory. The early experiments by Griggs *et al.* (1960) produced very slight plastic deformation even at a temperature of 800°C and 500 MPa confining pressure. An appreciable amount of plastic deformation of quartz was first obtained by Carter *et al.* (1961, 1964) at a confining pressure of about 2000 MPa and at temperatures up to 1000°C. In the absence of water, single crystals of quartz plastically deform at a stress larger than 1000 MPa at a temperature of about 1300°C and at a confining pressure of 300 MPa (Paterson & Bitmead, described by Doukhan & Trepied 1985). Even at such large stresses and high temperatures the dislocations (see section 7.8) were found to be limited in extent and were localized in sites of stress concentration; the major mechanism of deformation consisted of shear fracturing with formation of glass lamellae (Christie *et al.* 1973, Christie & Ardell 1974).

The exceedingly large resistance of quartz to plastic deformation in dry environment under laboratory conditions is in sharp contrast with the comparative ease with which quartz has been deformed under natural conditions. This paradox was resolved by the discovery of the phenomenon of water weakening of quartz by Griggs & Blacic (1964, 1965). They found that clear natural crystals of quartz can be plastically deformed at comparative ease, at about 800–1000°C under a confining pressure 1500 MPa (15 kilobar) and with a flow stress of about 200 MPa for a strain rate of 10^{-5} per second. The water in these experiments was made available by the dehydration of the talc pressure vessel. A similar *water weakening* or *hydrolytic weakening* was also subsequently seen during deformation of synthetic quartz crystals (Fig. 7.19a) which contained an appreciable amount of grown-in

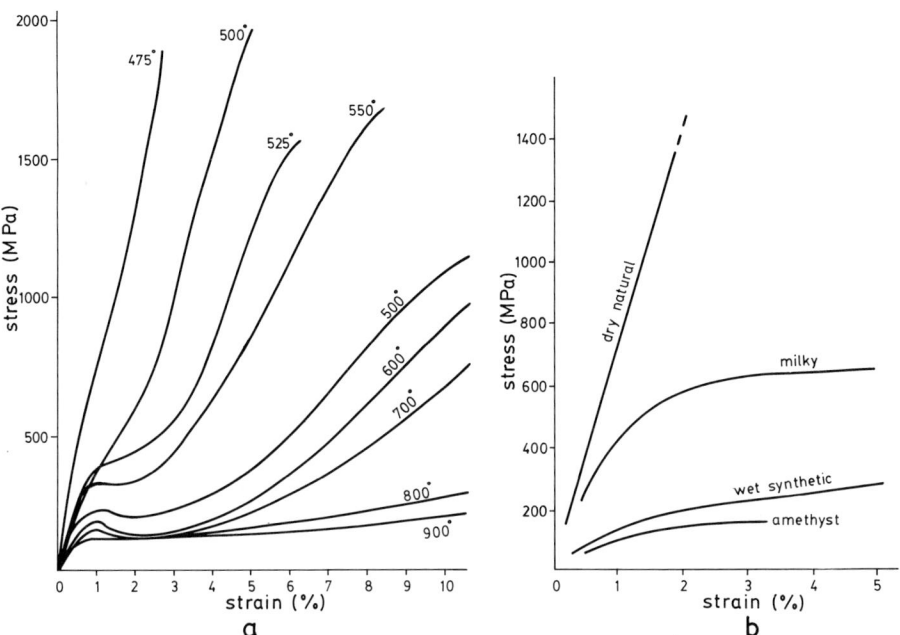

FIG. 7.19. (a) Stress–strain curves for specimens of "wet" synthetic quartz at 300 MPa confining pressure and at different temperatures (°C). After Morrison-Smith *et al.* 1976. (b) Stress–strain curves for "dry" natural quartz, natural milky quartz, amethyst and "wet" synthetic quartz, at 300 MPa confining pressure, 800°C and 10^{-5} s^{-1} strain rate. After Kekulawala *et al.* 1978.

water (Baëta & Ashbee 1967, 1970, Hobbs *et al.* 1972, Blacic 1975, Morrison-Smith *et al.* 1976). Water weakening was also found in natural amethyst crystals containing an appreciable amount of water (Kekulawala *et al.* 1978). Different aspects of the phenomenon of water weakening have been experimentally investigated by many other workers (e.g. Balderman 1974, Kirby & McCormick 1979, Kekulawala *et al.* 1981, Linker & Kirby 1981, Linker *et al.* 1984, Blacic & Christie 1984, Kronenberg & Tullis 1984, Jaoul *et al.* 1984, Mainprice & Paterson 1984, Doukhan & Trepied 1985, Ord & Hobbs 1986, Paterson 1989). The experiments have clearly demonstrated that water plays a decisive role in the plastic deformation of quartz. The experiments of Griggs & Blacic were carried out in a solid-medium apparatus. However, similar experiments with natural quartz crystals in a gas-medium apparatus failed to produce a significant weakening (Paterson & Kekulawala 1979). It has been suggested that the weakening observed in solid-medium apparatus was greatly assisted by introduction of water through microcracking and crack healing (Kirby & Kronenberg 1984, Kronenberg *et al.* 1986).

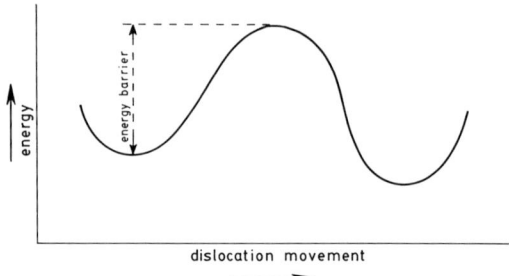

FIG. 7.20. Schematic representation of energy barrier which has to be overcome for moving a dislocation.

Although water weakening of quartz (Fig. 7.19a, b) has been demonstrated by experiments, the exact manner in which this weakening is brought about is not yet fully understood. Water may be present in quartz in different forms (e.g. Paterson 1989), i.e. as molecular aggregates (for example, as optical or smaller-scale bubbles or liquid inclusions), in solid solution and in dislocation cores (i.e. the regions near the dislocation lines where the crystal lattice is highly distorted). The experiments indicate that the rheology of quartzites and quartz crystals not only depends on the bulk concentration of water but also to a great extent on the scale of dispersion of water. Thus, milky natural quartz and many quartzites contain a larger amount of water than clear natural amethyst crystals or synthetic single crystals of quartz. Yet, the milky quartz and quartzites have generally a larger strength (Fig. 7.19b). This difference has been explained by different scales of dispersion of water in the two sets of samples (e.g. Paterson & Kekulawala 1979, Kronenberg & Tullis 1984, Paterson 1989).

Although the detailed mechanism of dislocation flow of quartz in a wet environment is not fully understood as yet, it has been tentatively suggested that there are three regimes of deformation (Doukhan & Trepied 1985, Paterson 1989).

(1) *Intrinsic regime*, at a relatively low temperature. As indicated in section 7.9, plastic deformation generally takes place by intracrystalline slip. In the region between the slipped and the unslipped portions there is a lattice distortion. This is the region of a *dislocation*. As the dislocation moves forward it leaves behind an undisturbed lattice. The movement of the dislocation is opposed by an energy barrier (Fig. 7.20). This is known as the *Peierls barrier*. At low to moderate temperatures the Peierls barrier of quartz is very high. This is the reason why, in the intrinsic regime, even in the presence of water, quartz remains extremely strong.

(2) *Hydrolytic glide control regime*, at an intermediate temperature. At such temperatures the presence of water enables the Peierls barrier

to be lowered so that it can be overcome and the dislocation can move at comparative ease. It has been suggested that the presence of water facilitates the movement of dislocations through nucleation and propagation of kinks (Griggs 1967, 1974). Some diffusion of water is also required for sites of kinks for continued nucleation. A steady state creep is unlikely to be achieved in this regime.

(3) *Recovery controlled regime*, at a relatively high temperature. In the presence of water at such high temperatures the Peierls barrier is negligible. Transmission electron microscope studies of plastically deformed synthetic quartz crystals (Morrison-Smith *et al.* 1976) shows that, while at low temperatures the dislocations are commonly straight and lie along simple crystallographic directions, in the high-temperature regime the dislocations are commonly curved, intertwined, with many small isolated loops and do not often lie in a single plane. The dislocation structures are typical of high temperature and imply dislocation climb. Both glide-controlled plastic deformation and recovery-controlled plastic deformation of quartz greatly depend on the diffusibility of water and on concomitant diffusion of oxygen and/or silicon (Paterson 1989).

Water weakening of olivine

The plastic deformation of olivine and olivine-rich rocks has received a great deal of attention in recent years. This is because of the realization that many global scale geological processes depend on the plastic deformation of the upper mantle in which olivine is the most dominant mineral constituent.

From a large number of experiments under controlled geologic conditions, we now have a fairly good idea about the rheology of olivine single crystals and of polycrystalline aggregates in dry conditions (e.g. Durham & Goetze 1977a, b, Karato *et al.* 1982, Kohlstedt & Ricoult 1984, Chopra & Paterson 1984). At a temperature of 1300°C and 300 MPa confining pressure single crystals and polycrystals of olivine show a steady state creep (Fig. 7.21) with the strain rate being approximately proportional to the cube power of stress. The flow stress also depends on the orientation of the slip systems. The flow stress is comparatively small when the planes of easiest slip are favourably oriented. Karato (1988) found that deformed polycrystalline aggregates of olivine show a strongly inhomogeneous strain distribution. Grains in which the easiest slip system is most favourably oriented have undergone the largest deformation.

The creep strength of olivine is markedly reduced by the presence of water (e.g. Carter & Ave Lallemant 1970, Blacic 1972, Post 1977, Chopra & Paterson 1981, 1984, Poumellec & Jaoul 1984, Mackwell

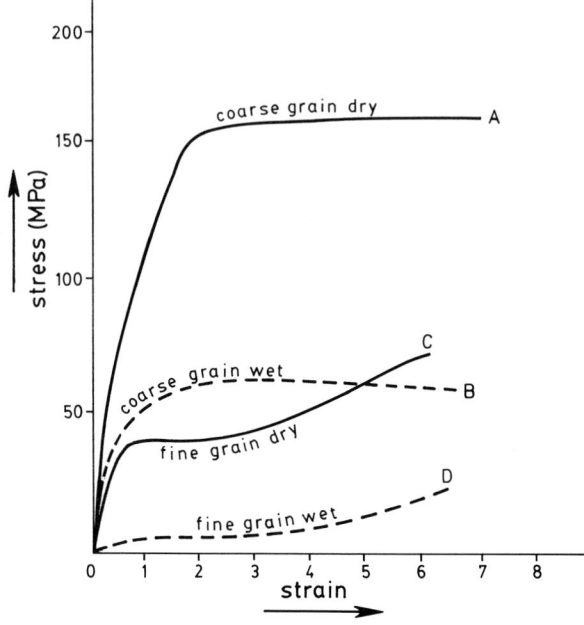

FIG. 7.21. Some typical stress–strain curves for plastic deformation of olivine. The curves *A* and *B* are for the coarse-grained specimens and the curves *C* and *D* are for the fine-grained specimens. The continuous and dashed lines are for "dry" and "wet" specimens, respectively. After Karato *et al.* 1986.

et al. 1985). The experiments of Chopra & Paterson showed that the fine-grained dunite specimens are weaker than the coarser grained specimens. In their experiments a small amount of melt phase appeared at the grain boundaries and the deformation was influenced by grain boundary processes. To avoid this effect Karato *et al.* (1986) conducted experiments with hot-pressed crushed olivine crystals in which the effect of partial melting was negligible. Two deformation regimes were recognized from these experiments. A steady state flow stress was achieved at 2 per cent strain in the coarse-grained regime. For these coarse-grained samples the flow stress was insensitive to grain size and the stress exponent was 3–4. With a fine-grain size a steady state flow stress was attained at smaller strains, about 0.2–0.5 per cent. This stage was followed by some hardening due to concurrent grain growth. In the fine-grained regime the flow stress is very sensitive to grain size. The flow stress increases with increasing grain size. The stress exponent is about 1. Water weakening effect is present in both the regimes. The weakening effect is more prominent in the fine-grained specimens than in the coarse-grained ones. According to Karato *et al.* (1986, also Karato 1989), in the coarse-grained regime or grain-size insensitive regime, the deformation takes place mostly by dislocation creep while in the fine-grained regime or grain-size sensitive regime diffusion creep is the dominant mechanism. In both cases the creep strength is significantly lowered by the presence of trace amounts of water (Fig. 7.21). Although the experimental deformations have given us a rough picture of the rheology of olivine, our knowledge about the deformation mechanism in olivine is still rather poor (Poirier 1985). In addition, there are many uncertainties in extrapolating the laboratory rheological data to the actual geological situation in which the strain rate (about 10^{-14} per second) is much smaller and the flow stress, roughly between 0.1 and 10 MPa, is much lower. An approximate idea of the behaviour of the upper mantle can be obtained from the tentative extrapolation by Paterson (1987).

Influence of confining pressure

The confining pressure has a relatively small influence on the flow stress or the creep rate. However, its effect may not be negligible if the laboratory data are extrapolated to regions of very high pressure in the earth's interior, especially to the deeper part of the mantle (Poirier 1985, pp. 149–150). Confining pressure is also significant in hydrolytic weakening of quartz. The water weakening effect of quartz was observed by Griggs (1967) at 15 kb confining pressure. The weakening effect was absent at low confining pressures (Kekulawala et al. 1978). This suggests that confining pressure plays an important role in hydrolytic weakening (Poirier 1985, p. 158).

Influence of grain size

Recent experimental studies have clearly demonstrated that grain size is an important variable in flow of rocks in certain deformation regimes in which the temperature is relatively high and the strain rate and flow stress are low. Thus, fine-grained Solenhofen limestone is stronger than the coarser-grained Carrara and Yule marbles at temperatures up to 500–600°C. On the other hand, at higher temperatures the limestone is weaker; the relative weakening of the limestone at the higher temperature was influenced by its fine-grain size (Schmid et al. 1977, 1980). A similar influence of grain size is also clearly brought out by the results of experimental deformation of anhydrite rocks by Müller et al. (1981). Figure 7.22 shows the strength at 10 per cent strain as a function of initial grain size at two different strain rates. The figure shows that above 300°C the strength of the fine-grained rocks is strongly dependent on the temperature and strain rate. The coarse-grained rock remains relatively insensitive to temperature and strain

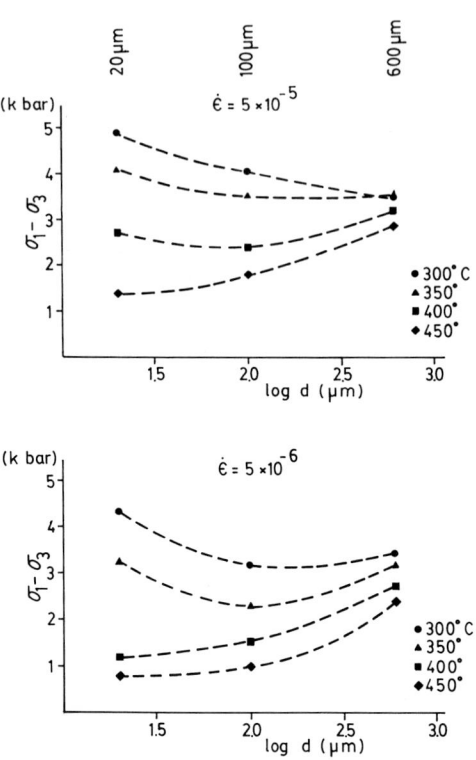

FIG. 7.22. Influence of grain size on the flow stress in experimental deformation of anhydrite rocks at different temperatures and at two different strain rates. After Müller *et al.* 1981.

rate even at higher temperatures. A striking example of the role of grain size is seen in experimental deformation of synthetic polycrystalline olivine crystals (Karato *et al.* 1986). These experiments show the presence of two different flow regimes. The flow of the coarse-grained olivine aggregate is not much dependent on grain size and has a fairly high stress exponent, $n \approx 3$ or more. The finer-grained material shows a strong grain size dependence, with n slightly above 1.

7.8. Dislocation and related phenomena

The variety of ductile behaviour of rocks and their dependence on pressure, temperature, strain rate and other factors is generally explained in terms of the deformation mechanisms which operate in the rocks. Before considering these agents of deformation, it will be helpful to have an elementary idea about dislocation and related phenomena. We shall discuss the deformation mechanisms in section 7.9.

Real crystals show defects or imperfections in the periodic arrangement of the lattice. The deviation from the periodic arrangement may appear as a *point defect*, localized in the neighbourhood of a few atoms, as a *line defect* and as a *surface* or *plane defect*. A *vacancy* is a type of point defect caused by the missing of an atom from its lattice position. Vacancies can migrate by exchange of ions from adjoining sites and can therefore be an agent of deformation through diffusion or atomic transfer of matter. The most important line defect is *dislocation* which is responsible for deformation of crystals by intracrystalline slip. Surface defects form when line defects are clustered on a surface. Grain boundaries, sub-grain boundaries and mechanical twins are examples of surface defects.

Dislocation is one of the most important agents of ductile deformation. There are two basic types of dislocation, *edge dislocation* and *screw dislocation* (Fig. 7.23). Figure 7.23a is a schematic representation of an edge dislocation with *ABCD* as the slip plane. The block above this plane has moved towards the left with respect to the block below the plane. The slip vector is from right to left. *AB*, the line separating the slipped and unslipped regions, is an edge dislocation perpendicular to the slip vector or Burger's vector *b*. A slip plane in the crystal is generally a plane which has the greatest atomic density along it. The movement of a dislocation requires much less shear stress than the theoretical shear stress. A characteristic feature of dislocation is that slip need not occur along a plane across an entire crystal but slip can occur in a part of a plane. The lattice arrangement is distorted only near the boundary of the slip; elsewhere it is undisturbed. In other words, the movement of the atoms is restricted only to the neighbourhood of the dislocation. As the dislocation moves forward, it leaves behind an undisturbed lattice. Figure 7.24a shows eight vertical planes in an undisturbed lattice. In Fig. 7.24b the left edge of the upper part has moved towards the right by one lattice spacing, while the right part of the lattice is still undisturbed. Above the potential slip plane the vertical planes 8 and 8′, 7 and 7′, etc., on the right part of the lattice are connected together. On the left side, however, the half plane 1 above the slip plane is connected with the half plane 2′ of the lower block. Consequently, there must be a half plane in the upper block which is not connected with one in the lower block. In Fig. 7.24b this extra half plane is at 5. The edge dislocation is on the slip plane below 5. The extra half plane 5 is now situated between 5′ and 6′ of the lower block. When the upper block is further sheared towards the right, the half plane 5 lines up with 6′ and thereby cuts off the half plane 6 from its continuity with 6′. The edge dislocation has thereby moved towards the right and is situated between 6′ and 7′ (Fig. 7.24c). The dislocation moves towards the right in this manner. When it reaches a free surface it gives rise to a step-like slip equal to the Burger's vector;

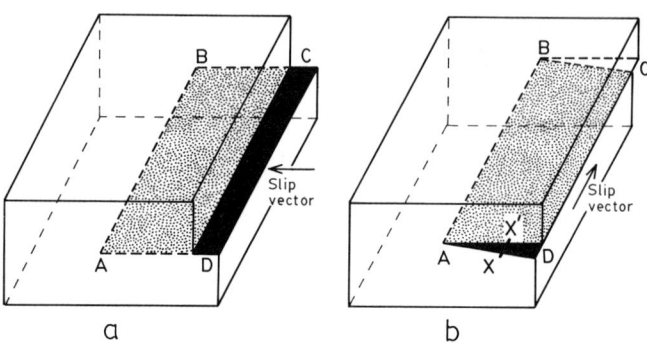

Fɪɢ. 7.23. Schematic representation of (a) edge dislocation and (b) screw dislocation.

the interior of the crystal is now dislocation-free (Fig. 7.24d). The slip plane and the slip direction together form a slip system.

Screw dislocation is a line defect which is parallel to the Burger's vector (Fig. 7.23b). In Fig. 7.23b, if we move from a point X' at the top to the point X along a circuit around the line of screw dislocation, we come down to a lower level on a spiral path (in the present case in the form of a right-handed screw). Unlike the edge dislocation, a screw dislocation does not have preferential slip planes.

In actual crystals, dislocation lines are rarely straight. They generally form curves or loops, the *dislocation loops*. When a small segment of the loop is either perpendicular or parallel to the Burger's vector, it is either a pure edge dislocation or a pure screw dislocation. Otherwise it has a mixed character of both edge and screw components.

A pure edge dislocation can move out of its plane and can climb up or down to a parallel slip plane. In order to climb, it has either to remove atoms from the extra half plane or to add atoms to the extra half plane. Thus, *dislocation climb* is diffusion-controlled and hence is likely to be effective at a relatively higher temperature. Dislocation climb does not take place in screw dislocation. In screw dislocation the slip can, however, change from one plane to another inclined plane, provided the inclined plane contains the common slip direction. This process, associated with screw dislocation, is known as *cross-slip*.

Crystals generally contain a number of dislocations. The number increases with increasing strain. Dislocation may pile up (forming a network) in front of a barrier or may interact with each other in such a manner as to impede deformation. This is the main reason for the phenomenon of strain hardening.

Apart from intracrystalline slip or translation glide, a shearing movement of layers of atoms may produce *deformation twins*. In intra-crystalline slip there is movement by an integral number of lattice spacings. The movement to cause deformation twins is much less than

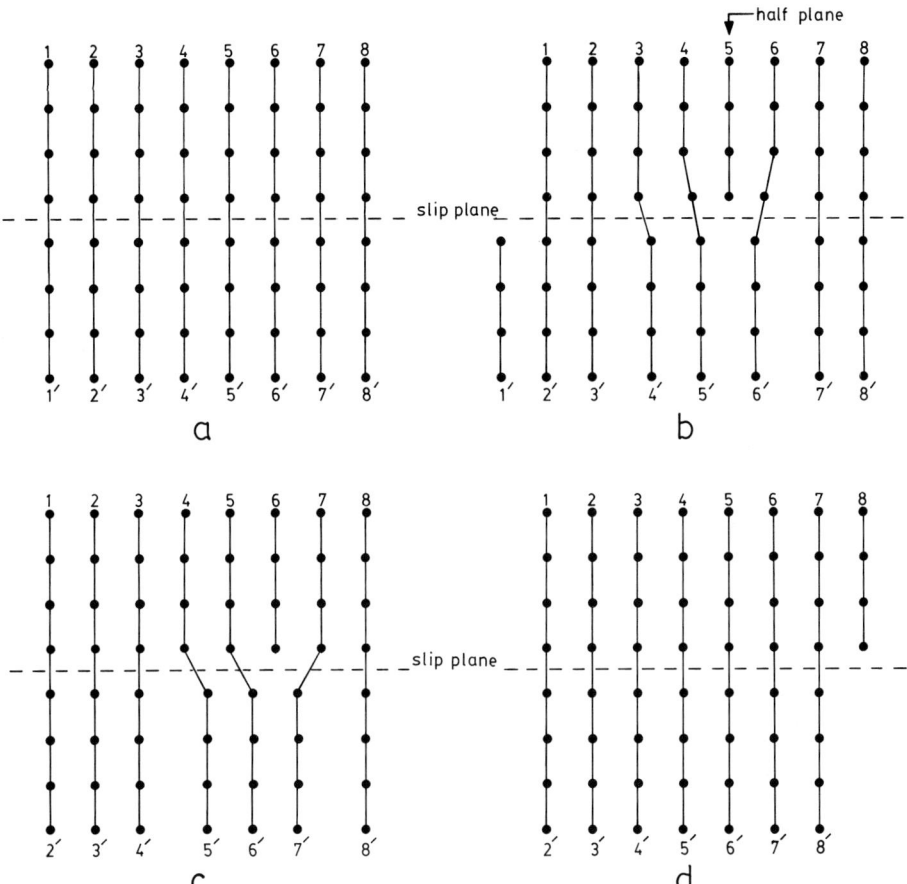

FIG. 7.24. Schematic representation of movement of edge dislocation through a crystal lattice.

that of lattice spacing. In translation gliding the crystal lattice has the same orientation above and below the slip plane. In twinning the lattice orientation is changed to a mirror image of the parent crystal. Unlike intracrystalline slip, mechanical twinning causes a relatively small amount of total deformation. However, it is often an important agent of plastic deformation.

7.9. Deformation mechanisms

Types of deformation mechanism

Leaving aside the flow of defect-free crystals, plastic deformation of polycrystalline aggregates may take place by at least five independent

113

and distinguishable deformation mechanisms (Ashby 1972). These are (1) *glide-controlled creep* caused by glide motion of dislocations, (2) *dislocation creep* involving both glide and dislocation climb, (3) *Nabarro creep* involving diffusion resulting from movement of point defects through the grains, (4) *Cobble creep* involving diffusion along grain boundaries and (5) *twinning*. In addition, grains or groups of grains may move relative to each other and cause permanent deformation. This last process may include a number of quite different mechanisms among which we may distinguish at least two categories, which are generally referred to as *cataclastic flow* and high temperature *grain boundary sliding*. Although both these involve relative movement between grains, they operate in quite different deformation regimes. These various processes are generally considered under three broad headings:

(1) Cataclastic flow.
(2) Crystal plasticity, including glide-controlled creep, dislocation creep and twinning.
(3) Diffusional flow and grain boundary sliding.

There are generally more than one of these processes operating during ductile flow. Thus, during mylonitization under ductile conditions quartz may deform by intracrystalline gliding along with some pressure solution (Mitra 1976). However, we can often distinguish deformation regimes in which one particular mechanism is dominant.

Cataclastic flow

Recent studies of both natural and experimentally deformed rocks indicate that cataclastic flow is an important deformation mechanism. In cataclastic flow the rock behaves in a macroscopically ductile manner but deformation is achieved by fracturing of grains and by rotation and sliding of the fragments. The early experimental studies on brittle–ductile transition (Heard 1960, Byerlee 1968) were in fact a transition from localized cataclasis (leading to brittle faulting) to a distributed microcracking, leading to macroscopically ductile flow (Rutter 1986). An essential requirement for cataclastic flow is the growth of uniformly distributed microcracks in a stable manner so that the cracks are neither localized nor are propagated rapidly to form a brittle fault before significant permanent deformation is achieved.

Cataclastic flow is sensitive to confining pressure. This is because cracking of the grains and the relative movement of the fragments tend to increase the volume of the rock. Secondly, an increase in confining pressure causes an increase in the normal stress across the microcracks and thereby resists frictional sliding along them. In general, cataclastic flow is not markedly sensitive to temperature and strain rate (Donath & Fruth 1971).

Cataclastic flow is often easy to identify from microscopic observation of abundance of grain-scale faults, offset twin lamellae, development of microcrush zones with angular fragments of a wide range of grain size along with a general absence of intracrystalline deformation. However, in many cases, cataclastic flow may not be easily identifiable from observations in an optical microscope. This is partly because some of the optical microstructures resulting from cataclasis may superficially resemble those produced by crystal plastic processes. Tullis & Yund (1987) have shown that feldspar aggregates experimentally deformed in a cataclastic flow regime show a flattening of original grains, patchy undulatory extinctions and sub-grain-like structures. Observations in transmission electron microscope (TEM), however, show that the grains are extensively cracked and faulted along cleavage planes in a more or less penetrative manner and this has caused a flattening of the grains and the development of a flattening foliation in the rock. TEM studies also show that the patchy undulatory extinction and the sub-grain-like structure have resulted from microcrack arrays and microcrush zones (Tullis & Yund 1977, 1987).

Cataclastic flow does not necessarily lead to the development of shear zones. Cataclasis may cause a distributed deformation without localization (e.g. Stearns 1969, Hadizadeh & Rutter 1983, Blenkinsop & Rutter 1986), although the deformation within certain low-temperature shear zones may indeed take place dominantly by cataclastic flow. The transition from brittle faulting to cataclastic flow at room temperature and with increasing confining pressure is only one mode of the brittle – ductile transition (Paterson 1978, Rutter 1986). In other physical conditions brittle – ductile transition may involve a transition from brittle faulting to plastic flow.

Crystal plasticity

Deformation by crystal plastic processes can usually be identified by optical microstructures (e.g. Carter *et al.* 1964, Christie *et al.* 1964, White 1973b), e.g. deformation twins, undulatory extinctions, deformation bands, deformation lamellae, Boehm lamellae and presence of equiaxed grains and subgrains. Although deformation twins may form in different minerals, they are particularly important in deformation of limestones and marbles. *Deformation bands*, frequently seen in quartz, calcite and other minerals, are somewhat irregular but roughly tabular zones in which the extinction is slightly different from that of the parent grain. *Deformation lamellae* are fine sub-parallel planes in quartz with a slightly different refractive index from that of the host grain. *Boehm lamellae*, also occurring in quartz, are similar to deformation lamellae in appearance but are marked by trails of bubbles. In certain cases evidence of plastic deformation in quartz grains is also found from trails of broken rutile needles or folding of boudinaged rutile needles included in grains of quartz (Mitra 1976).

Rocks deformed at high and moderate temperatures generally show equiaxed sub-grains and recrystallized grains. *Sub-grains* are groups of equiaxed polygonal units, forming a part of a grain, and with small misorientation among them. The misorientation may range from less than a degree to about 10°. The boundary between sub-grains is a *low angle boundary*. The blocks or polygons representing sub-grains contain few dislocations. The formation of sub-grains is associated with the process of redistribution of dislocations. This process of redistribution of dislocation resulting in the formation of sub-grains is known as *polygonization*. It is likely that the redistribution of dislocations in polygonization takes place by a combination of different mechanisms, slip, climb of edge dislocations and cross-slip of screw dislocations. Since the peripheral region of a grain, because of interaction with the deformation of adjoining grains, is strained to a greater extent, polygonization sometimes gives rise to a core-and-mantle structure, with the core of a grain free from subgrains and with a polygonal network of equiaxed sub-grains in the mantle. With progressive deformation or at higher temperatures the sub-grain structure is seen over the entire grain. With increase in strain, the misorientation of the sub-grain increases; when the misorientation is sufficiently large ($> 10°$), the equiaxed blocks are separated by high angle boundaries or grain boundaries (Hobbs 1968, Poirier & Nicolas 1975). This process, by which strain-hardened grains are replaced by unstrained grains in the course of deformation, is known as *dynamic recrystallization*. Polygonization, which also causes a softening of the grains during deformation at elevated temperatures, is a recovery structure; this process is known as *dynamic recovery*. Because of dynamic recovery and dynamic recrystallization, the dislocation pile-ups are eliminated and the grains can undergo further deformation, sometimes leading to an elongation of grains and sub-grains. Thus, the sub-grain and grain structures continuously undergo an evolution, one set of sub-boundaries or boundaries being replaced by a new set. Means & Ree (1988) have shown from directly observed microstructural evolution of octachloropropane that sub-grain boundaries may form in different ways. Some of these different types look alike under the optical microscope. Means's studies suggest that the occurrence of sub-grains in recrystallized grains is not always an evidence of dynamic recrystallization.

Dynamic recrystallization may take place by *rotation recrystallization* and by *migration recrystallization*. Rotation recrystallization is achieved by progressive misorientation of sub-grains. Owing to migration recrystallization the dislocation-free nuclei grow in size by migration of high angle boundaries. The increase in dislocation density impedes the motion of dislocations through the lattice. As a result, a larger stress is required to cause plastic deformation. This process is known as *strain hardening* or *work hardening*. This process is counteracted by the softening caused by dynamic recovery and dynamic recrystallization. A steady state creep is achieved when there is a balance between the rates of these two competing processes.

The deformation mechanisms described in this section may cause different types of creep, with different flow laws. When features produced by dislocation glide (undulatory extinction, deformation bands, etc.) are prominent, we generally have a *glide-controlled creep*. If, on the other hand, recovery structures are prominent, we have the case of *dislocation creep* (Ashby 1972) or recovery-controlled creep (Poirier 1985). In this type of creep, dislocation glide is associated with the climb of the edge dislocations and cross-slip of screw dislocations. As we saw in section 7.8, climb involves the diffusion of point defects; hence, when climb is an important mechanism, the creep is also diffusion-controlled. The flow law of recovery-controlled creep is a power law with $\dot{\varepsilon}$ proportional to σ^n. Recovery-controlled creep is also sometimes described as *power-law creep* (Poirier 1985, p. 103).

Dislocation creep may be associated with both dynamic recovery and dynamic recrystallization.

Diffusion flow

The change in shape in polycrystalline aggregates can also be achieved by transfer of matter by diffusion. The diffusion can take place by atomic transport of matter through the grains or along the grain boundaries. When *bulk diffusion* or diffusion through the grain dominates, the process of flow is called *Nabarro–Herring creep* or *Nabarro creep*. When *grain boundary diffusion* is dominant, the creep is known as *Cobble creep*. For diffusional creep of both these types the strain rate is proportional to the differential stress. In that way the behaviour is similar to that of a Newtonian liquid. However, there is generally a threshhold stress below which diffusional creep does not occur; hence, the creep behaviour is similar to that of a Bingham substance. Both Nabarro creep and Cobble creep are strongly dependent on the grain size. While Nabarro creep is inversely proportional to the square of the grain size ($\dot{\varepsilon} \propto 1/d^2$), Cobble creep is inversely proportional to the cube of the grain size ($\dot{\varepsilon} \propto 1/d^3$). Diffusional creep is expected in finer-grained rocks. It also occurs in deformation regimes in which steady state flow can take place at a relatively low differential stress. For the same grain size and temperature, diffusional flow occurs at a smaller flow stress than what is required for dislocation glide and power-law creep. The flow law for pressure solution is somewhat similar to that of Cobble creep (Rutter 1976).

Grain-boundary sliding and superplastic deformation

Deformation by solid state diffusion tends to create voids between grains. To prevent the creation of voids, diffusional flow has to be associated with some grain-boundary sliding. In the same way, deformation by grain-boundary sliding is accommodated by diffusional creep. Thus, diffusional flow and grain-boundary sliding are closely associated. It may, however, be possible to distinguish between two end members in which either diffusional creep or grain-boundary sliding is dominant.

In geological literature the dominant role of grain-boundary sliding is generally considered in connection with the behaviour known as *superplasticity*. Many metals and alloys can undergo very large strain in excess of 1000 per cent, at small stresses under elevated temperatures, without premature rupture or necking. The creep rate or strain rate is generally much larger than that of diffusional creep. Thus, 3×10^{-7} per second is a representative strain rate for Nabarro–Herring creep, while superplastic flow of metals often take place at 10^{-3} or 10^{-4} per second. Hence, unlike diffusional creep, superplastic deformation (SPD) can cause a very large strain at a relatively short time. The flow law of superplastic deformation is similar to that of power-law creep, i.e. $\dot{\varepsilon} \propto \sigma^n$. However, compared to dislocation creep the stress exponent, n, in SPD is very low and generally ranges between 1.4 to 2. Although the deformation mechanisms entering into superplastic deformation are still under investigation, it is generally believed that grain-boundary sliding is dominant in it.

Superplastic behaviour is displayed only by very fine-grained materials, with grain size less than $10\,\mu m$. The initial (before SPD) grains in the material may be elongated or equiaxed. After a certain amount of strain in SPD, all grains become more or less equiaxed. After this stage is reached the equiaxed microstructure is retained even after hundreds or thousands of per cent of strain (see Kashyap *et al.* 1985 for a lucid review of microstructural aspects of superplasticity).

Although grain-boundary sliding is the most dominant deformation mechanism in SPD, it is associated with various other phenomena, such as grain growth, grain-boundary migration, grain rotation, grain

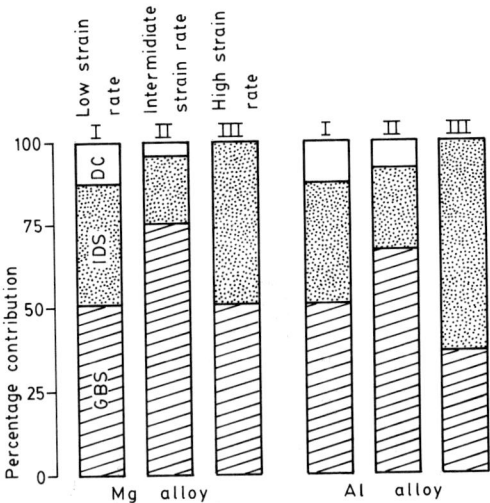

Fig. 7.25. Percentage contribution of grain-boundary sliding (GBS), intragranular dislocation slip (IDS) and diffusion creep (DC) in deformation of Mg alloy and an Al alloy. After fig. 8 of Kashyap *et al*. 1985 (based on Kaibyshev 1981).

rearrangement, diffusion and dislocation gliding. The relative contribution of grain-boundary sliding, intragranular dislocation slip and diffusion creep varies from material to material (e.g. Kashyap *et al*. 1985). Figure 7.25 illustrates the relative proportions of these deformation mechanisms in a magnesium alloy and an aluminium alloy in three regimes I, II, III, in low, intermediate and high strain rates respectively. The figure shows that the contribution of grain-boundary sliding is largest in intermediate strain rates.

In deformation resulting in dislocations, the individual grains become elongated. This is followed by or accompanied by polygonization and recrystallization and the grains are again deformed by dislocations. The bulk change of shape of the material is thus continually associated with elongation of the grains. How then can we explain the maintenance of an equiaxed structure while the material as a whole is being strongly strained by superplastic deformation? There are different microscopic models to explain this process. Evidently, the bulk deformation must be brought about by extensive *grain rearrangement*. Experimental deformation of metals show that there is a large amount of grain rearrangement during SPD; grains in contact with each other in an early stage may, at the end of deformation, be separated by many grains. The amount of separation of the grains is evidently larger in the principal direction of bulk extension (Kashyap *et al*. 1985). Clearly, such extensive grain rearrangement must be associated with sliding

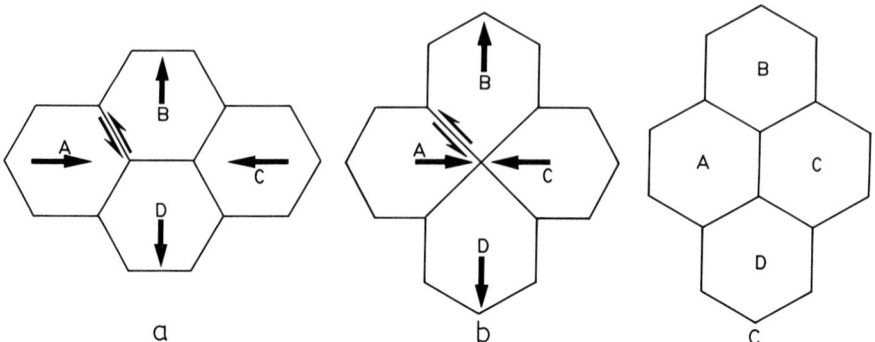

FIG. 7.26. Model of grain neighbour switch. During the shortening in a horizontal direction the grains *A* and *C* have become neighbours and the grains *B* and *D* have moved apart. After Ashby & Verrall 1973.

between the grains and rotation of grains. One of such models of grain rearrangement, the model of *grain-neighbour switch* (Ashby & Verrall 1973), has been verified in experimental deformation of some metal alloys. Figure 7.26a shows four grains *A, B, C* and *D*, with *A* and *C* separated from each other, and *B* and *D* in contact with each other. By the neighbour-switch mechanism *A* and *C* have come to lie side by side, while *B* and *D* have been separated. As a result there has been a bulk extension in the direction of separation of *B* and *D* (Fig.7.26c). The grains remain equiaxed and unstrained in the final stage, though there was some local strain at an intermediate stage (Fig.7.26b).

Grain-boundary sliding and grain rotation invariably weaken the crystallographic preferred orientation. In certain cases a new element of preferred orientation may be added during SPD.

Experimental deformation of fine-grained Solenhofen limestone (Schmid *et al.* 1977) at high temperatures (600–900°C) shows that there is a large contribution of grain-boundary sliding. The rheological data show that the stress exponent *n* is very low (about 1.7). It has therefore been concluded that Solenhofen limestone shows super-plastic behaviour at this deformation regime. Superplastic deformation is more difficult to identify from the microstructures of naturally deformed rocks. However, the following combination of characters generally suggests that the rock was deformed by superplastic flow:

(1) During deformation the rock had a very fine grain size ($< 10\,\mu m$).
(2) The temperature of deformation was high (i.e. greater than half the melting point of the rock).
(3) The rock has an equiaxed structure.
(4) The rock has undergone very large strain.

(5) The dislocation density in the rock is low to moderate.
(6) The preferred orientation is rather weak.

On the basis of such characters it has been suggested that some mylonites have developed by superplastic deformation (e.g. Boullier & Gueguen 1975, Behrman 1985).

Deformation mechanism maps and flow regimes

We have seen in the foregoing section that the rheological behaviour of a crystalline material depends on the predominant deformation mechanism. The latter in its turn will depend on the strain rate, temperature and the differential stress. This dependence can be clearly represented by preparing a *deformation mechanism map* for that material (Ashby 1972, Frost & Ashby 1982). The map is prepared in the stress – temperature space. The stress (σ), plotted as the ordinate, is normalized by dividing it by the shear modulus (μ). In a similar way the temperature (T), plotted along the abscissa, is divided by the melting point of the material (T_m); this ratio is called homologous temperature. The deformation mechanism map divides this stress (σ/μ) – temperature (T/T_m) space into separate fields in each of which a single deformation mechanism is dominant. It is essential to know from experimental data the exact form of the constitutive equation, $f(\sigma, \dot{\varepsilon}, T) = 0$, for the material for each field in which one deformation mechanism is predominant. The boundary between two fields is obtained by equating two corresponding equations and solving for the differential stress as a function of temperature. Any point in the map gives a value of stress and temperature for a particular deformation mechanism. By using the corresponding constitutive equation, the strain rate for that particular point can be plotted so as to enable us to draw contours of constant strain rates on the map. Such deformation mechanism maps have proved to be of great importance in metallurgy. However, the experimental data for rocks are still insufficient to prepare reliable deformation mechanism maps for most rocks. Figure 7.27 shows an initial attempt of preparing a deformation mechanism map for calcite (Rutter 1976) with a grain size of 100 μm. Although the extrapolation of the fields towards geologically realistic strain rates is conjectural, this map at least shows in a qualitative way the relative positions of the fields of dislocation glide, dislocation creep, Nabarro–Herring creep, Cobble creep and pressure solution.

The methods of identification of a deformation mechanism for rocks under experimental conditions are still controversial. Geologists therefore prefer to rely more on rheological measurements to depict the different regimes of flow behaviour of a rock. These are called *flow regimes* (Schmid *et al.* 1977, 1985, Paterson 1987). Each flow regime,

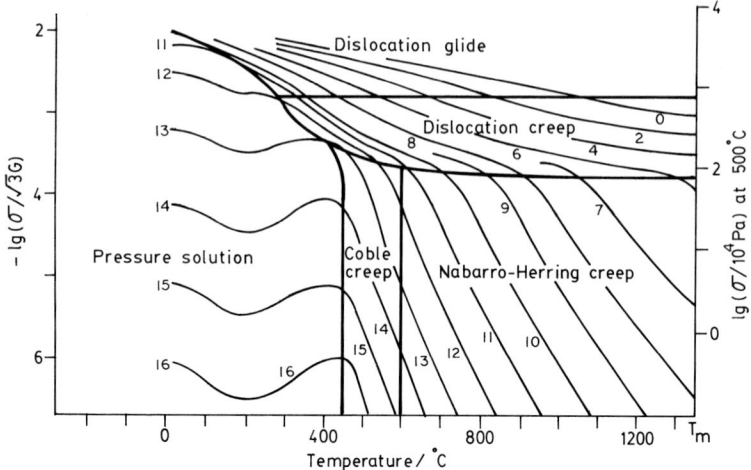

FIG. 7.27. An initial attempt of preparing a deformation mechanism map for calcite with grain size of 100 μm. The figures on the graphs give negative values of strain rates. Thus 7 indicates a strain rate of 10^{-7}. We can locate a point in the map if we know any two of the three variables of stress, strain rate and temperature. After Rutter 1976.

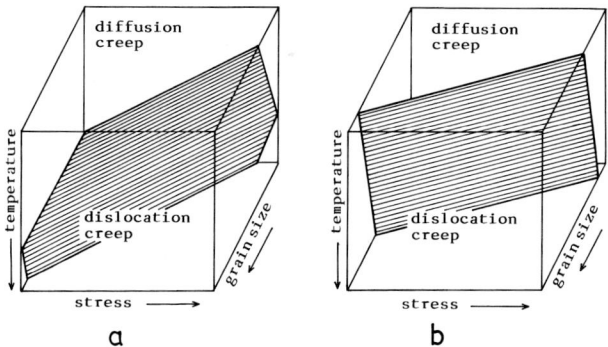

FIG. 7.28. Schematic deformation mechanism maps for olivine in stress–temperature–grain size space in dry (*left*) and wet (*right*) conditions. Simplified from Karato 1989, fig. 11.12.

distinguished by its flow law, has its own distinctive deformation mechanism; however, for the identification of a flow regime it is not necessary to establish exactly its dominant deformation mechanism. Experimental deformation of Solenhofen limestone between 400° and 900°C (Rutter 1974, Schmid *et al.* 1977, 1985) shows three deformation regimes, 1, 2 and 3.

Regime 1 is restricted to lower temperatures (< 500°C) and large flow stress (> 1900 bars) and is characterized by an exponential flow law. The deformation mechanism is dominantly *dislocation glide*.

Regime 2 is characterized by power-law creep with a comparatively large stress exponent ($n = 4.7$). The flow is mainly by dislocation creep.

Regime 3 also shows a power-law flow but has a low stress exponent ($n = 1.7$). The dominant deformation mechanism is grain-boundary sliding. Thus, regime 3 corresponds with superplastic deformation.

A somewhat similar result was also obtained by Karato (1989) for deformation of olivine and by Tullis & Yund (1991) for deformation of feldspar. The deformation mechanism map for olivine (Karato 1989), constructed in the three-dimensional space of stress–temperature–grain size, shows that dislocation creep dominates at relatively high stresses, coarse grain sizes and high temperatures (Fig. 7.28). There is a transition from dislocation creep to diffusion creep, with decreasing grain size and/or decreasing stress. Dislocation creep may continue to be active when dynamic recrystallization takes place at high temperatures and/or low stress. On the other hand, if there is a grain size reduction during dynamic recrystallization associated with dislocation creep at relatively low temperatures (and/or high stress), the deformation mechanism may switch over to diffusion creep. The schematic deformation maps for fine-grained (2–10 μm) feldspar aggregates (Tullis & Yund 1991) show that with increasing temperature there is a transition from cataclastic flow to recrystallization-accommodated dislocation creep at low to moderate metamorphic grades. On the other hand, the deformation mechanism map for the fine-grained feldspar aggregates in a wet environment suggests that, at natural strain rates, there is a direct transition from cataclastic flow to fluid-assisted grain-boundary diffusion creep.

7.10. Estimation of palaeostress

The flow stress in dislocation creep is related to certain microstructural parameters such as recrystallized grain size, sub-grain size and dislocation density. This raises the possibility that the palaeostress of rocks can be estimated from these parameters (e.g. Twiss 1977, Mercier *et al.* 1977, McCormick 1977, Ross *et al.* 1980, Kohlstedt *et al.* 1979, White 1979a, Weathers *et al.* 1979, Kohlstedt & Weathers 1980, Christie & Ord 1980, Ord & Christie 1984).

Dynamically recrystallized grain size is related to the flow stress by an equation of the form

$$\sigma = Ad^{-m} \tag{7.13}$$

where σ is the differential stress in megapascals, d is the grain size in μm and A and m are constants. The values of A and m for quartz, either theoretically or empirically determined, by different authors are

A	m	References
381	0.71	Mercier *et al.* 1977
603	0.68	Twiss 1977
4090	1.11	Christie *et al.* 1980, Ord & Christie 1984

There is evidently a large difference in the palaeostress estimates by the different authors. Thus, for a recrystallized grain size of 20 µm, the flow stress is 45, 79 and 147 MPa, depending on whether we choose the values of A and m given by Mercier *et al.* (1977), Twiss (1977) or Christie *et al.* (1980).

The general form of the equation relating stress and sub-grain size is

$$\sigma = l\mu b\, d^{-1} \qquad (7.14)$$

where μ is shear modulus, b the Burgers vector and l is a constant. l for minerals is between 25 and 80 (Takeuchi & Argon 1976). The shear modulus of quartzite is about 42 or 44×10^3 MPa and b for quartz is 5×10^{-4} µm. With a rather small value of l, Twiss calculated palaeostress from the formula $\sigma = 200\, d^{-1}$, where σ is in megapascals and d in µm. For a sub-grain size of 10 µm, for instance, $\sigma = 20$ MPa, according to Twiss (1977). If we take $l = 25$, we get a somewhat larger value of stress, i.e. $\sigma = 55$ MPa. The relation between dislocation density (N) and the flow stress (σ) is

$$\sigma = \alpha\mu b\, N^p \qquad (7.15)$$

where α and p are constants. It has been suggested that α has a value of 3 (Goetze 1975). p is usually taken as 0.5 (Goetze 1975, Twiss 1977, Weathers *et al.* 1979) or as 0.66 (McCormick 1977), 0.63 (Kohlstedt *et al.* 1979) or 0.67 (Kohlstedt & Weathers 1980). There are at present several expressions for estimating palaeostress from the dislocation density of quartzites:

$$\sigma = 2.47 \times 10^{-3}\, N^{0.5}\ \text{(Twiss 1977)}$$
$$\sigma = 1.64 \times 10^{-4}\, N^{0.66}\ \text{(McCormick 1977)}$$
$$\sigma = 6.6\ \ \times 10^{-3}\, N^{0.5}\ \ \text{(Weathers *et al.* 1979)}$$
$$\sigma = 2.89 \times 10^{-4}\, N^{0.67}\ \text{(Kohlstedt & Weathers 1980)},$$

where σ is in megapascals and the dislocation density, N, is per cm^2. For instance, if $N = 10.4 \times 10^8$ cm^{-2}, the stress is estimated as 79.7, 146.6, 212.8 and 317.9 MPa according to the formulae given above.

The values of stress inferred from grain size and dislocation density do not tally in general. Moreover, there are, at present, several problems in using these parameters as palaeostress indicators. For instance, the dislocation density may change after the main phase of deformation or the recrystallized grain size may be influenced by presence of fine mica along grain borders and hence may be unrelated to stress (White 1979b, Poirier 1985). However, each of these parameters may be used in certain situations in a semi-quantitative manner to see whether there is a spatial variation in the magnitude of stress, say, as we move towards a shear zone (e.g. White 1979b, Etheridge & Wilkie 1981).

CHAPTER 8

Finite Homogeneous Deformation

8.1. Homogeneous deformation

In structural geology we are often concerned with finite homogeneous strain. The strain is considered to be finite when it is not infinitesimally small. It is regarded as homogeneous if within a certain element of the body the state of strain (i.e. the principal strains and the orientations of their axes) is the same at all points of the element. By homogeneous deformation a point in the undeformed state with coordinates x_1, x_2, x_3 is shifted to the position (x_1', x_2', x_3') according to the equations

$$
\begin{aligned}
x_1' &= a_{11}x_1 + a_{12}x_2 + a_{13}x_3 \\
x_2' &= a_{21}x_1 + a_{22}x_2 + a_{23}x_3 \\
x_3' &= a_{31}x_1 + a_{32}x_2 + a_{33}x_3
\end{aligned}
\qquad (8.1)
$$

or,

$$
x_i' = a_{ij}x_j.
$$

This transformation is known as *linear transformation* or *affine transformation*. The following properties of affine transformation can be easily established: (i) a straight line is deformed to a straight line, (ii) parallel straight lines are deformed to parallel straight lines, (iii) a plane is deformed to a plane, (iv) parallel planes are deformed to parallel planes, (v) a circle or a sphere is deformed to an ellipse or an ellipsoid. Finite deformation is concerned with the initial and the final states and does not specify the nature of intermediate stages of deformation.

In general, a homogeneous deformation may not preserve lengths and angles. However, a characteristic feature of homogeneous deformation is that there is always a particular triplet of lines which preserves the angles. This triplet of lines is represented by the principal axes of an ellipsoid which is obtained by homogeneous deformation of a sphere. The axes of the ellipsoid are mutually perpendicular. In the undeformed state they form a triplet of mutually perpendicular diameters of the sphere.

By solving eqns. (8.1) we can express each of x_i in terms of x'_i:

$$\begin{aligned}
x_1 &= b_{11}x'_1 + b_{12}x'_2 + b_{13}x'_3 \\
x_2 &= b_{21}x'_1 + b_{22}x'_2 + b_{23}x'_3 \\
x_3 &= b_{31}x'_1 + b_{32}x'_2 + b_{33}x'_3
\end{aligned} \tag{8.2}$$

or,

$$x_i = b_{ij}x'_j$$

where b_{ij} is the inverse of the matrix (a_{ij}), i.e.

$$b_{ij} = A^{-1}$$

where the matrix A is:

$$\begin{bmatrix}
a_{11} & a_{12} & a_{13} \\
a_{21} & a_{22} & a_{23} \\
a_{31} & a_{32} & a_{33}
\end{bmatrix}. \tag{8.3}$$

There are several methods of obtaining the matrix of A^{-1} (see, for instance, Finkelbeiner 1966). One way of obtaining an element, say b_{ij}, is to delete the row and column passing through the element a_{ji} and find the cofactor of a_{ji}, with the proper sign, $(-1)^{i+j}$. Thus,

$$b_{21} = (-1)^{2+1}(a_{21}a_{33} - a_{31}a_{23}) = a_{31}a_{23} - a_{21}a_{33}.$$

8.2. Strain ellipsoid

When a sphere of unit radius is deformed by homogeneous deformation, the resulting ellipsoid is known as the strain ellipsoid. To obtain the equation of the strain ellipsoid and the lengths and orientations of their principal axes we proceed in the following manner.

(1) Start with the unit sphere

$$x_1^2 + x_2^2 + x_3^2 = 1. \tag{8.4}$$

(2) Substitute the expression for x_i from (8.2) into (8.4) and obtain the equation

$$ax_1'^2 + bx_2'^2 + cx_3'^2 + 2fx_2'x_3' + 2gx_1'x_3' + 2hx_1'x_2' = 1 \tag{8.5}$$

where the coefficients a, b, c, etc., are known in terms of the elements of the matrix b_{ij}. (8.5) is the equation of the strain ellipsoid with its centre at the origin of the coordinate system.

(3) Form the symmetric matrix

$$\begin{bmatrix} a & h & g \\ h & b & f \\ g & f & c \end{bmatrix} \tag{8.6}$$

and determine its three invariants, just as in the cases of the stress matrix and the infinitesimal strain matrix.

$$I_1 = a + b + c$$
$$I_2 = (ab + bc + ca) - (h^2 + f^2 + g^2) \tag{8.7}$$
$$I_3 = abc + 2fgh - af^2 - bg^2 - ch^2.$$

(4) Write out the *characteristic equation*

$$s^3 - I_1 s^2 + I_2 s - I_3 = 0 \tag{8.8}$$

and determine the three roots s_1, s_2, s_3 of this cubic equation. These are known as *characteristic roots*, *latent roots* or *eigenvalues*.

(5) If we choose the coordinate axes along the principal axes of the ellipsoid represented by (8.5), the equation of the ellipsoid will become

$$s_1 x_1'^2 + s_2 x_2'^2 + s_3 x_3'^2 = 1. \tag{8.9}$$

The lengths of the semi-axes of the ellipsoid are, therefore,

$$R_i = \frac{1}{\sqrt{s_i}}. \tag{8.10}$$

Let us choose R_1, R_2, R_3 so that $R_1 > R_2 > R_3$. The *finite principal strains* are then

$$e_1 = R_1 - 1, e_2 = R_2 - 1, e_3 = R_3 - 1$$

and the quadratic elongations are

$$\lambda_1 = R_1^2, \lambda_2 = R_2^2, \lambda_3 = R_3^2.$$

To determine the *orientation of the principal axes of the ellipsoid*, proceed in the following manner. Take three equations

$$(a - s_1) l_1' + h l_2' + g l_3' = 0$$
$$h l_1' + (b - s_1) l_2' + f l_3' = 0 \tag{8.11}$$
$$g l_1' + f l_2' + (c - s_1) l_3' = 0$$

where l_1', l_2', l_3' are the direction cosines of a semi-axis of the ellipsoid. Take any two of these equations and divide them by, say, l_2'. We then get two equations with two unknown quantities l_1'/l_2' and l_3'/l_2'. We find these quantities by solving the two equations and getting the direction ratios. The direction cosines are then obtained by the usual procedure. In the same way, by replacing s_1 by s_2 or s_3 in eqn. (8.11), we can find the direction cosines of the other axes of the ellipsoid.

8.3. Principal axes of strain, rotational and irrotational deformation

A consequence of the linear transformation (8.1) is that the principal axes of the strain ellipsoid form a unique triplet of material lines perpendicular to one another both before and after the deformation. Their *initial* orientations are defined as the *principal axes of strain*.

Let the direction cosines of one of the semi-axes of the strain ellipsoid, say of R_1, be l_1', l_2', l_3'. Let the semi-axis be represented by the line segment OP'. Let P be the position of P' before deformation. Let l_1, l_2, l_3 be the direction cosines of OP. Since the strain ellipsoid was obtained by transformation of a unit sphere, $OP = 1$. Let x_i and x_i' be the coordinates of P and P'. Then

$$l_1 = x_1, l_2 = x_2, l_3 = x_3 \text{ and}$$
$$l_1' = x_1'/R_1, l_2' = x_2'/R_1, l_3' = x_3'/R_1.$$

Substituting these expressions of x_i and x_i' in eqn. (8.1), we obtain three equations

$$\begin{aligned}
a_{11}l_1 + a_{12}l_2 + a_{13}l_3 &= l_1'R_1 \\
a_{21}l_1 + a_{22}l_2 + a_{23}l_3 &= l_2'R_1 \\
a_{31}l_1 + a_{32}l_2 + a_{33}l_3 &= l_3'R_1.
\end{aligned} \tag{8.12}$$

We have seen that if we know the coefficient matrix a_{ij}, then we can determine the value of R_1 from eqn. (8.10) and the direction cosines of R_1 from eqn. (8.11). Thus, eqns. (8.12) have three unknowns l_1, l_2, l_3. Since these are related by the equation $l_1^2 + l_2^2 + l_3^2 = 1$, we can determine the ratio $l_1 : l_2 : l_3$ from any two of the eqns. (8.12) and can then determine direction cosines l_i. These give the orientations of one of the principal axes of finite homogeneous strain. The orientations of the other two principal axes of strain can be found in a similar manner.

If the orientations of the principal axes of strain coincide with the orientations of the principal axes of the strain ellipsoid, then the deformation is *irrotational*. The deformation is considered to be *rotational* when the orientation of the two sets of axes are different. The reader should be cautioned that these definitions of the principal axes of strain and of rotational and irrotational deformation are not followed by all structural geologists.

8.4. Plane strain, flattening and constriction

If the volume remains constant during straining, then

$$\lambda_1 \lambda_2 \lambda_3 = 1$$

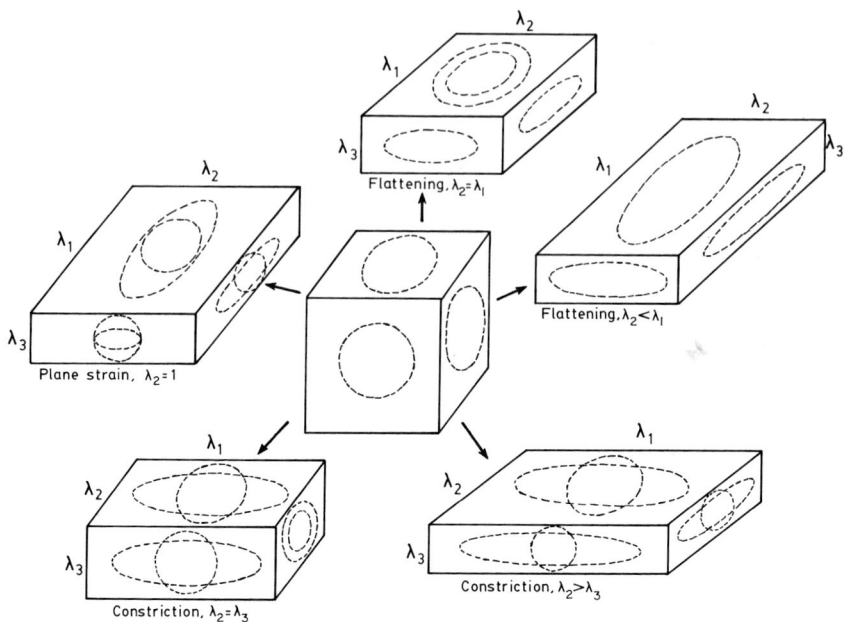

FIG. 8.1. Deformation of a cube resulting from five types of strain. The sections through the strain ellipsoid are shown on the faces of the deformed body.

where λ_i are the quadratic elongations along the principal axes. It is obvious that the three principal quadratic elongations are not independent. It is evident that $\lambda_1 > 1$ and $\lambda_3 < 1$. λ_2 may be equal to or greater than or less than 1. If $\lambda_2 = 1$, the deformation is of *plane strain* type. If $\lambda_2 > 1$, the deformation is of the *flattening* type. If $\lambda_2 < 1$, the deformation is of *constriction* type. In the flattening type of deformation we may have two cases, (a) $\lambda_2 = \lambda_1$ and (b) $\lambda_2 < \lambda_1$. If $\lambda_2 = \lambda_1$, there is an equal extension in all directions in the $\lambda_1 \lambda_2$ plane. The resulting strain ellipsoid is shaped like a pancake and is called an oblate ellipsoid. In a similar way, we may have two cases in the constriction type of deformation with (a) $\lambda_2 = \lambda_3$ and (b) $\lambda_2 > \lambda_3$. If $\lambda_2 = \lambda_3$, there is an equal shortening in all directions on the $\lambda_2 \lambda_3$ plane. The resulting strain ellipsoid is cigar-shaped and is called a prolate ellipsoid. For these different cases, the deformation of a cube and the characteristic sections through the strain ellipsoids are depicted in Fig. 8.1.

Thus, depending on the nature of strain along the λ_2-axis, we may have five types of strain:

1. Flattening, with $\lambda_2 > 1$ and $\lambda_2 = \lambda_1$.
2. Flattening, with $\lambda_2 > 1$ and $\lambda_2 \neq \lambda_1$.
3. Plane strain, with $\lambda_2 = 1$.

4. Constriction, with $\lambda_2 < 1$ and $\lambda_2 \neq \lambda_3$.
5. Constriction, with $\lambda_2 < 1$ and $\lambda_2 = \lambda_3$.

8.5. Finite strain along a line

Longitudinal strain

Once the principal quadratic elongations λ_i and the principal axes of the finite strain ellipsoid are determined, the equation of the strain ellipsoid can be expressed in a simpler form by choosing the coordinate axes along its principal directions:

$$\frac{x_1'^2}{\lambda_1} + \frac{x_2'^2}{\lambda_2} + \frac{x_3'^2}{\lambda_3} = 1. \tag{8.13}$$

We have used here the symbol x_i' to indicate that the strain ellipsoid can be obtained from the unit sphere

$$x_1^2 + x_2^2 + x_3^2 = 1, \tag{8.14}$$

the transformation from (8.14) to (8.13) having taken place by the formulae

$$\begin{aligned} x_1' &= \sqrt{\lambda_1}\, x_1 \\ x_2' &= \sqrt{\lambda_2}\, x_2 \\ x_3' &= \sqrt{\lambda_3}\, x_3. \end{aligned} \tag{8.15}$$

From deformed objects such as pebbles and other strain gauges, it is often possible to determine the values of λ_i and the orientations of the λ_i-axes. Once these are known, we can then calculate the longitudinal and shear strains in any direction.

Let P be a point on the unit sphere (8.14), with coordinates x_i and let the direction cosines of the line OP be l_i. Let P' be the new position of the point on the ellipsoid (8.13). Its coordinates are x_i' and the direction cosines of OP' are l_i'. Then

$$\begin{aligned} l_i &= x_i/OP = x_i \\ l_i' &= x_i'/OP'. \end{aligned} \tag{8.16}$$

From (8.15) and (8.16) we find

$$l_1' = \frac{\sqrt{\lambda_1}\, l_1}{OP'} = \frac{\sqrt{\lambda_1}\, l_1}{\sqrt{\lambda}}$$

$$l_2' = \frac{\sqrt{\lambda_2}\, l_2}{\sqrt{\lambda}} \tag{8.17}$$

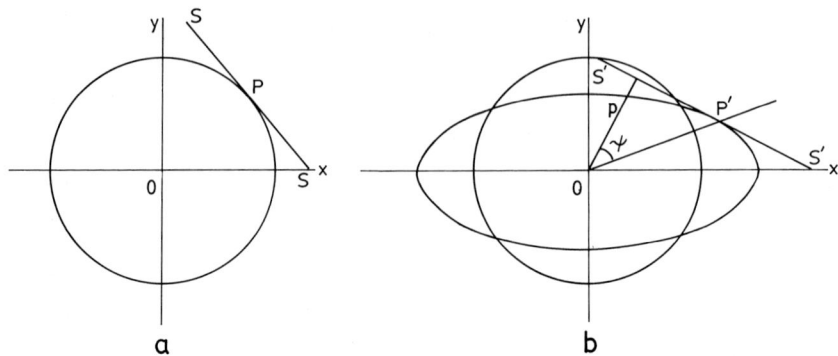

FIG. 8.2. (a) SS represents a plane touching a sphere at P. (b) When the sphere is deformed to an ellipsoid the deformed tangent plane $S'S'$ touches the sphere at point P'. OP' is no longer a normal to the tangent plane. p is now the normal to the tangent plane. ψ is the angle between OP' and p. The shear strain is $\gamma = \tan \psi$.

$$l_3' = \frac{\sqrt{\lambda_3}\, l_3}{\sqrt{\lambda}}.$$

It is evident that the quadratic elongation λ along a line can be expressed either in terms of l_i or of l'_i. Thus,

$$\lambda = (x_1'^2 + x_2'^2 + x_3'^2) = l_1^2 \lambda_1 + l_2^2 \lambda_2 + l_3^2 \lambda_3 \qquad (8.18)$$

or, expressing l_i in terms of l'_i with the help of eqn. (8.17) and substituting them in the equation $l_1^2 + l_2^2 + l_3^2 = 1$, we find

$$\frac{1}{\lambda} = \frac{l_1'^2}{\lambda_1} + \frac{l_2'^2}{\lambda_2} + \frac{l_3'^2}{\lambda_3}. \qquad (8.19)$$

Shear strain

The equation of the tangent plane S at the point P on the unit sphere is

$$l_1 x_1 + l_2 x_2 + l_3 x_3 = 1 \qquad (8.20)$$

where l_i are the direction cosines of the line OP. OP is normal to S. After deformation, OP is transformed to OP' and S is transformed to the tangent plane S' at P'. OP' is not, in general, normal to S. Thus, the change in the right angle is measured by the angle between OP' and the normal to the tangent plane S'. Let p be the length of the perpendicular from the origin O to the plane S' (Fig. 8.2). Then the shear angle ψ is given by the equation

$$\cos \psi = p/OP'. \qquad (8.21)$$

The equation of the deformed tangent plane S' is obtained from eqns. (8.15) and (8.20)

$$\frac{l_1 x_1}{\sqrt{\lambda_1}} + \frac{l_2 x_2}{\sqrt{\lambda_2}} + \frac{l_3 x_3}{\sqrt{\lambda_3}} = 1. \tag{8.22}$$

The length of the perpendicular p on this plane is given by the formula

$$p = \pm \frac{1}{\sqrt{\dfrac{l_1^2}{\lambda_1} + \dfrac{l_2^2}{\lambda_2} + \dfrac{l_3^2}{\lambda_3}}}. \tag{8.23}$$

From (8.21) and (8.23) we then find

$$\sec^2 \psi = \frac{OP'^2}{p^2} = \lambda \left(\frac{l_1^2}{\lambda_1} + \frac{l_2^2}{\lambda_2} + \frac{l_3^2}{\lambda_3} \right).$$

Since

$$\gamma^2 = \tan^2 \psi = \sec^2 \psi - 1,$$

$$\gamma^2 = \lambda \left(\frac{l_1^2}{\lambda_1} + \frac{l_2^2}{\lambda_2} + \frac{l_3^2}{\lambda_3} \right) - 1, \tag{8.24}$$

where λ is given by eqn. (8.18) in terms of l_i and λ_i. Equation (8.24) gives the shear strain along a line in terms of the principal quadratic elongations and the orientation of the *undeformed line*.

To obtain an expression for the shear strain in terms of the direction cosines of the deformed line OP', we substitute the expressions for l_i from eqns. (8.17) into eqn. (8.24):

$$\gamma^2 = \lambda^2 \left(\frac{l_1'^2}{\lambda_1^2} + \frac{l_2'^2}{\lambda_2^2} + \frac{l_3'^2}{\lambda_3^2} \right) - 1. \tag{8.25}$$

8.6. Strain ellipse on an inclined plane

Consider a plane inclined to the principal axes of the strain ellipsoid. The variation of longitudinal strain in different directions on this plane can be determined by constructing a strain ellipse on this plane. The strain ellipse is a central section through the strain ellipsoid. The equation of the strain ellipsoid, as given by (8.13) is

$$\frac{x_1'^2}{\lambda_1} + \frac{x_2'^2}{\lambda_2} + \frac{x_3'^2}{\lambda_3} = 1.$$

The equation of a plane passing through the centre of the ellipsoid is

$$l_1 x_1' + l_2 x_2' + l_3 x_3' = 0$$

where l_i are the direction cosines of the normal to the plane. The lengths of the semi-axes (r_i) of the strain ellipse are obtained from the equation (McCrea 1953, p. 109)

$$\frac{\lambda_1 l_1^2}{r^2 - \lambda_1} + \frac{\lambda_2 l_2^2}{r^2 - \lambda_2} + \frac{\lambda_3 l_3^2}{r^2 - \lambda_3} = 0. \tag{8.26}$$

Or,

$$Pr^4 - Qr^2 + 1 = 0, \tag{8.27}$$

where $P = \lambda_1 l_1^2 + \lambda_2 l_2^2 + \lambda_3 l_3^2$,

$$Q = (\lambda_1 \lambda_2 + \lambda_3 \lambda_1) l_1^2 + (\lambda_2 \lambda_3 + \lambda_1 \lambda_2) l_2^2$$
$$+ (\lambda_3 \lambda_1 + \lambda_2 \lambda_3) l_3^2.$$

The length of the semi-axes of the strain ellipse are obtained from the equation

$$r^2 = \frac{1}{2P} [Q \pm (Q^2 - 4P)^{\frac{1}{2}}]. \tag{8.28}$$

The direction ratios of the semi-axes are

$$a = \frac{\lambda_1 l_1}{r^2 - \lambda_1}, b = \frac{\lambda_2 l_2}{r^2 - \lambda_2}, c = \frac{\lambda_3 l_3}{r^2 - \lambda_3}.$$

The corresponding direction cosines are

$$l_1' = \frac{a}{\sqrt{a^2 + b^2 + c^2}}, l_2' = \frac{b}{\sqrt{a^2 + b^2 + c^2}}, l_3' = \frac{c}{\sqrt{a^2 + b^2 + c^2}}. \tag{8.29}$$

8.7. Lines of no finite longitudinal strain

Along certain directions of a constant-volume strain ellipsoid a line is neither lengthened nor shortened. These are the directions of no finite longitudinal strain. The distribution of such lines within a strain ellipsoid has a different pattern for the five types of strain ellipsoids mentioned in section 8.4. These patterns are best represented in stereographic or equal area projections.

(1) For plane strain the lines of no finite longitudinal strain lie on two circular sections through the strain ellipsoid which intersect along the λ_2-axis. Let us consider the strain ellipse on the $\lambda_1 \lambda_3$-plane (Fig. 8.3).

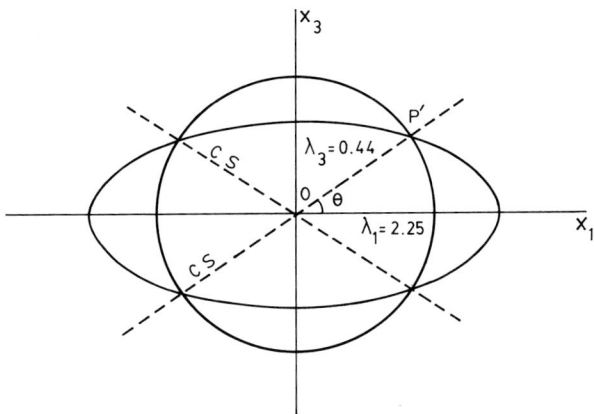

FIG. 8.3. Strain ellipse derived from a circle in the $\lambda_1\lambda_3$-plane for plane strain. The lines of no finite longitudinal strain lie on the two circular sections (CS).

The equation of the strain ellipse is

$$\frac{x_1'^2}{\lambda_1} + \frac{x_3'^2}{\lambda_3} = 1. \tag{8.30}$$

Let OP', the trace of the *circular section* on this ellipse, make an angle of θ with the λ_1-axis. Then, since $OP' = 1$,

$$\frac{\cos^2\theta}{\lambda_1} + \frac{\sin^2\theta}{\lambda_3} = 1. \tag{8.31}$$

Now, since $\lambda_1 \lambda_2 \lambda_3 = 1$ and, for plane strain, $\lambda_2 = 1, \lambda_3 = 1/\lambda_1$. Substituting this expression for λ_3 in (8.31), we find after simplification

$$\cos^2\theta = \frac{\lambda_1}{1 + \lambda_1},$$

or,

$$\tan^2\theta = 1/\lambda_1. \tag{8.32}$$

Hence the circular sections in a plane-strain ellipsoid (Fig. 8.4a) are parallel to the λ_2-axis and are inclined with the λ_1-axis at angles

$$\theta = \tan^{-1}\left(\pm\frac{1}{\sqrt{\lambda_1}}\right). \tag{8.33}$$

For infinitesimally small deformation λ_1 tends to unity and θ tends to $\pm 45°$. For finite deformation the absolute value of θ is less than $45°$.

(2) For a flattening type of strain with $\lambda_2 = \lambda_1$, all sections along the λ_3-axis will have an identical shape of the strain ellipsoid. Therefore, the lines of no finite longitudinal strain will plot in equal area projection on a small circle around the λ_3-axis. The strain ellipse on the $\lambda_1 \lambda_3$-plane

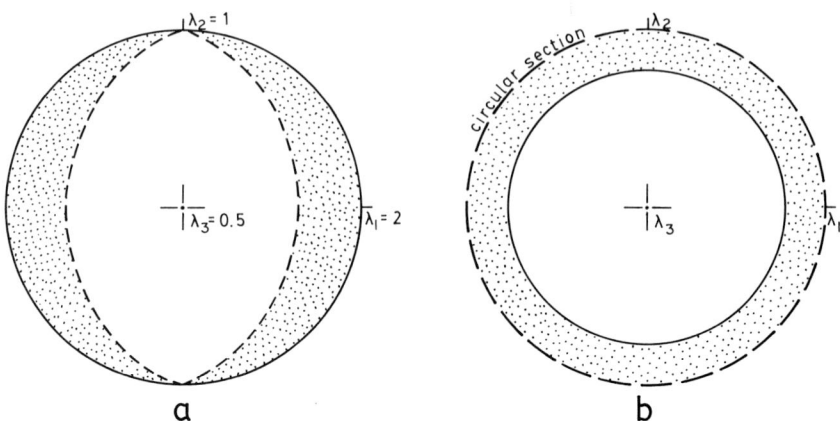

FIG. 8.4. (a) Equal area projection of circular sections for plane strain with $\lambda_1 = 2, \lambda_2 = 1, \lambda_3 = 0.5$. Lines which plot as points in the dotted area undergo finite extension and those which lie in the blank area are shortened. (b) For a flattening type of strain with $\lambda_2 = \lambda_1$, the lines of no finite longitudinal strain lie on a small circle around the λ_3-axis in equal area projection. The dotted area shows the field of extension.

is given by eqn. (8.30) and the condition in which the length of $OP' = 1$ is given by eqn. (8.31):

$$\frac{\cos^2\theta}{\lambda_1} + \frac{\sin^2\theta}{\lambda_3} = 1. \tag{8.34}$$

Substituting the condition $\lambda_2 = \lambda_1$ in the equation $\lambda_1 \lambda_2 \lambda_3 = 1$, we find

$$\lambda_3 = \frac{1}{\lambda_1^2}.$$

θ can then be expressed in terms of λ_1 only:

$$\tan^2\theta = \frac{1}{\lambda_1 (1 + \lambda_1)}. \tag{8.35}$$

The small circle will lie at an angle of $90° - \theta$ with the λ_3-axis (Fig. 8.4b). If the deformation is infinitesimally small, the small circle will lie at angle of $54°.7$ with the λ_3-axis. For finite deformation the angle will be larger than $54°.7$.

(3) For a flattening type of deformation with $\lambda_2 > 1$ but with $\lambda_2 \neq \lambda_1$, the following method may be adopted to plot the locus of the lines of no finite longitudinal strain. Let us first consider the strain ellipse on the $\lambda_1 \lambda_2$-plane (Fig. 8.5a). Since both λ_1 and λ_2 are greater than unity, all lines on this plane will be extended. Consider the elongation of a line OQ' at an angle of α with the λ_1-axis. The equation of the strain ellipse is $(x_1'^2/\lambda_1) + (x_2'^2/\lambda_2) = 1$. Now, $x_1' = OQ' \cos\alpha$ and $x_2' = OQ' \sin\alpha$.

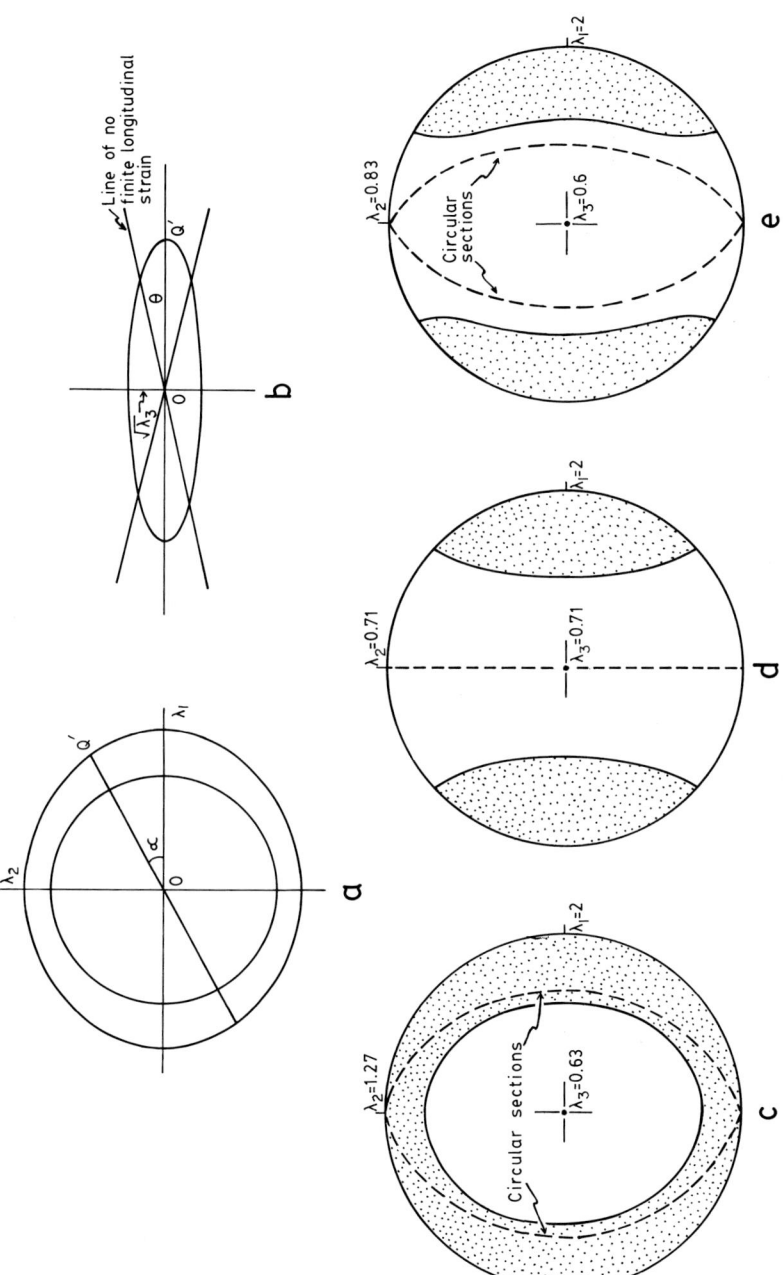

FIG. 8.5. (a) Strain ellipse derived from a circle in the λ_1, λ_2-plane for a flattening type strain with $\lambda_1 > \lambda_2$. All lines on this plane are extended. OQ' is a line which makes an angle α with the λ_1-axis. (b) Strain ellipse on a plane parallel to OQ' and λ_3-axis. The line of no finite longitudinal strain makes an angle θ with OQ'. (c) Locus of lines of no finite longitudinal strain (the inner elliptical curve) for $\lambda_1 = 2, \lambda_2 = 1.27, \lambda_3 = 0.63$ in equal area projection. Lines plotting within the dotted area are extended. The circular sections lie within the field of extension. (d) Locus of lines of no finite elongation in equal area projection for a constriction type of deformation with $\lambda_2 = \lambda_3$. The circular section (dashed line), perpendicular to the λ_1-axis, lies within the shortening field. (e) Locus of lines of no finite longitudinal strain for constriction type of deformation, with $\lambda_2 > \lambda_3$. Circular sections lie within the shortening field.

Substituting these expressions for x'_i in the equation of the strain ellipse, we find

$$\frac{\cos^2\alpha}{\lambda_1} + \frac{\sin^2\alpha}{\lambda_2} = \frac{1}{(OQ')^2} = \frac{1}{\lambda_{xy}}, \tag{8.36}$$

where λ_{xy} is the quadratic elongation along the line OQ'. We next construct a strain ellipse (Fig. 8.5b) on a plane passing through OQ' and the λ_3-axis. The equation of this strain ellipse is

$$\frac{\zeta^2}{\lambda_{xy}} + \frac{\xi^2}{\lambda_3} = 1. \tag{8.37}$$

Since $\lambda_3 < 1$, there will be some point P' on this ellipse with coordinates (ζ, ξ) and with length $OP' = 1$. If the angle between OP' and the long axis of the ellipse is θ, then $\zeta = \cos\theta$ and $\xi = \sin\theta$. Therefore,

$$\frac{\cos^2\theta}{\lambda_{xy}} + \frac{\sin^2\theta}{\lambda_3} = 1, \tag{8.38}$$

from which we obtain the relation

$$\tan^2\theta = \frac{\lambda_3(\lambda_{xy} - 1)}{\lambda_{xy}(1 - \lambda_3)}. \tag{8.39}$$

For different values of α, λ_{xy} is first calculated from eqn. (8.36) and then θ is determined with the help of eqn. (8.39). For a particular combination of α and θ, a line of no finite longitudinal strain can then be plotted in equal area projection (Fig. 8.5c). The locus of all such lines for a certain strain state is shown in Fig. 8.5c.

(4) For a constriction type of deformation, with $\lambda_2 = \lambda_3$, the lines of no finite longitudinal strain are found from eqn. (8.34):

$$\frac{\cos^2\theta}{\lambda_1} + \frac{\sin^2\theta}{\lambda_3} = 1.$$

Since in this case $\lambda_2 = \lambda_3$, we have $\lambda_3 = 1/\sqrt{\lambda_1}$ when the volume remains constant during deformation. The value of θ, the angle between the λ_1-axis and a line of no finite elongation, can then be found as

$$\tan^2\theta = \frac{1 + \sqrt{\lambda_1}}{\lambda_1}. \tag{8.40}$$

In equal area projection (Fig. 8.5d) the lines lie on a small circle at an angle of θ with the λ_1-axis. For an infinitesimally small deformation $\theta = 54°.7$. For finite deformation θ must be smaller than this value.

(5) For a constriction type of deformation with $\lambda_2 \neq \lambda_3$, the locus of lines of no finite longitudinal strain is plotted in equal area projection (Fig. 8.5e) in the same manner as in (4).

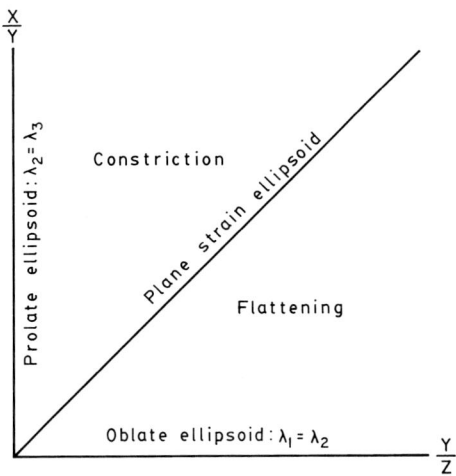

F<small>IG</small>. 8.6. Fields of constriction, flattening and plane strain in a Flinn diagram in which the shape of an ellipsoid can be represented graphically by plotting X/Y against Y/Z.

8.8. Characteristic features of five types of strain ellipsoid

The semi-axes of the strain ellipsoid are usually referred to as X, Y and Z parallel to λ_1, λ_2 and λ_3 respectively. A strain ellipsoid can be plotted as a point in a diagram with X/Y as ordinate and Y/Z as abscissa (Fig. 8.6). Any point which falls on the abscissa represents an oblate type of ellipsoid of revolution. A point on the ordinate represents a prolate type of ellipsoid of revolution. A point on the line at an angle of 45° with the abscissa represents a plane-strain ellipsoid. A point below this line represents a flattening type of strain ellipsoid and a point lying above this line represents a constriction type of strain ellipsoid. Instead of taking X/Y and Y/Z, it is sometimes convenient to take $\ln X/Y$ and $\ln Y/Z$ as the ordinate and the abscissa.

The strain in different directions on a plane can be conveniently represented by equal area projection of the plane (Figs. 8.4, 8.5). The characteristic features of the strain ellipsoids as given by such diagrams are summarized below.

(1) The loci of lines of no finite longitudinal strain in equal area projection divide the projection area into fields of shortening and fields of extension. In Figs. 8.4 and 8.5 a line which projects in the dotted area is extended and a line plotted in the blank area is shortened. Similarly, a plane which lies entirely within the dotted area has undergone extension in all directions and a plane which lies entirely within the blank area has undergone a shortening in all directions.

(2) Each strain ellipsoid has one or two circular sections. These are shown in Figs. 8.4 and 8.5. On such planes the elongation is equal in all directions. For an oblate type of uniaxial ellipsoid the circular section is perpendicular to the Z-axis (Fig. 8.4b). All lines are equally extended on this plane. In a flattening type of ellipsoid with $X \neq Y$, there are two circular sections parallel to the Y-axis, symmetrically oriented with respect to the YZ-plane (Fig. 8.5c). The circular sections lie entirely within the extension field. In plane strain the two circular sections coincide with the loci of lines of no finite longitudinal strain (Fig. 8.4a). In the prolate type of ellipsoid (with $\lambda_2 \neq \lambda_3$), the two circular sections are parallel to the Y-axis and lie entirely within the constriction field (Fig. 8.5e). Lastly, in the prolate type of ellipsoid, with $Y = Z$, the circular section is perpendicular to the X-axis (Fig. 8.5d). Along any line on this plane the quadratic elongation is

$$\lambda = \lambda_2 = \lambda_3.$$

The angle, β, between the Z-axis and a circular section is given by the equation

$$\tan^2 \beta = \frac{\lambda_1 (\lambda_2 - \lambda_3)}{\lambda_3 (\lambda_1 - \lambda_2)}. \tag{8.41}$$

(3) The fold axis on a bed is generally initiated parallel to the long axis of the strain ellipse (with semi-axes X_A and X_B) on the plane of the bed. However, folding can take place only if X_B is less than 1. A single direction of folding may occur only on those planes which intersect the locus of the lines of no finite longitudinal strain. On planes which lie entirely within the field of shortening, both X_A and X_B are less than unity. Hence, such planes may show domes and basins or crossing trends of folds. The planes which lie entirely within the extension field may show crossing trends of boudin axes.

(4) In general, the long axis of the strain ellipse does not coincide with the line of intersection of the bed and the XY-plane. This point is of great importance because it indicates that the fold axis may not lie on the XY-plane. However, for certain special orientations of the bed, X_A may lie on the XY-plane. (a) In an oblate type of uniaxial ellipsoid, X_A, and hence the fold axis, is always parallel to the XY-plane. (b) For all types of strain ellipsoids, a bed which is parallel to either the X-axis or the Z-axis will have X_A parallel to the XY-plane. (c) On beds parallel to the Y-axis, X_A will lie on the XY-plane only when the angle between the bed and the Z-axis is smaller than the angle between the Z-axis and the circular section. This angle is given by eqn. (8.41).

(5) It will be seen later that the schistosity is initiated parallel to the XY-plane while a stretching lineation, usually a mineral lineation, is

initiated parallel to the X-axis. Both a schistosity and a stretching lineation, giving rise to what is sometimes called an L–S tectonite, may form, unless the strain ellipsoid is of uniaxial oblate or prolate type. When the ellipsoid is of the uniaxial oblate type, a schistosity but no stretching lineation can develop. When the ellipsoid is of the uniaxial prolate type, a stretching lineation but no schistosity can form.

8.9. Linear transformation in two dimensions

The equations for finite homogeneous deformation are considerably simplified when the analysis is carried out for two dimensions only. A point in the undeformed state with coordinates (x_1, x_2) is shifted to the position (x_1', x_2') according to the equation

$$x_1' = a_{11}x_1 + a_{12}x_2$$
$$x_2' = a_{21}x_1 + a_{22}x_2. \tag{8.42}$$

By solving these equations we can express x_i in terms of x_i':

$$x_1 = \frac{a_{22}}{d} x_1' - \frac{a_{12}}{d} x_2'$$

$$x_2 = \frac{-a_{21}}{d} x_1' + \frac{a_{11}}{d} x_2', \tag{8.43}$$

where d is the determinant of the coefficient matrix of (8.42)

$$d = a_{11}a_{22} - a_{12}a_{21}. \tag{8.44}$$

Let us take a unit circle

$$x_1^2 + x_2^2 = 1 \tag{8.45}$$

and substitute in it the expressions for x_1' and x_2' from eqn. (8.43). We then obtain the following equation of the strain ellipse

$$ax_1'^2 + bx_2'^2 - 2hx_1'x_2' = 1 \tag{8.46}$$

where

$$a = \frac{a_{21}^2 + a_{22}^2}{d^2}, b = \frac{a_{11}^2 + a_{12}^2}{d^2}, h = \frac{a_{11}a_{21} + a_{12}a_{22}}{d^2}. \tag{8.47}$$

One method of determining the lengths of the semi-axes of the strain ellipse is to form the characteristic equation

$$s^2 - I_1 s + I_2 = 0 \tag{8.48}$$

where, as given by (8.7),

$$I_1 = a + b,$$

$$I_2 = ab - h^2. \tag{8.49a}$$

I_1 and I_2 are considered to be invariants in the sense that, if by any change of axes, without change of origin, the quantity $ax^2 + 2hxy + by^2$ becomes $a'x'^2 + 2h'x'y' + b'y'^2$, then

and
$$\begin{aligned} a + b &= a' + b' \\ ab - h^2 &= a'b' - h^{2'}. \end{aligned} \tag{8.49b}$$

Equation (8.48) has two roots, s_i, which are known as the eigenvalues of the characteristic equation.

$$s_i = \tfrac{1}{2}(I_1 \pm \sqrt{I_1^2 - 4I_2}), \tag{8.50}$$

or,

$$s_i = \tfrac{1}{2}(a + b) \pm \tfrac{1}{2}\sqrt{(a - b)^2 + 4h^2}. \tag{8.51}$$

The semi-axes of the strain ellipse are

$$X = \sqrt{\lambda_1} = \frac{1}{\sqrt{s_1}},$$

$$Y = \sqrt{\lambda_2} = \frac{1}{\sqrt{s_2}}. \tag{8.52}$$

The orientations of X and Y are given by the equation

$$\tan 2\theta' = \frac{2h}{b - a} \tag{8.53}$$

where a, b and h are given by (8.47).

A characteristic feature of homogeneous deformation is that the principal axes of the strain ellipse constitute a unique pair of material lines which are mutually perpendicular before and after deformation. The *initial* orientations of these lines are defined as the *principal axes of finite strain*. It can be shown that the orientations of the principal axes are given by the equation

$$\tan 2\theta = \frac{2(a_{11} a_{12} + a_{21} a_{22})}{a_{11}^2 + a_{21}^2 - a_{12}^2 - a_{22}^2} \tag{8.54}$$

and

$$\tan(\theta' - \theta) = \frac{a_{21} - a_{12}}{a_{11} + a_{22}}. \tag{8.55}$$

$(\theta' - \theta)$ is defined as the rotation (Fig. 8.7). Thus, if $a_{21} = a_{12}$, the deformation is *irrotational*, while if $a_{21} \neq a_{12}$, the deformation is *rotational*.

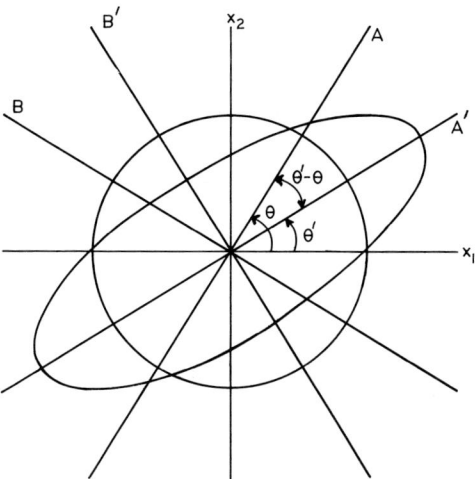

FIG. 8.7. *A* and *B* are two mutually perpendicular lines before deformation. After deformation they have been deformed to *A'* and *B'* while the angle between them has remained a right angle. *A* and *B* are the principal axes of strain. *A'* and *B'* are parallel to the principal axes of the strain ellipse. The angle between *A'* and *A* is defined as the rotation.

For two-dimensional deformation the elongation of a line making an initial angle α with the x_1-axis is obtained from the following equation:

$$\lambda = (a_{11}^2 + a_{21}^2) \cos^2\theta + (a_{12}^2 + a_{22}^2) \sin^2\theta$$
$$+ 2(a_{11}a_{12} + a_{21}a_{22}) \sin\theta \cos\theta. \qquad (8.56)$$

8.10. Some simple types of finite homogeneous deformation

Pure shear

In experiments with soft models we often use a pure shear apparatus. In this apparatus a rectangular block of the model is placed between two parallel walls (Fig. 8.8a). The walls are fixed in position so that the distance between them remains constant. In Fig. 8.8a the walls are shown as vertical. A rigid plate is placed at the top of the model and the model is shortened by a vertical compression. As a result the model is extended in a horizontal direction parallel to the fixed walls while the dimension of the model remains fixed in a direction normal to the walls. The strain ellipsoid in this case is of the plane-strain type, with $\lambda_2 = 1$. Hence, if the volume of model remains constant, $\lambda_3 = 1/\lambda_1$, where λ_3 is in the vertical direction and λ_1 is in the horizontal direction parallel to the fixed walls. If we choose the coordinate axes along these principal

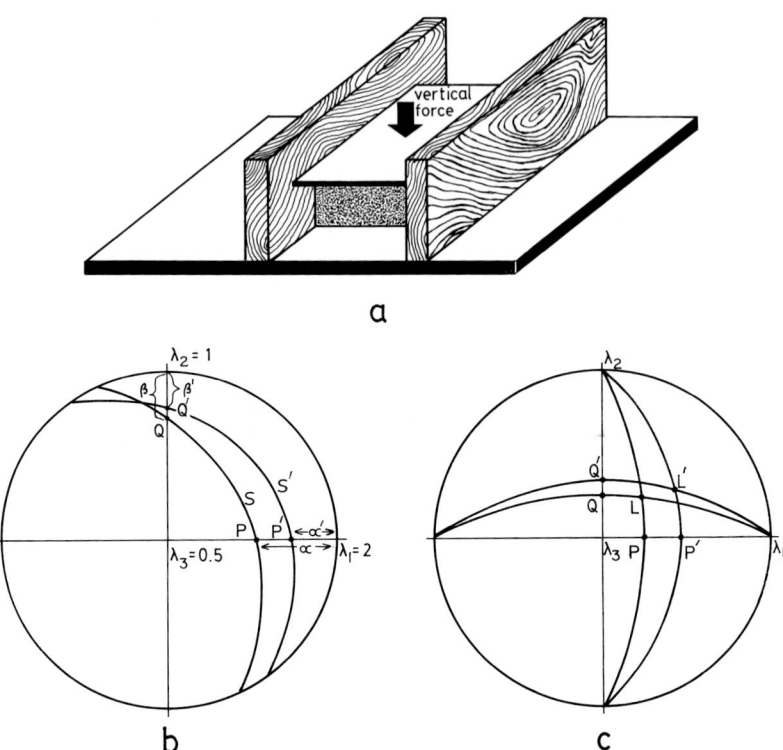

FIG. 8.8. (a) Pure shear apparatus. (b) Equal area projection method of determining the attitude of a planar structure after it is deformed by pure shear. (c) Equal area projection method of determining the attitude of a linear structure deformed by pure shear.

directions, a point (x_1, x_2, x_3) will be shifted to the position (x'_1, x'_2, x'_3) according to the formula

$$x'_1 = \sqrt{\lambda_1} x_1$$
$$x'_2 = x_2 \qquad\qquad (8.57)$$
$$x'_3 = \frac{1}{\sqrt{\lambda_1}} x_3.$$

In most cases the second equation of (8.57) is implied but not written out.

The equations (8.57) represent a pure shear transformation. It is an irrotational deformation of the plane-strain type. From (8.57) we find that

$$x_1 = \frac{1}{\sqrt{\lambda_1}} x'_1,$$

$$x_3 = \sqrt{\lambda_1}\, x_3'.$$

(8.58)

Consider the unit circle

$$x_1^2 + x_3^2 = 1.$$

Substituting the expressions of x_i from (8.58) in this equation we find

$$\frac{x_1'^2}{\lambda_1} + \frac{x_3'^2}{\lambda_3} = 1.$$

(8.59)

This is the equation of the strain ellipse on the $\lambda_1 \lambda_3$-plane with $\sqrt{\lambda_1}$ and $1/\sqrt{\lambda_1}$ as the semi-axes. On this plane, let a line make an initial angle α and the final angle α' with the λ_1-axis. Then, from (8.57) or (8.58) we find that

$$\tan\alpha' = \frac{x_3'}{x_1'} = \frac{1}{\lambda_1}\left(\frac{x_3}{x_1}\right) = \frac{1}{\lambda_1}\tan\alpha$$

(8.60a)

and

$$\tan\alpha = \lambda_1 \tan\alpha'.$$

(8.60b)

To find the deformed position of a planar or linear structure *inclined to all the principal axes*, it is convenient to adopt the following procedure.

(1) Plot the planar structure, S, in equal area or stereographic projection.

(2) Locate its intersections P and Q with the $\lambda_1 \lambda_3$-plane and the $\lambda_2 \lambda_3$-plane (Fig. 8.8b).

(3) Determine the angle α between P and the λ_1-axis and the angle β between Q and λ_2-axis. The transformed angle α' is found from eqn. (8.60). β' is found from the second and the third equations of (8.57):

$$\tan\beta' = \frac{x_3'}{x_2'} = \frac{1}{\sqrt{\lambda_1}}\frac{x_3}{x_2} = \frac{1}{\sqrt{\lambda_1}}\tan\beta.$$

(8.61)

(4) The positions of P' and Q' are then plotted, with P' at an angle of α' on the $\lambda_1 \lambda_3$-plane and Q' at an angle of β' on the $\lambda_2 \lambda_3$-plane. The new position of the planar structure S' is given by the great circle passing through P' and Q'.

To find the deformed position of a lineation (L), (1) draw a great circle through L and the λ_2-axis and locate the point of intersection P of the great circle and the $\lambda_1 \lambda_3$-plane (Fig. 8.8c). (2) Draw a great circle through L and the λ_1-axis and locate its point of intersection Q with the $\lambda_2 \lambda_3$-plane. Determine the positions of P' and Q' from equations (8.60a) and (8.61). Draw great circles through P' and λ_2-axis and through Q' and λ_1-axis. The point of intersection of these two great circles gives the position of L'.

The longitudinal strain and shear strain along a line not necessarily parallel to the $x_1 x_3$-plane, are obtained from equations (8.18), (8.19) (8.24) and (8.25), by taking $\lambda_2 = 1$ and $\lambda_3 = 1/\lambda_1$

$$\lambda = l_1^2 \lambda_1 + l_2^2 + l_3^2/\lambda_1,$$

$$\gamma^2 = \lambda\left(\frac{l_1^2}{\lambda_1} + l_2^2 + l_3^2 \lambda_1\right) - 1 \tag{8.62a}$$

where l_i are the direction cosines of the line before deformation, and

$$\frac{1}{\lambda} = \frac{l_1'^2}{\lambda_1} + l_2'^2 + l_3'^2 \lambda_1$$

$$\gamma^2 = \lambda^2\left(\frac{l_1'^2}{\lambda_1^2} + l_2'^2 + l_3'^2 \lambda_1^2\right) - 1 \tag{8.62b}$$

where l_i' are the direction cosines of the line after deformation. If the linear structure occurs on the $x_1 x_3$-plane, $l_1 = \cos\alpha$ and $l_3 = \sin\alpha$ and the equations are simplified to:

$$\lambda = \lambda_1 \cos^2\alpha + \frac{1}{\lambda_1}\sin^2\alpha$$

$$\gamma^2 = \lambda\left(\frac{\cos^2\alpha}{\lambda_1} + \lambda_1 \sin^2\alpha\right) - 1 \tag{8.63}$$

$$\frac{1}{\lambda} = \frac{\cos^2\alpha'}{\lambda_1} + \sin^2\alpha' \lambda_1$$

$$\gamma^2 = \lambda^2\left(\frac{\cos^2\alpha'}{\lambda_1^2} + \lambda_1^2 \sin^2\alpha'\right) - 1.$$

Simple shear

The apparatus for deforming a model by simple shear consists of two parallel walls one of which is fixed in position. The other wall can move parallel to itself while keeping a constant distance from the fixed wall (Fig. 8.9a). The coordinate axis x_1 is chosen parallel to direction of movement of the moving wall. The x_3-axis is chosen normal to the walls. The x_3-axis is perpendicular to the $x_1 x_2$-plane. A point with coordinates x_i is shifted to the position x'_i according to the equations

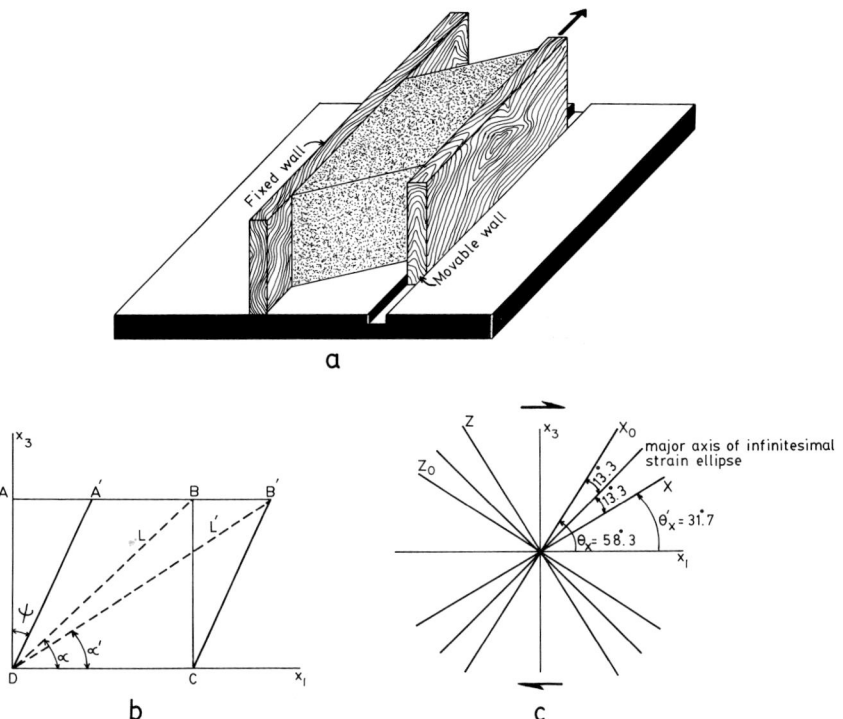

FIG. 8.9. (a) Simple shear apparatus. (b) Reorientation of a lineation (L to L') by simple shear. (c) X_0 and Z_0 are the initial orientations of two mutually perpendicular lines. X and Z are their positions after a deformation by simple shear of $\gamma = 1$. X and Z are also mutually perpendicular. X_0, Z_0 are the principal axes of strain while X and Z are the principal axes of the strain ellipse. The major axis of infinitesimal strain ellipse bisects the angle between X and X_0.

$$x'_1 = x_1 + \gamma x_3,$$
$$x'_2 = x_2, \qquad\qquad (8.64)$$
$$x'_3 = x_3.$$

The second equation can be omitted when we concentrate on deformation on x_1x_3-plane, noting that along the x_2-axis both the longitudinal strain and the shear strain are zero. On the x_1x_3-plane a rectangle $ABCD$ with sides parallel to the x_1- and x_3-axes (Fig. 8.9b) is deformed to a parallelogram $A'B'CD$, with ψ as the angle between DA and DA'. The shear strain corresponding to the lines AB and AD is $\gamma = \tan\psi$. γ is regarded as the amount of simple shear.

To obtain the equation of the strain ellipse on the x_1x_3-plane we start with a unit circle

$$x_1^2 + x_3^2 = 1$$

147

and express x_1 and x_3 in terms of x_1' and x_3' with the help of eqn. (8.64):

$$x_1'^2 + (\gamma^2 + 1)\, x_3'^2 - 2\gamma x_1' x_3' = 1. \tag{8.65a}$$

Comparing eqn. (8.65) with eqn. (8.46) we find that

$$a = 1,\, b = \gamma^2 + 1,\, h = \gamma. \tag{8.65b}$$

Therefore, from eqn. (8.51)

$$s_i = \tfrac{1}{2}(\gamma^2 + 2) \pm \tfrac{1}{2}\sqrt{\gamma^4 + 4\gamma^2}. \tag{8.66}$$

Since in the present case $\lambda_1\, \lambda_3 = 1$, the axes of the simple shear strain ellipse are given by the equations (8.52) as

$$
\begin{aligned}
X^2 &= 1/Z^2 = 1/s_1 = \tfrac{1}{2}(\gamma^2 + 2) + \tfrac{1}{2}\sqrt{\gamma^4 + 4\gamma^2} \\
Z^2 &= 1/X^2 = 1/s_2 = \tfrac{1}{2}(\gamma^2 + 2) - \tfrac{1}{2}\sqrt{\gamma^4 + 4\gamma^2}.
\end{aligned} \tag{8.67}
$$

The expressions for the semi-axes of the strain ellipse can be made simpler by making the right-hand side of eqn. (8.66) a perfect square. Let $\gamma/2 = s$. Then eqn. (8.66) can be written as

$$
\begin{aligned}
s_i &= (2s^2 + 1) \pm \sqrt{4s^4 + 4s^2} \\
&= (2s^2 + 1) \pm 2s\sqrt{s^2 + 1} \\
&= (s^2 + 1) + s^2 \pm 2s\sqrt{s^2 + 1} \\
&= [\sqrt{s^2 + 1} \pm s]^2
\end{aligned}
$$

or,

$$
\begin{aligned}
X &= \sqrt{s^2 + 1} + s \\
Z &= \sqrt{s^2 + 1} - s
\end{aligned} \tag{8.68}
$$

(Jaeger 1964).

The orientations of the strain ellipse axes are obtained from the relations (8.53) and (8.65b):

$$\tan 2\theta' = \frac{2}{\gamma}. \tag{8.69}$$

By comparing eqn. (8.64) with eqn. (8.42) we find that for simple shear $a_{11} = 1,\, a_{12} = \gamma,\, a_{21} = 0$ and $a_{22} = 1$. The orientations of the principal axes of finite strain are found by substituting these values in eqn. (8.54):

$$\tan 2\theta = -\left(\frac{2}{\gamma}\right). \tag{8.70}$$

The amount of rotation is found by substituting the values of a_{ij} into eqn. (8.55):

$$\tan(\theta' - \theta) = -\left(\frac{\gamma}{2}\right). \tag{8.71}$$

Simple shear is, thus, a plane-strain type of *rotational* deformation.

If the amount of simple shear is infinitesimally small, i.e. if γ tends to 0, eqn. (8.69) shows that θ' tends to $\pm 45°$. Thus, the principal axes of the infinitesimal simple shear strain ellipse are oriented at $\pm 45°$ with the x_1-axis.

Let us consider a numerical example with $\gamma = 1$. From eqns. (8.69), (8.70) and (8.71) we find that

$$\theta'_x = 31°.7, \; \theta'_z = 31°.7 + 90°,$$
$$\theta_x = 58°.3, \; \theta_z = 58°.3 + 90°,$$

where θ'_x and θ'_z give the orientations of the X- and Z-axes, the semi-major and semi-minor axes of the strain ellipse, and θ_x and θ_z give the orientations of the principal axes of strain, X_0 and Z_0, of finite simple shear. When these orientations are drawn in a diagram (Fig. 8.9c), from the origin, we find that X and X_0 are symmetrically situated on either side of the line $\theta = 45°$ the position of the major axis of the infinitesimal strain ellipse. *This is a characteristic feature, not only of simple shear, but of all types of rotational plane strain, i.e. the angle between X and X_0 is bisected by the major axis of the infinitesimal strain ellipse.* This means that the orientations of the X- and the X_0- axes can also be expressed by the following equations (Jaeger 1964):

$$\theta' = 45° - \tfrac{1}{2}\tan^{-1}(\gamma/2),$$
$$\theta = 45° + \tfrac{1}{2}\tan^{-1}(\gamma/2), \tag{8.72}$$

where $\tfrac{1}{2}\tan^{-1}(\gamma/2)$ is half the absolute value of the angle of rotation as given by eqn.(8.71).

The reorientation of planar and linear structures by finite simple shear can be determined in the following manner. From (8.64) we find that

$$\frac{x'_1}{x'_3} = \frac{x_1 + \gamma x_3}{x_3} = \frac{x_1}{x_3} + \gamma,$$

or,

$$\cot \alpha' = \cot \alpha + \gamma. \tag{8.73}$$

where α and α' are the initial and final angles between a linear structure and the x_1-axis (Fig. 8.9b). If the linear structure is at an angle to $x_1 x_3$-plane, the deformed line OP' must then lie on the plane defined by the direction of simple shear PP' and the undeformed line OP (Fig. 8.10c). This is an important relation and is often used in structural studies

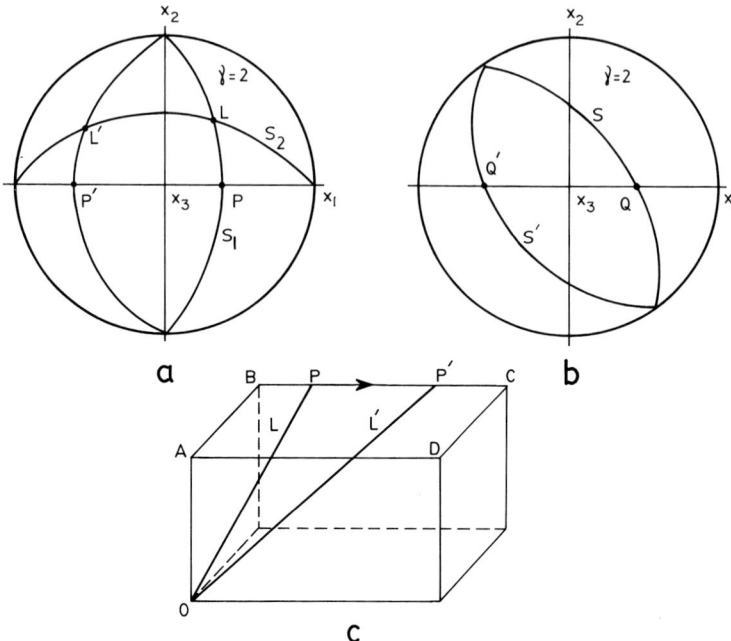

FIG. 8.10. (a) Equal area projection method of determination of the attitude of a linear structure after it is deformed by simple shear. (b) Equal area projection method of determining the attitude of a planar structure after it is deformed by simple shear. (c) Block diagram showing the change in orientation of a linear structure by simple shear.

(Turner & Weiss 1963, p. 484) in which an early lineation is deformed by slip-folding. Let L and L' be the initial and deformed positions of a linear structure in equal area projection (Fig. 8.10a). Draw a great circle S_1 through L and the x_2-axis, noting that the x_2-axis is parallel to the planes of simple shear. Draw another great circle S_2 through L and the x_1-axis. Again, note that the x_1-axis is the direction of simple shear. Let P be the intersection of S_1 and the x_1x_3-plane. The deformed position P' of the line of intersection is determined with the help of eqn.(8.73). L' will be situated at the intersection of S_2 and a great circle passing through P' and the x_2-axis (Fig. 8.10a).

Again, let S and S' be the initial and deformed positions of a planar structure in equal area projection (Fig. 8.10b). To determine the attitude of S' the following procedure is convenient. Let Q be the intersection of S and the x_1x_3-plane. Determine the position of Q' (i.e. the deformed position of Q) from eqn.(8.73), noting that the angle between Q' and the x_1-axis is α'. The projection of S' is obtained as the great circle passing through Q' and the line of intersection of S with the x_1x_2-plane.

150

Let a line OP, inclined to all the coordinate axes be of unit length and have direction cosines l_i. Then the coordinates of the point P are $x_i = l_i$. The quadratic elongation along the deformed line OP' is

$$\lambda = (OP')^2 = x_1'^2 + x_2'^2 + x_3'^2$$

or,

$$\lambda = 1 + \gamma^2 l_3^2 + 2\gamma l_1 l_3. \tag{8.74}$$

Now,

$$x_1' = (OP')l_1'$$
$$x_2' = x_2 = l_2 = (OP')l_2'$$
$$x_3' = x_3 = l_3 = (OP')l_3'.$$

Substituting these values in eqn. (8.64) we have

$$l_1 = \sqrt{\lambda}\,(l_1' - \gamma\, l_3')$$
$$l_2 = \sqrt{\lambda}\, l_2' \tag{8.75}$$
$$l_3 = \sqrt{\lambda}\, l_3'$$

where $\sqrt{\lambda} = OP'$.

Squaring and adding and noting that $l_i l_i = 1$, we find

$$1/\lambda = 1 - 2\gamma l_1' l_3' + \gamma^2 l_3'^2. \tag{8.76}$$

If the linear structure occurs on the $x_1 x_3$-plane, $l_1 = \cos\alpha$, $l_3 = \sin\alpha$, $l_1' = \cos\alpha'$, $l_3' = \sin\alpha'$ and eqns.(8.74) and (8.76) can be written as

$$\lambda = 1 + \gamma^2 \sin^2\alpha + 2\gamma \sin\alpha \cos\alpha$$
$$\frac{1}{\lambda} = 1 + \gamma^2 \sin^2\alpha' - 2\gamma \sin\alpha' \cos\alpha'. \tag{8.77}$$

To determine the shear strain along a line not necessarily lying on the $x_1 x_3$-plane consider a tangent plane at a point P on the unit sphere. OP is the normal to the tangent plane. When the sphere is deformed to an ellipsoid by a simple shear the orientation of OP and of the tangent plane will both change. In general, OP', the deformed position of OP, will not be a normal to the deformed plane. The angle between OP' and the normal to the deformed tangent plane gives us the shear angle ψ. Let the tangent plane to the unit sphere be

$$l_1 x_1 + l_2 x_2 + l_3 x_3 = 1,$$

where l_i are the direction cosines of OP. Expressing x_i in terms of x_i' with the help of eqn.(8.64) we obtain the equation of the tangent plane on the strain ellipsoid

$$l_1 x_1' + l_2 x_2' + (l_3 - l_1 \gamma)x_3' = 1.$$

The length of the normal from the origin to this plane is

$$p = \pm \frac{1}{\sqrt{1 + l_1^2 \gamma^2 - 2\gamma l_1 l_3}}$$

so that

$$\sec^2 \psi = \frac{(OP')^2}{p^2} = \lambda(1 + l_1^2 \gamma^2 - 2\gamma l_1 l_3)$$

and

$$\gamma_a^2 = \lambda(1 + l_1^2 \gamma^2 - 2\gamma l_1 l_3) - 1 \tag{8.78}$$

where γ_a is the shear strain along OP'. If the line lies on the $x_1 x_3$-plane we have

$$\gamma_a^2 = \lambda(1 + \gamma^2 \cos^2 \alpha - 2\gamma \sin\alpha \cos\alpha) - 1, \tag{8.79}$$

or eliminating λ with the help of the first equation of (8.77) we obtain, after simplification (Nadai 1963, p. 59)

$$\gamma_a^2 = \tfrac{1}{2}\gamma^2 \sin 2\alpha + \gamma \cos 2\alpha. \tag{8.80}$$

Note that γ is the amount of simple shear while γ_a is the shear strain along a line which was initially at an angle of α with the x_1-axis. If α' instead of α is known, we have first to determine α from eqn.(8.73) and then determine γ_a from (8.80). The condition in which γ_a has a maximum value is obtained by putting $\delta\gamma_a/\delta\alpha = 0$.

$$2\gamma_a \frac{\delta\gamma_a}{\delta\alpha} = \gamma^2 \cos 2\alpha - 2\gamma \sin 2\alpha = 0,$$

or, if $\gamma \neq 0$,

$$\tan 2\alpha = \frac{\gamma}{2}. \tag{8.81}$$

For example, if $\gamma = 1$, $\alpha = 13.3$, and from eqn. (8.73), $\alpha' = 10.8$. The corresponding value of γ_a is 1.06. Thus, in simple shear deformation *the maximum shear strain is not along the direction of simple shear.*
  Both pure shear and simple shear deformations are of the plane-strain type with $\lambda_2 = 1$. Pure shear deformation is irrotational while simple shear deformation is rotational. In pure shear, the axes of the strain ellipsoid have the same orientation irrespective of the magnitude of deformation. In simple shear, the long axis of the strain ellipsoid makes a smaller angle with the direction of simple shear with increasing value of γ. However, the geometry of simple shear deformation can be obtained by a two stage deformation, first a pure shear deformation with λ_1 and λ_2 given by eqns.(8.67) and with the λ_i-axes oriented along

θ_1 and θ_2 as given by eqn.(8.70) and second, by a rigid rotation through an angle $\tan^{-1}(-\gamma/2)$ as indicated by eqn.(8.71).

Combined pure and simple shears

Ramberg (1975) has considered a complex type of plane strain by simultaneous combination of pure shear and simple shear (Fig. 9.14) with constant strain rates. The equation for transformation of a point (x_1, x_3) to the point (x_1', x_3') for this case cannot be obtained simply by adding eqns.(8.57) and (8.64). Instead, we have to add the rates of displacement for pure shear and for simple shear and then integrate the combined rate-of-displacement equation. The rate of displacement equation for pure shear (Ramberg 1975, p. 6) is

$$\dot{x}_1 = \dot{\varepsilon}_1 x_1 \tag{8.82}$$
$$\dot{x}_3 = -\dot{\varepsilon}_1 x_3 \tag{8.83}$$

where the dots indicate differentiation with respect to time t and the rate-of-displacement equation for simple shear is

$$\dot{x}_1 = \dot{\gamma} x_3$$
$$\dot{x}_3 = 0. \tag{8.84}$$

For combined pure shear and simple shear with constant values of $\dot{\varepsilon}_1$ and $\dot{\gamma}$, the rate-of-displacement equation will be

$$\dot{x}_1 = \dot{\varepsilon}_1 x_1 + \dot{\gamma} x_3$$
$$\dot{x}_3 = -\dot{\varepsilon}_1 x_3. \tag{8.85}$$

From this system of differential equations we obtain

$$x_1 = \exp(\dot{\varepsilon}_1 t) x_1^\circ + \frac{\dot{\gamma}}{\dot{\varepsilon}_1} \sinh(\dot{\varepsilon}_1 t) x_3^\circ$$
$$x_3 = \exp(-\dot{\varepsilon}_1 t) x_3^\circ \tag{8.86a}$$

where (x_1°, x_3°) are the coordinates of a point on the $x_1 x_3$-plane at $t = 0$. Rewriting this equation in the notation followed earlier, we have

$$x_1' = \exp(\bar{\varepsilon}_1) \cdot x_1 + \frac{1}{s_r} \sinh(\bar{\varepsilon}_1) \cdot x_3$$
$$x_3' = \exp(-\bar{\varepsilon}_1) \cdot x_3, \tag{8.86b}$$

where $s_r = \dot{\varepsilon}_1/\dot{\gamma}$, the ratio of the rates of pure shear and simple shear (Ramberg 1975, Ghosh & Ramberg 1976). Since $\varepsilon_1/\gamma = \dot{\varepsilon}_1\dot{\gamma} = s_r$, the coefficients of eqn.(8.86b) can also be expressed in terms of γ and s_r by replacing ε_1 by γs_r. It should be noted that $\bar{\varepsilon}_1$ in eqn.(8.86b) is the natural strain, so that $\exp(\bar{\varepsilon}_1) = \sqrt{\lambda_1}$.

From eqns.(8.46) and (8.47) we find that the equation of the strain ellipse is

$$ax_1'^2 + bx_2'^2 - 2hx_1'x_2' = 1 \tag{8.87}$$

where

$$a = \exp(-2\bar{\varepsilon}_1)$$

$$b = \exp(2\bar{\varepsilon}_1) + \frac{1}{s_r^2}\sinh^2(\bar{\varepsilon}_1) \tag{8.88}$$

$$h = \frac{1}{2s_r}[1 - \exp(-2\bar{\varepsilon}_1)].$$

Equations (8.53), (8.54) and (8.55) give the orientations of the axes of the strain ellipse, the orientations of the principal axes of finite strain and the rotation. The lengths of the semi-axes of the strain ellipse are given by eqns.(8.51) and (8.52). The orientations of the axes of the infinitesimal strain ellipse (Ghosh 1982) are given by the following equation

$$\tan 2\theta = \frac{1}{2s_r}. \tag{8.89}$$

The equation shows that if $s_r \; (=\dot{\varepsilon}_1/\dot{\gamma})$ tends to zero, i.e. for a simple shear deformation, the major axis of infinitesimal strain ellipse lies at an angle of 45° with the simple shear direction. As s_r increases, this angle decreases. For example, if $\dot{\varepsilon}_1/\dot{\gamma} = 1$, $\theta = 13°.3$ and if $\dot{\varepsilon}_1/\dot{\gamma} = 2$, $\theta = 7°$. Thus, in combined pure shear and simple shear the XY-plane of the strain ellipsoid may lie at a small angle with the direction of simple shear even if the deformation is small.

8.11 Mohr diagram for strained and unstrained states

For plane strain, eqns. (8.18) (8.24), (8.19) and (8.25) can be written as

$$\lambda = \lambda_1 \cos^2\alpha + \lambda_2 \sin^2\alpha, \tag{8.90a}$$

$$\gamma = \frac{\lambda_1 - \lambda_2}{\sqrt{\lambda_1 \lambda_2}} \sin\alpha \cos\alpha, \tag{8.90b}$$

$$\lambda' = \lambda_1' \cos^2\alpha' + \lambda_2' \sin^2\alpha', \tag{8.90c}$$

$$\gamma' = (\lambda_2' - \lambda_1') \sin\alpha' \cos\alpha', \tag{8.90d}$$

where $\lambda' = 1/\lambda$, $\lambda_1' = 1/\lambda_1$, $\lambda_2' = 1/\lambda_2$, $\gamma' = \gamma/\lambda$ (Jaeger 1964, pp. 35–38, Ramsay 1967, pp. 65–68). α and α' are the angles between the λ_1-axis

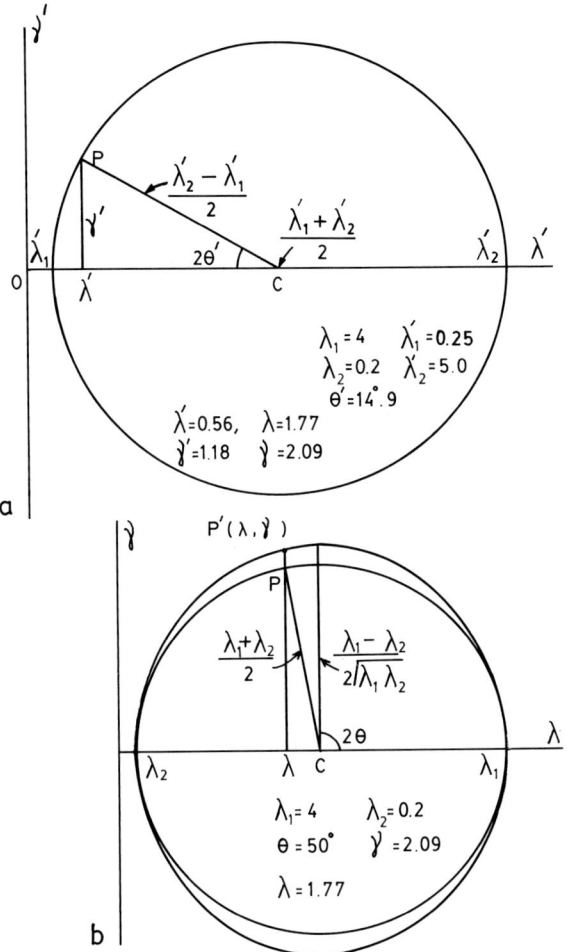

FIG. 8.11. (a) Mohr diagram for the strained state. The Mohr circle has centre C at $(\lambda_1' + \lambda_2'/2, 0)$ and with radius $\frac{1}{2}(\lambda_2' - \lambda_1')$. The angle $2\theta'$ at the centre of the circle is measured clockwise from λ_1'. (b) Mohr diagram for unstrained state for the same deformation as in (a). The Mohr diagram is an ellipse with semi-axes $\frac{1}{2}(\lambda_1 - \lambda_2)$ and $(\lambda_1 - \lambda_2)/2\sqrt{\lambda_1\lambda_2}$ along the λ- and γ-axes, respectively. If there is no area decrease, i.e. $\lambda_1\lambda_2 = 1$, the Mohr diagram would have been a circle with radius $\frac{1}{2}(\lambda_1 - \lambda_2)$. P is a point on such a circle with CP at an angle of 2θ measured anticlockwise from λ_1. A line through P parallel to the γ-axis intersects the Mohr ellipse at P'. The coordinates of P' give us λ and γ.

and a line before and after deformation. These equations can be expressed in terms of the double angles 2α and $2\alpha'$ by following the same procedure as in the case of the Mohr diagram for stress:

$$\lambda = \tfrac{1}{2}(\lambda_1 + \lambda_2) + \tfrac{1}{2}(\lambda_1 - \lambda_2)\cos 2\alpha, \qquad (8.91a)$$

$$\gamma = \frac{1}{2\sqrt{\lambda_1 \lambda_2}} (\lambda_1 - \lambda_2) \sin 2\alpha, \tag{8.91b}$$

$$\lambda' = \tfrac{1}{2}(\lambda'_1 + \lambda'_2) + \tfrac{1}{2}(\lambda'_1 - \lambda'_2) \cos 2\alpha', \tag{8.91c}$$

$$\gamma' = \tfrac{1}{2}(\lambda'_2 - \lambda'_1) \sin 2\alpha'. \tag{8.91d}$$

The Mohr diagram for the strained state is drawn with the help of eqns. (8.91c & d), with λ' as abscissa and γ' as ordinate. The Mohr diagram in this case is a circle. If P is a point on it with CP making an angle $2\alpha'$ with the λ'-axis (Fig. 8.11), the coordinates of this point give us the values of λ' and γ' for the deformed line. The Mohr diagram for the unstrained state is an ellipse with centre at $[(\lambda_1 + \lambda_2)/2, 0]$ and with the semi-axes of lengths $\tfrac{1}{2}(\lambda_1 - \lambda_2)$ and $\tfrac{1}{2}(\lambda_1 - \lambda_2)(\lambda_1 \lambda_2)^{-\frac{1}{2}}$ parallel to the λ- and γ-axes, respectively (Fig. 8.11b). If there is no area change, i.e. if $\lambda_1 \lambda_2 = 1$, then the Mohr diagram is a circle.

CHAPTER 9

Progressive Deformation

9.1. Introduction

Finite deformation considers only the initial shape of the body and the final state of deformation. To study the evolution of structures, however, we need to consider *progressive deformation* which specifies the intermediate stages of deformation. Thus, for example, the folding of a layer involves a layer-parallel shortening; boudinage, on the other hand, requires a layer-parallel extension. Hence, the initiation of folding in a layer and the initiation of boudinage in the profile plane of the same layer could not have taken place at the same time; they must have formed in successive stages. In this chapter we shall consider some simple types of progressive deformation.

9.2. Progressive pure shear

The transformation equation for pure shear is [eqn. (8.57)]

$$x' = \sqrt{\lambda_1}\, x$$
$$y' = (1/\sqrt{\lambda_1})y,$$

from which we obtain the relation

$$x'y' = xy.$$

In other words, the product of $x'y'$ is a constant, say $\frac{1}{2}a^2$:

$$xy = \tfrac{1}{2}a^2. \tag{9.1}$$

This is the equation of a rectangular hyperbola with x- and y-axes as asymptotes and the vertices situated at a distance of a from the origin at an angle of $\pm 45°$ with x- or the y-axis (Fig. 9.1). Depending on the initial position, each particle moves in pure shear along a hyperbolic *particle path* (Fig. 9.2) when λ_1 is gradually increased. Consider a line segment OA, with O at the origin and the point A moving along a

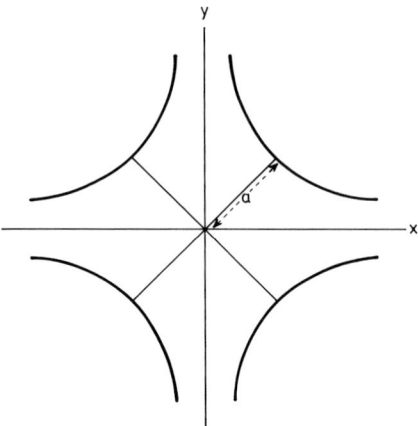

FIG. 9.1. Hyperbola, represented by the equation $xy = \frac{1}{2}a^2$.

hyperbolic particle path (Fig. 9.3a). OA initially makes an angle of $70°$ with the x-axis. With progressive deformation the line segment will be progressively shortened. The shortening along the line will be a maximum at $\lambda_1 = 2.6$, when the point A has moved to the position B, with OB making an angle of $45°$ with the x-axis. With further deformation the line will start to lengthen but will still be shorter than the initial length OA. It will be equal to its original length at $\lambda_1 = 7.6$, when it reaches the position OC at an angle of $20°$ with the x-axis. It will become longer than its initial length only if the progressive deformation goes beyond this stage (say, the line OF). On the other hand, if a line (Fig. 9.3b) was initially parallel to OB (at angle of $60°$ with the x-axis), it will first shorten and then lengthen and will regain its initial length at $\lambda_1 = 3$, in the position OD (at an angle of $30°$ with the x-axis). Beyond this stage of deformation it will enter into the field of finite extension.

We can draw the following conclusions from the above discussion:

(1) In pure shear deformation the $45°$ lines (Fig. 9.4) divide the $\lambda_1 \lambda_2$-plane into fields of instantaneous shortening (dotted area) and fields of instantaneous extension (blank area).

(2) Any line starting from the origin and going into the dotted area of Fig. 9.4 undergoes a shortening at each instant. We may say that the line undergoes an *incremental shortening* or shortening in a small increment of deformation.

(3) Any line which lies in the blank area of the figure undergoes an instantaneous extension or *incremental extension*.

(4) A line whose *final orientation* is in the dotted area must have undergone a finite (or total) shortening. It has also undergone a shortening throughout the entire history of progressive deformation.

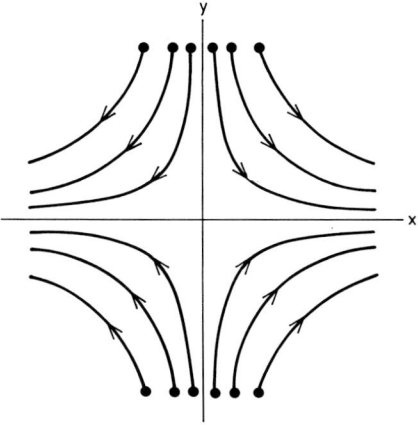

FIG. 9.2. Hyperbolic particle paths in pure shear.

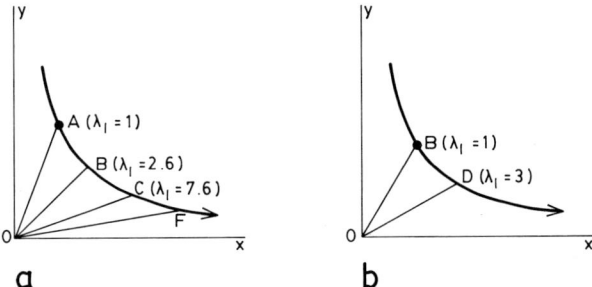

FIG. 9.3. (a) In pure shear deformation the 45° lines divide the $\lambda_1\lambda_2$-plane into fields of instantaneous shortening and extension. A line OA at an angle of 70° with the λ_1-axis first undergoes shortening till it comes to the position OB. It undergoes instantaneous extension between OB and OC, where $OC = OA$; however, there has been no finite extension of the line. A finite extension takes place when the line moves beyond the position of OC. (b) A line OB making an initial angle of 60° with the λ_1-axis will undergo finite extension at a much smaller value of λ_1.

(5) On the other hand, if a line lies in the blank area, we cannot say whether it has undergone a finite extension or not. It may have any one of the following histories: (a) it might have been extended throughout the entire history of the deformation (line OA in Fig. 9.5); (b) it might have been first shortened and then extended to come to lie parallel to the line of no finite longitudinal strain (line OB in Fig. 9.5); (c) it may have been first shortened and then extended and the finite strain along it is an extension (line OC in Fig. 9.5).

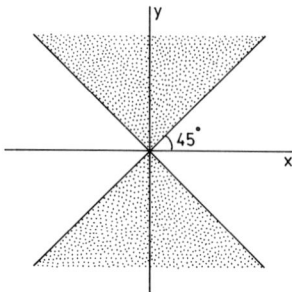

Fɪɢ. 9.4. Fields of instantaneous shortening (dotted) and fields of instantaneous extension (blank) in pure shear.

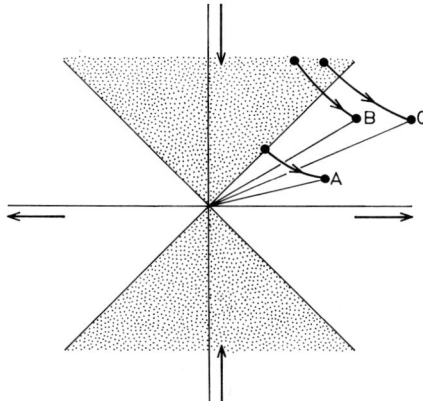

Fɪɢ. 9.5. Shortening and extension of differently oriented lines in progressive pure shear.

(6) In progressive pure shear, with a constant orientation of the principal axes of strain, all oblique lines rotate towards the direction of the λ_1-axis. A line may pass from the shortening field to the extension field but it cannot go from the extension field to the shortening field. The lines parallel to the principal axes will not change orientation.

Figure 9.6a shows three veins A, B, C with initial inclinations of $\theta = 90°$, $\theta = 63°$, and $\theta = 10°$ with the λ_1-axis. In an early stage of deformation (Fig. 9.6b) the vein A has undergone buckle folding without a change in orientation. Similarly, the vein C has undergone boudinage with a slight change in orientation. The vein B has undergone folding and has rotated as a whole towards the λ_1-axis. Figure 9.6c shows an advanced stage of deformation in which the vein B has entered into the field of finite longitudinal extension and has under-

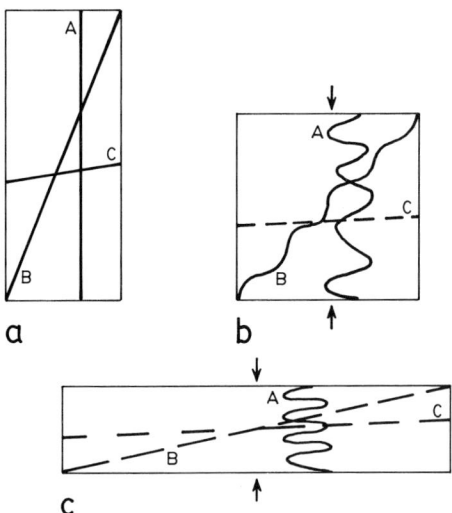

FIG. 9.6. Buckling and boudinage of differently oriented veins in pure shear.

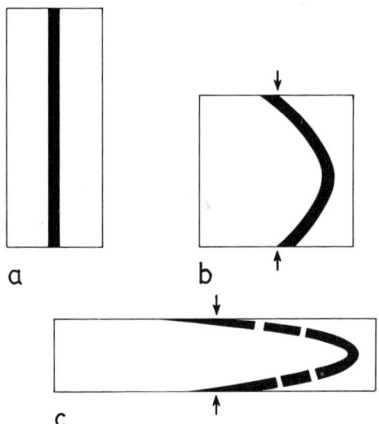

FIG. 9.7. In a layer undergoing buckle folding, boudinage can take place in the profile plane only at a late stage when the fold limbs have entered into the extension field.

gone boudinage. Figure 9.7a shows a competent layer parallel to the direction of shortening. In the early stage of deformation the layer undergoes folding. Boudinage cannot take place because the limbs are still in the compression field (Fig. 9.7b). The limbs are boudinaged only at a later stage (Fig. 9.7c) when they have rotated to the field of extension.

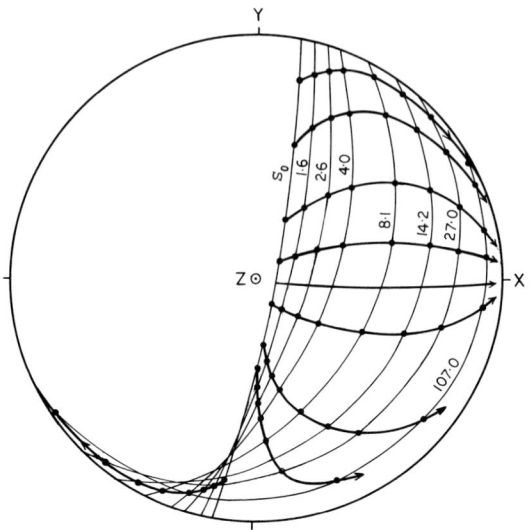

FIG. 9.8. Stereographic projection showing the change in orientation of an S-plane and lineations lying on it with progressive pure shear. S_0 is the initial position of the plane at $\lambda_1 = 1$. The figures represent values of λ_1 with progressive deformation.

Figure 9.8 shows how a plane (S_0) and several lines (L_0) lying on it deform in progressive pure shear. Initially the plane and the lines lying on it are oblique to all the principal axes of strain. With progressive deformation the S-plane rotates towards the XY-plane and the lineations move towards the X-axis. The path of a lineation is neither a great circle nor a small circle. The method of obtaining the position of the S-plane and the lineation for the successive values of λ_1 has been described in section 8.10.

9.3. Progressive simple shear

The transformation equation for simple shear is (eqn. 8.64):

$$x' = x + \gamma y,$$
$$y' = y.$$

When γ is gradually increased each particle with initial coordinates (x, y) moves along a straight line parallel to the x-axis (Fig. 9.9). In Figs. 9.10a and 9.10b all lines (such as OA) passing through the origin and occurring in the dotted area undergo an instantaneous shortening, while all lines (such as OB) lying in the blank area undergo an instantaneous extension. A line rotates clockwise in dextral simple shear and anticlockwise in sinistral simple shear. In dextral simple shear, an initial line OA has successive positions OB, OC, OD, etc. (Fig. 9.10c). The line

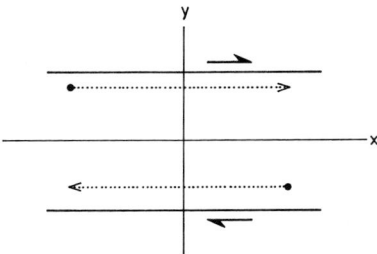

FIG. 9.9. Straight line particle paths in simple shear.

undergoes a longitudinal shortening in all positions between OA and OB, where OB is parallel to the y-axis. With progressive simple shear it starts to extend beyond this position but still remains shorter than its original length. It becomes equal to its original length when it reaches the position OD, with the angle DOB equal to the angle AOB. An actual extension of the line takes place only if the deformation progresses beyond this stage. On the other hand, if a line has an initial orientation within the blank area of Fig. 9.10 (say the line OB or OC), it will continue to undergo an extension with progressively increasing simple shear. Thus, just as in the case of pure shear, a layer initially in the field of shortening may undergo folding and then be boudinaged when the layer as a whole enters the field of extension. However, a layer which has undergone boudinage by simple shear cannot undergo folding in continuation of the same simple shear movement provided the nature of the instantaneous simple shear remains the same throughout the entire history.

It will be seen in Chapter 14 that the schistosity in rocks either tracks the XY-plane of the strain ellipsoid or remains roughly parallel to this plane. If the deformation is by progressive simple shear, the XY-plane of the infinitesimal strain ellipsoid makes an angle of 45° with the direction of simple shear. This angle is reduced with progressive deformation (Fig. 9.11). Hence the schistosity cannot make an angle greater than 45° with the direction of simple shear. When this angle has become rather small, a large increment of simple shear is required to make a small change in the orientation of the schistosity. Thus, at very large values of simple shear, the orientation of the schistosity virtually remains unchanged with progressive deformation.

We have considered so far the evolution of structures on a plane parallel to the xy coordinate plane. This plane contains the major and the minor principal axes of the strain ellipsoid. The normal to this plane is parallel to the intermediate axis of the strain ellipsoid. A bed inclined to all the principal axes of the strain ellipsoid will show on it an elliptical section of the strain ellipsoid. The principal axes of this *strain ellipse* will not coincide with those of the strain ellipsoid. Folding of the

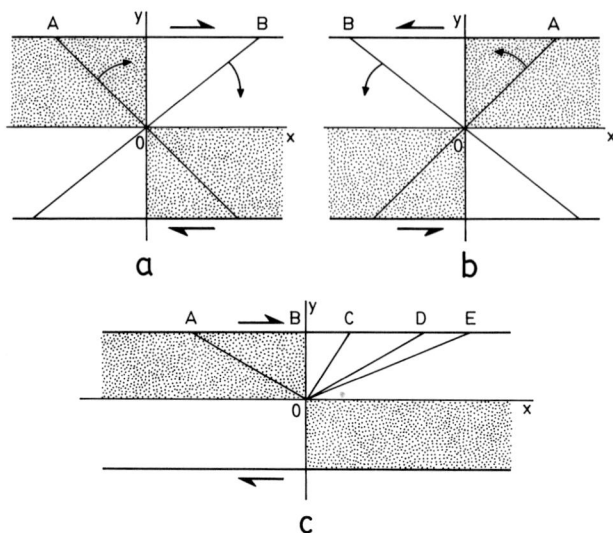

FIG. 9.10. Fields of instantaneous shortening (dotted) and extension (blank) in simple shear. In dextral simple shear (a) a line rotates clockwise and in sinistral simple shear (b) it rotates anticlockwise towards the direction of simple shear (x-axis). In progressive simple shear a line can move from the field of shortening to the field of extension, but an extended line does not undergo shortening at any other stage (c).

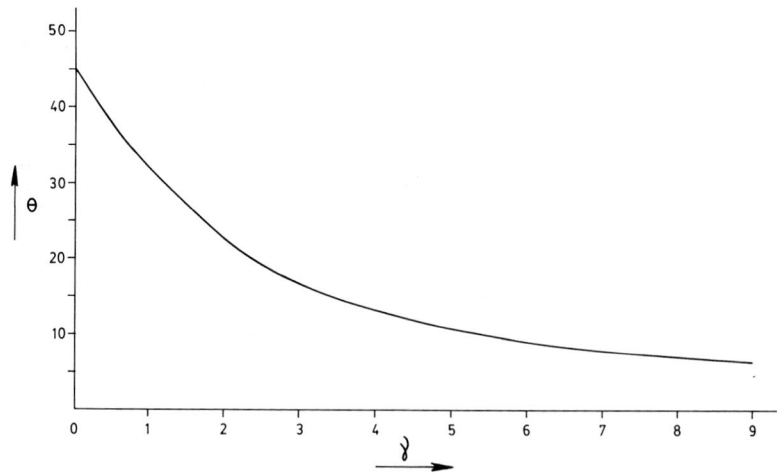

FIG. 9.11. Change in the angle (θ) between the direction of simple shear and the X-axis of the strain ellipsoid in progressive simple shear.

164

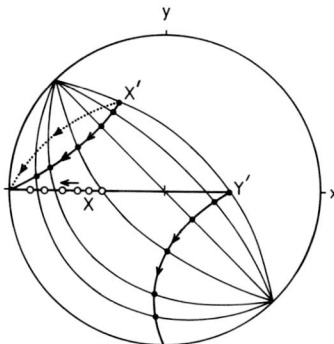

FIG. 9.12. Paths of the long (X') and the short (Y') axes of the strain ellipse on an inclined plane with progressive simple shear. The path does not coincide with that (dotted curve) of a marker line which initially coincided with the X'-axis. The successive positions of the major axis (X) of the strain ellipsoid are shown by open circles.

inclined bed will occur only if there has been a shortening along the minor axis of the strain ellipse. In that event the fold axis will be parallel to the major axis of the ellipse. Since the bed undergoes at the same time an extension parallel to the fold axis, boudinage may take place, with the boudin lines perpendicular to the fold axis, even when the folds are quite open. Boudin lines subparallel to the fold axis may develop only when the folds have tightened. Figure 9.12 shows how the major and the minor axes of strain ellipse on an inclined plane change orientations in the course of progressive simple shear. Once formed, the fold axis may continue to remain parallel to the major axis of the strain ellipse at all stages of deformation. Alternatively, it may passively rotate as a material line with progressive deformation. In general, these two alternative paths remain close to each other but do not coincide (Fig. 9.12). For both cases, the orientation of the fold axis does not coincide with the line of intersection of bed and the XY-plane of strain ellipsoid although the angle between the two lines remains rather small.

Figure 9.13 shows the manner in which an S-plane, oblique to all the coordinate axes and lineations lying on the plane, change orientations with progressive deformation. Each lineation moves towards the direction of simple shear (x-direction) along a plane which passes through the direction of simple shear and the initial lineation. In stereographic or equal area projection the locus of each lineation in progressive simple shear is therefore a great circle.

9.4. Progressive plane strain of a more complex type

In steady progressive plane deformation, in which the orientation of the axes of the instantaneous strain ellipse remains constant throughout the history of deformation, the path of a particle is not necessarily a

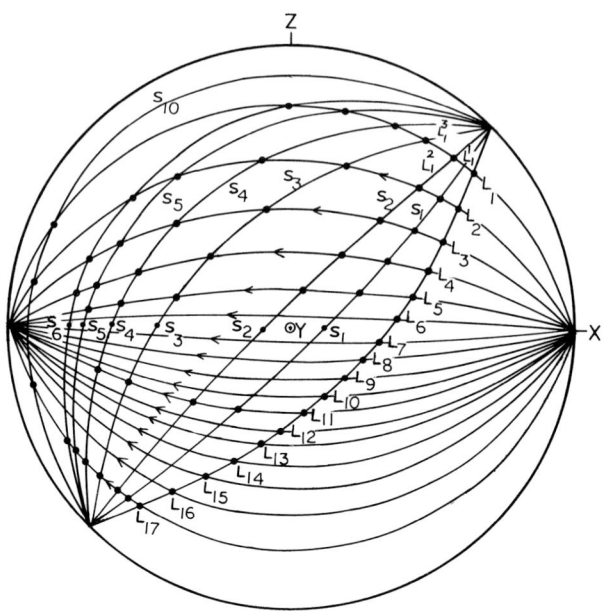

FIG. 9.13. Change in orientation of an *S*-plane and lineations lying on it with progressive simple shear. Each lineation locus is a great circle passing through the initial orientation of the lineation and the direction of simple shear.

straight line (as in progressive simple shear) or a rectangular hyperbola (as in progressive pure shear). Ramberg (1975) has considered a type of plane deformation by a combination of pure shear and simple shear in which a particle moves along a non-rectangular hyperbola. Let us choose a set of rectangular coordinate axes, x and y with the direction of simple shear along the x-axis and the planes of shear parallel to the xy-plane. The x-axis is also parallel to the direction of maximum stretching of the pure shear part of deformation. The maximum shortening in pure shear is along the y-axis (Fig. 9.14). $\dot{\gamma}$ is the rate of simple shear and $\dot{\varepsilon}_x$ and $\dot{\varepsilon}_y$ are the principal strain rates of pure shear. The ratio $\dot{\varepsilon}_x/\dot{\gamma}$ has been designated s_r. For a steady progressive deformation the strains ε_x and γ increase in course of time but s_r remains constant. A particle with initial position (x, y) changes its position to (x', y') in accordance with eqn. (8.86) (Ghosh 1982, eqn.7):

$$x' = \exp(\varepsilon_x)x + \frac{1}{s_r}(\sinh \varepsilon_x)y$$

$$y' = \exp(-\varepsilon_x)y$$

where ε_x is the natural strain for pure shear. The particle path can be obtained by increasing ε_x in successive steps and calculating x', y' for

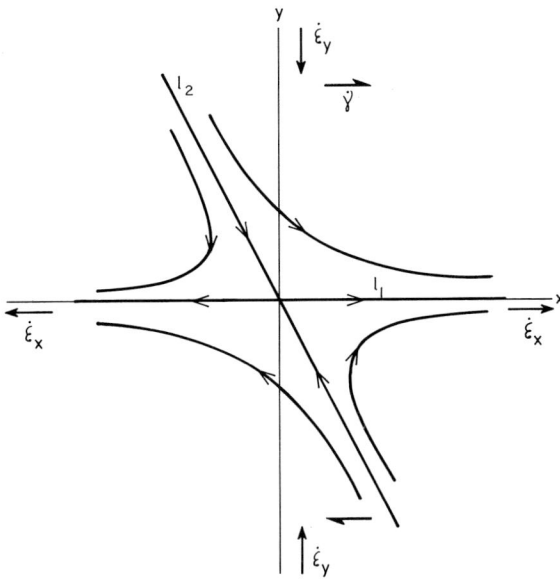

FIG. 9.14. Hyperbolic particle paths in combined pure shear and simple shear. l_1 and l_2 are the two apophyses; along each of them a particle moves in a straight line. Lines parallel to l_1 and l_2 do not rotate. Along l_1 a line is extended, along l_2 a line is continuously shortened.

any initial particle of coordinates x and y. For the equation given above it has been assumed that there is no area change in the xy-plane. Hence, the sum of natural strains along the x- and y-directions is zero. In case of a change in volume, $(-\varepsilon_x)$ in the second equation should be replaced by ε_y.

The hyperbolic particle paths are shown in Fig. 9.14. The hyperbolic curves in this figure have two apophyses. One of these, say l_1, is parallel to the x-axis and the other one, say l_2, is oriented at an angle of θ, with

$$\tan\theta = -2s_r. \qquad (9.2)$$

The equation shows that, as long as the ratio $\dot{\varepsilon}_x/\dot{\gamma}$ remains constant, the orientation of the two apophyses do not change during progressive deformation. The angle between the two apophyses decreases with a decrease in s_r.

Material lines parallel to the two apophyses do not change their orientations. However, material lines along one of the apophyses, viz. l_1, undergo extension while lines parallel to l_2 undergo a shortening throughout the entire history of the deformation. All other lines rotate towards the extensional apophysis, l_1. However, lines lying within the obtuse angles between l_1 and l_2 rotate forward, i.e. in a sense

167

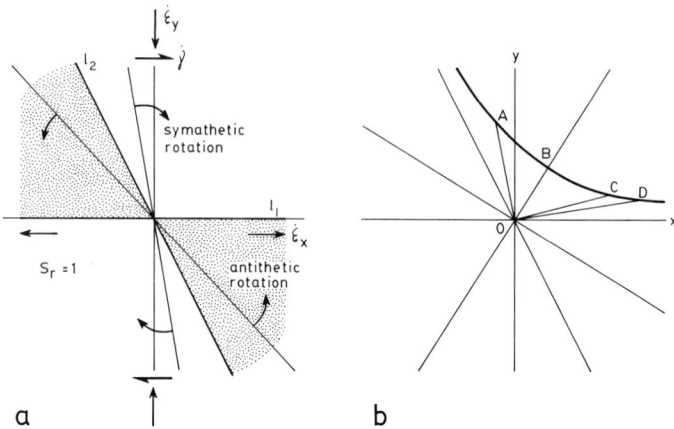

FIG. 9.15. (a) l_1 and l_2 are the two apophyses of particle paths in a progressive non-coaxial deformation, with $S_r = 1$. A line occurring within the dotted area shows antithetic rotation and a line in the blank area shows sympathetic rotation. (b) A line OA moving from field of instantaneous shortening to the field of instantaneous extension. It undergoes a finite extension when it has moved to the position OD, with angle $BOD > AOB$.

sympathetic to the sense of shear while lines lying within the acute angle rotate backward or antithetically (Fig. 9.15a).

The acute and the obtuse bisectrices between the two apophyses divide the xy-plane into fields of instantaneous shortening and fields of instantaneous extension (Fig. 9.16). Any line starting from the origin and lying within the extension field will continue to extend throughout the entire history of deformation. A line (say, OA in Fig. 9.15b) lying in the shortening field will continue to shorten till it becomes parallel to a bisectrix (OB). Beyond this position it will continue to extend but will not show a finite extension till it has passed beyond the position of OC where the angle AOB = the angle BOC (Fig. 9.15b). The orientations of the principal axes of the strain ellipse are given by the following equation (Ghosh 1982, eqn.8):

$$\tan 2\theta' = \frac{\dfrac{1}{s_r}\left[1 - \exp\left(-2\gamma\, s_r\right)\right]}{\left[\dfrac{1}{s_r}\sinh\left(\gamma s_r\right)\right]^2 + \left[\exp\left(2\gamma s_r\right) - \exp\left(-2\gamma s_r\right)\right]} \tag{9.3}$$

where θ' is the angle between a principal axis of the strain ellipse and the extensional apophysis. Figure 9.17 shows how the major axis of the strain ellipse changes orientation in course of progressive deformation. The figure shows that when s_r increases, i.e. when a pure shear is added

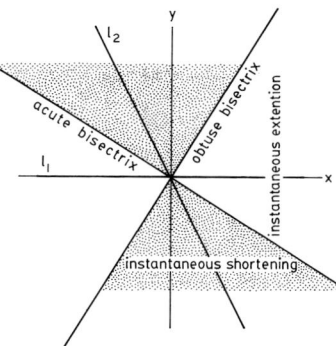

FIG. 9.16. The acute and the obtuse bisectrices between l_1 and l_2 dividing the xy-plane into a field of instantaneous shortening and fields of instantaneous extension.

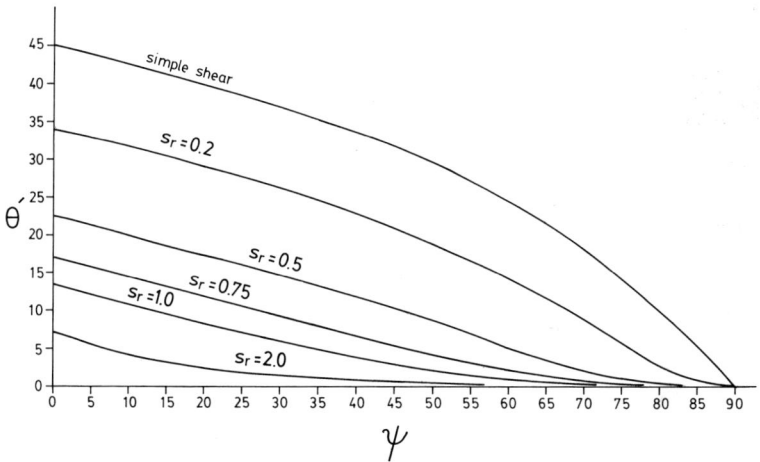

FIG. 9.17. Change in orientation of major axis of strain ellipse on the xy-plane in non-coaxial deformation with combined pure shear and simple shear. The value of simple shear is tan ψ. θ' is the angle between the major axis of the strain ellipse and the extensional apophysis.

to the simple shear, the major axis of the strain ellipse makes a smaller angle with the direction of shearing motion. With progressive deformation it comes close to this direction; hence, the maximum range of rotation of the strain ellipse also becomes smaller with an increase in the value of $s_r = (\dot{\varepsilon}_x/\dot{\gamma})$.

We shall see in Chapter 14 that the cleavage in rocks forms approximately parallel to the XY-plane of the strain ellipsoid. If the deformation in a ductile shear zone takes place by a combination of wall-parallel shear

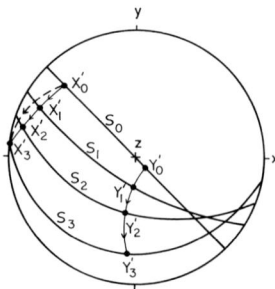

FIG. 9.18. Changing orientations of the major (X') and minor (Y') axes of the strain ellipse on a layer inclined to all the principal axes of the strain ellipsoid. The dashed line shows the locus of a material line which initially coincided with X'. S_1, S_2, etc., are the successive positions of the layer in progressive deformation with combined pure shear and simple shear with $S_r = 1$. Lower hemisphere stereographic projection.

and a shortening across the shear zone, the cleavage in the shear zone can then initiate at a fairly low angle with the shear zone walls.

Figure 9.18 shows the changing orientations of the principal axes of the strain ellipse on a layer inclined to all the principal axes of the strain ellipsoid. With progressive deformation the layer rotates towards the xy-coordinate plane, the major axis of the strain ellipse rotates towards the x-coordinate axis (parallel to the extensional apophysis) and the minor axis of the strain ellipse rotates towards the y-coordinate axis or the intermediate axis of the strain ellipsoid (see Flinn 1962, Ramberg & Ghosh 1977, Treagus & Treagus 1981 for a more detailed discussion). Folding of the inclined layer can occur only if there is a finite shortening along the minor axis of the strain ellipse. If folds do develop, the fold axis is initiated parallel to the major axis of the strain ellipse on the bed. With progressive deformation it either tracks this axis or rotates as a material line along a slightly different path (Fig. 9.18). Boudinage may occur on the inclined layer, with boudin lines initially perpendicular to the fold axis. With progressive deformation the boudin lines will also rotate towards the x-axis; as a consequence the angle between the fold axis and the boudin lines will no longer remain at a right angle.

9.5. Shortening of extended layers

In the type of deformation considered so far, a line which is once extended cannot undergo a shortening with progressive deformation. As a consequence, a cleavage which forms parallel to the XY-plane of the strain ellipsoid cannot be crenulated in continuation of the same deformation. In a similar way, the axial plane of a fold cannot be folded in continuation of the same deformation. Yet, in nature, the crenulation of a cleavage or the folding of an axial plane cleavage occasionally takes place in the course of a progressive non-coaxial deformation. This mode of structural evolution is possible if the deformation history

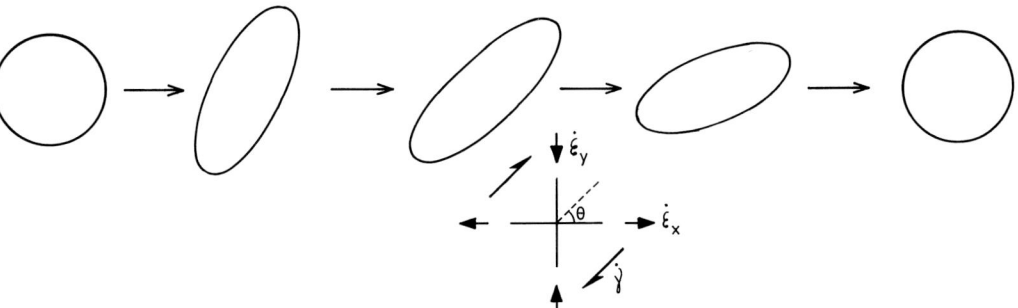

F<small>IG</small>. 9.19. Pulsating strain ellipse in non-coaxial deformation with closed particle paths.

involves *pulsating strain ellipsoids* or *closed particle paths* (Ramberg 1975). It is more likely, however, that in the course of a prolonged period of deformation the orientations of the principal axes of the infinitesimal strain ellipsoid do not remain the same because of changes in the boundary conditions. As a result of such changes, an extended line may come to lie in the field of instantaneous shortening. Under such conditions it is possible to have the folding of a cleavage in a continuous deformation or the folding of a boudinaged layer by a shortening across the boudin lines. As yet, there is no satisfactory theoretical analysis which derives this kind of structural evolution as a necessary consequence of a specific set of geological conditions.

A history of pulsating strain ellipsoids is obtained when the kinematical vorticity number (see Chapter 11) is greater than 1. Ramberg (1975) has derived this kind of deformation history by combining pure shear and simple shear, with the principal axes of the pure shear part parallel to the coordinate axes x and y, and the direction of simple shear at an angle of θ with the x-axis (Fig. 9.19). Under this condition a point (x, y) changes its position to (x', y') according to the equations

$$x' = a_{11}x + a_{12}y$$
$$y' = a_{21}x + a_{22}y \tag{9.4}$$

where

$$a_{11} = \cos(bt) + \frac{1}{b}(s_r - \sin\theta\cos\theta)\sin(bt)$$

$$a_{12} = \frac{1}{b}\cos^2\theta \sin(bt)$$

$$a_{21} = -\frac{1}{b}\sin^2\theta \sin(bt)$$

$$a_{22} = \cos(bt) - \frac{1}{b}(s_r - \sin\theta\cos\theta)\sin(bt),$$

171

with $b = (2s_r \sin\theta \cos\theta - s_r^2)^{\frac{1}{2}}$ and t representing time. The pulsating strain ellipsoids are obtained when

$$\sin2\theta - s_r > 0. \tag{9.5}$$

Since the maximum value of $\sin2\theta$ is 1, s_r must be less than 1. As the time t is increased, each particle moves along an *elliptical path*.

In progressive pure shear, all material lines rotate towards the direction of principal extension, and in simple shear all lines rotate towards the direction of shear. Similarly, in combined pure shear and simple shear with hyperbolic paths, the lines rotate towards the direction of the extensional apophysis. On the other hand, for elliptical particle paths a material line continues to rotate in the same sense; it does not approach a stable position. Similarly, the length of the major axis of the strain ellipse cannot go on increasing indefinitely. After reaching a maximum value the axial ratio of the strain ellipse starts to decrease (Fig. 9.19). Thus, in this kind of deformation, a bed which has undergone boudinage can once again be folded in the same continuous deformation. Similarly, the development of a cleavage and its folding can take place in the same deformation.

9.6. Deformation path

The finite strain ellipsoid, as reconstructed from deformed objects tells us nothing about the history of deformation. The deformation path concept (Elliott 1972), first introduced by Flinn (1962), describes the manner in which the strain ellipsoids of intermediate stages change their shape and orientation. The deformation path may be coaxial or non-coaxial. In the case of progressive coaxial deformation the orientation of the principal axes of finite strain remains the same at all stages of deformation. For such a situation the manner in which the strain in the intermediate stages changes is sufficient to describe the progressive deformation. The resulting path is called a *strain path*. A strain path may also be reconstructed for non-coaxial deformation in which the orientation of strain axis cannot be determined.

A method of deriving the strain path utilizes the heterogeneous character of strain within a small scale. Thus, the ooids in a deformed oolitic limestone may show different intensities of strain resulting from slight variations of ductility. A basic assumption of this method is that the strain history is the same for each component although the finite strain at each stage is variable among the different components (Wood 1974, Donath & Wood 1976). The strain path can be determined if the degree of heterogeneity is large.

A convenient way of representing the strain path is to plot the strain ellipsoids in a Flinn diagram (Flinn 1956, see also Flinn 1962, 1978). In

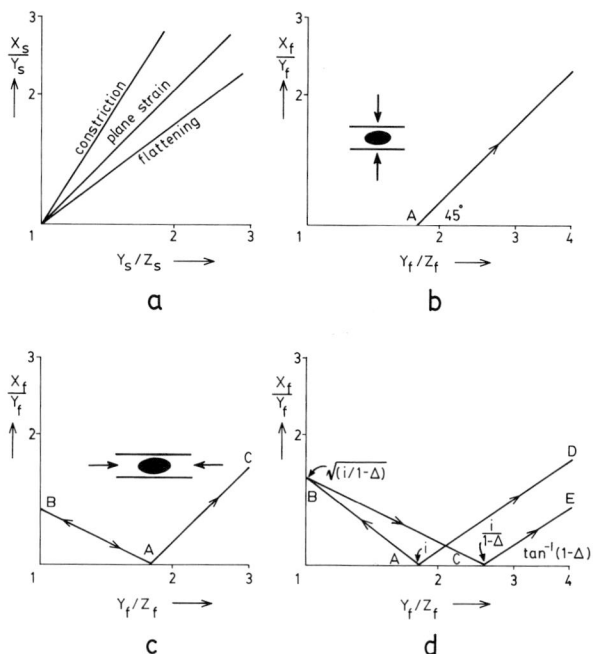

FIG. 9.20. Logarithmic Flinn plot of strain paths. (a) Tectonic strain paths without volume loss. X_s, Y_s and Z_s are lengths of principal semi-axes of the strain ellipsoid. A straight line path making an angle (θ) of $< 45°$ with the Y/Z-axis indicates progressive deformation in the flattening field. Paths of plane strain and constriction make angles of $\theta = 45°$ and $\theta > 45°$. (b) and (c) Progressive shape change of initially spherical marker which had first undergone a bedding normal compaction and then a tectonic plane strain. The principal shortening for tectonic strain is normal to bedding in (b) and parallel to bedding in (c). (d) Same as in (b) and (c) but with volume loss Δ. The line AD shows the shape change when tectonic shortening is normal to bedding. When the shortening is parallel to bedding the path is from A to B, then B to C and finally from C to E.

this diagram the ratio X/Y is plotted against the ratio Y/Z. The plotting is often done on a logarithmic scale. If the deformation is by plane strain, the strain path is along a line passing through the origin and making an angle of $45°$ with the ordinate or the abscissa (Fig. 9.20a). If the k value remains constant during deformation the plots of the strain ellipsoids of successive stages will lie along a straight line passing through the origin. It may be recalled that

$$k = \frac{(X/Y) - 1}{(Y/Z) - 1}.$$

For a flattening type of deformation, $k < 1$; for each value of k the strain path will be a straight line making an angle of less than $45°$ with the Y/Z-axis. For a constriction type of deformation the straight line strain path will make an angle of greater than 45° with Y/Z-axis (Fig. 9.20a). The strain path will be curved if the k value changes in successive stages of deformation.

Hobbs *et al.* (1976, p. 43) have suggested a modified form of the Flinn diagram in which a third axis has been added along which we can plot the angle θ between the layering and the major principal axis of strain. Each strain ellipsoid is plotted as a point with its three coordinates $(X/Y, Y/Z$ and θ). The line connecting these points will represent a deformation path.

The final shape of deformed objects, such as ooids or lapilli, is a result of a tectonic strain superimposed on a pretectonic shape. If the pretectonic shape is spherical, the ellipsoid resulting from the tectonic strain immediately gives us the shape of the strain ellipsoid. In many cases, the pretectonic shape is approximately ellipsoidal. This may either be due to initial variation in shape or due to diagenetic compaction (e.g. Cloos 1947b, Ramsay 1967, Oertel 1970, Ramsay & Wood 1973, Sanderson 1976, Bell 1985). The following examples show how the shape of the deformation path is controlled by the different factors.

(1) *Tectonic plane strain with no volume loss, superimposed on diagenetic compaction*

Let us choose a set of coordinate axes x, y and z, with z normal to the bedding. These directions are to be distinguished from the directions of the tectonic strain axes X, Y, Z. In the initial stage, let there be a compaction along the z-axis (Fig. 9.20b). After compaction of a spherical body it has a uniaxial oblate or pancake-like shape, with $X/Y = 1$ and $Y/Z > 1$. The axis of revolution of the oblate strain ellipsoid is normal to the bedding. In a Flinn diagram such an ellipsoid plots as the point A on the Y/Z-axis. For the tectonic strain following the stage of compaction, consider two situations of plane strain, (i) with the maximum shortening normal to the bedding and (ii) with maximum shortening parallel to the bedding. It is assumed that the tectonic deformation is coaxial. In the first situation the ratio of X/Y and Y/Z will progressively increase and the successive stages of the ellipsoid will plot on a straight line starting from the point A and making an angle of 45° with the Y/Z-axis (Fig. 9.20b). In the second situation there will be a gradual lengthening along the layer-normal (Fig. 9.20c) with increasing values of X/Y along with decreasing values of Y/Z, till the ellipsoid attains a prolate or cigar-like shape. At this stage (point B in Fig. 9.20c) of progressive deformation, $Y/Z = 1$ and the value of X/Y is equal to the value of $\sqrt{Y/Z}$ for the point A. With continued shortening, the ellipsoid again becomes uniaxial oblate type. The deformation path traces back from B to A (Fig. 9.20c). At this stage the ellipsoid has the same shape as it initially had after compaction; however, the axis of revolution of the ellipsoid is now parallel to the bedding. Beyond this stage the deformation path (AC in Fig. 9.20c) is a straight line at an angle of 45° with Y/Z-axis.

The strain path considered above is for the total ellipsoid combining compaction and tectonic strain. If we consider tectonic strain alone, the strain path will be a straight line starting from the origin and making an angle of 45° with Y/Z-axis.

(2) Tectonic plane strain with incremental volume loss superimposed on diagenetic compaction

Let the ellipsoid after compaction be represented by the point A in the Flinn diagram (Fig. 9.20d). As in the first case, this is an oblate spheroid with the axis of revolution normal to the bedding. If the tectonic plane strain causes a further shortening along this normal, along with volume loss (Δ), we have

$$X_s Y_s Z_s = (1 - \triangle) \tag{9.6}$$

(Ramsay & Wood 1973, eqn. 9),where X_s, Y_s, Z_s are the ellipsoid axes for the tectonism alone. For the combined effects of compaction and tectonism, there may be two alternative strain paths depending on whether (i) X_s is parallel and Z_s is normal to bedding or (ii) X_s is normal and Z_s is parallel to bedding. In the first situation (Fig. 9.20d) the strain path is a straight line starting from A and making an angle of $\tan^{-1}$ $(1 - \Delta)$ with the Y/Z-axis. In the second situation, the strain path starts from the point A to the point B. The point B represents the stage when the oblate ellipsoid of stage A has become a uniaxial prolate ellipsoid, with $Y/Z = 1$ and $X/Y = \sqrt{i/(1 - \Delta)}$, where i is equal to the value of Y/Z for the point A. Beyond this stage of deformation the strain path descends along a straight line BC. At C, lying on the Y/Z-axis, the value of Y/Z is $i/(1 - \Delta)$. With further deformation the strain path is along a straight line at an angle of $\tan^{-1} (1 - \Delta)$ with the Y/Z-axis.

One of the most important methods of determining the deformation history is to measure the incremental strains and incremental rotation from the length and orientation of fibrous crystals growing in veins and in pressure shadows (e.g. Choukroune 1971, Durney & Ramsay 1973, Wickham 1973, Ramsay 1980b, Casey *et al.* 1983, Ramsay & Huber 1983, Etchecopar & Malavieille 1987). The changing orientations of the fibrous crystals (quartz, calcite, chlorite, etc.) record the changing orientations of the direction of principal extension in the rock, while the length of the fibres in a particular orientation record the incremental displacement. The detailed methods of measurement and of reconstruction of the deformation history are given in Ramsay & Huber (1983).

CHAPTER 10

Measurement of Strain

10.1. Introduction

Finite strain in rocks can be measured from the deformed shapes of certain geological objects. These are known as *strain markers* or *strain gauges*. To determine strain geologists have used a variety of strain markers, e.g. fossils, conglomerates, breccias, oolites, clastic grains, grain aggregates, reduction spots, amygdales, lapilli, boudins, burrows, dessication cracks, etc. In addition, buckle folded layers and rotated planar markers have also been used to determine the bulk strain in rocks. Depending on the initial geometry of the strain marker, different techniques of strain measurement have been devised (Cloos 1947b, 1971, Flinn 1962, Ramsay 1967, Dunnet & Siddans 1971, Matthews *et al.* 1974, Shimamoto & Ikeda 1976, Mimran 1976, Robin 1977, Lisle 1977a, 1979, Oertel 1970, Fry 1979a & b, Hanna & Fry 1979, Ramsay & Huber 1983). The most comprehensive study on the techniques of strain determination is by Ramsay (1967) and by Ramsay & Huber (1983). Some of the methods of strain determination are described in the following sections.

When pebbles or other deformed objects occur in a loose or friable matrix, the lengths of their principal axes can be directly measured in the field. In most cases, however, the initial measurements are done on three orthogonal sections cut through an oriented specimen. A strain ellipse is then reconstructed for each of these sections. From these data we can then calculate the orientations and axial ratios of the principal axes of the strain ellipsoid. The first step in this technique is therefore to reconstruct the strain ellipse on a section through the rock.

10.2. Spherical markers

The simplest method of strain determination is to use strain markers which are approximately spherical in the initial state. In a section through the undeformed rock these appear as circular markers. If the rock is homogeneously deformed the circles are transformed into

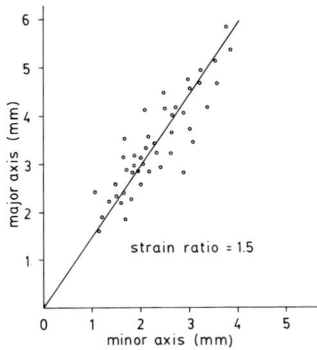

FIG. 10.1. For initially spherical strain markers the lengths of major (X) and minor (Z) axes of the deformed objects are plotted in a diagram with X-axis as ordinate and Z-axis as abscissa. Slope of the line which best fits the cluster of points gives the average strain ratio.

ellipses. Although we cannot determine the magnitude of strain if we do not have any knowledge of the size of the initial circular markers, the orientations of the principal axes of the strain ellipse can be directly obtained from the orientations of the principal axes of the deformed markers. The ratio of the major and the minor axes of an elliptical marker gives us the ratio of the principal axes of the strain ellipse.

In the usual case, three orthogonal sections are cut through an oriented specimen of the rock. If the rock is foliated it is preferable, though not essential, to make one section parallel to the foliation. If there is also a lineation, the second section may be made parallel to it and perpendicular to the foliation. The third section may be made perpendicular to the other two sections. The lengths of the principal axes of a number of elliptical markers are then measured, directly from the rock surface, or from a tracing, or from an enlarged photograph. If the strain markers are small, the lengths of their axes can be measured from thin sections by using an ocular micrometer. The strain ratio is given by the arithmetic mean of the axial ratios of the elliptical markers. Alternatively, the data can be plotted as a graph, with the length of the major axis as ordinate and the length of the minor axis as abscissa. All the points should lie on or close to a line passing through the origin. The slope of the line is equal to the average strain ratio (Fig. 10.1).

10.3. Fry method for isotropic initial fabric

This elegant method of determining the strain ratio (Fry 1979a & b, Hanna & Fry 1979) assumes that before deformation the centres of the marker objects had an isotropic and uniform distribution, i.e. the distances between the neighbouring centres were statistically uniform. The method has the advantage that it can be used even if there is a ductility contrast between the strain markers and their matrix. Moreover, the method does not make any assumption about the initial shape of the strain markers. The method fails if the centres of the objects in the undeformed state had a Poisson distribution.

The Fry method is based on the argument that, when a set of points with statistically uniform distribution is deformed, the average distance between the neighbouring points in any direction increases or decreases in the same ratio as the length of a marker line in that direction. The maximum increase takes place in the direction of the long axis of the strain ellipse. The average distance between the points decreases the most in a direction parallel to the short axis of the strain ellipse. The axial ratio of the strain ellipse can be determined by the following graphical method.

1. Place a sheet of tracing paper either directly on the surface of the slab of rock or on a photograph of the surface and mark out the centres of the strain markers. To avoid confusion in plotting, it is preferable to number the points as 1, 2, 3, etc. Draw a set of parallel reference lines on the tracing sheet, preferably parallel to the cleavage trace or bedding trace or any other line whose orientation is known.

2. On another tracing sheet mark out a central point and draw a set of parallel reference lines

3. Place the second sheet on the first, bring its central point on any one (say, number 1) of the marked points of the first sheet and make the two sets of reference lines parallel.

4. Mark on the second sheet the positions of all the points (say, 2, 3, etc.) of the first sheet except the point lying at the centre point.

5. Shift the second sheet while keeping the reference lines in alignment and bring the centre point on another marked point (say, point number 2). Mark on the second sheet the positions of all other points (1, 3, 4, etc.) of the first sheet. Repeat this operation till the central point has been placed on all the points.

When all the points are plotted in this manner, there will appear an elliptical area around the central point. The elliptical area can be identified by the absence of any points or by a low concentration of points. At the fringe of this area there will be an elliptical ring of high density of points (Fig. 10.2). The strain ratio is given by the ratio of the lengths of the long and the short axes of the ellipse which fits the area of low concentration around the central point or the fringe area of high concentration. The orientation of this ellipse will also give us the orientation of the strain ellipse.

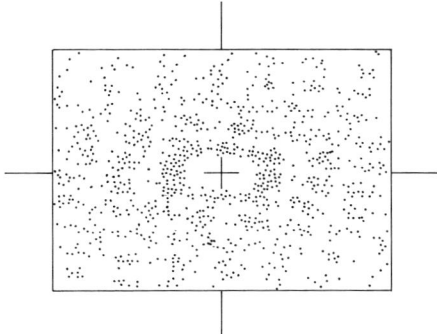

FIG. 10.2. Fry method of determining strain ratio.

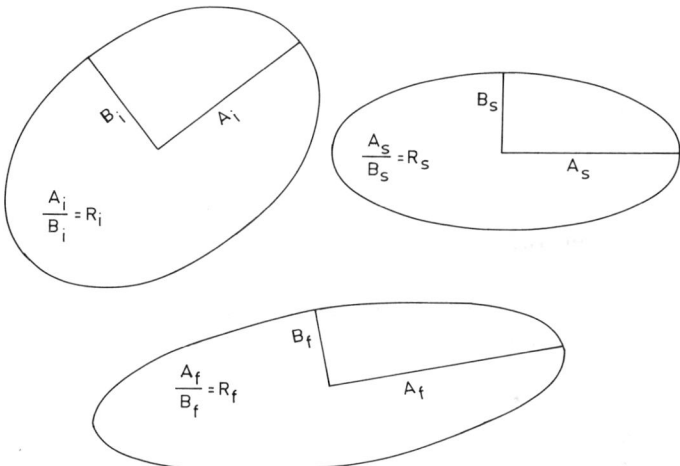

FIG. 10.3. *Upper left*: initial elliptical marker with semi-axes A_i and B_i, with axial ratio R_i. *Upper right*: strain ellipse with axial ratio R_s. *Lower*: Deformed marker with axial ratio R_f.

10.4. Randomly oriented elliptical markers

The initial shapes of many strain markers, such as pebbles and clastic grains, can be approximated by ellipsoids. In a section through the rocks these appear as elliptical markers. By homogeneous strain these are deformed to ellipses of other shapes and orientations. Ramsay devised a method of determining the principal strain ratio from such elliptical markers. In its simplest form this method depends upon the assumption that the elliptical markers were randomly oriented in the undeformed state (Ramsay 1967).

Let A_s and B_s be the lengths of the major and the minor axes of the strain ellipse on a plane. Let the *x*- and the *y*-coordinate axes be chosen

179

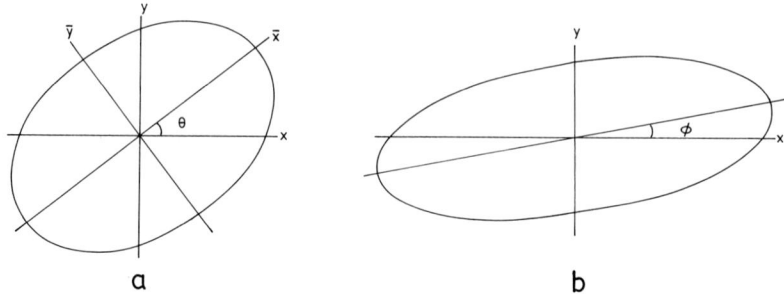

FIG. 10.4. (a) Initial elliptical marker with major axis at an angle θ with the x-coordinate axis. (b) Deformed marker with major axis at an angle Φ with the x-coordinate axis. x- and y-axes are parallel to principal axes of strain ellipse.

parallel to A_s and B_s. Consider an undeformed elliptical marker with the major and the minor axes as A_i and B_i, the i in the subscript indicating the initial or undeformed state of the marker. Let θ be the initial angle between the x-axis and the long axis of the marker. In the final or deformed state, the long and the short axes of the elliptical marker are A_f and B_f, with the long axis A_f at an angle Φ with the x-axis. The axial ratios of the long and the short axes of these different ellipses (Fig. 10.3) are designated in the following way:

$$R_i = A_i/B_i, \text{ the initial axial ratio,}$$
$$R_s = A_s/B_s, \text{ the strain ratio } (\sqrt{\lambda_x}/\sqrt{\lambda_y}),$$
$$R_f = A_f/B_f, \text{ the final axial ratio.}$$

To understand the basic principle underlying Ramsay's method of strain determination we should first find out how Φ can be expressed in terms of θ, R_i and R_s and also how R_s can be expressed in terms of θ, R_i and Φ.

Figure 10.4a shows an undeformed elliptical marker with its long axis at an angle θ with the x-coordinate axis. Let us also choose $\bar{x}$- and $\bar{y}$-coordinate axes parallel to the long and short axes of the ellipse. In the $\bar{x},\bar{y}$-coordinate system the equation of this ellipse is

$$\frac{\bar{x}^2}{A_i^2} + \frac{\bar{y}^2}{B_i^2} = 1. \tag{10.1}$$

The $\bar{x}$- and $\bar{y}$-coordinate axes are then rotated clockwise through an angle $-\theta$ so that they coincide with the x- and y-axes. For such a rotation

$$\begin{aligned} x &= \bar{x}\cos(-\theta) + \bar{y}\sin(-\theta), \\ y &= -\bar{x}\sin(-\theta) + \bar{y}\cos(-\theta) \end{aligned} \tag{10.2a}$$

and

$$\bar{x} = x\cos(-\theta) - y\sin(-\theta),$$
$$\bar{y} = x\sin(-\theta) + y\cos(-\theta). \tag{10.2b}$$

Substituting $\bar{x}$ and $\bar{y}$ in eqn. (10.1) by x and y with the help of eqn. (10.2b), we have

$$\left(\frac{\cos^2\theta}{A_i^2} + \frac{\sin^2\theta}{B_i^2}\right)x^2 + \sin2\theta\left(\frac{1}{A_i^2} - \frac{1}{B_i^2}\right)xy +$$

$$\left(\frac{\sin^2\theta}{A_i^2} + \frac{\cos^2\theta}{B_i^2}\right)y^2 = 1. \tag{10.3}$$

This is the equation of the undeformed elliptical marker with reference to the x,y-coordinate system. When this elliptical marker is deformed, a point (x,y) in its periphery is transformed to a point (x',y'), where

$$x' = \sqrt{\lambda_x}\, x = A_s\, x$$
$$y' = \sqrt{\lambda_y}\, y = B_s\, y \tag{10.4}$$

and the equation of the final ellipse becomes

$$\left(\frac{\cos^2\theta}{A_i^2} + \frac{\sin^2\theta}{B_i^2}\right)\frac{x'^2}{A_s^2} +$$

$$\sin2\theta\left(\frac{1}{A_i^2} - \frac{1}{B_i^2}\right)\frac{x'y'}{A_s B_s} +$$

$$\left(\frac{\sin^2\theta}{A_i^2} + \frac{\cos^2\theta}{B_i^2}\right)\frac{y'^2}{B_s^2} = 1. \tag{10.5}$$

It may be recalled that the long axis of an ellipse, $ax^2 + 2hxy + by^2 = 1$, makes an angle Φ with the x-coordinate axis (Fig. 10.4b), with the angle Φ given by the equation

$$\tan2\Phi = \frac{2h}{a-b}. \tag{10.6}$$

Hence, for the ellipse given by eqn. (10.5),

$$\tan2\Phi = \frac{\sin 2\theta\left(\dfrac{1}{A_i^2} - \dfrac{1}{B_i^2}\right)\dfrac{1}{A_s B_s}}{\left(\dfrac{\cos^2\theta}{A_i^2} + \dfrac{\sin^2\theta}{B_i^2}\right)\dfrac{1}{A_s^2} - \left(\dfrac{\sin^2\theta}{A_i^2} + \dfrac{\cos^2\theta}{B_i^2}\right)\dfrac{1}{B_s^2}}.$$

Multiplying the numerator and denominator by $A_i^2 A_s^2 / \cos^2\theta$ we find, after simplification,

$$\tan 2\Phi = \frac{2\sin 2\theta \, (R_i^2 - 1) \, R_s}{(R_s^2 - 1)(R_i^2 + 1) + (R_s^2 + 1)(R_i^2 - 1)\cos 2\theta} . \qquad (10.7)$$

This is an important equation, first derived by Ramsay (1967, eqn. 5-22), for calculating Φ when θ, R_i and R_s are given.

Equation (10.3) shows the general form of the equation of an ellipse, where the coefficients are given in terms of the lengths of its principal axes and the angle between the long axis and the x-coordinate axis. The equation of the final ellipse, with A_f and B_f as principal axes, and with an angle Φ between A_f and the x-coordinate axis can also be written in the same way:

$$\left(\frac{\cos^2\Phi}{A_f^2} + \frac{\sin^2\Phi}{B_f^2} \right) x^2 +$$

$$\sin 2\Phi \left(\frac{1}{A_f^2} - \frac{1}{B_f^2} \right) xy +$$

$$\left(\frac{\sin^2\Phi}{A_f^2} + \frac{\cos^2\Phi}{B_f^2} \right) y^2 = 1. \qquad (10.8)$$

This ellipse is identical with the final ellipse given by the eqn. (10.5). Hence, the corresponding coefficients of the two equations should be identical. Equating the ratio of the coefficients of x^2 and y^2 of eqn. (10.8) with the ratio of the corresponding coefficients of eqn. (10.5), we have

$$\frac{\dfrac{\cos^2\Phi}{A_f^2} + \dfrac{\sin^2\Phi}{B_f^2}}{\dfrac{\sin^2\Phi}{A_f^2} + \dfrac{\cos^2\Phi}{B_f^2}} = \frac{\dfrac{\cos^2\theta}{A_i^2 A_s^2} + \dfrac{\sin^2\theta}{B_i^2 A_s^2}}{\dfrac{\sin^2\theta}{A_i^2 B_s^2} + \dfrac{\cos^2\theta}{B_i^2 B_s^2}} .$$

Multiplying the numerator and the denominator of the left-hand term by $A_f^2 / \cos^2\Phi$ and those of the right-hand term by $A_i^2 A_s^2 / \cos^2\theta$, we find, after simplification,

$$\frac{1 + R_f^2 \tan^2\Phi}{R_f^2 + \tan^2\Phi} = \frac{1 + R_i^2 \tan^2\theta}{R_s^2 \tan^2\theta + R_i^2 R_s^2} \qquad (10.9)$$

or,

$$R_f^2 = \frac{\tan^2\Phi\,(1 + R_i^2\tan^2\theta) - R_s^2\,(R_i^2 + \tan^2\theta)}{R_s^2\tan^2\Phi\,(R_i^2 + \tan^2\theta) - (1 + R_i^2\tan^2\theta)}. \qquad (10.10)$$

This is another important equation which was derived by Ramsay (1967, eqn. 5-27) to determine the strain ratio from deformed elliptical markers.

Consider a set of randomly oriented initial elliptical markers with a constant R_i and let these be deformed up to a certain strain ratio R_s. For a particular value of θ, the value of Φ is first determined from eqn. (10.7). R_f can then be calculated from eqn. (10.10). By regularly varying θ from 0 to $\pm 90°$, we can get a number of pairs of Φ and R_f values for constant R_i and R_s. Each of these deformed elliptical markers is plotted graphically, say with R_f as ordinate and Φ as abscissa. We may also eliminate θ from eqns. (10.7) and (10.10) and obtain a single equation (Dunnet 1969, eqn. 16, Lisle eqn. 2.6) which relates R_f, R_i, R_s and Φ.

$$\cos 2\Phi = \frac{\left(R_f + \dfrac{1}{R_f}\right)\left(R_s + \dfrac{1}{R_s}\right) - 2\left(R_i + \dfrac{1}{R_i}\right)}{\left(R_f - \dfrac{1}{R_f}\right)\left(R_s - \dfrac{1}{R_s}\right)}. \qquad (10.11)$$

Either with the help of eqns. (10.7) and (10.10) or from eqn. (10.11), we can construct a number of R_f/Φ curves for different values of R_s. Figure 10.5 shows three such families of curves for $R_s = 1.5$, $R_s = 2$ and $R_s = 4$. A number of such diagrams can be prepared, either with the help of a computer or with a programmable calculator. There are three types of R_f/Φ curves depending on the value of R_i. A closed curve is obtained when $R_i < R_s$. The maximum and the minimum of the curve both occur on the line $\Phi = 0$. When $R_i = R_s$, the curve remains between $\Phi = \pm 45°$ and meets the Φ-axis at two points. For $R_i > R_s$, the curves do not close and the value of Φ ranges between $\pm 90°$; the maximum of R_f of each of these curves lies on the line $\Phi = 0$ and the minimum of R_f lies on $\Phi = 90°$. For closed curves ($R_i < R_s$), the range or fluctuation of Φ decreases with increasing R_s. For a low value of the initial shape factor (R_i) and a fairly large strain ratio, the fluctuation of Φ is quite small.

To determine the strain ratio, each pair of values of R_f and Φ is plotted as a point in a diagram with $\ln R_f$ as ordinate and Φ as abscissa. To give reproducible strain values Dunnet (1969) suggests a minimum of about 40 R_f/Φ plots for ooids or pisolites and between 60 to 100 plots for pebbles and clastic grains. If the assumption of random orientation of the elliptical markers is justified, the scatter diagram of R_f/Φ plots should be symmetrical about a line parallel to the R_f-axis. The position of this line may also be located by taking the arithmetic mean of the Φ values of the data points. The minimum and maximum R_f should also lie on this line. The Φ value of this line gives the orientation of the major axis of the strain ellipse.

The two-dimensional strain ratio, R_s, is generally determined by a combination of different methods (cf. Ramsay 1967, Dunnet 1969, p. 128, Lisle 1985, Ramsay & Huber 1983).

(1) The strain ratio can be determined from the maximum and minimum values of R_f (Ramsay 1967). For randomly oriented elliptical markers with a constant axial ratio, the maximum value of R_f should be given by those markers whose long axis is parallel to the long axis of the strain ellipse. The minimum R_f will be given by the marker whose long axis in the initial state coincides with the short axis of the strain ellipse. In other words, the maximum of R_f is obtained when $\theta = 0°$ and minimum of R_f is obtained when $\theta = 90°$.

Consider an initial elliptical marker, with $\theta = 0$ and with the principal axes A_i and B_i parallel to the strain axes A_s and B_s, respectively (Fig. 10.6). After deformation the principal axes of the elliptical marker will be

$$A_f = A_i A_s,$$
$$B_f = B_i B_s,$$

so that

$$R_f(\text{maximum}) = \frac{A_f}{B_f} = \frac{A_i}{B_i} \cdot \frac{A_s}{B_s},$$

or

$$R_f(\text{maximum}) = R_i R_s. \tag{10.12}$$

If as in Fig. 10.7 the long axis of the undeformed elliptical marker is oriented parallel to the short axis of the strain ellipse, the lengths of the principal axes of the deformed elliptical marker will be $B_i A_s$ along the x-direction and $A_i B_s$ along the y-direction. If $R_i > R_s$, the long axis of the final ellipse will be parallel to the y-axis; in this case (Fig. 10.7)

$$A_f = A_i B_s,$$
$$B_f = B_i A_s,$$

$$R_f(\text{minimum}) = \frac{A_f}{B_f} = \frac{A_i}{B_i} \cdot \frac{B_s}{A_s},$$

or

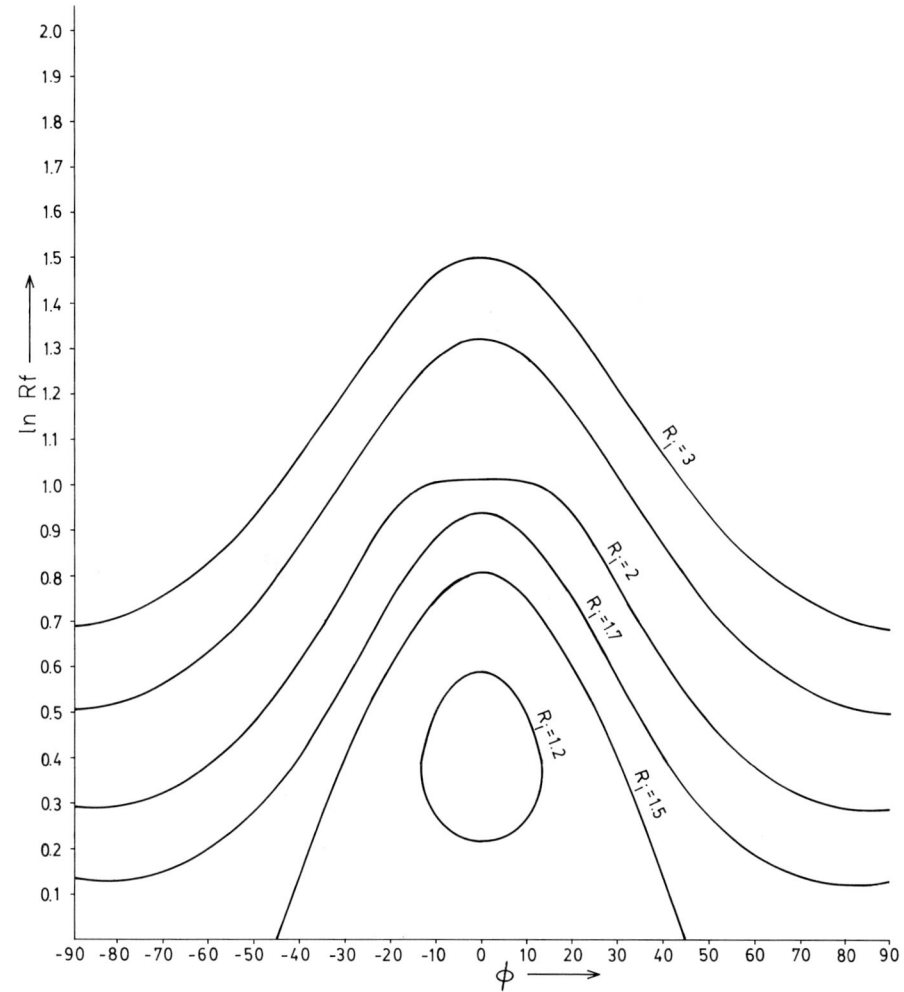

FIG. 10.5a.

$$R_f(\text{minimum}) = \frac{R_i}{R_s}. \qquad (10.13a)$$

If $R_i < R_s$ (Fig. 10.7), the lengths of the principal axes of the final ellipse will be

$$A_f = B_i A_s,$$
$$B_f = A_i B_s,$$

or,

185

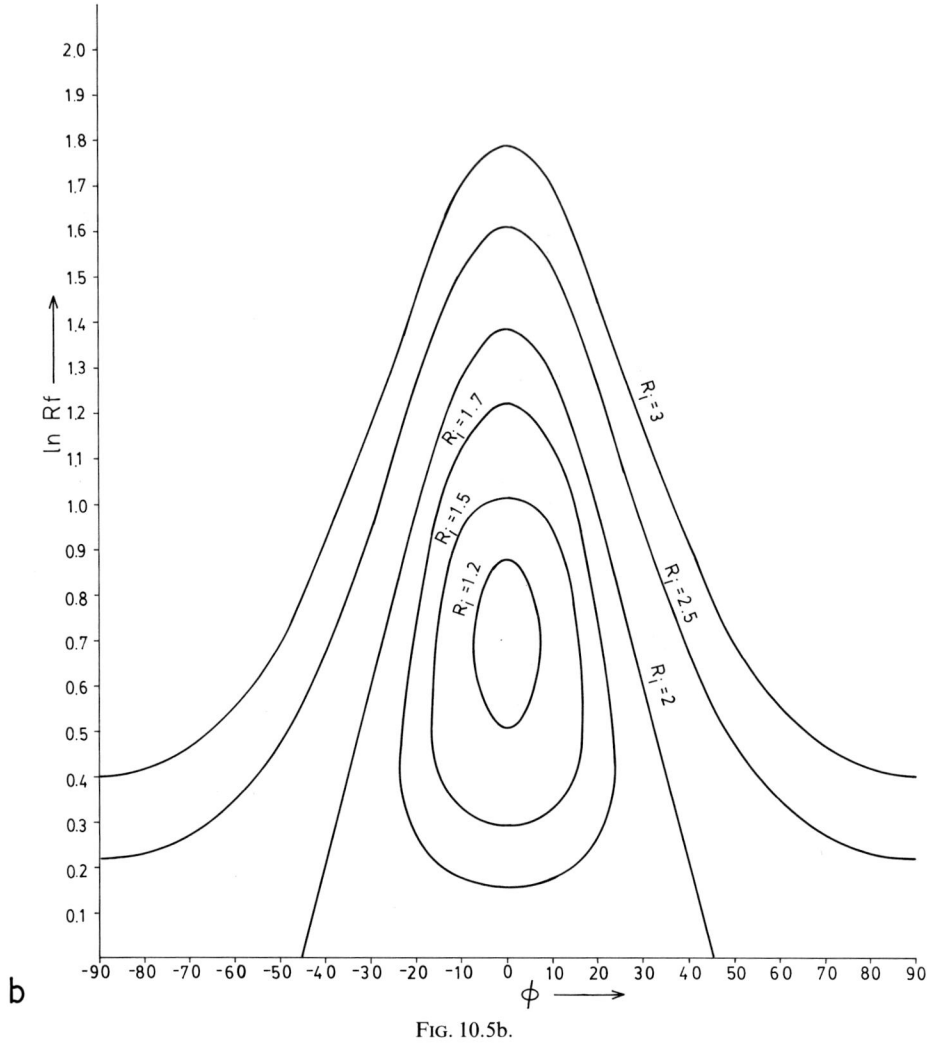

FIG. 10.5b.

$$R_{f(\text{minimum})} = \frac{B_i}{A_i} \cdot \frac{B_s}{B_i},$$

or,

$$R_{f(\text{minimum})} = R_s / R_i. \qquad (10.13b)$$

Thus, the strain ratio can be determined from the following equations (Ramsay 1967)

$$R_s^2 = \frac{R_{f(\text{max})}}{R_{f(\text{min})}}, \text{(when } R_i > R_s),$$

186

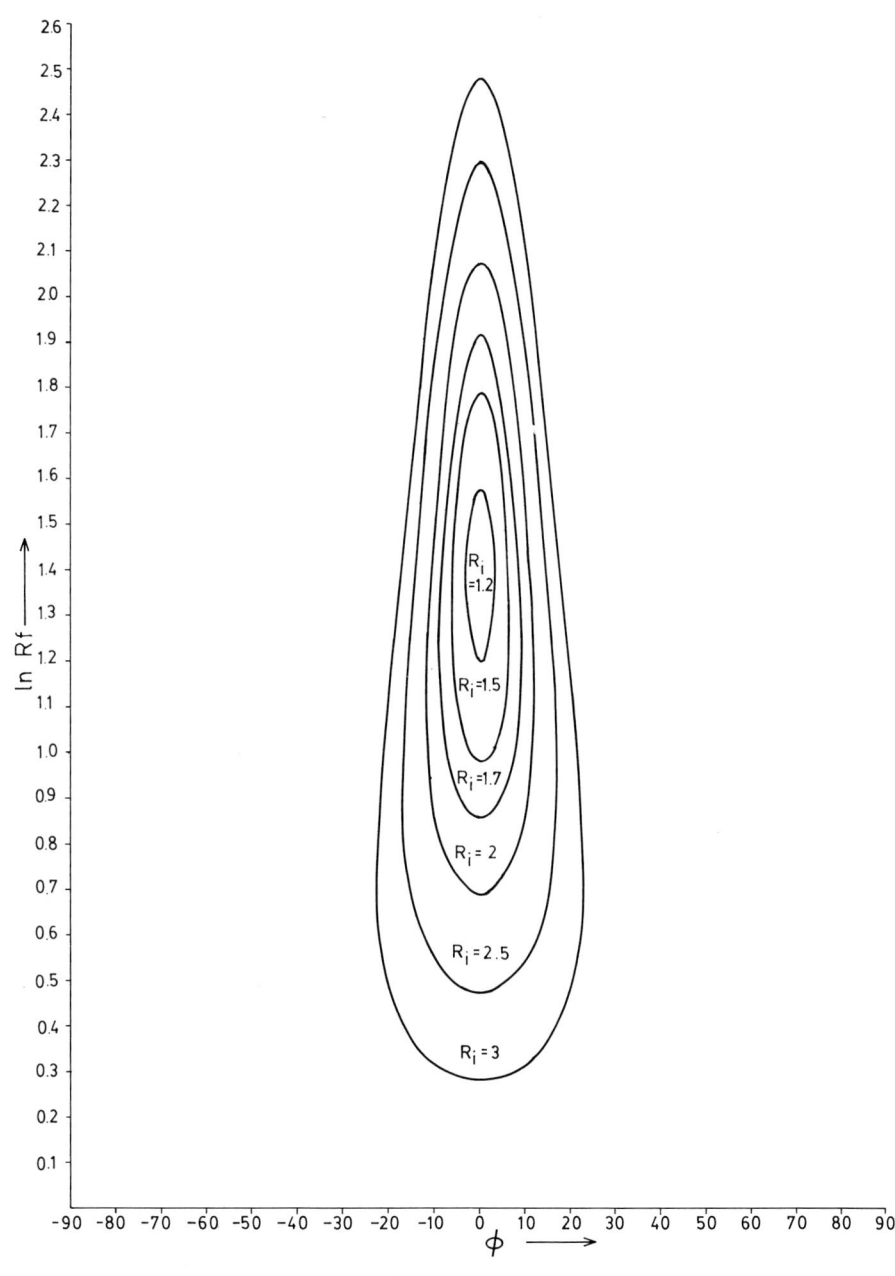

C

Fɪɢ. 10.5. (a) R_f/Φ curves for $R_s = 1.5$. (b) R_f/Φ curves for $R_s = 2$. (c) R_f/Φ curves for $R_s = 4$.

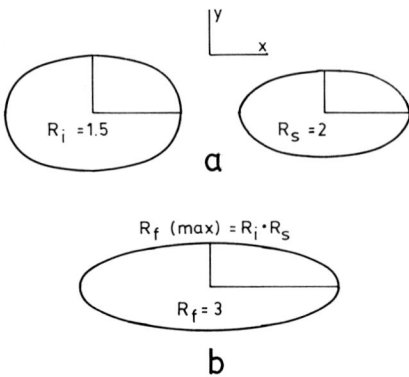

FIG. 10.6. Deformation of an elliptical marker with $R_i = 1.5$, with A_i and B_i parallel to A_s and B_s. In this case $R_f = R_i \times R_s$.

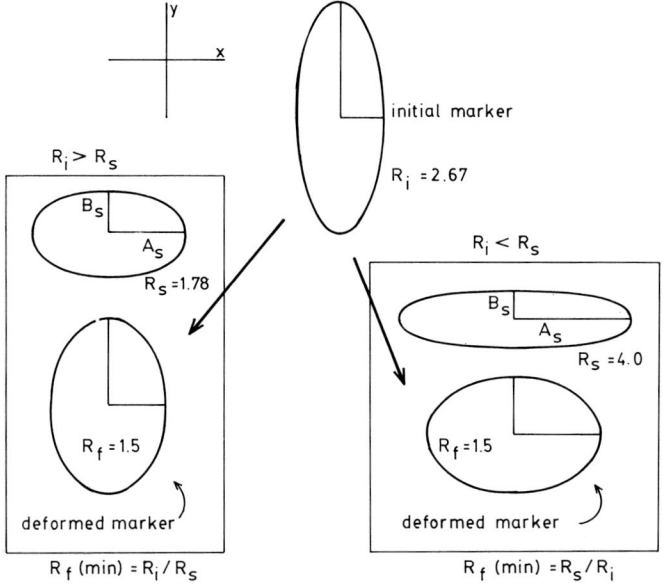

FIG. 10.7. Deformation of an elliptical marker with A_i parallel to B_s and B_i parallel to A_s. If $R_i > R_s$, as in the left-hand figure, we have $R_f = R_i/R_s$. If $R_i < R_s$, $R_f = R_s/R_i$.

$$R_s^2 = R_{f(max)} \times R_{f(min)}, \, (R_i < R_s). \qquad (10.14)$$

(2) An approximate value of the strain ratio can be obtained from the harmonic mean of all the R_f values. The harmonic mean, H, can be calculated in the following way:

$$H = n/(R_{f1}^{-1} + R_{f2}^{-1} + R_{f3}^{-1} + \ldots + R_{fn}^{-1}).$$

MEASUREMENT OF STRAIN

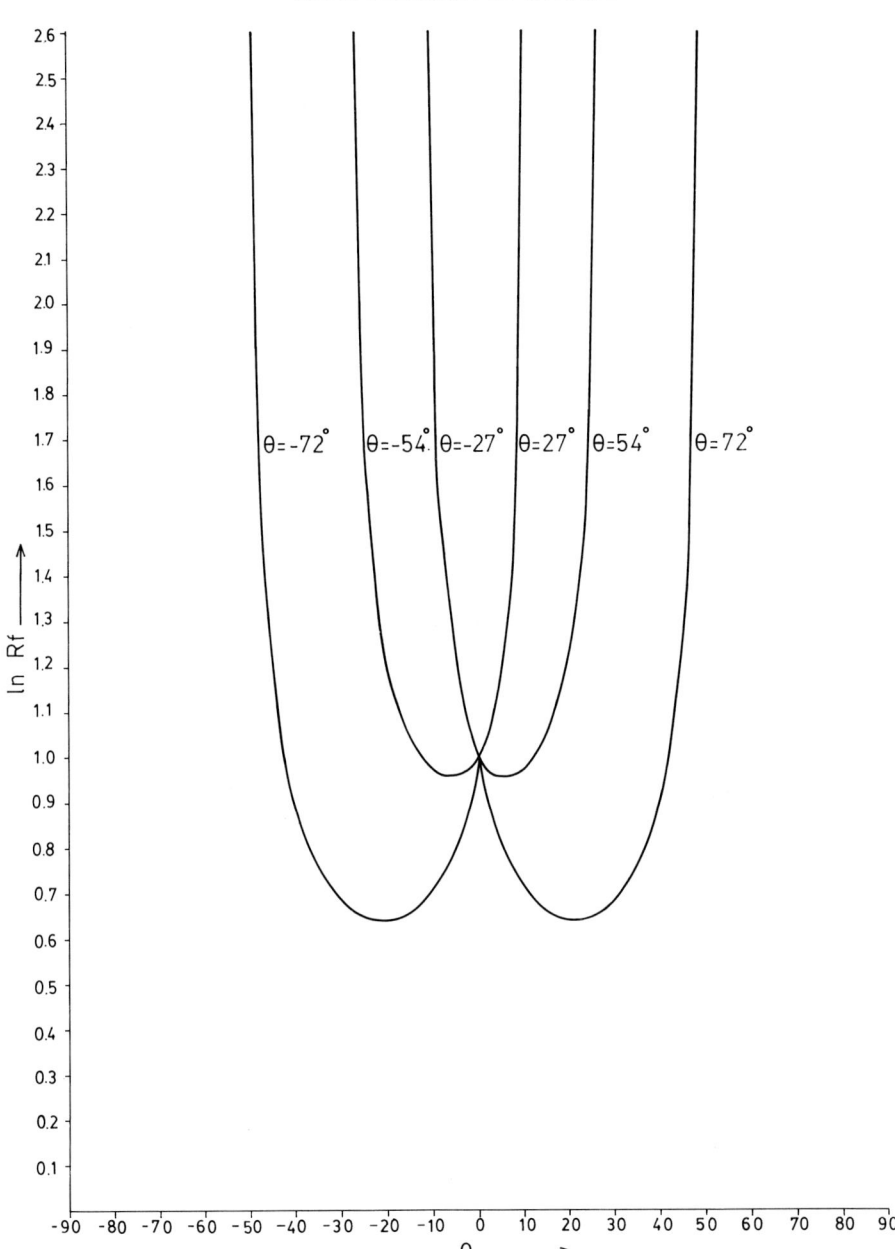

FIG. 10.8. θ-curves for $R_s = 2.7$.

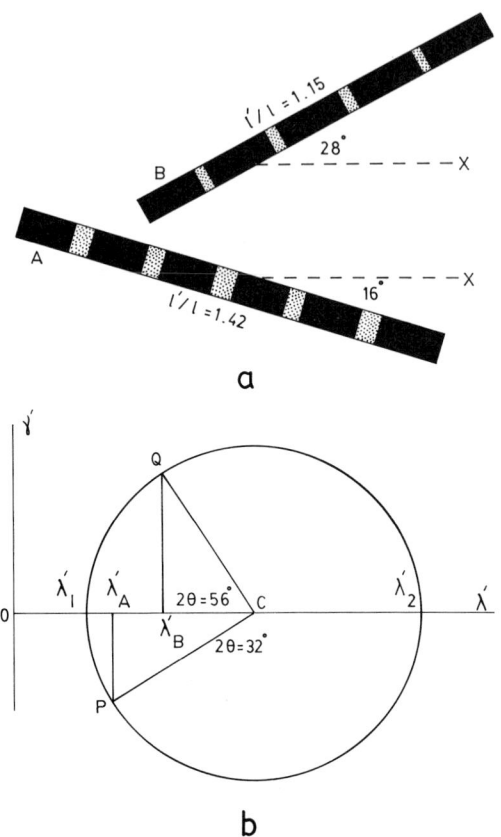

FIG. 10.9. Determination of principal strains from two boudinaged veins A and B. From (a) the stretch is measured for both vein A and vein B. Then as in (b), a circle is drawn and the lines CP and CQ are drawn at angles $2\theta' = 56°$ and $2\theta' = 32°$. The perpendiculars from P and Q on the λ'-axis give the positions of λ'_A ($= 0.5$) and λ'_B ($= 0.75$). From these positions the scale of the λ'-axis is determined and the position of the origin O is located. λ'_1 and λ'_2 are then read off from the diagram. λ_1 and λ_2 are found to be 2.67 and 0.44, respectively.

It has been shown by Lisle (1977a, 1979) that the R_s value is slightly smaller than the harmonic mean.

(3) After an approximate value of R_s is obtained (by either of the two methods given above), R_f/Φ families of curves of different R_s values are placed one after another over the plotted data till we can select the set of curves of visual best fit. R_s and R_i are given by the curve which best fits the envelope of the data points.

Evidently, this method of determination of the strain ratio can be used only if the assumptions of homogeneous deformation and randomness of initial orientation are justified.

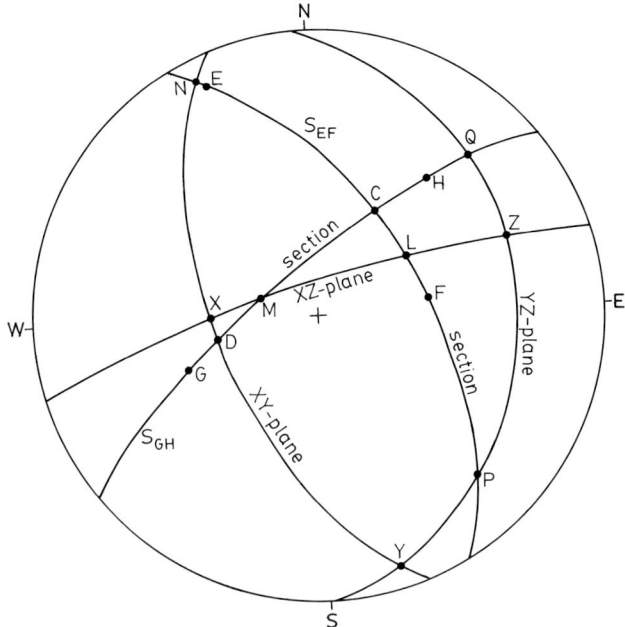

Fig. 10.10. Method of determining strain ratio in three dimensions from data on two section planes S_{EF} and S_{GH}. A_E and B_F are semi-major and semi-minor axes of S_{EF}. A_G and B_H are the axes of the strain ellipse on S_{GH}. L is the intersection of the XZ-plane and S_{EF} while M is the intersection of the XZ-plane and S_{GH}. P and Q are the corresponding intersections with the YZ-plane. N and D are the intersections of S_{EF} and S_{GH} with the XY-plane.

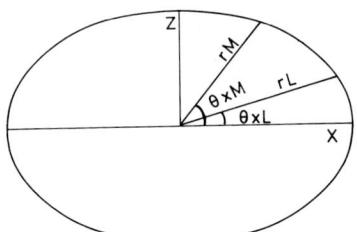

Fig. 10.11. Strain ellipse on the XZ-plane with radii r_M and r_L. r_M and r_L are at angles of θ_{XM} and θ_{XL} with the X-axis.

10.5. Non-random initial fabric

An asymmetry of the R_f/Φ plots generally indicates a non-random initial fabric. The strain ratio for such rocks can be determined from the R_f/Φ plots provided we can make an assumption regarding the nature of initial preferred orientation. The method of finding R_s in this case is much more difficult than that for an initial random fabric. Some of these methods have been described by Dunnet & Siddans (1971) and Lisle (1977b, 1985).

Let us consider here a relatively simple situation in which the XY-plane (say parallel to the cleavage) is identifiable and it can be assumed that the initial sedimentary fabric was symmetrical about the bedding. A technique of finding R_s in this case is to use the θ curves of Lisle (1977b, 1985). For any specific value of R_s, these are the loci of elliptical markers with variable R_i but with the same initial orientation θ. The different sets of θ curves can be constructed either from eqns. (10.7) and (10.10), or by eliminating R_i from these equations and combining them into a single equation (Lisle 1977b, p. 385)

$$R_f^2 = \frac{\tan2\theta(R_s^2 - \tan^2\Phi) - 2R_s\tan\Phi}{\tan2\theta(1 - R_s^2\tan^2\Phi) - 2R_s\tan\theta}.$$

Figure 10.8 shows, for example, some of the θ curves for $R_s = 2.7$.

For finding R_s, several sets of θ curves of different values of R_s are overlain, one after another, on the R_f/Φ points until a symmetrical θ distribution is brought about. If the cleavage trace on the R_f/Φ diagram is taken to be the trace of the XY-plane, the cleavage trace is aligned with the line $\Phi = 0$ of the θ curve diagram.

An alternative technique described by Dunnet & Siddans (1971) uses a computer to "destrain" the bedding trace and each individual marker in successive steps till a symmetrical distribution of the R_f/Φ points about the bedding trace is obtained.

10.6. Other methods of finding strain ratio in two dimensions

Buckle-folded veins

In slates and schists we often get thin diversely oriented veins of quartz, carbonate or other materials. Similar diversely oriented veins of pegmatite also occur in the gneisses. If the veins are buckle folded and if the geometry of the folds indicates that there is a large competence contrast with the host rock, the strain ratio can be determined from the buckle shortening of the veins (e.g. Talbot 1970). If there is a large competence contrast, the shortening of a vein takes place essentially by folding and there is very little thickening of the vein by homogeneous strain. The initial length of a vein between any two points can be determined from a photograph by measuring the length of a curved line along the fold-arcs. The stretch is given by the ratio of the final length and the initial length. To determine the strain ratio from a Mohr diagram we require the stretches of at least any three non-parallel veins or the stretches of any two veins and their angles with a principal direction of the strain ellipse (say the trace of the cleavage on a surface perpendicular to it).

Boudinage structures

This method of finding the strain ratio can also be applied if the diversely oriented veins have undergone boudinage instead of buckle folding (Fig. 10.9). The method is applicable only if the veins have behaved in a brittle manner as indicated by straight sharp edges and no barrelling of the individual boudins. The stretch of such a layer is obtained from the ratio of the length of the boudinaged layer between any two points and the sum of the boudin lengths on the plane of section.

Deformed fossils

Two-dimensional strain ratios have been determined from the distorted shapes of different types of fossils, e.g. belemnites, crinoid stems, brachiopods, trilobites, ammonites, corals and graptolites. The methods of strain determination from the distorted shapes of each of these fossils have been described by Ramsay (1967, pp. 228–249) and by Ramsay & Huber (1983, pp. 127–148). Elongation in different directions can sometimes be determined from boudinage of belemnites and crinoid stems. In most cases, however, the deformation of a fossil is analysed from the change in the angle between two linear elements. Fossils like trilobites and brachiopods have a bilateral symmetry. From the shapes of these fossils it is possible to identify two perpendicular lines in the undeformed state. In the general case these lines do not remain perpendicular when the fossils are deformed and the shear strain can be calculated from the change in the right angle. The strain ratio can then be determined with the help of a Mohr diagram from the shear strains in differently oriented fossils lying on the same plane (e.g. Ramsay 1967, pp. 236–237).

10.7. Determination of three-dimensional strain

In rocks which contain a prominent cleavage and a stretching lineation, the cleavage is usually taken to be parallel to the XY-plane and the lineation parallel to the X-axis. This assumption considerably simplifies the problem of determination of the principal strain ratios from the two-dimensional data (Ramsay 1967). However, this assumption should not be made when the problem itself is to find out whether the cleavage is exactly parallel to the XY-plane of the strain ellipsoid. If the sections are cut parallel to XZ-plane (parallel to lineation and perpendicular to cleavage), the YZ-plane (perpendicular to both the cleavage and the lineation) and the XY-plane (parallel to the cleavage), the strain ratios on these sections directly give us the values of X/Z, Y/Z and X/Y. It is not essential to have the data of the third section since any one of these ratios can be obtained from the other two. The final results are usually presented by giving the X/Z and Y/Z ratios or by the ratio $X:Y:Z$.

The principal strain ratios can also be determined from observations of any two mutually perpendicular sections, provided the orientations of the XY-plane and of the X-axis can be measured in the field from the orientations of the cleavage and the stretching lineation. The following procedure may be adopted to find out the ratios X/Y, Y/Z and X/Z.

Plot the two section planes S_{EF} and S_{GH}, the principal axes of strain ellipses and the X, Y, Z axes in stereographic projection. Let R_1 and R_2 be the principal axes of the strain ellipse on S_{EF} and let R'_1 and R'_2 be the corresponding axes on S_{GH}, with $R_1 > R_2$ and $R'_1 > R'_2$. Measure the angles of R_1, R_2, R'_1 and R'_2 from each of X, Y and Z axes and calculate the direction cosines. Let l_1, l_2, l_3 be the direction cosines of R_1; m_1, m_2, m_3 be the direction cosines of R_2; l'_1, l'_2, l'_3 be the direction cosines of R'_1 and m'_1, m'_2, m'_3 be the direction cosines of R'_2.

If we choose the coordinate axes x, y, z along the principal axes of the strain ellipsoid, $(x^2/X^2)+(y^2/Y^2)+(z^2/Z^2)=1$, we have,

$$(l_1^2/X^2) + (l_2^2/Y^2) + (l_3^2/Z^2) = 1/R_1^2,$$
$$(m_1^2/X^2) + (m_2^2/Y^2) + (m_3^2/Z^2) = 1/R_2^2,$$

so that,

$$\frac{m_1^2 Y^2 Z^2 + m_2^2 X^2 Z^2 + m_3^2 X^2 Y^2}{l_1^2 Y^2 Z^2 + l_2^2 X^2 Z^2 + l_3^2 X^2 Y^2} = \frac{R_1^2}{R_2^2} = \text{say}, R_{12}^2,$$

or,

$$(Y^2/X^2)(m_1^2 - l_1^2 R_{12}^2) + (Y^2/Z^2)(m_3^2 - l_3^2 R_{12}^2) + (m_2^2 - l_2^2 R_{12}^2) = 0.$$

By following the same procedure for the section plane S_{GH}, we have

$$(Y^2/X^2)(m_1^2 - l_1^2 R_{12}^2) + (Y^2/Z^2)(m_3^2 - l_3^2 R_{12}^2) + (m_2^2 - l_2^2 R_{12}^2) = 0.$$

By solving these two equations we have

$$Y^2/Z^2 = (P - S)/(PT - QS),$$
$$Y^2/X^2 = (T - Q)/(PT - QS),$$

where

$$P = (m_1^2 - l_1^2 R_{12}^2)/(l_2^2 R_{12}^2 - m_2^2),$$
$$Q = (m_3^2 - l_3^2 R_{12}^2)/(l_2^2 R_{12}^2 - m_2^2),$$
$$S = (m_1^7 - l_1^7 R_{12}^7)/(l_2^7 R_{12}^7 - m_2^7),$$
$$T = (m_3^7 - l_3^7 R_{12}^7)/(l_2^7 R_{12}^7 - m_2^7).$$

An alternative method involves the calculation of rM and rL and to reconstruct the strain ellipse on the XZ-plane (Figs. 10.10, 10.11).

As shown by Ramsay (1967, pp. 142–147), when the orientations of X, Y and Z are unknown, the method of determination of the principal strain ratios becomes much more complex.

CHAPTER 11

Rotation of Structural Elements

11.1. Rotation of different geometric elements

In structural geology we often consider the rotations of different types of structural elements. Thus, a rigid porphyroblast may rotate without any internal deformation. Evidence of such rigid body rotation of porphyroblasts is furnished, in case of postcrystalline deformation, by the discordance between the trails of inclusions in the porphyroblasts and the schistosity of the matrix (Fig. 11.1a) and in case of paracrystalline rotation by the presence of symmetrical S- or Z-shaped trails of inclusions (Fig. 11.1b). In Sander's terminology this type of rotation is known as *external rotation* (Turner & Weiss 1963, p. 275). Buckle folding of a competent layer also involves external rotation or body rotation. On the other hand, a layer may undergo rotation, i.e. a change in orientation, by homogeneous deformation (Fig. 11.1c). In this case the deformations in the layer and in the matrix are identical. Rotation of this type is described as *internal rotation* or as *rotation of a kinematically passive element* (Weiss 1955, Turner & Weiss 1963). We may also consider sometimes the rotation of a *geometrically defined structural element* such as the long axis of a strain ellipse or the normal to an S surface. As the S surface rotates passively in response to a homogeneous deformation, the change in orientation of a *geometrically defined* normal may be quite different from that of a *material line* which was initially normal to the S surface (Fig. 11.2). Lastly, we may also speak about the part of rotation involved in rotational or non-coaxial deformation.

In the general case, for a particular value of finite deformation, the angles of rotation of these unlike structural elements differ among themselves. As an example we may consider the well-known model of simple shear. Let us assume that on a plane normal to rotational axis of simple shear, a rigid ellipsoidal porphyroblast with aspect ratio 2, is oriented in such a way that its longest axis makes an angle of 45° with the direction of simple shear. Let there also be a kinematically passive lineation in the same direction in the matrix. Then, for a simple shear of

$\gamma = 1$, the porphyroblast will rotate through an angle of 21.9° while the lineation will rotate through an angle of 18.4° only. For the same value of simple shear, the major axis of the strain ellipse in the matrix will rotate 13.3° and the rotational component of deformation will be 26.6°. The amount of rotation will depend on two factors, viz. the initial orientation of a structural element and the amount of simple shear. For the rigid porphyroblast the aspect ratio is an additional controlling factor. Thus, if the porphyroblast were spherical it would have rotated through an angle of 28.7°. The equations which describe these different kinds of rotation have been discussed by several authors (e.g. Jeffery 1922, Taylor 1923, Mason & Manley 1957, Bretherton 1962, Ramberg 1959, Ramsay 1967, Rosenfeld 1970, Gay 1968, Ghosh 1975, Ghosh & Ramberg 1976, Ghosh & Sengupta 1973, Ramberg & Ghosh 1977).

A study of these different kinds of rotation is essential for an understanding of the course of evolution of many types of geologic structures such as the drag patterns of foliation around porphyroblasts or porphyroclasts, the patterns of inclusion trails in paratectonic porphyroblasts, the *en echelon* arrangement of boudins and the development of different patterns of porphyroclast-tail systems.

11.2. Vorticity

In structural geology we are mostly concerned with finite strain and finite rotations. Nevertheless, the equation for the instantaneous rates of rotation can also be very useful in certain situations. Let us consider a plane deformation defined by the equations

$$u = a_{11}x + a_{12}y$$
$$v = a_{21}x + a_{22}y \qquad (11.1)$$

where u and v are the velocities along x- and y-axes, respectively. Depending upon the coefficients a_{ij}, the motion represented by eqn. (11.1) may or may not cause a local rotation of the material around the z-axis, i.e. the axis normal to the xy-plane. This rate of rotation (ω_z) is given by the equation

$$\omega_z = \tfrac{1}{2}\left(\frac{\delta v}{\delta x} - \frac{\delta u}{\delta y}\right). \qquad (11.2)$$

If the deformation is three-dimensional we shall, in addition, have rotations around the x- and y-axes. These components of the rate of rotation are

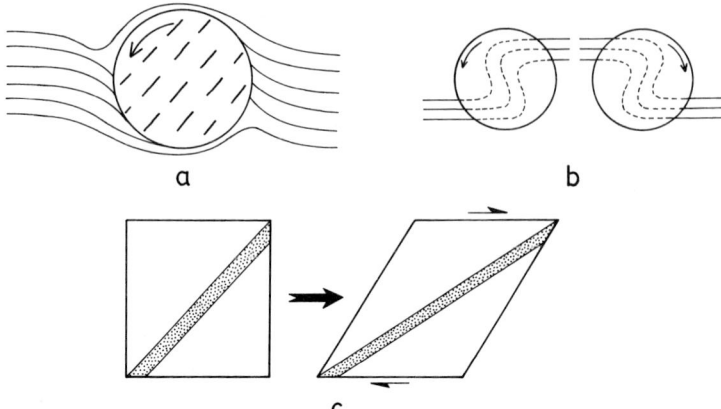

F<small>IG</small>. 11.1. Rigid body rotation and internal rotation. (a) Porphyroblast with a straight trail of inclusion discordant with the outer schistosity. The microstructure indicates postcrystalline rigid body rotation of the porphyroblast with respect to its matrix. (b) Paracrystalline rigid body rotation of porphyroblast. The sense of rotation is given by the S- or Z-shaped trails of inclusion in the porphyroblast. (c) Internal rotation of a layer by simple shear.

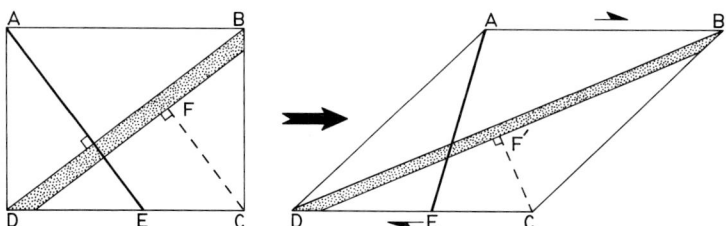

F<small>IG</small>. 11.2. Distinction between rotation of a geometrically defined line and of a material line. In the left-hand figure, DB is a layer and AE is a material line normal to the layer. FC is the geometrically defined normal. The right-hand figure shows that after a simple shear deformation both DB and AE have changed their orientations and the angle between them is no longer a right angle. $F'C$ is now a geometrically defined normal.

$$\omega_x = \tfrac{1}{2}\left(\frac{\delta w}{\delta y} - \frac{\delta v}{\delta z}\right),$$

$$\omega_y = \tfrac{1}{2}\left(\frac{\delta u}{\delta z} - \frac{\delta w}{\delta x}\right). \tag{11.3}$$

ω_x, ω_y and ω_z are the components of the *rate of rotation vector*. The vector, with components $2\omega_x$, $2\omega_y$ and $2\omega_z$ is called *vorticity*. Thus for the two-dimensional flow represented by (11.1) the vorticity is:

$$W = 2\,\omega_z = \frac{\delta v}{\delta x} - \frac{\delta u}{\delta y}. \qquad (11.4)$$

If the vorticity vanishes, the motion is *irrotational* or *coaxial*; if it does not vanish, the motion is *rotational* or *non-coaxial*. Thus, the magnitude of vorticity is equal to twice the magnitude of local rate of rotation.

In what kind of a flow do we have a local rotation and how do we measure this rate of rotation? Consider a vanishingly small spherical element within the flowing liquid. Imagine that this spherical element has suddenly solidified. If this rigid spherical body does not rotate, the flow is irrotational; if it does rotate, the flow is rotational. The magnitude of vorticity is equal to twice the rate of rotation of this spherical body (Truesdell 1954). Thus, the magnitude of vorticity or the local rate of rotation can be obtained from the rotation of a rigid equant grain, such as a porphyroblast of garnet, embedded in a much softer matrix.

As an example, let us consider progressive pure shear and progressive simple shear. For pure shear the velocities in the x- and y- directions are

$$u = \dot{\varepsilon}_x \cdot x,$$
$$v = \dot{\varepsilon}_y \cdot y, \qquad (11.5)$$

where $\dot{\varepsilon}_x$ and $\dot{\varepsilon}_y$ are the strain rates in the x- and the y-directions. Comparing with eqn. (11.1) we find that $a_{11} = \dot{\varepsilon}_x$ and $a_{22} = \dot{\varepsilon}_y$ while $a_{12} = a_{21} = 0$. From eqn. (11.4) we find that, for pure shear, $\omega = 0$. Thus, progressive pure shear is irrotational.

In simple shear the velocities in the x- and the y-directions are

$$u = \dot{\gamma}y,$$
$$v = 0. \qquad (11.6)$$

Therefore, $\omega_z = -\dot{\gamma}/2$, and $W = -\dot{\gamma}$. Progressive simple shear is, thus, rotational. The rate of rotation of a rigid equant inclusion is half the rate of simple shear.

Simple shear is not the only type of plane non-coaxial deformation. How do we decide whether a particular type of non-coaxial deformation has a large or a small degree of rotationality? Truesdell (1954, p. 107) has proposed that the degree of rotationality or the degree of non-coaxiality can be measured by the *kinematical vorticity number*

$$W_k = \frac{W}{[2\,(\dot{\varepsilon}_1^2 + \dot{\varepsilon}_2^2 + \dot{\varepsilon}_3^2)]^{\frac{1}{2}}} \qquad (11.7)$$

where $\dot{\varepsilon}_i$ are the principal linear strain rates and W is the magnitude of the vorticity vector. Alternatively, for the two-dimensional case, and in

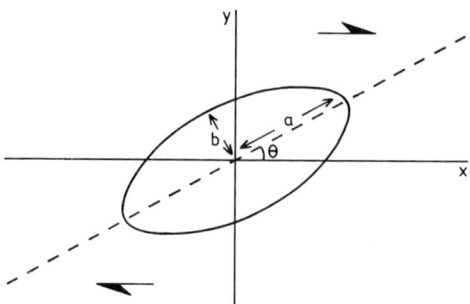

FIG. 11.3. Rigid elliptical inclusion with the semi-major axis a at an angle θ with the direction of simple shear.

terms of the coefficients of eqn. (11.1),

$$W_k = \frac{a_{21} - a_{12}}{\sqrt{2}[a_{11}^2 + a_{22}^2 + \frac{1}{2}(a_{12} + a_{21})^2]^{\frac{1}{2}}}. \tag{11.8}$$

From eqns. (11.1) and (11.8) we find that for simple shear the absolute value of the kinematical vorticity number is 1. The degree of rotationality or non-coaxiality increases with an increase in the kinematical vorticity number. For irrotational flow $W_k = 0$.

11.3. Rate of rotation of a rigid ellipsoidal inclusion and of a passive marker plane in simple shear

Let us consider a rigid ellipsoidal body embedded in a ductile matrix with one of the symmetry axis of the ellipsoid parallel to the z-axis of progressive simple shear. This is the axis of rotation of the ellipsoidal inclusion. Let the two other axes of the inclusion, lying on the xy-plane, be a and b, with $a > b$, and with $a/b = R$. Let the angle between the a-axis of the inclusion and the direction of simple shear be θ (Fig. 11.3). The rate of rotation of the inclusion is then given by the equation

$$\omega = \frac{-\dot{\gamma}(R^2 \sin^2\theta + \cos^2\theta)}{R^2 + 1} \tag{11.9}$$

(Jeffery 1922), where $\dot{\gamma}$ is the rate of simple shear. The negative sign on the right-hand side is because of the convention followed in this book that a positive shear strain γ_{xy} causes a reduction in the angle between lines parallel to the positive x- and the positive y-axes. On the other hand, following the convention of kinematics, an anticlockwise rotation has been taken as positive.

For a spherical inclusion, $R = 1$, and from eqn. (11.9) we find that

$$\omega = -\frac{\dot{\gamma}}{2}. \tag{11.10}$$

For an elliptical inclusion of a certain value of R, the rate of rotation will depend on the orientation of the a-axis. The absolute value of the rate of rotation is largest when $\theta = 90°$, i.e. when the a-axis is normal to the shearing plane. The rate of rotation is smallest when the a-axis is parallel to the shear direction (Fig. 11.4). It should be noted that the long axis of the incremental strain ellipse of simple shear is oriented at an angle of $\theta = 45°$. At this orientation the rigid ellipsoids with different values of R all rotate at the same rate, i.e. at the rate of rotation of a spherical inclusion. At $\theta > 45°$ a longer inclusion rotates faster while at $\theta < 45°$ a shorter inclusion rotates faster.

When the ratio a/b is very large, the ellipsoidal inclusion has a tapering sheet-like shape. It then rotates more or less as a passive marker plane. When a passive marker plane is deformed by finite simple shear the change in orientation, as given by eqn. (8.73), is:

$$\cot\theta = \cot\theta_0 + \gamma$$

where θ_0 and θ are the initial and the final orientations of the plane. Differentiating with respect to time t we have

$$-\csc^2\theta \, . \, \dot{\theta} = \dot{\gamma},$$

or,

$$\omega = \dot{\theta} = -\sin^2\theta \, . \, \dot{\gamma}. \tag{11.11}$$

The equation shows that the absolute value of the rate of rotation of a marker plane is a maximum when $\theta = 90°$. Its rate of rotation vanishes when it is parallel to the shear direction ($\theta = 0$).

11.4. Rate of rotation of a rigid ellipsoidal inclusion and of a passive marker plane in pure shear

Let us now consider a rigid ellipsoidal object embedded in a ductile matrix with one of the symmetry axis of the inclusion parallel to the intermediate axis of pure shear deformation. Let the xy-plane be perpendicular to this axis, with the x-axis as the direction of maximum stretching. Let the angle between the x-axis and the long axis (a) of the elliptical section of the rigid inclusion be θ. The rate of rotation of the inclusion is given by the equation (Ghosh & Ramberg 1976, p. 4)

$$\omega = -\dot{\varepsilon}_x \frac{R^2 - 1}{R^2 + 1} \sin 2\theta \tag{11.12}$$

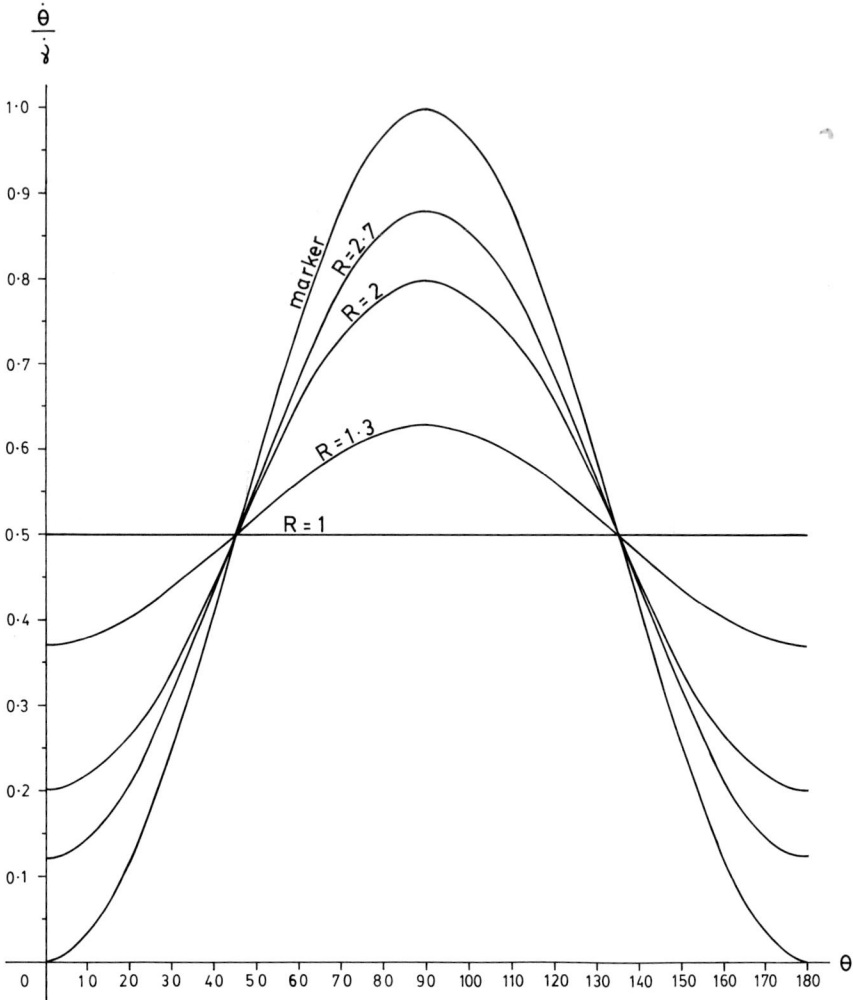

FIG. 11.4. Rate of rotation ($\dot{\theta}$) relative to the rate of simple shear ($\dot{\gamma}$) of rigid inclusions and a passive marker at different orientations under bulk simple shear. At $\theta < 45°$, $\dot{\theta}/\dot{\gamma}$ is largest for a spherical inclusion and is smallest for a marker line. At $\theta = 45°$, the rigid inclusion and the marker line all rotate at half the rate of simple shear. At $\theta > 45°$ but $< 135°$, the marker line rotates faster than an elliptical inclusion and a longer inclusion rotates faster than a shorter inclusion.

where $\dot{\varepsilon}_x$ is the rate of natural strain in the x-direction. The equation shows that the rotation can be either positive or negative depending on whether the a-axis lies on the left or on the right side of the positive y-axis. The absolute value of the rate of rotation is a maximum when

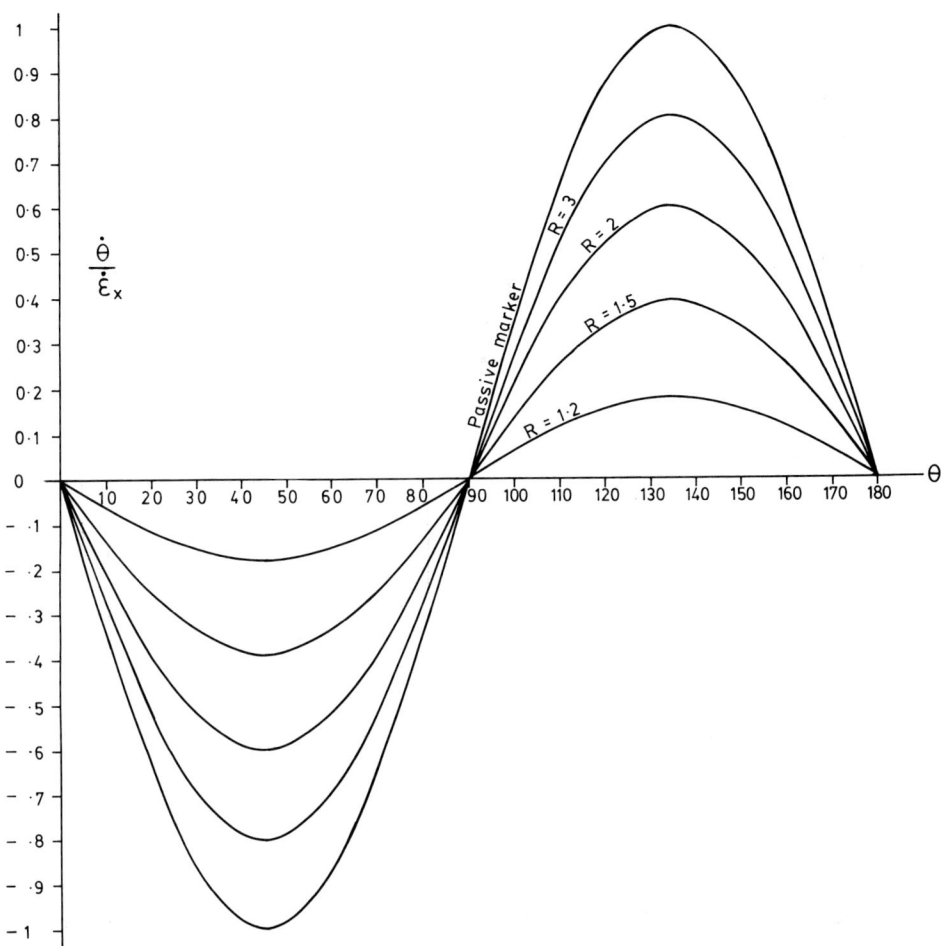

FIG. 11.5. Rates of rotation ($\dot{\theta}$) of rigid inclusions and of a passive marker line in pure shear. $\dot{\theta}$ is expressed relative to the rate of extension ($\dot{\varepsilon}_x$) of pure shear along the x-axis. Negative values of $\dot{\theta}$ indicates clockwise rotation and a positive value indicates anti-clockwise rotation. Note that the sense of rotation is opposite on either side of the y-axis ($\theta = 90°$). The absolute value of the rotation rate is larger for a longer inclusion and is largest for the passive marker. The rate of rotation is 0 for a circular inclusion.

$\theta = 45°$ or $135°$ (Fig. 11.5). The rate of rotation vanishes when the inclusion is spherical, i.e. $R = 1$.

In order to determine the rate of rotation of a passive marker plane in pure shear let us start with the particle path equation for progressive pure shear

$$x = x_0 \exp(\dot{\varepsilon}_x t)$$
$$y = y_0 \exp(-\dot{\varepsilon}_x t) \qquad (11.13)$$

(Ramberg 1975). Thus

$$\frac{x}{y} = \frac{x_0}{y_0} \exp(2\dot{\varepsilon}_x t)$$

or,

$$\cot\theta = \cot\theta_0 \exp(2\dot{\varepsilon}_x t). \tag{11.14}$$

Differentiating with respect to time, we get

$$-\csc^2\theta \cdot \dot{\theta} = \cot\theta_0 \exp(2\dot{\varepsilon}_x t) \cdot 2\dot{\varepsilon}_x.$$

Substituting the expression of $\cot\theta_0$ in terms of $\cot\theta$ from eqn. (11.14) we have, after simplification

$$\omega = \dot{\theta} = -\dot{\varepsilon}_x \sin 2\theta. \tag{11.15}$$

By comparing eqns. (11.12) and (11.15) we find that the two rates of rotation differ by a factor $(R^2-1)/(R^2+1)$. Since this factor is always less than 1, the absolute value of the rate of rotation of a passive marker is always larger than that of an elongate rigid inclusion at the same orientation. When R is sufficiently large, $(R^2-1)/(R^2+1) \approx 1$, and the inclusion rotates more or less as a passive marker plane.

11.5. Rotation rates in more complex types of non-coaxial deformation

We have seen that progressive simple shear is a type of non-coaxial deformation with the kinematical vorticity number $W_k = 1$. We have also seen that a line parallel to the x-axis, i.e. the direction of simple shear, does not rotate with progressive deformation. Let us now consider a type of plane non-coaxial deformation in which $W_k < 1$. There are, in this type of deformation, and in a plane perpendicular to the vorticity vector, two directions, say l_1 and l_2, along which the material lines have zero angular velocity. These are the two apophyses of the hyperbolic particle paths (Fig. 11.6; Ramberg 1975). There is an extension parallel to one of these apophyses (extensional apophysis of Passchier 1987). Lines are shortened along the other apophysis. In the following discussion the x-axis has been chosen parallel to the extensional apophysis and the xy-plane is taken to be normal to the vorticity vector. Let θ be the angle between the x-axis and the long axis of the rigid inclusion. The rate of rotation of the inclusion is given by the equation (Ghosh & Ramberg 1976)

$$\dot{\theta} = W(A\sin^2\theta + B\sin 2\theta + C\cos^2\theta), \tag{11.16}$$

where W is the magnitude of vorticity and

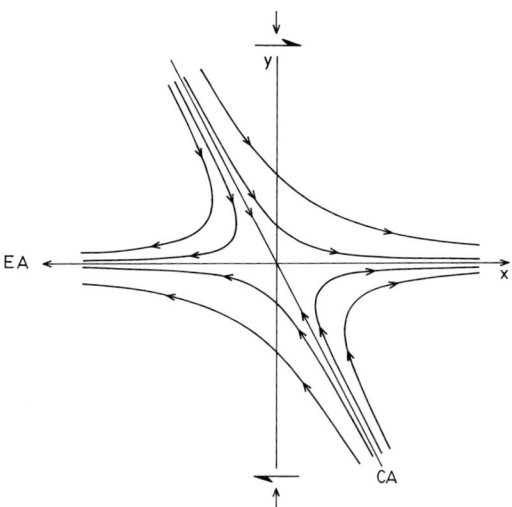

FIG. 11.6. Particle paths for a plane non-coaxial deformation with $W_k < 1$. Along the directions of the extensional apophysis (EA) and compressional apophysis (CA) the particles move in a straight line. The sense of rotation of a line is opposite on either side of CA.

$$A = \frac{R^2}{R^2 + 1}, \quad B = \frac{s_r(R^2 - 1)}{R^2 + 1}, \quad C = \frac{1}{R^2 + 1}.$$

R is the axial ratio of the inclusion on the xy-plane and s_r is a parameter dependent on the kinematical vorticity number W_k (Ghosh 1987, eqn. 9):

$$s_r = \frac{\sqrt{1 - W_k^2}}{2W_k}. \tag{11.17}$$

s_r may also be interpreted in the following way. Let the deformation be imagined as a combination of steady pure shear and steady simple shear. $\dot{\varepsilon}_x$ and $\dot{\varepsilon}_y$ are the principal strain rates of pure shear along the x- and y-axes. The direction of simple shear is also along the x-axis. S_r is equal to the ratio of pure shear and simple shear strain rates $\dot{\varepsilon}_x/\dot{\gamma}$ Since W, the vorticity, is equal to twice the rate of rotation of a spherical body, the rates of rotation of the rigid inclusions may be expressed, with the help of eqn. (11.16), in terms of the dimensionless entity $\dot{\theta}/W$ (Fig. 11.7).

Figure 11.7 shows that there are two orientations at which the inclusions with different values of R all rotate at the same rate, i.e. at the rate of rotation of a spherical inclusion. These two orientations are given by the equation

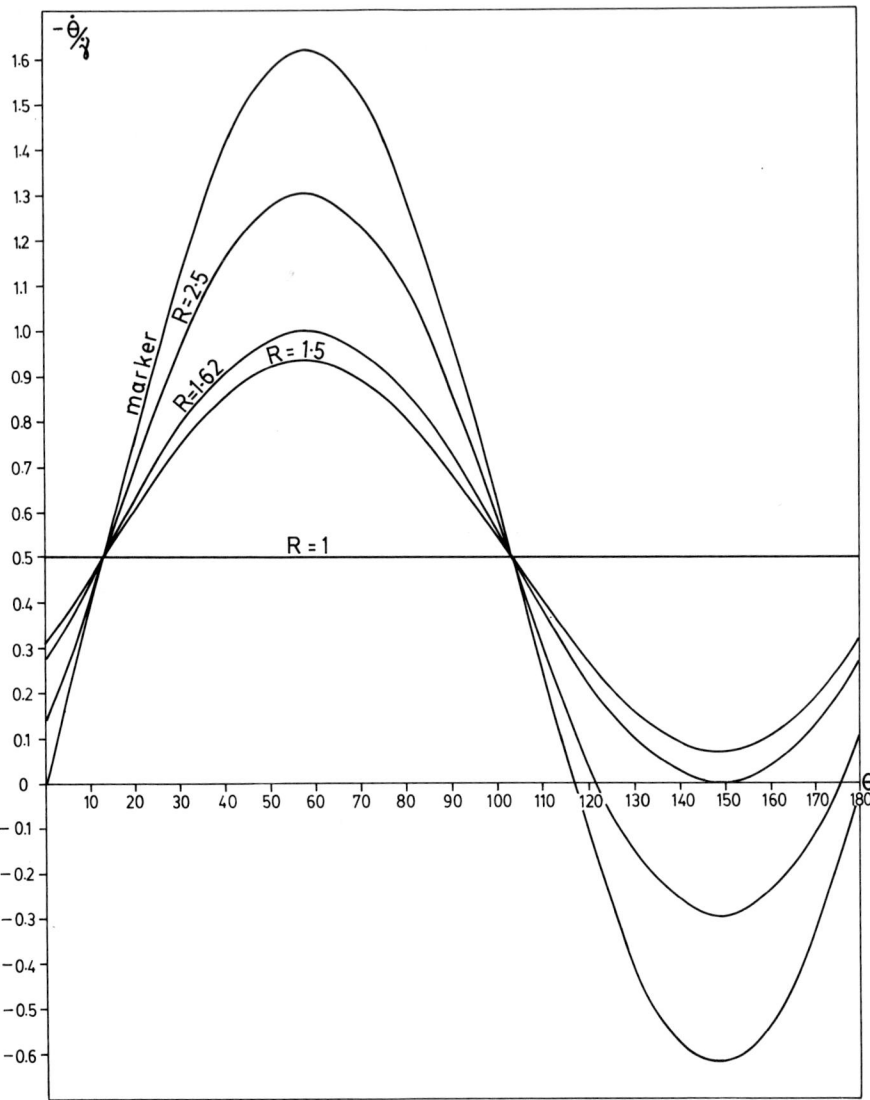

Fig. 11.7. Rates of rotation of rigid ellipsoidal inclusions and of a passive marker in plane non-coaxial deformation with $W_k = 0.45$.

$$\tan2\theta = \frac{1}{2s_r}. \tag{11.18}$$

These are also the principal directions of the incremental strain ellipsoid. For Fig. 11.7, $W_k = 0.45$ and, hence, by eqn. (11.17), $s_r = 1$.

From eqn. (11.18) we find that the long axis of the incremental strain ellipsoid makes an angle of 13.3° with the x-axis. An important aspect of this type of non-coaxial deformation is that, under certain conditions, the ellipsoidal inclusions may attain a stable orientation. Once the a-axis of the inclusion attains this orientation it does not rotate any more. A stable orientation can be attained only if

$$\frac{R^2 - 1}{R} \geqslant \frac{1}{s_r} \qquad (11.19a)$$

or,

$$\frac{R^2 - 1}{R} \geqslant \frac{2W_k}{\sqrt{(1 - W_k^2)}}. \qquad (11.19b)$$

In the numerical example considered above, with $W_k = 0.45$, a rigid ellipsoidal inclusion may have a stable orientation only if $R \geqslant 1.62$. These relations show that a stable position may be attained if either the axial ratio of the inclusion is large or the kinematical vorticity number is small. The stable orientation (Ghosh & Ramberg 1976, p. 13) is found from the equation

$$\cot\theta = [- s_r (R^2 - 1) - \sqrt{s_r^2 (R_2 - 1)^2 - R^2}] \qquad (11.20a)$$

or,

$$\cot\theta = \frac{- B - \sqrt{(B^2 - AC)}}{C}. \qquad (11.20b)$$

where A, B and C are given in connection with eqn. (11.16). For the same numerical example, with $W_k = 0.45$ the stable orientation for an inclusion with $R = 2.5$ is $\theta = -5.8°$.

The rate of rotation of a passive marker plane is the same as that of a rigid inclusion when R is infinitely large (Ghosh & Ramberg 1976, p. 9). When R tends to infinity, the coefficients of eqn. (11.16) can be obtained by using Hospital's rule:

$$A = 1, B = s_r, C = 0.$$

The rate of rotation of a passive marker plane is obtained by substituting these values in eqn. (11.16):

$$\dot{\theta} = W (\sin^2 \theta + s_r \sin 2\theta). \qquad (11.21)$$

Indeed, the difference between the rates of rotation of a marker plane and a rigid ellipsoidal inclusion is very small when $R > 5$.

The main conclusions of the sections 11.2 to 11.5 are summarized below:

1. A rigid spherical body embedded in a ductile matrix rotates only if the deformation is non-coaxial. Twice the rate of rotation of a rigid spherical inclusion is equal to the vorticity. The degree of non-coaxiality can be represented by the kinematical vorticity number which depends only on the vorticity and the principal stretching rates.

2. An ellipsoidal inclusion rotates in pure shear unless its symmetry axes are parallel to the principal axes of strain. The absolute value of the rate of rotation is always smaller than that of a passive marker plane of the same orientation. Inclusions with different orientations may have different senses of rotation.

3. A spherical inclusion rotates in simple shear at half the rate of simple shear. In simple shear an elongate inclusion always rotates in the same sense. Unless R is quite large, the rate of rotation of an inclusion does not vanish at any orientation. In simple shear the rate of rotation of an inclusion is a maximum when its long axis is normal to the shearing plane; it is a minimum when it is parallel to the direction of shear.

4. In other types of plane non-coaxial deformation, with $W_k < 1$, there is a small range of orientation in which an elongate inclusion rotates in a sense opposite to that of the vorticity. Depending on the kinematical vorticity number, a rigid ellipsoidal inclusion has a critical value of R. When R is below this critical value, the inclusion will rotate in all orientations. If R is equal to or is greater than this critical value, the inclusion cannot rotate indefinitely. There is a stable orientation at which the rotation of the inclusion vanishes.

5. In all types of deformation the rate of rotation of a marker plane is equal to that of a rigid inclusion with a very large value of R.

11.6. Finite rotations

In finite homogeneous deformation there is a set of three lines which remain perpendicular to each other both before and after the deformation. The angle between their final and the initial orientations is the finite rotation. If this angle is zero, the deformation is coaxial or irrotational. The initial positions of these lines are called the principal axes of finite deformation. The final positions of the lines are parallel to the principal axes of the finite strain ellipsoid. The principal axes of the infinitesimal or incremental strain ellipsoid bisect the angles between the principal axes of finite strain and the principal axes of the finite strain ellipsoid (Fig. 11.8).

The finite rotations of a marker plane and of a rigid inclusion are different from each other and from the rotation of the principal axes of the strain ellipsoid. Recognition of such dissimilar rotations is significant in the interpretation of many of the geological structures. As an example let us consider the rotation of a schistosity in a steady

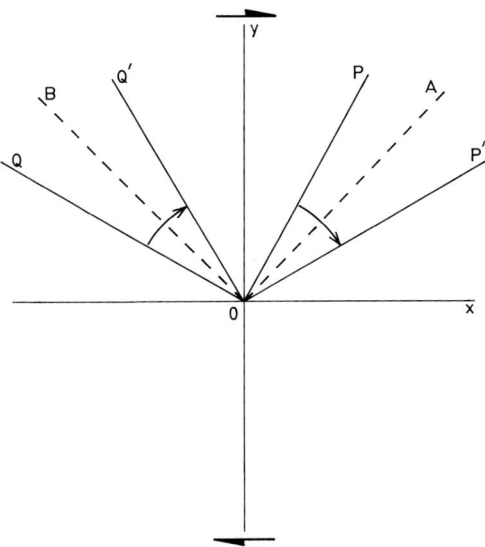

FIG. 11.8. P and Q are two mutually perpendicular material lines which, after deformation, have rotated to positions P' and Q' which are also mutually perpendicular. The deformation is rotational when the angle $POP' \neq 0$. The angle POP' is defined as the rotation. OA and OB, the bisectrices of the angles POP' and QOQ' are the principal axes of infinitesimal strain.

progressive deformation. The schistosity initiates parallel to the XY-plane of the strain ellipsoid. Once formed, the schistosity may continue to remain parallel to the XY-plane throughout the entire history of the deformation. In this case it is said to track the XY-plane. On the other hand, after its initiation the schistosity may rotate passively as a marker plane. Its orientation will then deviate from that of the XY-plane. Is this deviation large or small? Can we set a limit to this deviation? Let us first consider the case of steady simple shear deformation. In simple shear, the major axis of the strain ellipse on the xy-plane is initially at an angle of 45° with the direction of simple shear. Let us assume that the schistosity is initiated in this orientation. Now consider the following two cases: (1) The schistosity tracks the XY-plane. (2) The schistosity undergoes only a passive rotation during the ongoing deformation. In the first case the final orientation, represented by the angle between the schistosity and the x-axis (the direction of simple shear), is

$$\theta = 45° - \tfrac{1}{2}\tan^{-1}\left(\frac{\gamma}{2}\right).$$

In the second case this orientation is

$$\theta' = \tan^{-1}\left(\frac{1}{1+\gamma}\right).$$

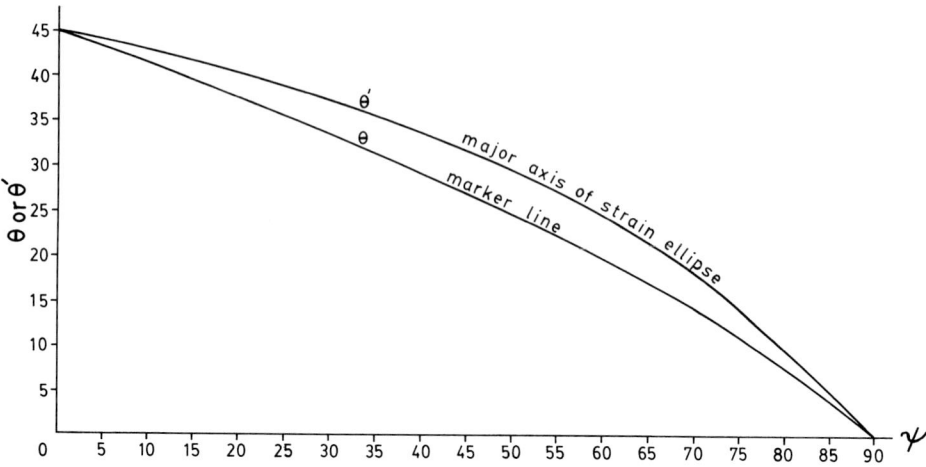

FIG. 11.9. Change in orientation of the major axis of the strain ellipse and a marker line initially at 45° with direction of simple shear. The maximum difference in angle between the two orientations is 5°.15 when the simple shear $\gamma = \tan \psi = 1$.

The difference between these two orientations is

$$\beta = \theta - \theta' = \frac{\pi}{4} - \tfrac{1}{2} \tan^{-1} \left(\frac{\gamma}{2} \right) - \tan^{-1} \left(\frac{1}{1 + \gamma} \right). \quad (11.22)$$

To find the maximum possible value of β we differentiate it with respect to γ and put $\delta\beta/\delta\gamma = 0$. From this relationship we find that β has the largest value when $\gamma = 1$. Substituting this value of γ in equation (11.22) we find that

$$\beta_{(max)} = 45° - \frac{3}{2} \tan^{-1} \left(\tfrac{1}{2} \right) = 5°.15.$$

This is the maximum possible deviation (Ghosh 1975) between a schistosity which tracks the XY-plane and that which rotates as a marker plane after its initiation. Figure 11.9 shows that the deviation is largest when $\gamma = 1$. The deviation is much smaller at the initial stages of deformation and at large values of γ. Evidently, the schistosity may not be initiated in the 45° position. If the initial angle is smaller, the maximum possible deviation is smaller than 5°.15. It has been shown (Ghosh 1982) that the maximum deviation is much smaller than 5°.15 in other types of steady, plane, non-coaxial deformations with $W_k < 1$. Thus, if W_k is 0.45, the maximum deviation between a schistosity tracking the XY-plane and a passively rotated schistosity is only 2°.2. Such deviations will be too small to be detected in the field; however, even

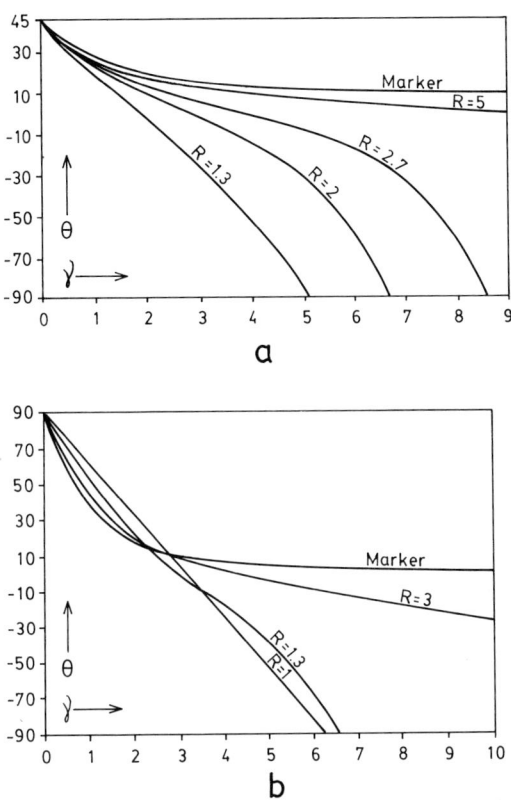

FIG. 11.10. (a) Change in orientation (θ) of rigid ellipsoidal inclusions, and of a passive marker in progressive simple shear. The long axes of the inclusions are initially at an angle of 45° with the direction of simple shear. (b) Change in orientation of an ellipsoidal inclusion and of a passive marker in plane non-coaxial deformation with $W_k = 0.93$. The long axes of inclusions were initially perpendicular to the extensional apophysis (l_1).

such small deviations may cause the development of a large shear strain parallel to the schistosity.

A spherical inclusion rotates through an angle of $\gamma/2$ in progressive simple shear. It is important to note that the angle in this case is given in radians. The rotation of the ellipsoidal inclusions in finite simple shear and in other types of steady, plane, non-coaxial deformations may be obtained by integrating the rotation rates as given by eqns. (11.9) and (11.16). The amount of rotation will depend upon the amount of deformation, the axial ratio R, the vorticity number and the initial orientation of the a-axis of the inclusion (Fig. 11.10). We shall consider here two situations: (1) The rotation of the rigid inclusions, such as porphyroclasts or porphyroblasts, takes place in the same deformation in which

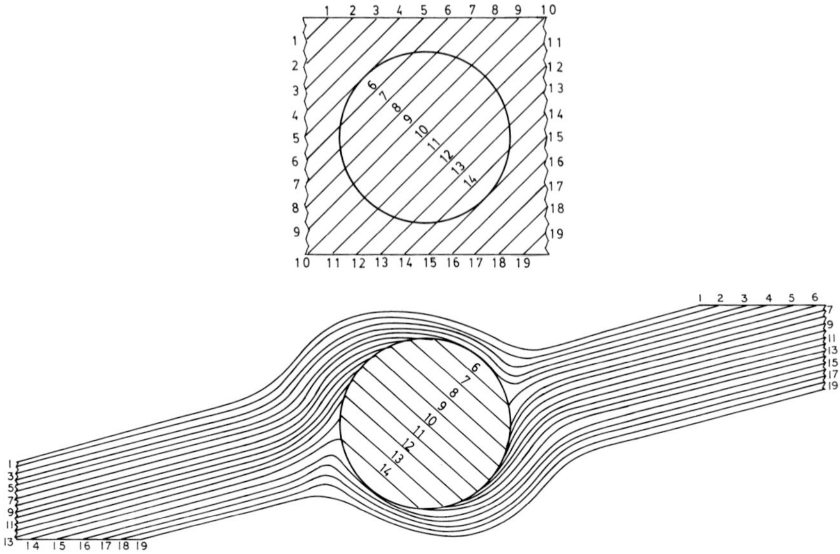

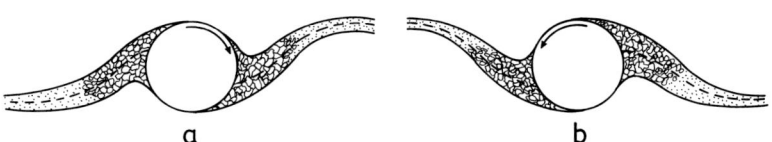

FIG. 11.11. Graphical method of reconstructing the drag pattern of foliation around a rigid spherical inclusion under simple shear of $\gamma = 3$. In the lower figure the porphyroblast has rotated clockwise through an angle of 86°. The foliation, initially at an angle of 45°, has undergone a clockwise rotation of 31°.

FIG. 11.12. δ-type of porphyroclast-tail system. The tail steps up to the right for a clockwise rotation as in (a) and steps up to the left (b) for an anticlockwise rotation.

the schistosity in the matrix of the rock is initiated and rotated. Once formed, the schistosity may either track the XY-plane or may rotate as a marker plane. (2) The schistosity developed in an earlier deformation and the major phase of rotation of the porphyroblast took place in a later deformation. The schistosity, in this case, does not necessarily have an orientation close to the XY-plane of the second deformation.

In the first situation an elongate porphyroblast or porphyroclast is likely to have its a-axis initially parallel to the schistosity. If the deformation is by progressive simple shear, the rigid inclusion will rotate through a larger angle than the schistosity. Along the periphery of the rigid grain the schistosity will be dragged in the same sense as the sense of rotation of the rigid grain. Thus, in dextral simple shear, with a

212

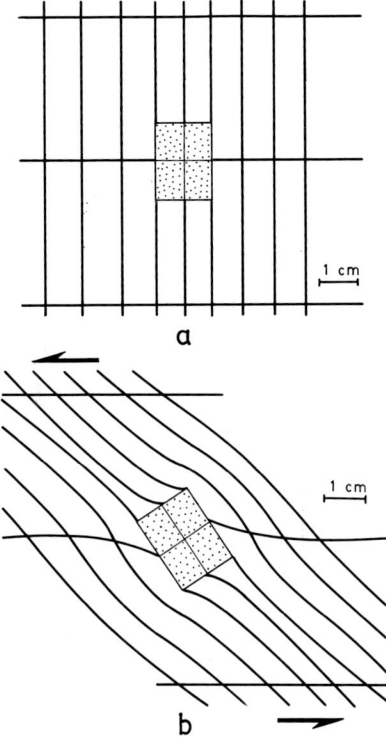

FIG. 11.13. (a) Rigid rectangular wooden block, 2 cm × 1.5 cm, embedded in silicone putty, with the long axis at a right angle to the direction of simple shear. (b) The same model after left-handed simple shear. The marker lines in the matrix have rotated through a larger angle than the rigid block. After fig. 20a of Ghosh & Ramberg 1976.

clockwise rotation of the porphyroblast the foliation on the right-hand side will be dragged downward near the periphery and the foliation on the left-hand side will be dragged upward (Fig. 11.11). A similar asymmetric pattern may also develop in the tails of certain types of porphyroclasts in mylonites. A rigid porphyroclast in this case develops a mantle of fine recrystallized grains of the same material. Since this fine-grained material is deformable, it is dragged out in the form of a tail. When the tail is sufficiently long, it becomes parallel to the mylonitic foliation in the matrix. With continued rotation of the porphyroclast the portion of the tail attached to the porphyroclast is rotated along with it, while the tail segment away from the porphyroclast is rotated as a passive marker and becomes parallel to the foliation. In such a porphyroclast-tail system (generally described as δ-type), the straight portion of the tail often shows a stair-stepping pattern (Fig. 11.12). Thus when there is a clockwise rotation of the porphyroclast the

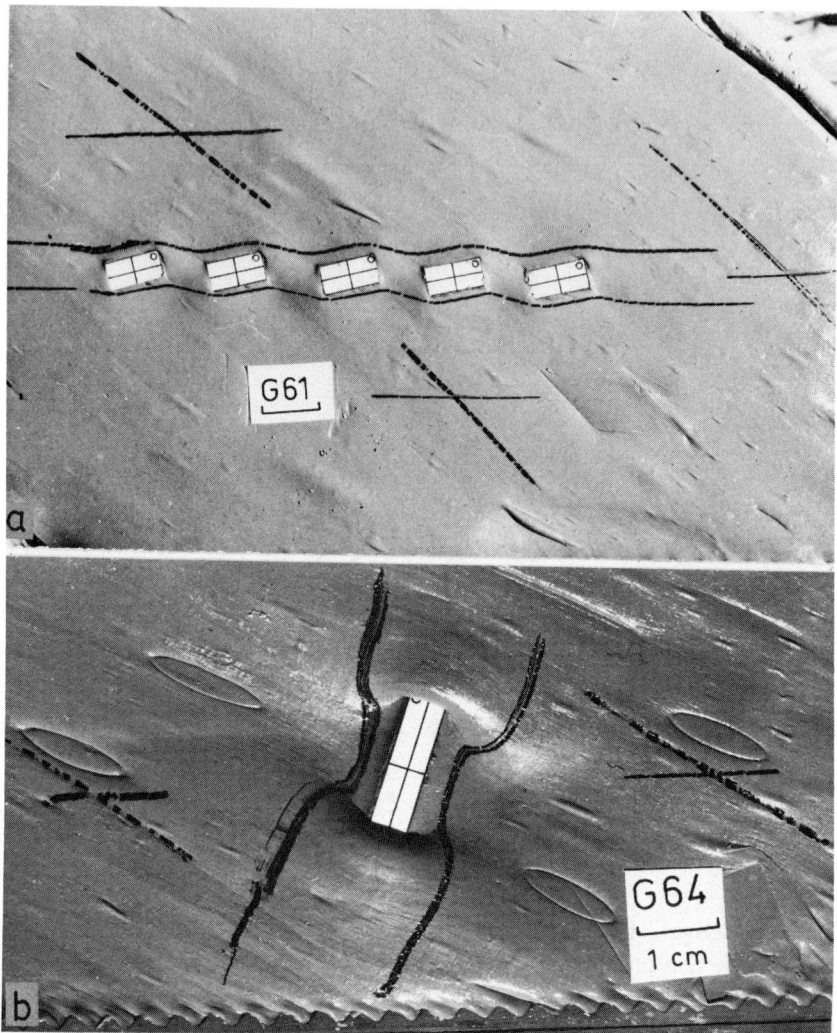

FIG. 11.14. (a) A row of rigid blocks, with axial ratio 2:1 and aligned parallel to the direction of left-handed simple shear. The blocks have rotated anticlockwise, while their centres still occur on a line parallel to the direction of simple shear. (b) Rectangular rigid block oriented at an angle of 135° with the direction of left-handed simple shear. The marker lines in the matrix, which were initially parallel to the long axis of the inclusion, are deformed to a millipede structure.

straight portion of the tail steps up to the right (Passchier & Simpson 1986).

The distorted pattern of the foliation around rigid objects may be different from what has been just described if, as in the second situation

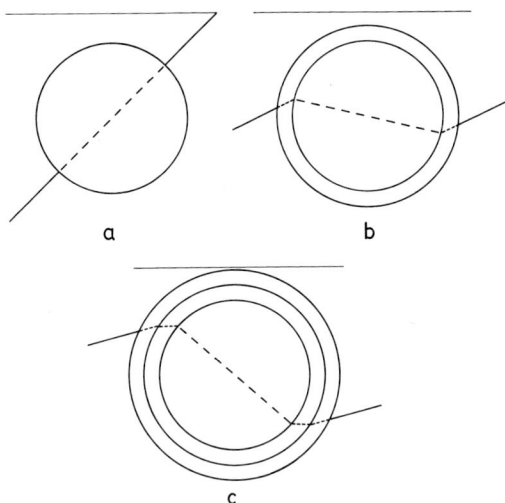

FIG. 11.15. (a) A spherical porphyroblast overgrowing a foliation. (b) The porphyroblast has undergone a clockwise rotation under dextral simple shear. The outer foliation, which has rotated through a smaller angle, is overgrown by the growing porphyroblast. (c) Another stage of rotation followed by growth. When the growth and rotation take place simultaneously, the internal trail has a smoothly curved shape.

mentioned above, the development of the matrix schistosity and rotation of the porphyroblast took place in separate deformations. Depending upon the orientation, the pre-existing schistosity may then rotate faster (Fig. 11.13) or slower (Fig. 11.14a) than the rigid grain. Moreover, a porphyroblast which rotates slower than the foliation at an early stage may rotate faster than it at a later stage (Fig. 11.10b). The sense of rotation of the porphyroblast may then be difficult to deduce from the pattern of the distorted schistosity around it. However, as Fig. 11.10b shows, if the deformation is very large, the total rotation of a spherical or a slightly oblong rigid body becomes much larger than that of the schistosity in the matrix. The asymmetry of the distorted S-surface can then give us the sense of rotation.

An unusual kind of drag pattern around porphyroblasts, with an inward bowing of the foliation, develops when the direction of shear is at a large obtuse angle to the foliation. Such structures were first recorded during experimental deformation (Fig. 11.14b) of the matrix material around rigid objects (Ghosh 1975, 1977, Ghosh & Ramberg 1976). The structures developed when the orientation of a pre-existing foliation is at a high angle to the direction of extension in the later deformation. The structure can initiate as a symmetrical feature but may become more and more asymmetrical with a progressive non-coaxial deformation. Similar structures from natural rocks have been described by Bell & Rubenach (1980, 1983) as *millipede structures*.

A porphyroblast may overgrow the bedding or the schistosity. The trace of these S-planes is sometimes retained within the porphyroblast as trails of inclusions (Fig. 11.15). This internal trace of the S surface is described as *Si* or as *trails of inclusions* to distinguish it from the *Se* or the external schistosity in the matrix of the rock. The trails of inclusions have a characteristic pattern when the porphyroblast grows as it rotates. Such porphyroblasts are called *paratectonic* and the rotation is then described as *paracrystalline rotation*. As the porphyroblast rotates the *Si* in it rotates through a larger angle than *Se* in the matrix. As a result the *Si* in the growing porphyroblast becomes curved (Fig. 11.15), a Z-shaped *Si* indicating clockwise rotation and a S-shaped *Si* indicating an anticlockwise rotation. (For other forms see Ghosh & Ramberg 1978.)

The dissimilar rotations of a rigid inclusion and a passive marker line may also cause the development of a side-stepping or en echelon pattern in a row of boudins. If the bulk deformation is homogeneous, a line joining the centres of the boudins rotates as a passive marker. The *en echelon* arrangement occurs because the rotation of the individual boudins is different from that of the line joining the centres of the boudins (Fig. 11.14a).

CHAPTER 12

Geometry of Folds

12.1. Definition of a fold

A fold is represented by a curved surface or a stack of curved surfaces whose initial curvature has increased by deformation. Since stratification in undeformed sedimentary beds is planar within a short distance, a distinctly curved or wavy stratification surface is commonly described as a fold. No doubt, cross-stratified beds may show initial curvatures of the foreset laminae. Such initially curved bedding laminae are said to be folded if the initial curvature is perceptibly increased by deformation.

There is no hard and fast definition about the limits of a single fold. Depending on the nature of enquiry we can speak about a fold on a single surface, a fold on a layer bounded by two surfaces or a fold in a stack of layers with several interfaces. For most purposes it is convenient to define the lateral limits of a fold (Ramsay 1967) on a single surface by two consecutive lines of inflection (Fig. 12.1). The definition may run into difficulties for certain special types of folds, as in zigzag folds where the points of inflection either do not exist or are difficult to locate, in certain types of arrow-head folds in which the limbs are sigmoidally curved and in cuspate folds in which the closures in one direction are smoothly curved but the closures of an opposite sense are sharp-hinged (Fig. 12.2).

12.2. Axis of folds

The geometry of an irregular wavy surface is extremely difficult to describe either in language or through diagrams. Fortunately, most folds, at least in small segments, have a sort of regularity, in the sense that their shapes do not show significant variation in one particular direction. In other words, in sections along a particular direction the trace of the folded surface appears as a straight line, while in all other sections the trace appears as a wavy line. This particular direction is

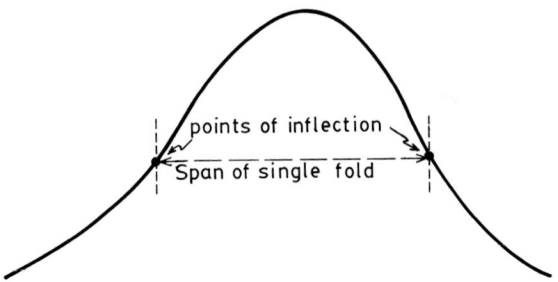

FIG. 12.1. Span of a single fold in transverse section.

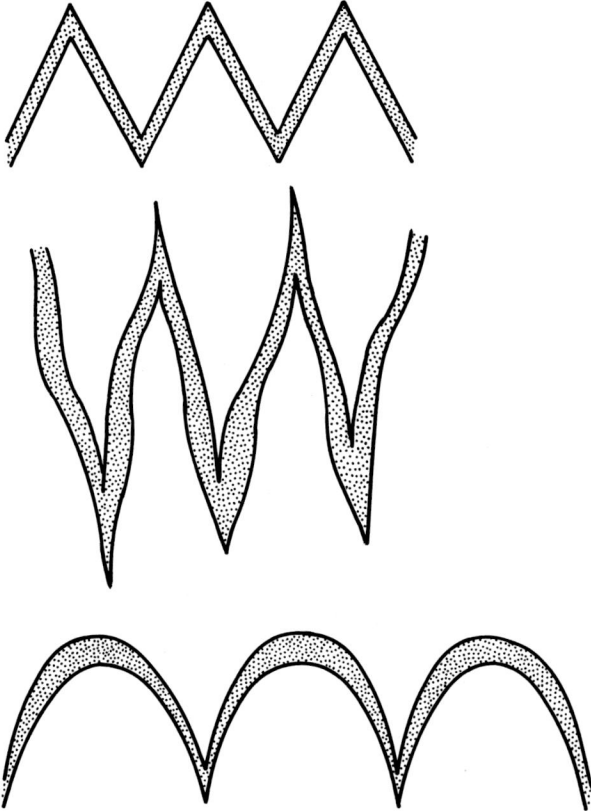

FIG. 12.2. Chevron folds, arrow-head folds and cuspate folds.

called the *fold axis*. A fold axis may, then, be defined as a line which, moving parallel to itself, generates the folded surface (Clark & McIntyre 1951, Weiss 1959a). The resulting fold is known as a *cylindrical fold.* The fold axis does not have a fixed position in space; it has only a constant orientation throughout a volume of rock within which the fold is cylindrical.

The fold axis is the most important structural element of a fold because the structures show the maximum continuity in this direction. A folded surface shows the maximum curvature in a plane perpendicular to the fold axis. Hence, the geometry of a cylindrical fold is best described by the orientation of the fold axis along with a description of the section perpendicular to the fold axis. Such a section is called a *transverse profile* or simply a profile.

12.3. Structural elements of a folded surface

A folded surface may be the interface between two beds, the interface between gneissic bands or simply a cleavage surface. For convenience of description the surface on which the fold appears may be called the *form surface* (McIntyre & Weiss 1956). The geometry of the folded surface is described in terms of the relative positions or orientations of certain planes, lines or points. These are the *structural elements of a fold.* The most important of these, the *fold axis,* has already been defined in the previous section. The other important elements are defined below.

The trace of a folded surface appears as a wavy line on the plane of the transverse profile. Roughly speaking, a point which separates a convex and a concave segment of the wavy line is called an *inflection point.* In other words, the points of inflection separate, on the transverse profile, fold-segments of opposite senses of curvature (Fig. 12.3). The line joining the corresponding inflection points of successive transverse sections is called the *inflection line* (Fig. 12.4). As mentioned earlier, two adjoining inflection lines on a folded surface mark out the limits of a fold.

On the transverse section, the highest and the lowest points of a folded surface are known as the *crest point* and the *trough point* (Fig. 12.3). The line obtained by joining the corresponding crest points in serial cross-sections of a fold is known as a *crest line* (Fig. 12.4). Similarly, we can define a *trough line* as a line which joins the lowest points of a fold. The point at which the folded surface shows the maximum, absolute value of curvature is a *hinge point* (Fig. 12.3). The line joining the corresponding points on successive profiles is the *hinge line* (Fig. 12.4) of the fold. For cylindrical folds the hinge line is parallel to the fold axis.

On the profile plane, the wavy trace of the folded surface may be represented by a function $y = f(x)$, where the x-axis joins two consecutive inflection points and the y-axis (with positive y directed upward) is at a right angle to the x-axis. In that case, the inflection points are those

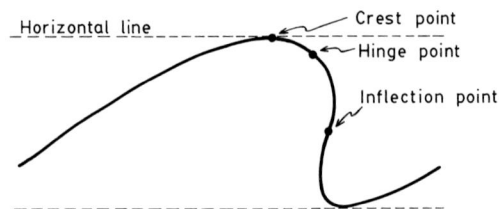

FIG. 12.3. Crest point, hinge point and inflection point on the transverse section of a fold.

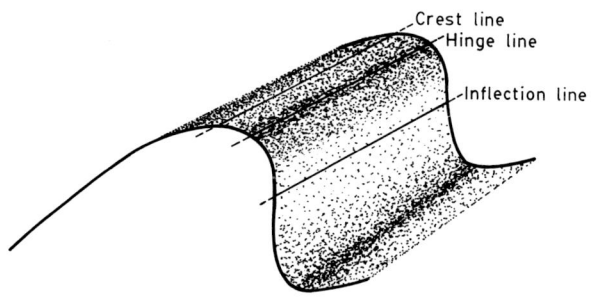

FIG. 12.4. Crest line, hinge line and inflection line on a folded surface.

where $d^2y/dx^2 = 0$. At a hinge point the absolute value of d^2y/dx^2 will be a maximum.

The surface which joins the adjoining inflection lines of a folded surface is called the *median surface* (Fig. 12.5). The two surfaces within which a train of folds rises and falls are known as *enveloping surfaces* (Fig. 12.6; Turner & Weiss 1963). The *amplitude* (Fig. 12.6) of a fold may be defined as the distance between the median surface and the enveloping surface. For folds in which the lines of inflection are not well defined, the median surface may be taken as the surface equidistant from the two enveloping surfaces.

In transverse profile, a folded surface may be periodic or non-periodic. For a more or less periodic fold-train the distance between alternate hinge points or alternate inflection points is known as the *wavelength*. The length of the curved trace of form surface between two alternate hinge points is the *arc length* of the fold (Fig. 12.7).

A *limb* of a fold may be defined as a segment of a fold between a hinge line and the adjacent inflection line (Fig. 12.8a). The lengths of the long and the short limbs of an asymmetric fold (Anthony & Wickham 1978) may be measured in this way. However, the shape of the folded surface depends to a large extent on the broadness or the sharpness of the fold hinge. To distinguish these different shapes we need to define the fold

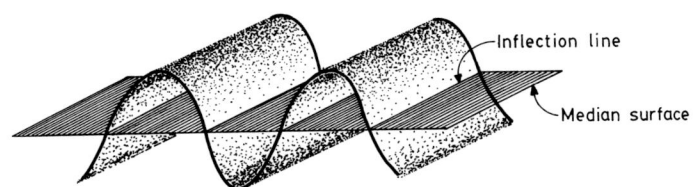

Fig. 12.5. The median surface of a fold passing through the lines of inflection.

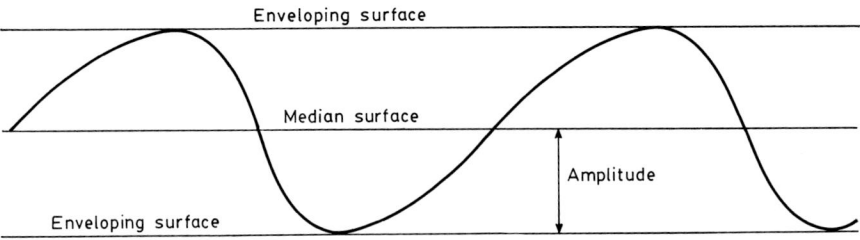

Fig. 12.6. The amplitude and the enveloping surfaces of a folded surface in transverse section.

limb in a different manner. Ramsay (1967) has defined a *hinge zone* (Fig. 12.8b) as that segment of a fold in which the radius of curvature is smaller than half the distance between inflection points at either end of a fold. The segment of arc between an inflection point and the border of the hinge zone then constitutes a fold-limb.

The *interlimb angle* of a fold (Fig. 12.8c) is the angle subtended by the tangents at two adjacent inflection points. The plane which passes through the hinge line and bisects the interlimb angle is called the bisecting plane (Haman 1961, Turner & Weiss 1963, pp. 108, 110).

12.4. Structural elements of folds on a single layer or on a stack of layers

The surface obtained by joining the corresponding inflection lines of successive interfaces is known as the *inflection surface* (Fig. 12.9a). In a similar way we can define a *crest surface* (Fig. 12.9b) and a *trough surface*. The surface or plane obtained by joining the successive hinge lines is called an *axial surface* (Fig. 12.9c) or *axial plane*. The line of intersection of the axial surface and any other plane or surface is known as the *axial trace*. The region towards the inner or concave side of a folded layer is described as the *core* of a fold. In certain cases, it is convenient to distinguish the boundary surfaces on the convex and the concave sides of a folded layer as the *extrados* and the *intrados* (Fig. 12.10), respectively (Goguel 1962, pp. 184–185).

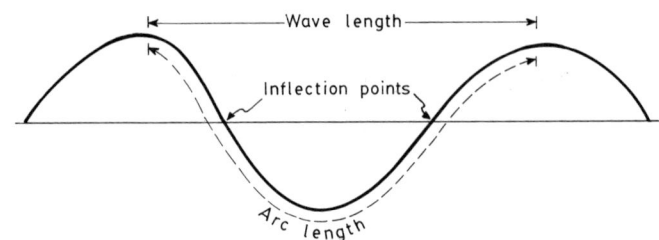

FIG. 12.7. The arc length and the wavelength of a folded surface.

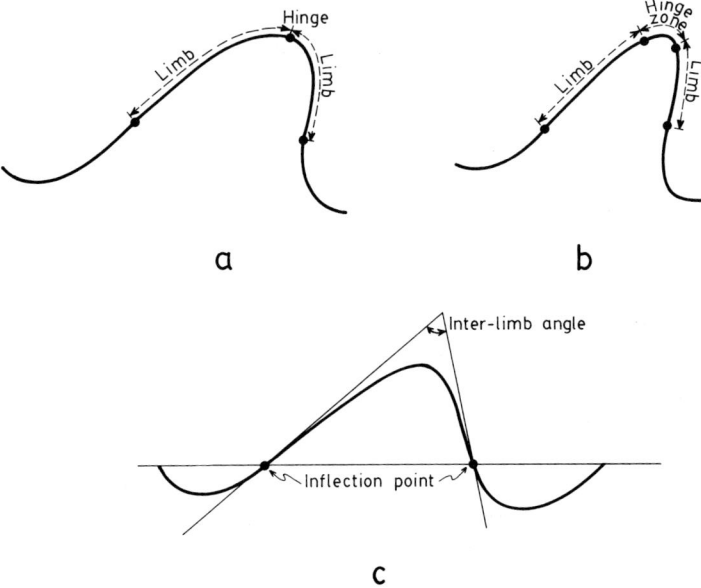

FIG. 12.8. (a) Hinge point and limbs of a fold. (b) The hinge zone of a fold. (c) The interlimb angle.

A useful structural element in folded layers is a *dip isogon* (Fig. 12.11; Elliott 1965) or a line of equal dip on successive interfaces on the transverse profile. The orientations of the dip isogons over a fold qualitatively describe the variation in thickness and the difference of curvatures between successive interfaces.

12.5. Qualitative description of fold geometry

An accurate description of the geometry of a fold is difficult and time-consuming. Hence it is often advantageous to describe a fold qualitatively by using a combination of certain terms which gives us a

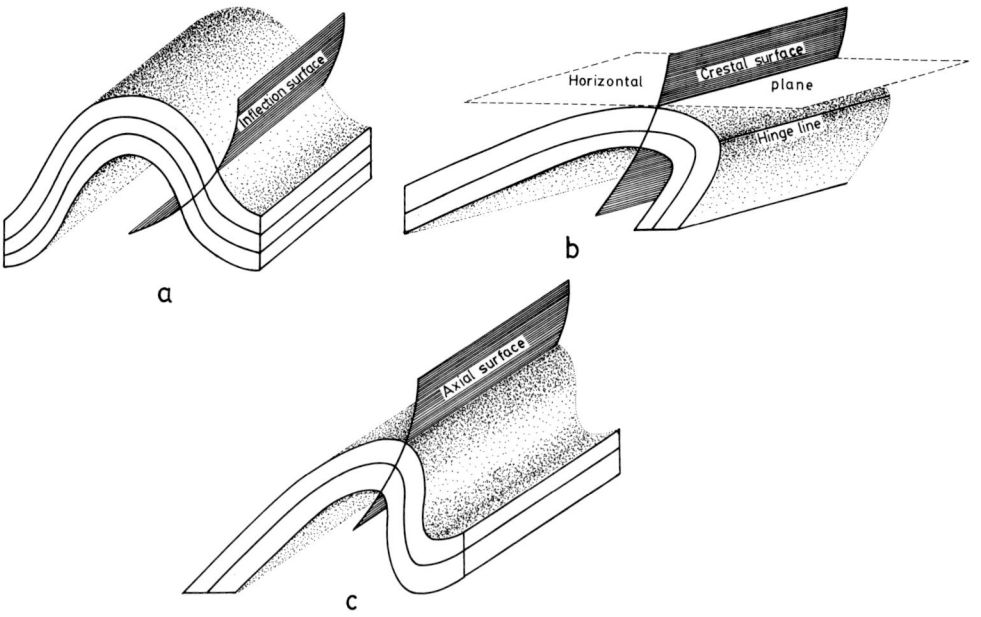

FIG. 12.9. Inflection surface, crestal surface and axial surface of folds.

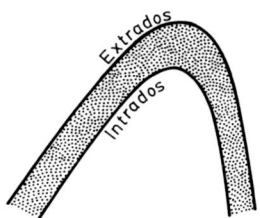

FIG. 12.10. Extrados and intrados or the convex and concave sides of a folded layer.

rough idea of the shape of a fold and of the orientations of its structural elements.

Terms which describe folds with dissimilar orientations of structural elements

(a) *Based on sense of curvature*

Antiform (Bailey *et al.* 1939): a fold that closes upward (Fig. 12.12a).

Synform (Bailey *et al.* 1939): a fold that closes downward (Fig. 12.12b).

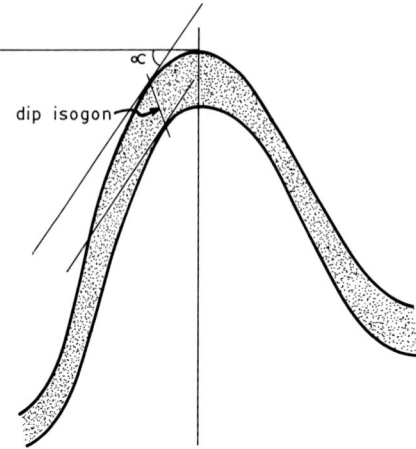

FIG. 12.11. Dip isogon, corresponding to the dip α, is determined by drawing on one surface of the folded layer a tangent dipping at an angle α. A parallel tangent touching the other surface is then drawn. The line joining the two points of tangency is the dip isogon or a line of equal dip on successive surfaces.

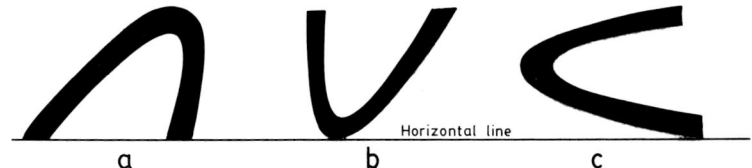

FIG. 12.12. (a) Antiform, (b) synform and (c) neutral fold in transverse profiles.

Neutral fold (Bailey & McCallien 1937): a fold that closes sidewise (Fig. 12.12c).

Dome: an antiformal structure with no distinct trend of the hinge line.

Basin: a synformal structure with no distinct trend of the hinge line.

(b) *Based on the plunge of the fold axis*

Horizontal fold: a fold whose axis is horizontal (Fig. 12.13a).

Plunging fold: a fold whose axis is inclined (Fig. 12.13b).

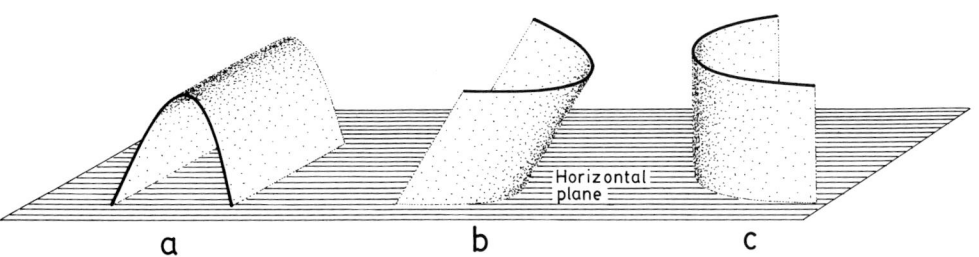

Fig. 12.13. Horizontal, plunging and vertical folds.

Vertical fold: a fold with vertical axis (Fig. 12.13c). Fleuty (1964) suggested the following classification on the basis of the amount of plunge of the fold axis:

Subhorizontal fold: plunge between 0 and 10°.

Gently plunging fold: plunge between 10 and 30°.

Moderately plunging fold: plunge between 30 and 60°.

Steeply plunging fold: plunge between 60 and 90°.

Sub-vertical fold: plunge between 80 and 90°.

(c) *Based on orientation of axial plane*

Upright fold (Willis & Willis 1929): with vertical or nearly vertical axial plane (Fig. 12.14a).

Recumbent fold: with axial plane dipping at an angle of 10° or less (Fig. 12.14b).

Inclined fold (Willis & Willis 1929): with inclined axial plane (Fig. 12.14c).

Reclined fold (Sutton 1960; distinguishing geometrical characters described by Naha 1959): inclined fold in which the pitch of the fold axis on the axial plane is between 80 and 100° (Fig. 12.14d).

Overturned fold: inclined fold in which both the limbs have the same sense of inclination (Fig. 12.14e).

Fleuty (1964) suggested the following definitions based on the amount of dip of the axial plane:

Upright fold: dip 90–80°.

Steeply inclined fold: dip 80–60°.

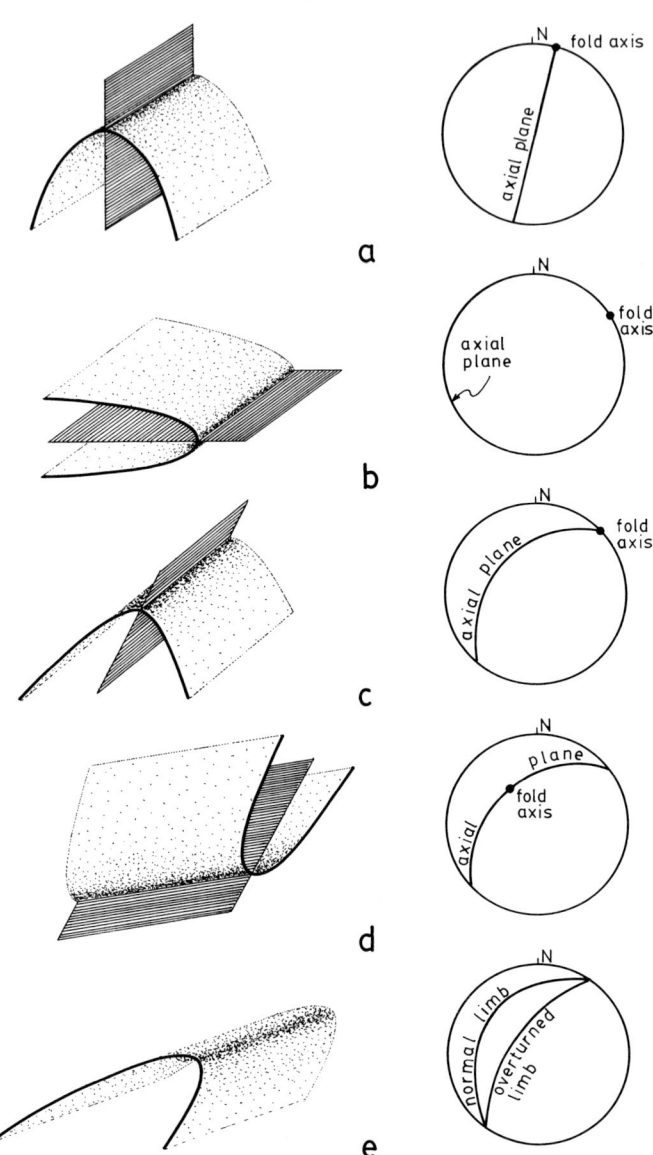

FIG. 12.14. (a) Upright fold, (b) recumbent fold, (c) inclined fold, (d) reclined fold and (e) overturned fold. (a), (b), (c) and (d) can be distinguished from the stereographic projections of the fold axis and the axial plane.

Moderately inclined fold: dip 60–30°.

Gently inclined fold: dip 30–10°.

Recumbent fold: dip 10–0°.

Terms which describe folds irrespective of absolute orientations of structural elements

(a) *Based on direction of younging relative to sense of fold closure*

Anticline: a fold in which direction of younging is away from the fold core (Fig. 12.15a).

Syncline: a fold in which direction of younging is towards the fold core (Fig. 12.15b).

Anticlinorium: a large anticline with many smaller folds on its back (Fig. 12.15c).

Synclinorium: a large syncline with many small folds on its back (Fig. 12.15c).

Synformal anticline: a fold that closes downward but with direction of younging away from the fold core (Fig. 12.15d).

Antiformal syncline: a fold that closes upward but in which the younging is towards the fold core (Fig. 12.15d).

(b) *Based on the symmetry of folds*

Symmetric fold: a fold in which the axial plane is a plane of symmetry (Fig. 12.16a). This is the most logical definition (Turner & Weiss 1963) although originally (Decker 1920, Willis & Willis 1929, Hills 1963) the term was used for folds with vertical axial planes. Symmetric folds are sometimes described as M-type folds (Ramsay 1967).

Asymmetric folds: a fold in which the axial plane is not a plane of symmetry (Fig. 12.16b). To represent the sense of asymmetry, two types, i.e. the S-type and the Z-type of folds, may be distinguished (Ramsay 1967).

(c) *Based on the nature of the hinge line*

Cylindrical fold: a fold which can be generated by moving a line parallel to itself. A cylindrical fold has a rectilinear hinge line parallel to the fold axis.

Non-cylindrical fold: a fold which cannot be generated by moving a line parallel to itself. The hinge line is either curved or the fold is conical.

Conical fold: a type of non-cylindrical fold whose shape approximates a part of a cone (Dahlstrom 1954, Haman 1961, Evans 1963).

(d) *Based on curvature of axial surface*

Plane fold (Turner & Weiss 1963): with planar axial surface. These are generally described as *plane cylindrical folds*, with planar axial surface and straight hinge line (Fig. 12.17a) and *plane non-cylindrical folds*, with planar axial surface and curved hinge line (Fig. 12.17b). The term *eyed fold* (Nicholson 1963) is sometimes used when a plane non-

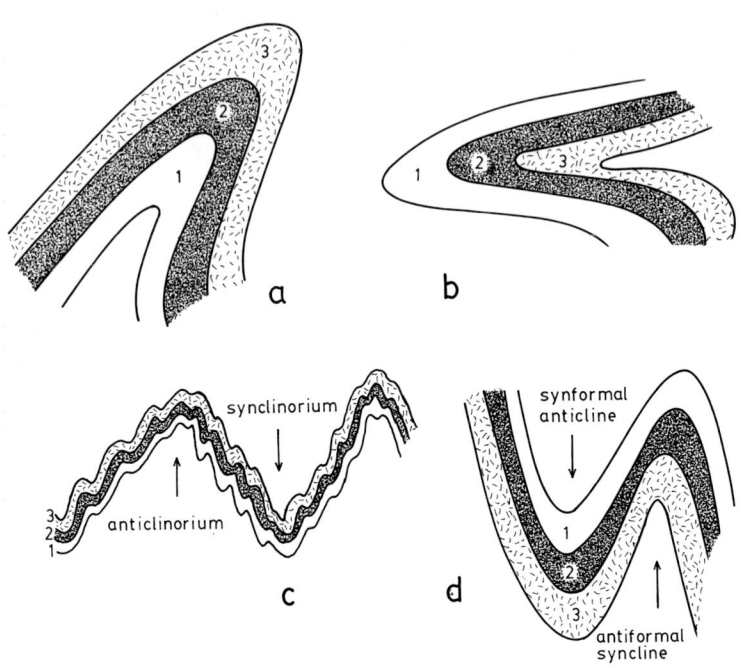

FIG. 12.15. Anticline, syncline, anticlinorium, synclinorium, synformal anticline and antiformal syncline.

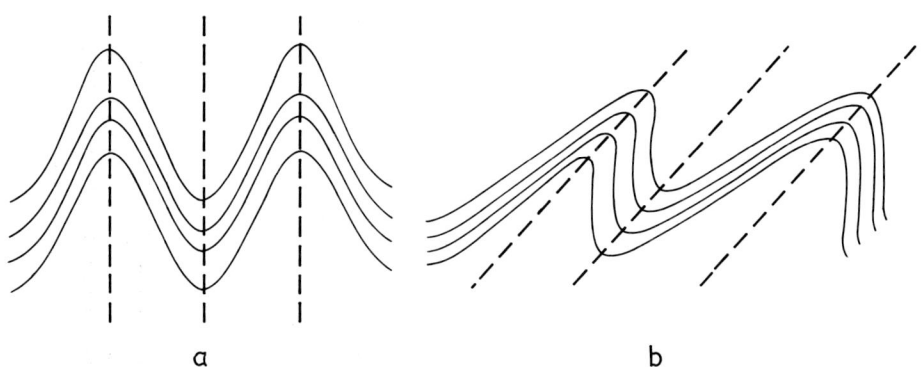

FIG. 12.16. (a) Symmetric and (b) asymmetric folds.

cylindrical fold forms an oval outcrop. It should be noted, however, that a plane non-cylindrical fold may not always form an eye-shaped outcrop (Fig. 12.17c). *Sheath fold* is a type of plane non-cylindrical fold in which the hinge line is strongly curved in a hairpin bend (Fig. 12.17d).

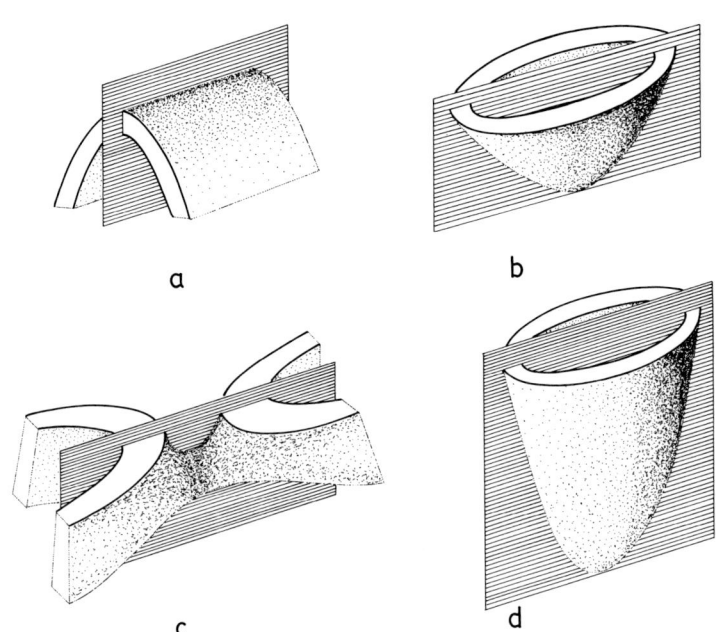

Fɪɢ. 12.17. (a) Plane cylindrical antiform, (b) plane non-cylindrical synform with an axial depression, (c) plane non-cylindrical antiform with axial depression, (d) a synformal sheath fold.

Non-plane fold: a fold with curved axial surface. Again, there are two sub-types, *non-plane cylindrical fold* (Fig. 12.18a) and *non-plane non-cylindrical fold* (Fig. 12.18b).

(e) *Based on interlimb angle* (Fleuty 1964)

Gentle fold: with interlimb angle between 180 and 120°.

Open fold: with interlimb angle between 120 and 70°.

Close fold: with interlimb angle between 70 and 30°.

Tight fold: with interlimb angle less than 30° and greater than 0°.

Isoclinal fold: with sub-parallel limbs (Fig. 12.20a).

Fan fold: with negative interlimb angle. The term *elastica* is sometimes uncritically used to describe a fan fold. However, all fan folds do not necessarily conform to the ideal shape of an elastica. Elastica is a mathematically well-defined curve; a low-amplitude undulating elastica is hardly distinguishable from a sine curve. Its fan-like form becomes obvious only when its amplitude becomes sufficiently large (Fig. 12.19).

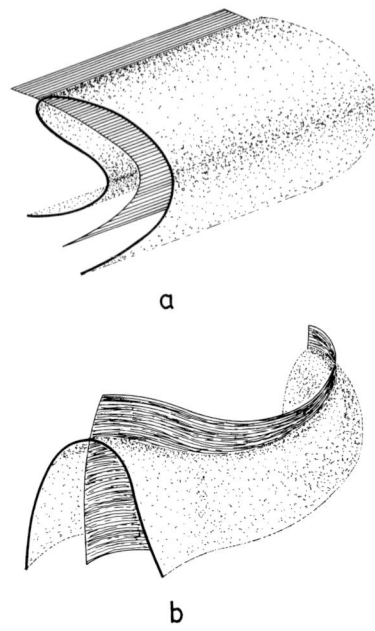

a

b

FIG. 12.18. (a) Non-plane cylindrical and (b) non-plane non-cylindrical folds.

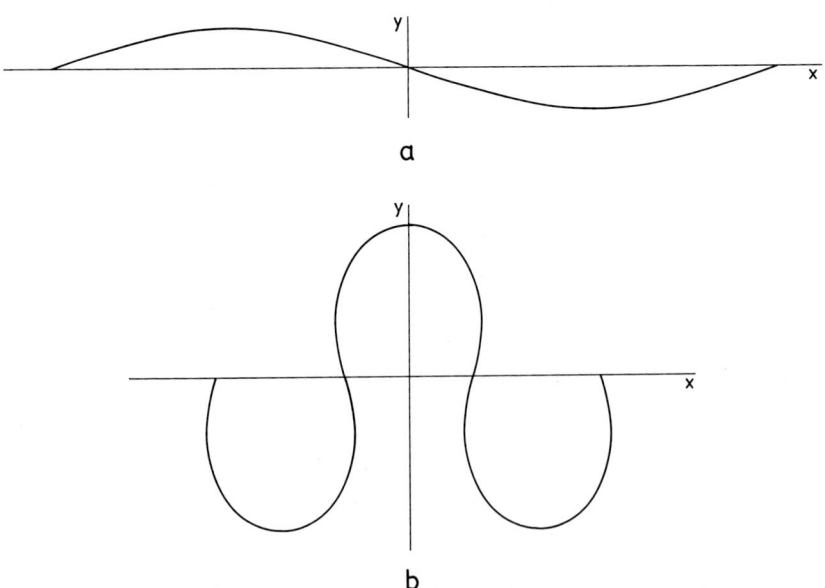

a

b

FIG. 12.19. (a) Low-amplitude elastica and (b) high-amplitude fan-shaped elastica.

(f) *Based on the shape of the hinge*

Round-hinged or *broad-hinged fold*: a fold with a broad hinge zone compared to the limb.

Chevron fold (zigzag fold, accordion fold, concertina fold): a fold with straight limbs and with a sharp hinge (Fig. 12.2).

Arrow-head fold (flame fold): a fold with a sharp hinge and with distinctly curved limbs (Kranck 1953). Often the folds are flame-shaped with sigmoidally curved limbs (Fig. 12.2).

Cuspate fold (Hills 1963): a train of fold with sharp hinges on one set of closures and with rounded hinges on the oppositely directed closures (Fig. 12.2).

(g) *Based on the number of hinges*

Single-hinged fold: a fold with a single hinge between two points of inflection.

Conjugate fold (Johnson 1956): a double-hinged fold with sharp hinges (Fig. 12.20b).

Box fold: a double-hinged fold with more or less rounded hinges. The term is generally restricted to folds with flat tops and steeply dipping limbs

(h) *Based on the geometrical relations among neighbouring structures*

Periodic folds: a train of fold with more or less the same geometry between alternate points of inflection (Fig. 12.21a).

Non-periodic folds: folds which are not periodic (Fig. 12.21b).

Polyclinal folds (Greenly 1919): a group of folds with non-parallel axial planes but with sub-parallel hinge lines (Fig. 12.21c).

Disharmonic folds: a group of folds in which the folds of one layer differ strongly in size or style from folds of an overlying or underlying layer (Fig. 12.21d).

Décollement: a process which gives rise to a train of folds in a layer which becomes detached from the adjacent layer. The adjacent layer remains more or less unfolded (Fig. 12.21e). The term is usually restricted to large-scale structures.

Intrafolial folds: an isolated, asymmetric, tight or isoclinal fold sandwiched between apparently unfolded foliation surfaces. Usually, the apparently unfolded layer forms the limbs of larger folds. When the

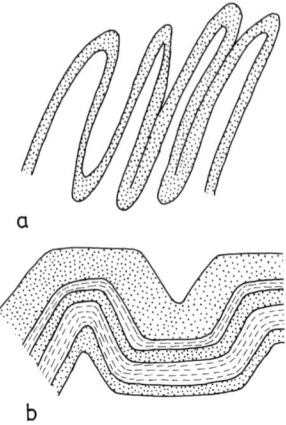

FIG. 12.20. (a) Isoclinal folds, (b) conjugate folds.

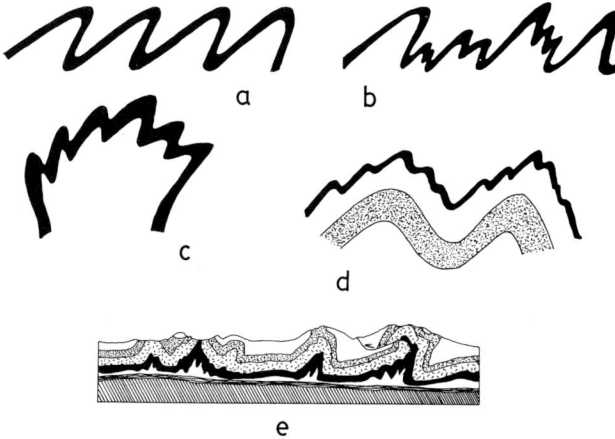

FIG. 12.21. (a) Periodic, (b) non-periodic, (c) polyclinal and (d) disharmonic folds. (e) *Décollement* of cover rocks over the basement, as illustrated by the Jura. After Buxtorf 1907.

intrafolial folds occur as tectonic inclusions they are described as *rootless intrafolial folds* (Turner & Weiss 1963).

Subsidiary folds (*minor folds*, *drag folds*, *parasitic folds*): smaller folds occurring over a larger fold, with sub-parallel hinge lines.

En echelon folds: folds arranged in a step-like or *en echelon* fashion.

(i) *Based on fold-trend with reference to orogenic trend*

Longitudinal folds: with axes sub-parallel to the orogenic trend.

Transverse folds: with axes transverse to the orogenic trend.

12.6. Morphological classification of folds

For the complete representation of the shape of a fold on a layer we need to know (1) the shape of one of the bounding surfaces of the layer in the transverse profile of the cylindrical fold and (2) the way in which the thickness of the layer varies along the fold profile. So far, these two aspects of the fold shape have been treated separately. We shall consider in this section the variation in layer thickness. The problem of representing the shape of a folded surface will be taken up in the next section.

The normal thickness of a layer cannot be defined for layers with non-parallel bounding surfaces. Consider the transverse profile of a folded layer (Fig. 12.22) with H and G as hinge points on its two bounding surfaces. Let us consider first a simple situation in which the tangents to the folded surfaces at H and G are parallel and are perpendicular to the axial surface trace HG. Take any point p on the outer surface and find a point q on the inner surface so that the tangents at p and q are parallel. Let α be the angle between the tangent at p and the folded surface at a hinge point. α may be called the dip angle at p with reference to the orientation of the folded surface at the hinge point. pq is then the dip isogon for the dip angle α. The *orthogonal thickness* (t_α) corresponding to the dip angle α is defined by Ramsay (1967) as the distance between the two tangents at p and q. The *axial plane thickness* (T_α) is the intercept, between the two tangents, of a line parallel to the axial surface trace on the profile plane. Evidently, as the figure shows

$$\frac{t_\alpha}{T_\alpha} = \cos \alpha. \tag{12.1}$$

The variation of thickness along the layer can be represented by showing the variation of t_α or T_α against α. In the shape analysis of folds we are not concerned with the absolute values of t_α or T_α. If we take a photograph of a fold and enlarge it to different sizes, the thickness will appear different in the different photographs but the shape of the fold will remain the same. For convenience of comparison between the shapes of different folds it is therefore essential to represent the thickness variation by a non-dimensional parameter, say, by $t'_\alpha = t_\alpha/t_0$ and $T'_\alpha = T_\alpha/T_0$, where t_0 and T_0 are the orthogonal and axial plane thicknesses at the hinge point, with $t_0 = T_0$. The shape of the fold can be represented by the variation of t'_α or T'_α against α (Ramsay 1967). Consider a fold in which the orthogonal thickness remains constant along the layer. For such a folded layer the radius of curvature of the outer arc (extrados) will be larger than that of the inner arc (intrados). The dip isogons at different parts of the fold will converge towards the fold core. As Ramsay (1967) has shown, the manner of variation of t'_α (or of T'_α) is

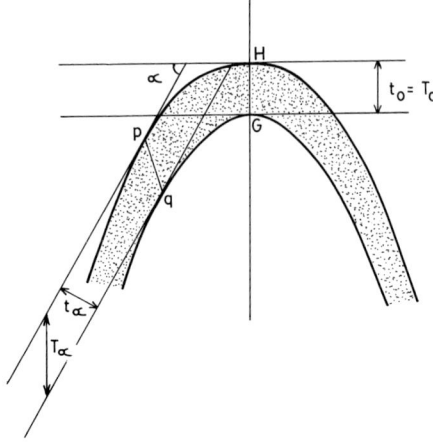

FIG. 12.22. Orthogonal thickness t_a and axial plane thickness T_a in a fold. At the hinge, $\alpha = 0$ and $t_0 = T_0$.

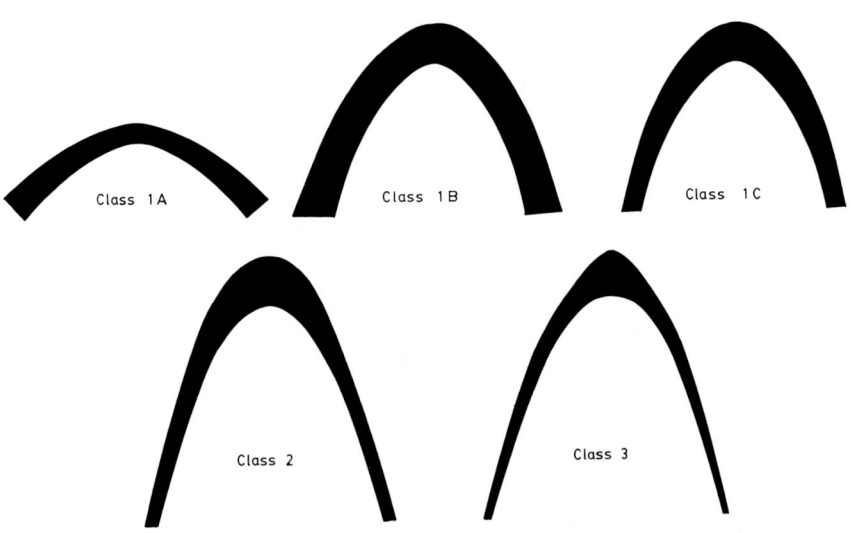

FIG. 12.23. Morphological classification of folds based on variation of t'_a or T'_a.

associated with a particular manner of variation of the dip isogon or of the radii of curvature of the outer and the inner arcs, and these parameters can be utilized for the following geometrical classification of folds (Figs. 12.23, 12.24).

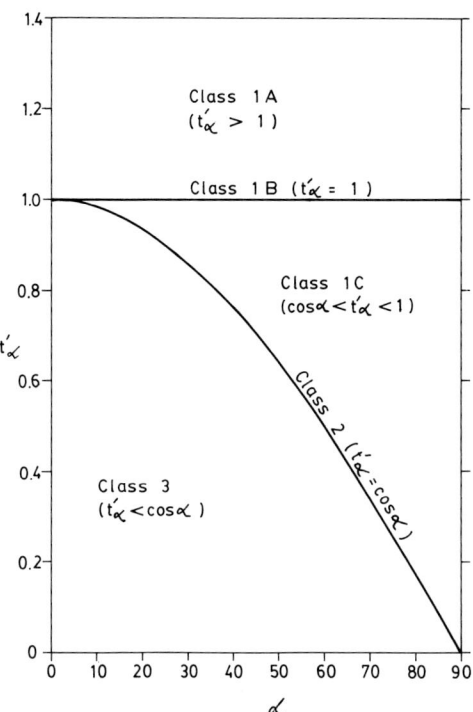

FIG. 12.24. Fields of morphological fold classes. A fold can be plotted in this diagram as a curve showing the variation of t'_α with α.

Class 1. Folds with convergent dip isogons. This necessitates that the radius of curvature of the outer arc (say, at the hinge zone) is larger than that of the inner arc. Ramsay (1967) has recognized three subclasses under this heading:

1A. Folds in which the orthogonal thickness is minimum at the hinge, i.e. $t'_a > 1$. Such a geometry implies that the dip isogons are strongly convergent. These were earlier described as supratenuous folds (Nevin 1931).

1B. Folds in which the orthogonal thickness is constant along the layer, i.e. $t'_a = 1$. These are the *parallel folds* of Van Hise (1896) or *concentric folds* of Leith (1923).

1C. Folds in which the orthogonal thickness is maximum at the fold hinge and decreases away from it. In class 1C folds $\cos \alpha < t'_a < 1$.

Class 2. Folds in which the dip isogons are parallel. This geometry can develop only if the axial plane thickness is constant all along the fold or,

235

in other words, the outer and the inner arcs have exactly the same shape. These are the *similar folds* of Van Hise (1896). In class 2 folds $t'_\alpha = \cos \alpha$ and $T'_\alpha = 1$.

Class 3. Folds in which the dip isogons diverge towards the fold core. For such folds the radius of curvature of the outer arc (say, at the hinge) is smaller than that of the inner arc. The orthogonal thickness is also a maximum at the fold hinge. In class 3 folds $t'_\alpha < \cos \alpha$.

The different categories of folds can be distinguished in most cases merely by eye inspection and without the drawing of dip isogons. We should first check whether the orthogonal thickness is constant (class 1B) or variable. A minimum thickness at the hinge (1A) is rarely found. Hence folds which are not parallel generally show a thickening of the hinge. For such folds we should see whether the inner arc is more strongly curved than the outer arc as in 1C folds. In the reverse case, i.e. if we find that the outer arc forms a sharper hinge we can identify the fold as a class 3 fold. If the inner and the outer arcs look more or less similar, we should consider the possibility that the fold may be a similar or class 2 fold. This possibility should then be tested by measuring the axial plane thickness in different parts of the fold.

Ramsay's classification is the one that is being currently used. It has several advantages over the earlier classifications. (1) The earlier classification by Van Hise (1896) only considered two fold types, the parallel and the similar (cf. Mertie 1959). The types, class 1C and class 3 folds, were not included in any of the earlier classifications. (2) In the current classification of Ramsay the morphological difference among folds of the same class or sub-class can be clearly brought out by plotting each fold by its t'_α/α or T'_α/α curve (fig. 12.24). (3) The classification is not arbitrary. Apart from the comparatively rare 1A folds, the other varieties generally appear in specific situations of fold genesis. Thus, during buckle folding of a multilayer consisting of alternate competent and incompetent layers, the incompetent layers invariably show class 3 folds while the competent layers generally show 1C folds. 1B folds occur in the competent layers only when the competence contrast is extremely large.

For certain types of non-parallel and non-similar folds the dip isogons cannot be drawn for certain segments of the folded layer; t_α and T_α remain undefined for these segments.

The variation of t'_α (or T'_α), which characterizes the morphology of a fold, depends on the choice of the reference line with respect to which α is measured. In the simplest and the most common case (Case 1), as described above, the trace of the axial surface is perpendicular to the trace of folded surface at both the hinge points on the profile plane. This condition may not hold good for all folds (Hudleston 1973a, p.11).

In certain folds (Case 2), the axial trace on the profile plane is not at a right angle to the folded surfaces at the hinge points, although the angle between the axial trace and the folded surface is the same at the hinges of both the inner and the outer arcs. In the third case considered by Hudleston, the axial trace makes different angles with the fold surfaces at the two hinge points. For the case 2, $t_0 \neq T_0$. Hudleston suggested that, for these folds, the reference line for measuring α on the profile plane should be chosen as the tangents to the folded surfaces at those points where t_α is a maximum and is equal to T_α, and where the dip isogon is normal to the folded surface. The case 3 folds, which are extremely rare, cannot be incorporated in this scheme of fold-classification.

12.7. The shape of a folded surface

If a folded surface in cross-section is represented by the function $y = f(x)$, then the shape of any periodic symmetric fold can be represented by taking a sufficiently large number of terms of a Fourier series of the form

$$y = a_1 \sin x + a_3 \sin 3x + a_5 \sin 5x + \ldots \qquad (12.2)$$

where $a_1 > a_3 > a_5 > \ldots$. For example, let six units of length represent half the wavelength of a fold and let a_1, a_3 and a_5 be 6.10, 0.77 and 0.22 units of length. The curve 1 in Fig. 12.25 has been drawn for the equation $y = 6.10 \sin x$. Similarly, curve number 2 and curve number 3 have been drawn from the equations $y = 0.77 \sin 3x$ and $y = 0.22 \sin 5x$. Curve 4 has been drawn by adding the y-coordinates of each of these curves. From the figure we can see that within the space of a single half-wave of the first harmonic represented by the first term of the series of eqn. (12.2), there are three half-waves for the third harmonic and five for the fifth. The changes brought about by the successive later terms of the series of eqn. (12.2) become smaller and smaller and, in practice, a curve can be represented with sufficient accuracy by taking a small number of terms.

To simplify the method Stabler (1968) suggested that we consider only a quarter wavelength (b) of a fold, i.e. the segment of a fold between an inflection point and the adjacent hinge point (Fig. 12.26) and measure the amplitude (a), the quarter wavelength (b) and the y-coordinates of two other points for $x = 1/3b$ and $x = 2/3b$. Substituting these values in eqn. (12.2) and taking only three terms of the series we get the following three equations with the three unknown quantities, a_1, a_3 and a_5:

$$a_1 \sin 30° + a_3 \sin\ 90° + a_5 \sin 150° = y_1$$
$$a_1 \sin 60° + a_3 \sin 180° + a_5 \sin 300° = y_2 \qquad (12.3)$$

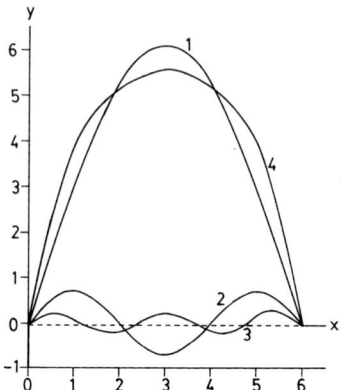

FIG. 12.25. Curve 1 represents the equation $y = a_1 \sin x$ with $a_1 = 6.1$ units of length and the scale is so chosen that a half wave of 6 units of length represents π. Curves 2 and 3 are drawn from the equations $y = 0.77 \sin 3x$ and $y = 0.22 \sin 5x$. Curve 4 is drawn from the equation $y = a_1 \sin x + a_3 \sin 3x + a_5 \sin 5x$.

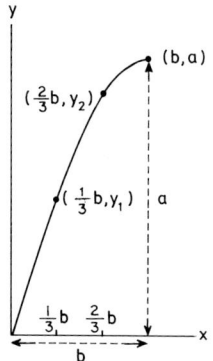

FIG. 12.26. A quarter wave of a folded surface between an inflection point and an adjacent hinge point. The two points lying on the folded surface have the coordinates $(1/3b, y_1)$ and $(2/3b, y_2)$, where b is the quarter wavelength.

$$a_1 \sin 90° + a_3 \sin 270° + a_5 \sin 450° = a$$

where the quarter wavelength of the fold is taken to be equal to $\pi/2$. Equation (12.3) can be simplified to

$$a_1 + 2a_3 + a_5 = 2y_1$$
$$a_1 - a_5 = (2/\sqrt{3})\, y_2 \qquad (12.4)$$
$$a_1 - a_3 + a_5 = a.$$

238

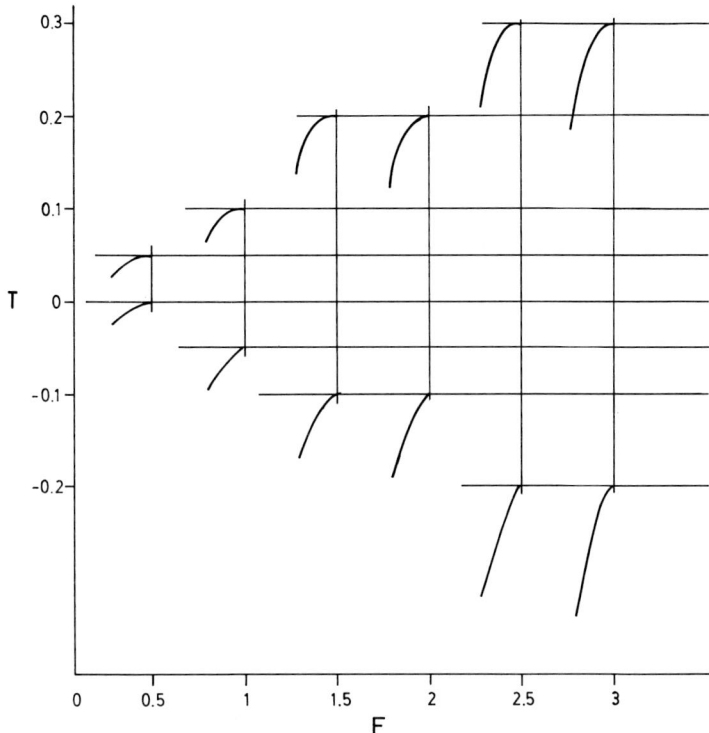

FIG. 12.27. $F-T$ diagram for plotting the shape of a folded surface. After Stabler 1968.

The solution of these equations gives us

$$a_1 = \tfrac{1}{3}(a + y_1) + 0.577\, y_2$$
$$a_3 = \tfrac{1}{3}(2y_1 - a) \qquad\qquad (12.5)$$
$$a_5 = \tfrac{1}{3}(a + y_1) - 0.577\, y_2.$$

Stabler suggested that most folds can be approximately represented by its first and third harmonics, viz. F and T values, where

$$F = a_1/b \text{ and } T = a_3/b.$$

By taking F and T as coordinate axes, a fold can then be plotted as a point on the F–T diagram (Fig. 12.27). Increasing F-values indicate increasing values of a/b. Comparatively larger positive values of T/F are shown by folds with comparatively rounded hinges. Negative values of T/F are shown by comparatively sharp-hinged folds.

Calculation of a_1 and a_3 by this method may cause significant errors. To reduce these errors we should have a larger number of sampling

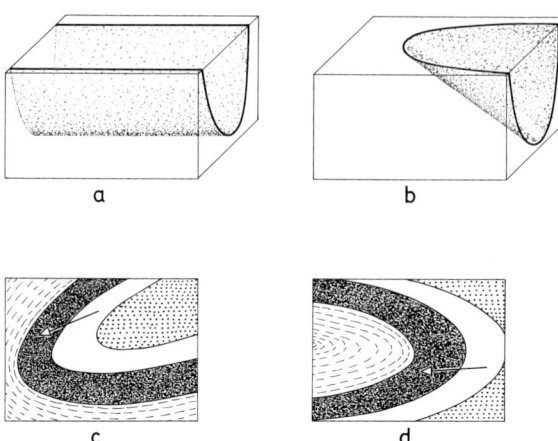

FIG. 12.28. (a) Block diagram of horizontal fold showing straight and parallel limbs on the horizontal face. (b) Plunging fold showing curved outcrop on the horizontal face. (c) Outcrop of plunging antiform, with plunge direction towards the convex side of the fold closure. (d) Outcrop of plunging synform, with plunge direction towards the concave side.

points and a larger number of terms of the Fourier series. For calculating a_1, a_3 etc., Hudleston (1973a) proposed equations which are based on the trapezoidal rule of numerical integration. According to Hudleston, the first and the third harmonics are sufficient to represent the gross characteristics of a folded surface although more than two harmonics may be required for a proper representation of the shape. It should also be noted that circular arcs and sharp-hinged folds require more than two harmonics for a fairly faithful representation (Hudleston 1973a, Stowe 1988). The Fourier series represented by eqn. (12.2), even with a large number of terms cannot be applied for folds with relative limb dips greater than 90°. The shapes of folds, can also be represented in terms of the arclength and the relative limb dip. This method (Chapple 1968) has the advantage that all folds, including fan folds, can be analysed. The only disadvantage of this method is that it is difficult to measure the arc length with sufficient accuracy.

12.8. Outcrop patterns of folds

Consider the traces of a stack of folded layers on a section through a fold. If the fold is cylindrical then the traces of all the folded surfaces will appear as straight lines on any section parallel to the fold axis. On a section oblique to the fold axis the traces of folded surfaces will be curved. For a horizontal fold, therefore, the outcrops of the limbs will be straight and parallel. If the fold plunges its outcrop will be curved (Fig. 12.28).

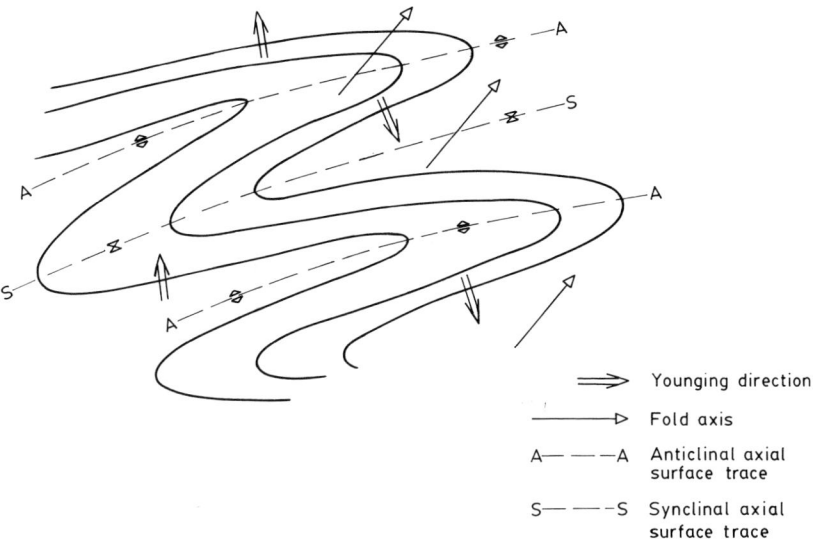

	Younging direction
	Fold axis
A— — —A	Anticlinal axial surface trace
S— — —S	Synclinal axial surface trace

FIG. 12.29. Plotting of younging direction to distinguish anticline and syncline.

The geometry of a cylindrical fold is completely defined by its map pattern and the orientation of the fold axis. The geometry can be accurately determined by constructing a transverse profile. We can obtain some important information even without such a construction. Thus, we can recognize an antiform when the fold axis plunges towards the convex side of a fold closure and a synform can be recognized when the fold axis plunges towards the concave side (Fig. 12.28c, d). Where the younging directions can be plotted in the map they enable us to distinguish anticlines and synclines depending on whether the young-ing is away from or is towards the core (Fig. 12.29). We should pay special attention to the angle between the axial trace and the trend of the fold axis (Fig. 12.30). The axial trace and the trend of the fold axis will be parallel in two situations:

1. The fold axis is horizontal (Fig. 12.30a).
2. The axial plane is vertical (Fig. 12.30b).

If the fold shows a closure on a horizontal section and shows at the same time a parallelism of the axial trace and the trend of the fold axis, we can then conclude that it is a plunging upright (with vertical axial plane) fold. In the same way, we can identify inclined folds when the fold trend is at an angle to the axial trace (Fig. 12.30c). In the map the large-scale reclined folds can be easily distinguished by the fact that in such folds the trend of the fold axis will be nearly at a right angle to the

241

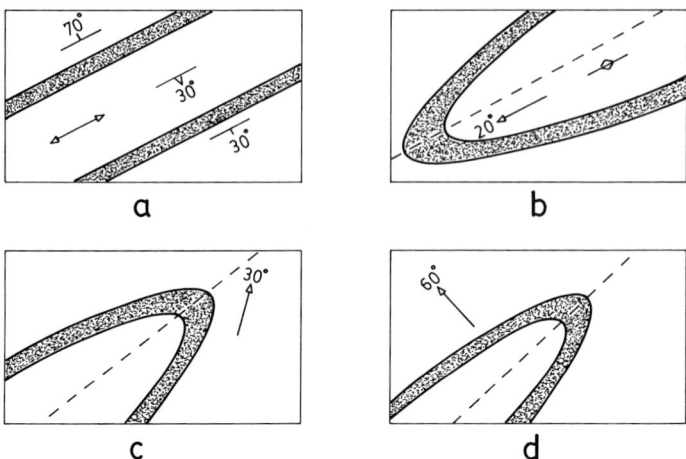

FIG. 12.30. The axial surface trace and the trend of the fold axis are parallel when (a) the fold is horizontal and (b) when the fold is upright. In a plunging inclined fold, as in (c), the axial surface trace is at an acute angle to the fold trend. (d) In a plunging neutral fold (reclined or recumbent) the fold trend is at a right angle to the axial surface trace.

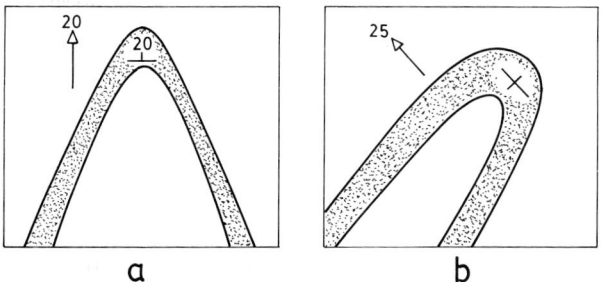

FIG. 12.31. Maps of a plunging upright fold and of a plunging reclined fold. (a) In the upright fold the dip is minimum at the hinge. (b) In the reclined fold the dip becomes vertical at or near the hinge.

axial trace (Fig. 12.30d). However, if the plunge of the fold axis $< 10°$, the fold should be called a recumbent fold.

In an upright fold the lowest dip is found at the hinge. Moreover, the dip amount will be roughly equal to the amount of plunge of the fold axis. In a reclined or a recumbent fold, on the other hand, there should be vertical dip in the neighbourhood of the hinge zone (Fig. 12.31).

A useful method of interpreting the large-scale structure from the geological map is that employed by Bailey & McCallien (1937) to interpret the "Schichallion twist" in Perthshire. The method (see also

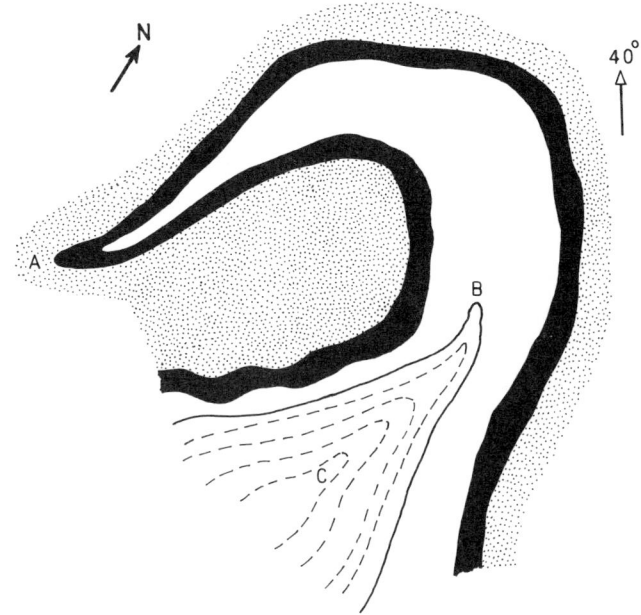

FIG. 12.32. Holding the map, with the plunge direction upward, the map gives a rough idea of the geometry of the fold. The map shows a non-plane cylindrical fold. The fold is reclined near *A*, upright near *B* and inclined near *C*. The fold is nearly isoclinal in most parts.

Mackin 1950) consists in holding the map while keeping the plunge direction upward in the map. The map then gives a somewhat distorted picture of the transverse profile. Although the exact forms of the folds are not obtained by this simple method, it can be employed to identify antiforms, synforms, neutral folds, as well as upright and inclined folds (Fig. 12.32). Evidently the method can be employed when the folds are plunging and cylindrical and when the effect of topography on the outcrop pattern is small.

The outcrop pattern of a fold changes with a change in the orientation of the axis in different parts of a fold. The fold may show an *arcuation* (Bucher 1933) of the fold axis, i.e. a change in the trend of the fold axis. The outcrop pattern then shows a curved trace of the axial surface (Fig. 12.33a). It should be noted, however, that a similar outcrop pattern may also be produced by non-plane cylindrical folds, i.e. in folds with a curved axial surface but with a more or less constant orientation of the fold axis (Fig. 12.33b). Figures 12.33c to 12.33f show some other types of fold outcrops in which the orientations of the axis is different in different parts, although in each of these cases the axial surface is planar and vertical. Figures 12.33c and d show the charac-

243

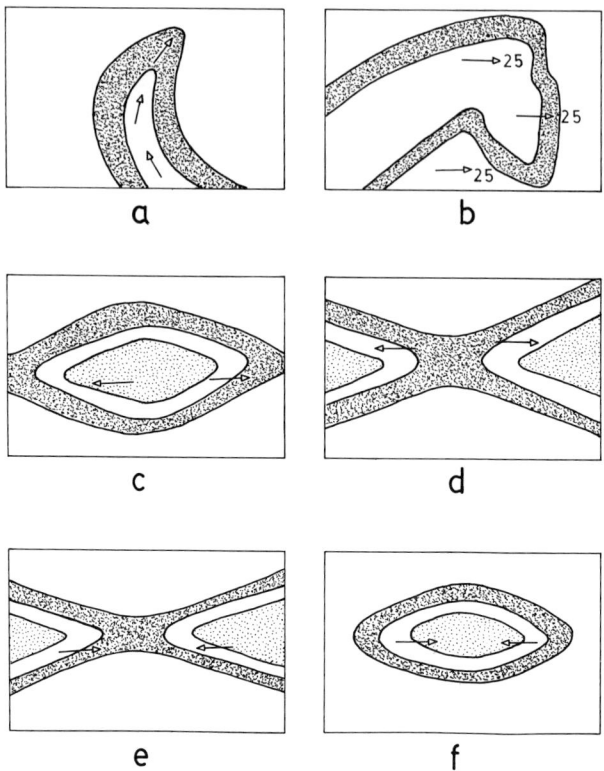

FIG. 12.33. Different types of outcrop patterns of folds on a flat topography. (a) Curved axial surface trace resulting from changing trend of fold axis. (b) Curved axial trace in a non-plane cylindrical fold. (c) Oval outcrop of antiform with axial culmination. (d) Outcrop of synform with axial culmination. (e) and (f) Outcrops of antiform and synform with axial depression.

teristic outcrops of an antiform and a synform with a zone of *axial culmination* (a zone of upward bending of the hinge line) running across each of them. Figures 12.33e and f show the outcrops of an antiform and a synform each traversed by a zone of axial depression (a zone of downward bending of the hinge line). Note that in each of these four outcrop patterns the sense of the fold closure is reversed when the plunge direction is reversed.

12.9. The axis of macroscopic folds

The attitudes of the axes of mesoscopic folds are generally determined by measuring the attitudes of the hinge lines of the folds (Fig. 12.34a). Where the bedding laminae and the axial plane cleavage are both prominent the fold axis can be conveniently determined by

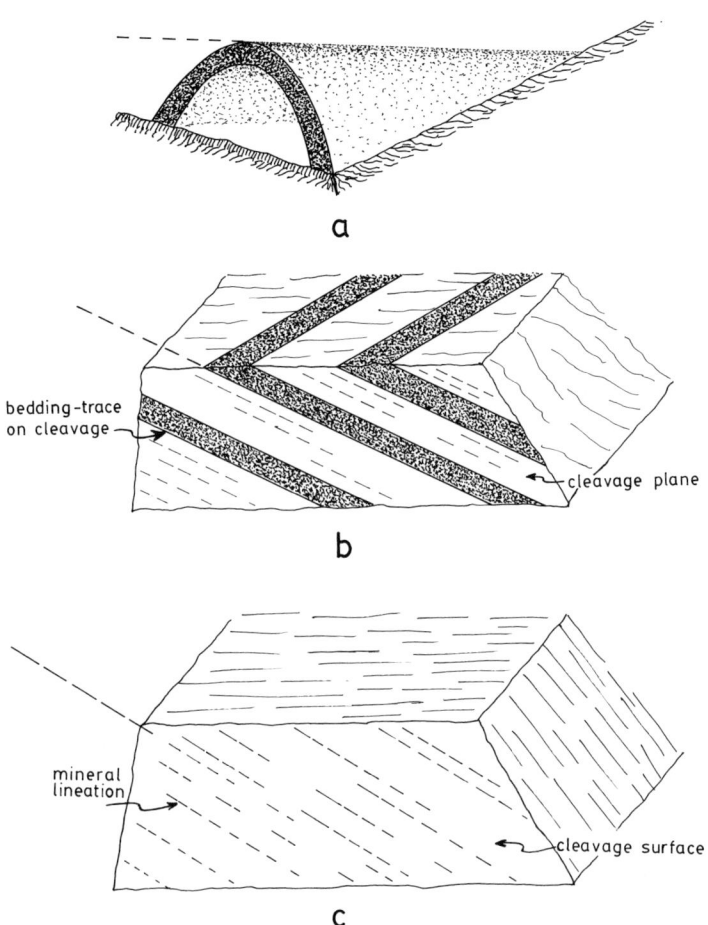

FIG. 12.34. Structural elements from which the attitude of fold axis can be measured. (a) Hinge line (dashed line) of mesoscopic fold, (b) trace of bedding on cleavage, (c) mineral lineation, if it is independently known from critical outcrops that the lineation is parallel to the fold axis.

measuring the bedding–cleavage intersection (Fig. 12.34b). The folded surfaces in many areas contain a prominent lineation (such as a mineral lineation or a striping lineation) parallel to the fold axis. The attitude of this lineation can then be taken as the attitude of fold axes even in outcrops where the fold hinges are not exposed or the folds themselves are not visible (Fig. 12.34c).

The attitude of the axis of a large-scale fold can be determined by either of the following two methods or preferably by a combination of them. The first method is based on the assumption that the axes of the mesoscopic folds are approximately parallel to the axes of the macro-scopic folds of the same generation. The axes of the mesoscopic folds are plotted in either stereographic or equal area projection. If the majority of the points fall close together, then the point of their maximum concentration can be taken as the axis of the large-scale folds of the same generation.

To apply the second method to determine the attitude of the axis of a macroscopic fold, we need to have a fairly large number of measure-ments of the attitude of the folded surface (bedding or foliation) from different parts of the fold. This S-plane, whose attitude we measure at any point of the fold, must be a tangent plane to the folded surface at that point. If the fold is cylindrical its tangent planes at different parts will intersect along lines parallel to the axis of the fold (Fig. 12.35). The normal to these tangent planes will be differently oriented but each will be at a right angle to the fold axis. We therefore have the following two methods of determining the fold axis from the data of S-plane orien-tations. In one of the methods we plot the S-planes as great circles in equal area projection (Fig. 12.36a). If there are n planes the number of intersection points will be $\frac{1}{2}n(n-1)$. The point of maximum concen-tration of these intersections is called the β-axis (Fig. 12.36b). A natural fold is never strictly cylindrical. The larger the fold the more does it deviate from the shape of an ideal cylindrical fold. A large-scale fold is cylindrical only in a statistical sense. Hence even a few aberrant data may give rise to a comparatively large number of spurious intersections. Thus, if we plot 30 readings of S-planes out of which 3 are slightly aberrant, these three planes may give rise to as many as 84 spurious points of intersection. For a large number of readings it is therefore a common practice to adopt the second method in which we plot the S-poles in equal area projection. If the fold is statistically cylindrical the poles will lie on or near a great circle (Fig. 12.36c). This is called a π-circle which represents a plane normal to the statistically defined axis of the large-scale fold. The pole to the π-circle is also called the β-axis which, in an ideal cylindrical fold, is geometrically equivalent to the β-axis obtained from intersections of S-planes plotted as great circles. It may happen that the π-poles are scattered in such a way that a clearly defined π-circle cannot be drawn. This will indicate that, for the domain as a whole, the large-scale fold is non-cylindrical.

12.10. Construction of a transverse profile

To represent the geometry of a large-scale cylindrical fold it is often required to construct a tectonic profile, i.e. a section perpendicular to the fold axis. For a plunging fold this can be done in the following way (Fig. 12.37).

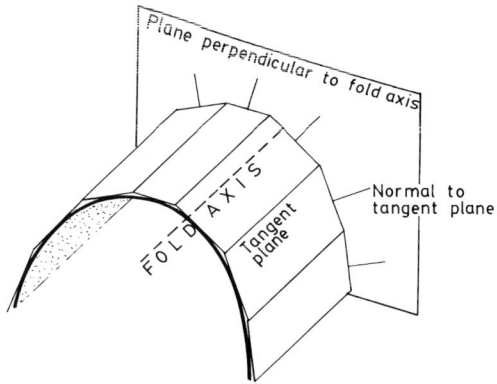

Fig. 12.35. The tangent planes in different parts of a cylindrical fold intersect along lines parallel to the fold axis. The normals to these planes are parallel to the transverse profile of the fold.

1. Take two mutually perpendicular lines OX and OY on the map perpendicular and parallel to the trend of the fold axis. It is convenient to place the lines outside the mapped region whose profile we are going to construct. Note that both OX and OY are on a horizontal plane.

2. From O draw a line OA making an angle (θ) with the line OY equal to the plunge of the fold axis. From any point B on OA draw a perpendicular and let it meet OY at C. OAC is then a vertical plane parallel to the fold axis. The line OC is its strike, the line OA on this plane is the fold axis and the line BC is the trace of the profile plane.

3. Take any point (P) on a folded surface in the map. If OX and OY are taken as the coordinate axes with the origin at O, then the perpendicular distances of P from OY and OX are the X and Y coordinates of P.

4. Let Q be the foot of the perpendicular from P on the axis OY and let P' be the foot of the perpendicular from Q on the line BC.

5. In a separate diagram take coordinate axes OX' and OY' lying on the profile plane. This is a plane which passes through the horizontal line OX of the map and is inclined to the horizontal plane at angle ($90°-\theta$) towards a direction opposite to the plunge direction of the fold axis. Plot a point P' with coordinates $X' = X$ and $Y' = BP'$. P' is the projection of the point P on the profile plane. Instead of the geometrical construction to obtain the value of Y' we can simply obtain it from the equation $Y' = Y \sin\theta$, where θ is the plunge of the fold axis. If the ground surface is uneven this equation has to be modified to

$$Y' = Y \sin\theta + h \cos\theta$$

where h is the altitude of the point P.

6. In a similar way take other points on the trace of the folded surface on the map and project each of them on the profile plane. The

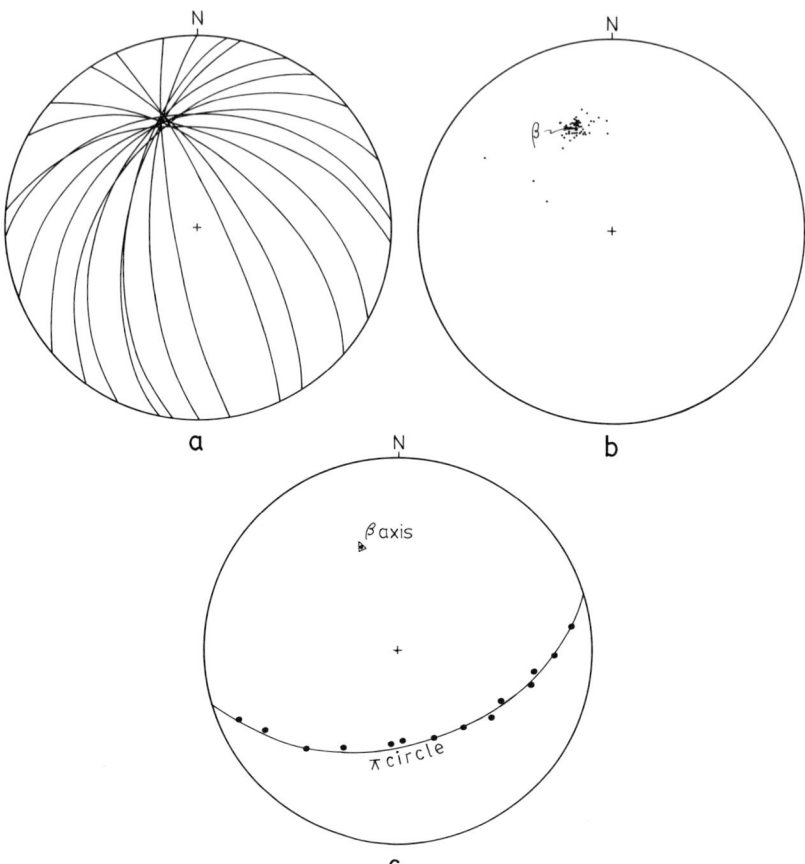

Fig. 12.36. (a) Great circle traces of beds from different parts of a fold in equal area projection. (b) The points of intersection from (a) are plotted in a separate diagram. The maximum concentration is taken as the β-axis. (c) The poles to the S-surfaces lying along a great circle (the π-circle). The normal to the π-circle is the β-axis.

true profile of the folded surface will be obtained by joining these points by a smooth curve.

12.11. Orientation of axial plane

In addition to noting down the general shape of a mesoscopic fold and the orientation of its hinge line, the orientation of its axial plane should be measured wherever possible. To measure the orientation of the axial plane we can adopt one of the following methods.

1. If the fold is isoclinal the orientation of its limb gives us the orientation of the axial plane.

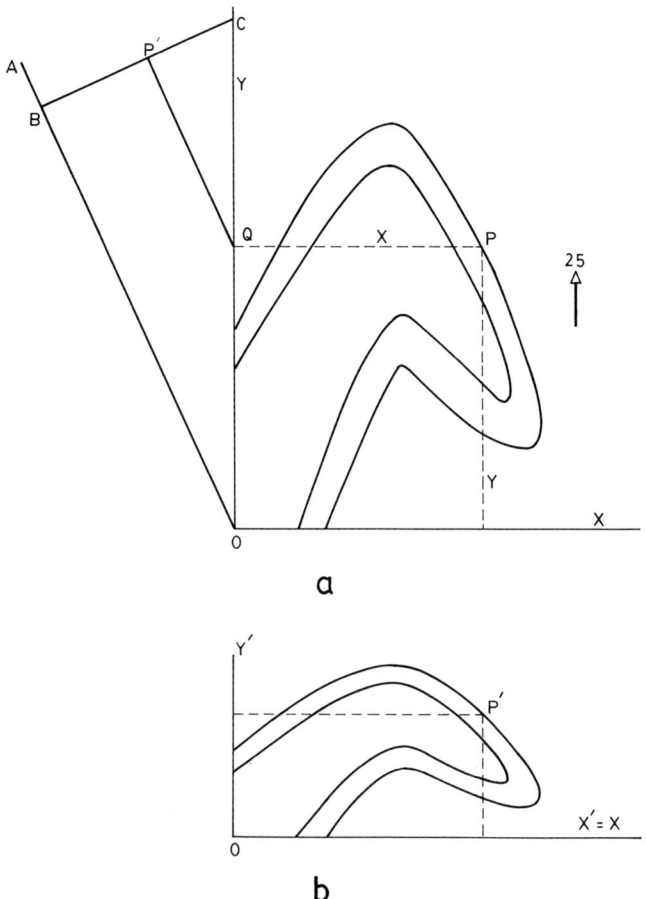

FIG. 12.37. Method of construction of transverse profile. (a) Map of folded bed. *COA* is a vertical section parallel to the fold axis. The angle *COA* is 25°, the plunge of the fold axis. *CB* is the trace of the transverse profile on the longitudinal section. (b) Transverse profile of the fold in (a).

    2. The axial plane of a fold can also be determined by measuring the attitude of the axial plane cleavage. If the cleavage shows a distinct fanning or refraction within the domain of a fold, then we should try to find an exposed cleavage surface at or near the hinge of the fold. Its attitude will be essentially parallel to the axial plane.

    3. For a mesoscopic fold with an exposed hinge line it is convenient to measure the attitudes of the hinge line and that of the axial surface trace on a more or less planar outcrop face. These are then plotted in stereographic projection. The attitude of the axial plane is obtained from the great circle passing through these two points.

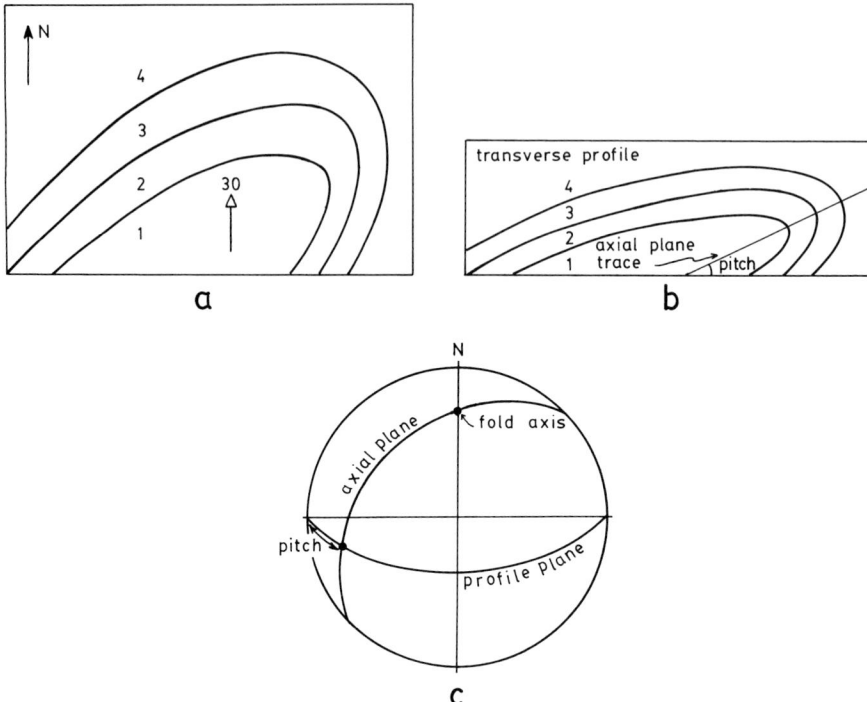

FIG. 12.38. Graphical method of determining the attitude of the axial plane from the attitude of the fold axis and the outcrop pattern of a plunging fold on a flat topography. (a) Map pattern. (b) Transverse profile with axial surface trace on it. (c) Plotting of axial plane in stereographic projection.

It is important to note that the line joining the points of maximum curvature in an outcrop of a fold may not lie on the axial surface (Badgley 1959, Schryver 1965). This line on the map may coincide with the axial surface trace in some special circumstances, e.g. when the axial plane is vertical or the fold is a sharp-hinged fold. In other situations the reconstruction of the axial plane of a large-scale fold from the orientation of the fold axis and of the line joining the points of maximum curvature in the map (on a flat topography) may involve some error. To avoid this we should prepare a transverse profile, measure the pitch of the axial surface trace on it and determine the attitude of the axial plane from stereographic projection (Fig. 12.38).

CHAPTER 13

Mechanism of Folding

13.1. Three types of folds

Depending on the mechanism of folding, the natural folds can be classified into three broad categories: (1) *buckle folds*, (2) *bending folds* and (3) *passive folds* (Ramberg 1963b, Hudleston 1986). Depending on the mechanism, folds can also be classified into two broad groups of *flexure folds* and *shear folds*. Buckle folds result from an instability when a layer or a stack of layers is subjected to a layer-parallel compression. A contrast in competence among the associated layers is essential for buckling. The presence of layers of contrasting competence produces a mechanical anisotropy. Buckle folds can also develop in a cleaved rock which has a planar anisotropy in the form of a cleavage but no distinct layering (Biot 1965a, Cobbold *et al.* 1971). Bending folds are produced by transverse forces acting across the layers. The layers undergoing bending folds may or may not have competence contrasts. In rocks showing passive folds the layers do not have significant competence contrasts.

The three mechanisms can be easily distinguished by performing the following three simple experiments (Fig. 13.1): (1) Put a long thin ruler vertically on a table and gently push it downward. When the pressure is small the ruler will remain straight. As the pressure is slowly increased it will be found that at one stage it suddenly becomes curved. This abrupt change from straight to a curved form at a particular pressure is due to the development of an elastic instability. (2) For the second experiment put the ruler horizontally between two stacks of books and put a paper weight or a piece of rock in the middle of the ruler. The ruler will evidently bend downward. The bending folds in nature develop by similar action of lateral forces. (3) To illustrate the principle of passive folding take a rectangular block of modelling clay and draw two steep, parallel, gently curved lines with a marking pen on one of its vertical faces. The curved lines will behave in the same manner as the traces of passive marker planes. Put a piece of plywood on the top of the model and press vertically downward. The curvatures of the lines drawn on the vertical faces will increase. Note that the folding of these lines takes place without any competence contrast in the model and without any forces acting across the lines. The folding here takes place in an entirely passive manner.

The following discussion will mainly be concerned with buckle folding because buckle folds are much more common than the other types of folds. Another reason for a greater emphasis on buckle folds is that there are very few detailed studies on bending and passive folds.

There is a large volume of literature on buckle folding. The problem has been attacked by using theoretical, experimental and finite element methods. The problem is quite complex and only an elementary form of the theory is presented in the following sections. Any one interested to go deep into this problem may first get acquainted with the theory of elastic stability. A clear exposition of this theory is given by Timoshenko (1936) or Timoshenko & Gere (1961). The papers published by Biot (1961, 1964, 1965b, c, d) and Biot *et al.* (1961) in the *Bulletin of the Geological Society of America* summarize the geologically relevant aspects of his studies on buckling. A more detailed account of these studies has also been put together in his book (Biot 1965a). Along with Biot's work, Ramberg's studies on buckling in viscous materials laid a firm theoretical basis for our understanding of the mechanics of folding in natural rocks (Ramberg 1959, 1961, 1963a, b, c, 1964, 1967, 1968, 1970). Biot's and Ramberg's studies have been extended by several workers to elaborate different aspects of the problem of buckle folding (e.g. Bayly 1964, 1970, 1971, 1974, Chapple 1968, 1969, Chapple & Spang 1974, Cobbold 1975, Cobbold *et al.* 1971, Currie *et al.* 1962, Dietrich 1970, Dietrich & Carter 1969, Donath 1968, Fletcher 1974, Ghosh 1966, 1968, 1970, 1974a, Ghosh & Ramberg 1968, Honea & Johnson 1976, Hudleston 1973a, b, Hudleston & Stephansson 1973, Hudleston & Holst 1984, Johnson 1969, Johnson & Ellen 1974, Johnson & Honea 1975, Johnson 1977, Latham 1985a, b, Parrish 1973, Paterson & Weiss 1966, Ramberg & Stephansson 1964, Ramberg & Stromgård 1971, Ramsay 1962a, 1974, Ramsay & Huber 1987, Sherwin & Chapple 1968, Smith 1975, 1977, Treagus 1973, 1981, 1983, Treagus & Treagus 1981).

13.2. Buckling of a slender bar

The simplest way to understand the process of buckling is to consider the deformation of a straight slender bar under an axial load. Let *P* be the axial load and *h* the thickness of the bar. It will be assumed that (1)

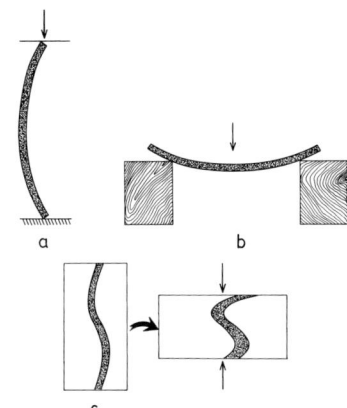

FIG. 13.1. Three mechanisms of folding: (a) buckling, (b) bending and (c) passive folding by accentuation of initial curvature.

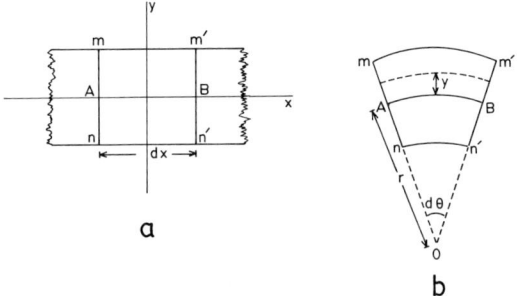

FIG. 13.2. (a) *mn* and *m'n'* are two transverse sections of a bar with *AB* as axis. (b) After buckling, *AB* remains a neutral surface. The strain along a fibre at a distance *y* from the neutral surface is *y/r*.

the material behaves elastically, (2) the amplitude of the buckles are very small and (3) there is no shear strain along the axis of the bar so that the cross-sections of the folded bar are everywhere perpendicular to the bar axis.

Consider a small segment separated by two cross-sections *mn* and *m'n'* at a distance *dx* apart, with the *x*-axis taken parallel to the bar axis and the *y*-axis perpendicular to it (Fig. 13.2). In the deformed state of the bar the cross-sections make an angle *dθ* with each other. If the radius of curvature of the middle surface *AB* is *r*, the arc length of *AB* is *rdθ*. The longitudinal strain of a fibre at a distance *y* from the middle surface is

$$\varepsilon_x = \frac{(r + y)d\theta - rd\theta}{rd\theta} = \frac{y}{r}.$$ (13.1)

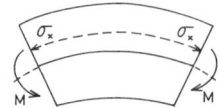

FIG. 13.3. Bending moment M on transverse sections of a bar.

The equation shows that on the outer arc of the curved bar, where y is positive, there will be an extension of the fibre, while on the inner arc, where y is negative, there will be a contraction. At $y = 0$, i.e. at the middle surface, the longitudinal strain will vanish. From eqn. (13.1) we have

$$\sigma_x = E\varepsilon_x = E . \frac{y}{r} \tag{13.2}$$

where σ_x is the local stress acting along the tangential direction and E is Young's modulus. From the distribution of the longitudinal stresses on one of the faces of the cut-out segment of the bar (Fig. 13.3) it is evident that there will be a bending moment (M) about the middle line AB. At any one point on this face the bending moment will be $M = \sigma_x . y$. The sum of the bending moments on the cross-section is

$$M = \int \sigma_x y da \tag{13.3}$$

where a is the area of the cross-section. From eqns. (13.2) and (13.3) we find that

$$M = \frac{E}{r} \int y^2 \, da.$$

The integral $\int y^2 da$, depending entirely on the shape of the cross-section, is represented by the symbol I and is known as the area moment of inertia. If b is the width of the bar $da = b . dy$. Therefore,

$$I = \int y^2 \, da = b \int_{-h/2}^{h/2} y^2 \, dy = \frac{bh^3}{12}. \tag{13.4}$$

The bending moment is then given by the expression

$$M = \frac{EI}{r}. \tag{13.5}$$

The quantity EI is known as the *flexural rigidity*.

Equation (13.5), usually attributed to Bernoulli and Euler, is the starting point for analyses of buckling. It states that at any point on the middle surface the bending moment is proportional to the curvature. In rectangular coordinates the curvature is given by the formula

$$\frac{1}{r} = \frac{-\dfrac{d^2y}{dx^2}}{\left[1 + \left(\dfrac{dy}{dx}\right)^2\right]^{3/2}}.$$ (13.6)

The negative sign before d^2y/dx^2 is justified when the slope dy/dx is positive and decreases with increasing x. Since, according to our initial assumption of small deflection, dy/dx is small, the square of dy/dx is negligible in comparison with unity and eqn. (13.5) can be written as

$$M = -EI \cdot \frac{d^2y}{dx^2}$$ (13.7)

where $y = f(x)$ represents the shape of the curved axis of the bar. Let us now specify that the bending moment is generated by an external force P acting along the bar axis. At any point on the bar axis, the bending moment is then

$$M = Py.$$ (13.8)

From eqns. (13.7) and (13.8) we have

$$Py = -EI\frac{d^2y}{dx^2},$$

or,

$$\frac{d^2y}{dx^2} + \frac{P}{EI}y = 0.$$ (13.9)

The general solution of this linear second order differential equation is

$$y = c_1 \sin\left(\sqrt{\frac{P}{EI}}x\right) + c_2 \cos\left(\sqrt{\frac{P}{EI}}x\right).$$ (13.10)

The particular shape of the curved axis of the bar will depend on how the ends of the bar are fixed. Let us consider the case in which the bar-ends are hinged, so that the cross-sections are free to rotate. Taking the origin of the coordinates at one end of the bar, and noting that there is no deflection at the bar ends, we find that the deflection curve (Fig. 13.4a) should conform with the following two conditions:

(1) $y = 0$ at $x = 0$,
(2) $y = 0$ at $x = L$,

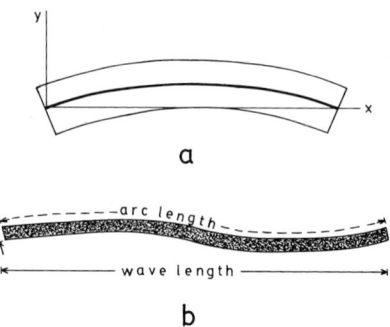

FIG. 13.4. Deflection curve of the middle line of a bar, with $y = 0$ at $x = 0$ and $y = 0$ at $x = L$, where L is a half-wave. (b) In a fold with very low amplitude the arc length is approximately the same as the wavelength.

where L is the length of the bar. Substituting the first condition in eqn. (13.10) we find that $c_2 = 0$. The second boundary condition is satisfied when

$$\sqrt{\frac{P}{EI}} \cdot L = n\pi \qquad (13.11)$$

where n is an integer. The shape of the deflection curve is thus

$$y = c_1 \sin\left(\sqrt{\frac{P}{EI}}x\right) \qquad (13.12)$$

where c_1 is the amplitude of the sine curve. The axial load, as given by eqn. (13.11), is

$$P = n^2 \frac{\pi^2 EI}{L^2}.$$

The minimum load at which buckling can take place is obtained when $n = 1$. This is known as the critical load

$$P_{\text{crit}} = \frac{\pi^2 EI}{L^2}. \qquad (13.13)$$

Below this critical load the straight form of the bar is stable. At or above this load the straight form becomes unstable. This simple derivation not only gives an insight into the phenomenon of buckling, it also shows why a symmetrical gently curved form derived by buckling of a long straight unit can be represented by a sine curve.

13.3. Differential equation of deflection of a thin elastic plate

Although the derivation of the buckling equation for a thin plate is somewhat more complex, the basic principles and the initial assumptions are the same as in the case of buckling of a slender bar. Let us consider a thin elastic rectangular plate of thickness h, with x- and y-coordinate axes chosen along the middle surface and with z-axis normal to it. Let the buckling of this plate take place under the action of a plate-parallel force, $P_x (= \sigma_x h)$, along with a lateral load q acting along the z-direction. It is assumed that P_x remains parallel to the middle surface even when the plate is deformed to a slightly curved cylindrical form. P_x is then no longer perpendicular to z-axis. Hence its projections along the z-axis will not vanish. For equilibrium the sum of all the forces along the z-direction must be equal to zero. The differential equation of deflection of the cylindrically curved middle surface of the plate is obtained from this condition:

$$D\frac{\delta^4 w}{\delta x^4} + \sigma_x h \frac{\delta^2 w}{\delta x^2} - q = 0 \qquad (13.14)$$

where w is the deflection of the middle surface of the plate in the z-direction and D is known as the flexural rigidity of the plate. For an incompressible material D is given by the following formula

$$D = \frac{1}{3}\mu h^3 \qquad (13.15)$$

where μ is the shear modulus of the plate.

Buckling of the plate can take place even without the lateral load q. As in the case of a bar, if the plate has hinged ends and is symmetrically folded its middle surface can be approximated by a sine curve

$$w = w_0 \sin lx \qquad (13.16)$$

where w_0 is the amplitude and $l = 2\pi/L$, L being the wavelength. Substituting the expression for w in eqn. (13.14), taking $q = 0$, and carrying out the differentiation we find

$$D . w_0 l^4 \sin lx - \sigma_x h w_0 l^2 \sin lx = 0$$

or,

$$Dl^2 - \sigma_x h = 0.$$

Replacing D by (13.15) we find

$$\sigma_x = \frac{1}{3}\mu\, h^2 l^2. \tag{13.17}$$

This is the critical buckling stress.

13.4. Buckling of an embedded layer

The problem of buckling of a thin layer embedded in another medium was solved by Biot (1957, 1961, 1965a) and Ramberg (1963a, 1964, 1967, p. 72). An approximate solution of this problem can be obtained directly from eqn. (13.14). When an elastic layer is embedded in another elastic medium, the deflection of the layer is affected by a reaction of the embedding medium. This reaction is oppositely directed to the deflection w. Biot has shown that this lateral resisting force, q, is proportional to the deflection and can be approximated by the expression

$$q = -\, 4\mu_2 l w \tag{13.18}$$

where μ_2 is the shear modulus of the embedding medium. Equation (13.14) then takes the form

$$\frac{1}{3}\mu_1 h^3 \frac{\delta^4 w}{\delta x^4} + \sigma_x h \frac{\delta^2 w}{\delta x^2} + 4\mu_2 l w = 0$$

where μ_1 is the shear modulus of the layer.

Substituting the expression (13.16) for w in this equation and carrying out the differentiation we find, after simplification,

$$\frac{1}{3}\mu_1 l^3 h^3 - \sigma_x l h + 4\mu_2 = 0$$

or,

$$\sigma_x = \frac{1}{3}\mu_1 (lh)^2 + \frac{4\mu_2}{lh}. \tag{13.19}$$

It may be recalled that $l = 2\pi/L$ and hence $lh = 2\pi h/L$. Thus, the magnitude of σ_x in eqn. (13.19) will depend upon wavelength/thickness ratio, L/h. Buckling will occur at that value of L/h which makes σ_x a minimum. This value of L/h is obtained from the condition

$$\frac{\delta\sigma_x}{\delta(lh)} = 0. \tag{13.20}$$

From eqns. (13.19) and (13.20) we find

$$\frac{2}{3}\mu_1 lh - \frac{4\mu_2}{(lh)^2} = 0$$

or,

$$(lh)^3 = 6\mu_2/\mu_1$$

or,

$$\frac{L}{h} = 2\Pi_3 \sqrt{\frac{1}{6}\frac{\mu_1}{\mu_2}} \tag{13.21}$$

(Biot 1957).

So far we have considered only elastic materials. Natural rocks, however, undergo permanent deformation during folding. Although the rheology of natural rocks is quite complex, we may, as a first approximation, consider them as viscous liquids. The viscosity of natural rocks is very large and hence the strain rates are very slow. As a consequence the acceleration terms in the equations of viscosity may be neglected for our purpose. Moreover, if we restrict our attention to mesoscopic structures the body force will be negligibly small in comparison with the viscous forces. Under such conditions the buckling equations for incompressible viscous materials may be derived from those of the elastic system by replacing the strains by the strain rates and the shear modulus by the coefficient of viscosity. With these changes eqn. (13.14) can be written as

$$\frac{1}{3}\mu_1 h^3 \frac{d}{dt}\left(\frac{\delta^4 w}{\delta x^4}\right) + \sigma_x h \frac{\delta^2 w}{\delta x^2} - q = 0 \tag{13.22}$$

where

$$q = -4\mu_2 l \frac{dw}{dt}$$

and

$$w = w_0 \sin lx.$$

μ_1 and μ_2 in this equation are the viscosities of the layer and the embedding medium respectively. Substituting the expressions for q and w in eqn. (13.22) and carrying out the differentiations we obtain

$$\frac{1}{3}\mu_1 l^4 h^3 \frac{dw}{dt} - \sigma_x l^2 hw + 4\mu_2 l \frac{dw}{dt} = 0$$

or,

$$\frac{1}{w}\frac{dw}{dt} = \frac{\sigma_x}{\frac{1}{3}\mu_1 l^2 h^2 + \frac{4\mu_2}{lh}}$$

or,

$$\frac{dw}{w} = \frac{\sigma_x \, dt}{\frac{1}{3}\mu_1 l^2 h^2 + \frac{4\mu_2}{lh}}. \tag{13.23}$$

The solution of this equation yields

$$w = C e^{\left(\frac{\sigma_x t}{Q}\right)} \tag{13.24}$$

where C is a constant and

$$Q = \frac{1}{3}\mu_1 l^2 h^2 + \frac{4\mu_2}{lh}. \tag{13.25}$$

For the elastic system it was argued that, of all possible wavelengths, that which minimizes σ_x will prevail in the buckling mode. For the viscous system the argument is somewhat different. Here we assume that the viscous layer contains minute initial irregularities which grow in amplitude with the application of σ_x. Of all possible waves, those which grow fastest will dominate and will mask other waves. Equation (13.24) shows that, for any value of σ_x, the deflection w is a maximum when Q is minimum. The minimum of Q is obtained from the condition

$$\frac{dQ}{d(lh)} = 0.$$

Substituting in this equation the expression for Q as given by eqn. (13.25) and carrying out the differentiation we find

$$(lh)^3 = 6\frac{\mu_2}{\mu_1},$$

or,

$$L = 2\Pi h_3 \sqrt{\frac{1}{6}\frac{\mu_1}{\mu_2}}. \tag{13.26}$$

Equation (13.26) is the same as eqn. (13.21). It should be noted, however, that μ_1/μ_2 in eqn. (13.26) represents the ratio of viscosities between the

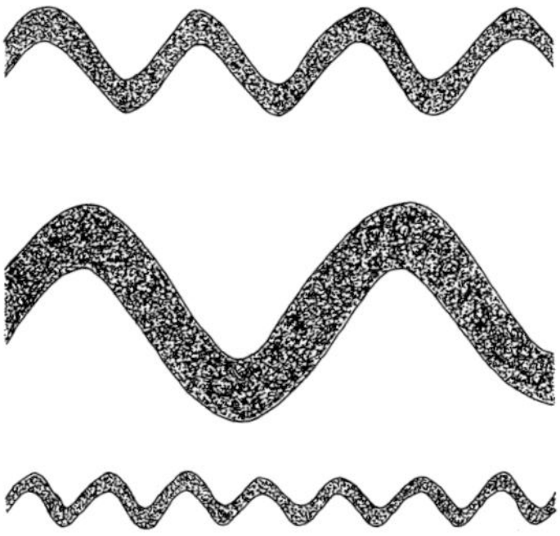

FIG. 13.5. Folds of different sizes in layers of different thicknesses in the same rock. In case of buckle folding, the ratio of arc length-to-thickness remains more or less the same.

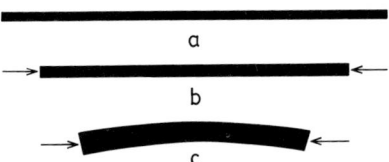

FIG. 13.6. (a) Undeformed layer. (b) Layer after layer-parallel homogeneous strain. (c) Layer after buckling. The total shortening is a result of both homogeneous strain and buckle shortening.

layer and its embedding material. The wavelength given by eqn. (13.26) is known as the *dominant wavelength* (Biot 1957) or *characteristic wavelength* (Ramberg 1964).

13.5. Geometry of single-layer buckle folds

The theory of single-layer buckling as given above is valid for a very small amplitude of the folds. L, as given by eqn. (13.26), does not therefore represent the wavelength of natural folds. For large values of μ_1/μ_2, the *arc lengths* of folds do not change much in course of progressive tightening of the folds. Since the arc length is very close to the wavelength in low-amplitude folds, the ratio of arc length-to-thickness (Fig. 13.4b) gives us an approximate estimate of the viscosity ratio as predicted by eqn. (13.26)

Single-layer folds in natural rocks are generally observed in quartz or calcite veins occurring in pelitic rocks or pegmatitic and aplitic veins in granitic terrains. The tight or isoclinal folds in the latter type of veins are sometimes described as ptygmatic structures. The arc lengths of the folds, as predicted by the theory, increases with increasing competence contrast as well as with increasing thickness of the vein. For the same pair of rocks, the ratio of arc length-to-thickness remains approximately the same (Figs. 13.5, 13.8a).

261

As compared to buckling folds in multilayers, single-layer folds, either in nature or in experiments, show a small range of shapes. For the same ratio of arc length-to-wavelength, the hinges of most single-layer folds are more rounded than that of a sine curve and less rounded than that of a true elastica. Chevron folds and conjugate folds do not normally develop by buckling of single layers.

Single-layer folds belong either to 1B or 1C class in Ramsay's fold classification. They do not show a prominent thickening of the hinge with reference to the limbs.

13.6. Single-layer folding with layer-parallel homogeneous strain

The theory of single-layer folding, as given above, was verified experimentally by Biot *et al.* (1961) for materials with high viscosity ratios. However, for low viscosity ratios there may be a significant departure from the theoretical results. Sherwin & Chapple (1968) measured the ratios of arc length-to-thickness from a large number of folds in thin quartz veins occurring in slates and phyllites. The mean relative arc lengths of these folds range between 4 to 6.8. For such small ratios of arc length-to-thickness, the Biot theory is not applicable. Buckling experiments with soft materials show that at low viscosity contrasts the competent layer undergoes a large amount of layer-parallel uniform shortening and a consequent thickening of the layer when the limb dip is very low (Fig. 13.6). When this layer-parallel homogeneous strain is taken into account the ratio of arc length-to-thickness of the final folds is given by the equation

$$\frac{L}{h} = 2\Pi_3 \sqrt[3]{\frac{1}{6}\frac{\mu_1}{\mu_2}\frac{\lambda_1 + 1}{2\lambda_1^2}} \tag{13.27}$$

where λ_1 is the quadratic elongation *in a direction perpendicular to the layering*. L and h are the arc length and thickness for the mature fold. It may be noted that eqns. (13.27) and (13.26) become identical when $\lambda_1 = 1$, i.e. when there is no layer-parallel strain. To apply this theory for specific field examples, λ_1 for the competent layer has to be independently determined. This is possible only in exceptional circumstances. For the population of folds studied by Sherwin & Chapple the viscosity ratios calculated from eqn. (13.27) ranged between 14 and 30. It should be noted that eqn. (13.27) cannot be applied to those folds in which there is a stretching or a shortening along the fold axis.

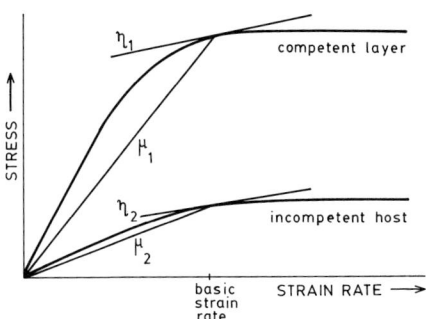

FIG. 13.7. Flow law for a non-Newtonian layer (upper curve) and its host (lower curve). μ and η are the two parameters which determine the non-linearity. After Smith 1979, fig. 2.

13.7. Single-layer folding in non-Newtonian materials

In the theory of single-layer folding, as given in the previous section, it was assumed that the layer and its embedding medium behaved as Newtonian liquids. However, there are reasons to believe that rocks often show a non-linear viscous rheology. Fletcher (1974) and Smith (1977, 1979) have shown that development of buckle folding is strongly influenced by the non-linear behaviours of the layer and its embedding medium. The behaviour of a Newtonian liquid depends upon a single parameter, the coefficient of viscosity μ. In the type of non-linear material considered by Fletcher and Smith, the behaviour of the material depends upon two parameters μ and η (Fig.13.7). The ratio μ/η ($= n$) is a measure of the non-linearity. For a Newtonian liquid $n = 1$. Let μ_1 and η_1 correspond to the competent layer while μ_2 and η_2 correspond to the incompetent material. Let $m = \mu_1/\mu_2$, $n_1 = \mu_1/\eta_1$ and $n_2 = \mu_2/\eta_2$. For large values of the viscosity ratio m, the ratio of dominant wavelength-to-thickness for a non-Newtonian material is (Smith 1979)

$$\frac{L}{h} = 2\Pi \left[\frac{1}{6} \frac{\mu_1/n_1}{\mu_2/\sqrt{n_2}} \right]^{\frac{1}{3}} \qquad (13.28a)$$

or,

$$\frac{L}{h} = 2\Pi \left(\frac{1}{6} \frac{\mu_1}{\mu_2} \right)^{\frac{1}{3}} \cdot \frac{n_2^{\frac{1}{6}}}{n_1^{1/3}}. \qquad (13.28b)$$

If $n_1 = n_2 = 1$, eqn. (13.28b) becomes the same as eqn. (13.26) for Newtonian liquids. Smith has shown that if μ_1/μ_2 is large and n_2 is not too large, the dominant wavelength increases with μ_1/μ_2 and eventually approximates the Biot wavelength. However, if n_2 is large and m is not too large, L/h lies between 4 and 6.

263

The main conclusions of this section may be summarized in the following way:

(1) Single-layer folds with large values of the viscosity ratio can be explained by the Biot–Ramberg theory of folding.

(2) Single-layer folds with short wavelengths (<8) are fairly common in nature. Such folds may have developed in Newtonian materials, provided, as shown by Sherwin & Chapple (1968), there is a very large amount of initial layer-parallel strain.

(3) Where L/h is small and there is very little layer-parallel shortening, the buckle folding must have been influenced by non-linear rheology.

13.8. Wavelength of large-scale buckle folds

During the development of large-scale folds, the upward movement of a layer in an anticline or the downward movement in a syncline must be affected by the pull of gravity. Under the effect of gravity the anticlines will tend to subside and the synclines will tend to rise. As a result the folds will tend to flatten out. This tendency will be opposed by the layer-parallel compressive force (Ramberg 1967, p. 65). Consider an elastic layer of thickness h floating on a viscous substratum of density ρ. Let $\Delta\sigma$ be the deviatoric stress necessary to produce stable elastic buckles in the layer. The buckling of the elastic layer under the joint action of the force of gravity and the lateral stress $\Delta\sigma$ was analysed by Smoluchowski (1909, 1910) and later by Ramberg & Stephansson (1964, also Ramberg 1967). The analyses of these authors show that the ratio, wavelength/thickness, is given by the following expression

$$\frac{L}{h} = \Pi \sqrt{\frac{2\triangle\sigma}{h\rho g}}, \tag{13.29}$$

while the deviatoric stress $\triangle\sigma$ is given by the expression

$$\triangle\sigma = \sqrt{\left[\frac{Eh\,\rho\,g}{3\,(1-v^2)}\right]}. \tag{13.30}$$

Here E and v are Young's modulus and Poisson's ratio of the elastic layer. Several experiments were performed by Ramberg & Stephansson (1964) to test this theory. The experiment consisted of buckling of elastic rubber sheets floating on either mercury or a saturated solution of potassium iodide. The experimental results agreed remarkably well with the theory.

The equations (13.29) and (13.30) show that a larger compressive stress is required to produce folds of greater wavelength. Since the deviatoric stress cannot exceed the strength of the rocks, equation (13.30) sets an upper limit to the size of buckle folds. Ramberg & Stephansson (1964) have considered the case of buckling of the earth's crust. A strong granitic rock has a strength of the order of 5×10^9 dynes/cm^2. Hence the maximum value of wavelength/thickness, according to eqn. (13.30), is

$$\frac{L}{h} = \Pi \sqrt{\left[\frac{5 \times 10^9}{3 \times 981 \times 15 \times 10^5}\right]} \approx 3.5,$$

with the crustal thickness as 30 km and substratum density as 3 g/cm^3. Thus, waves longer than 100 km cannot form by compression of the crust (Ramberg & Stephansson 1964, Ramberg 1967, pp. 75–76). This analysis is of considerable importance in theories of geotectonics, since it shows that the earth's crust cannot form very large buckle folds, i.e. folds of the order of the width of a geosyncline. The analysis of large-scale buckle folding has also been extended to the case of viscous layers and multilayers (Ramberg 1967).

13.9. Buckle folding of multilayers

There have been several theoretical and experimental studies on buckling of multilayers. As shown by these studies the wavelength of buckle-folded multilayers depends upon a number of factors, e.g. the number and thickness of competent layers, the spacing between the competent layers, the competence contrast among the layers and the competence of the medium confining the multilayer. Just as in the case of single embedded layers, the thicker and stiffer multilayers tend to produce larger folds (Fig. 13.8b).

Buckle-folded multilayers show a wider range of fold shapes than buckle-folded single layers. Thus, for example, multilayers, both in nature and in model experiments, may show round-hinged folds, chevron folds, kink bands and conjugate folds (Fig. 13.9).

When smoothly curved single-hinged folds develop in a multilayer composed of alternate competent and incompetent units, the competent units show either class 1B or class 1C fold geometry. In the incompetent units the folds show class 3 geometry (Fig. 13.10).

Experimental deformation of soft multilayered models enables us to identify some of the physical factors which control the shapes of folds. Depending on the spacing between the competent layers, two different buckling modes (Ramberg 1964) are possible. (i) If the spacing is large, each competent layer buckles more or less as an independent unit (Fig. 13.11a). The arc length of the folds is approximately the same as that of a single-layer fold. Such independent folding can take place only if the

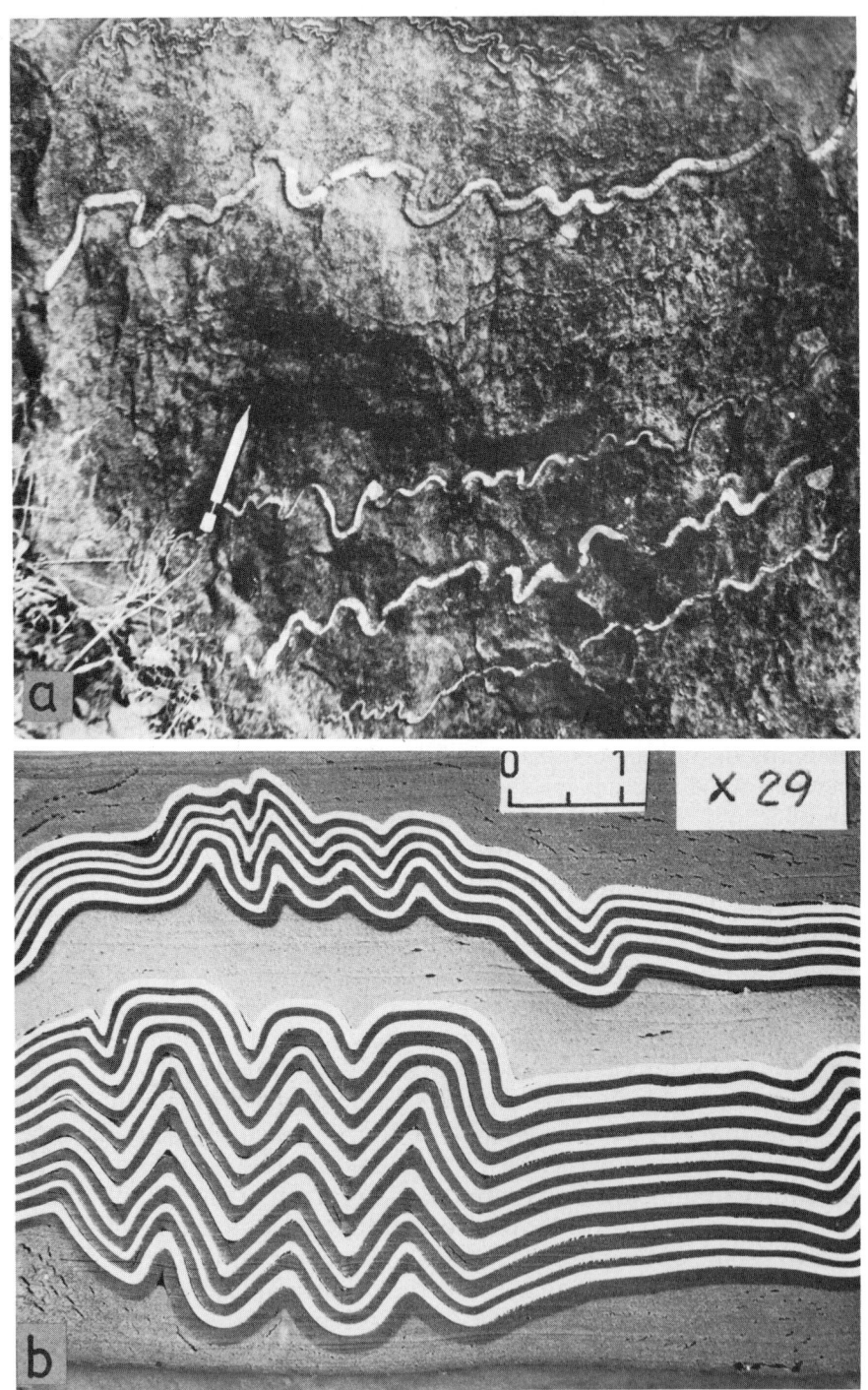

Fig. 13.8.

spacing between the competent units is considerably larger than the arc lengths of single-layer folds. (ii) For a close spacing of the competent units the folding of the multilayer is harmonious (Fig. 13.11b). For the rest of the discussion we shall be concerned with such harmonious folding in close-spaced multilayers.

A multilayer may be composed of alternate competent and incompetent layers or may have only competent layers separated by surfaces of discontinuity. In natural rocks a surface of discontinuity may, for example, be represented by a thin veneer of mica between layers of quartzite. The mechanical effect of a surface of discontinuity is that it enables the competent layers on either side of it to slide past each other. Thus, a multilayer may have a large, moderate or small competence contrast among the different layers or it may have greater or smaller ease of gliding.

The shape of folds in buckled multilayers is essentially controlled by two factors: (i) the competence contrast or ease of gliding within multilayer and (ii) the nature of the medium within which the multilayer is confined. As shown by experiments with analogue models, a variety of fold shapes (e.g. Figs. 13.8b, 13.12, 13.13) may develop under different combinations of these two factors (Ghosh 1968).

The arc lengths of folds are large (compared with the total thickness of the multilayer) when the confining medium is weak or very incompetent. With stronger confinement the arc lengths decrease. If the confining material is very competent the folds are much smaller than the total thickness of the multilayer (Fig. 13.12a). The amplitudes of the folds die out at the contact with the confining medium. This type of folding belongs to the category of *internal buckling* of Biot (1964, 1965a). Internal buckling may also take place if the total thickness of a multilayer is much larger than its individual members and is not too small with respect to the length of the multilayer. Biot has described such a situation as self-confinement.

Depending on the mechanical property of a multilayer and the nature of its confining medium the following cases of buckle folding may be distinguished.

(1) *Large competence contrast or great ease of gliding in multilayer*

(a) *Weak confinement*

The multilayer buckles into smooth single-hinge folds with arc lengths much larger than the total thickness (Fig. 13.13a). The

---

FIG. 13.8 (*opposite*). (a) Buckle-folded quartz veins of different arc lengths in phyllite. (b) Buckle folding of a thin and a thick multilayer embedded in painter's putty. The multilayers are composed of layers of modelling clay with greased interfaces. The thicker multilayer produces larger folds. From Ghosh 1968.

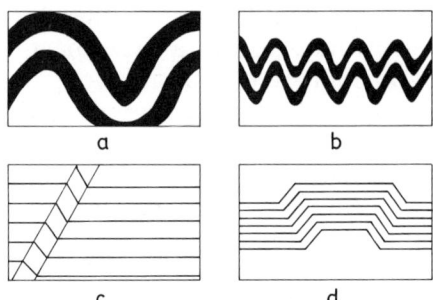

FIG. 13.9. A variety of fold shapes in buckle-folded multilayers: (a) round-hinged fold, (b) chevron fold, (c) kink band and (d) conjugate folds.

FIG. 13.10. Buckling folds in multilayers with class 1B or class 1C fold geometry in competent layers (black) and class 3 geometry in incompetent layers (white).

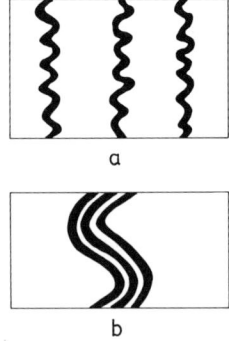

FIG. 13.11. (a) Buckling of multilayer with a wide spacing between the competent units; each layer buckles more or less independently. (b) Buckling of close-spaced multilayer with harmonious folding of each layer.

roundness of the fold hinges is generally retained even when the folds become tight or isoclinal. In case of an extreme tightening the hinges in the fold core may become sharp. However, this kind of folding generally does not give rise to a regular train of chevron folds.

(b) *Strong or moderately strong confinement*
The multilayer buckles into smooth single-hinged folds with arc lengths smaller than the total thickness. With progressive tightening the folds in the central zone of the multilayer become chevron-like with long straight limbs and narrow hinge zones (Fig. 13.8b). Towards the outer parts of the multilayer, i.e. on either side of the zone of chevron, folding the folds become cuspate with alternate roundish and sharp hinges. If the confining medium is very competent the folds die out at its margins.

(2) *Small competence contrast or restricted ease of gliding in multilayers*

(a) *Weak confinement*
Under a layer-parallel shortening the multilayer develops conjugate folds, the initial wavelengths of the conjugate folds being larger than the total thickness of the multilayer (Fig. 13.13b).

(b) *Strong confinement*
Under layer-parallel shortening the multilayer develops small kink bands or conjugate folds. With continued shortening the kink bands are progressively replaced by chevron folds (Fig. 13.12a).

There have been several experiments with a wide variety of materials to demonstrate that smooth-hinged folds form in multilayers with high competence contrasts or low friction between layers and that kinks and conjugate folds develop when the competence contrast is low or the interfacial friction is large. In Ghosh's (1968) experiments the modelling clay multilayers with oiled interfaces produced smooth single-hinged folds while those with greased interfaces gave rise to conjugate folds. Greszczuk (1974) performed buckling experiments with lamina-reinforced composites. The composites consisted of aluminium laminae interlayered with resin. When the Young's modulus of the resin was comparatively small (2500 psi and 22,900 psi) the folds had smoothly curved single hinges. However, when the resin had a large Young's modulus (62,380 psi) the specimens failed by development of kink bands. Similar results were also obtained by Honea & Johnson (1976) in rubber multilayers. When the dry rubber layers were in strong frictional contact with each other, the multilayer was deformed into kink bands. Smooth single-hinged folds were produced when the layer interfaces were lubricated.

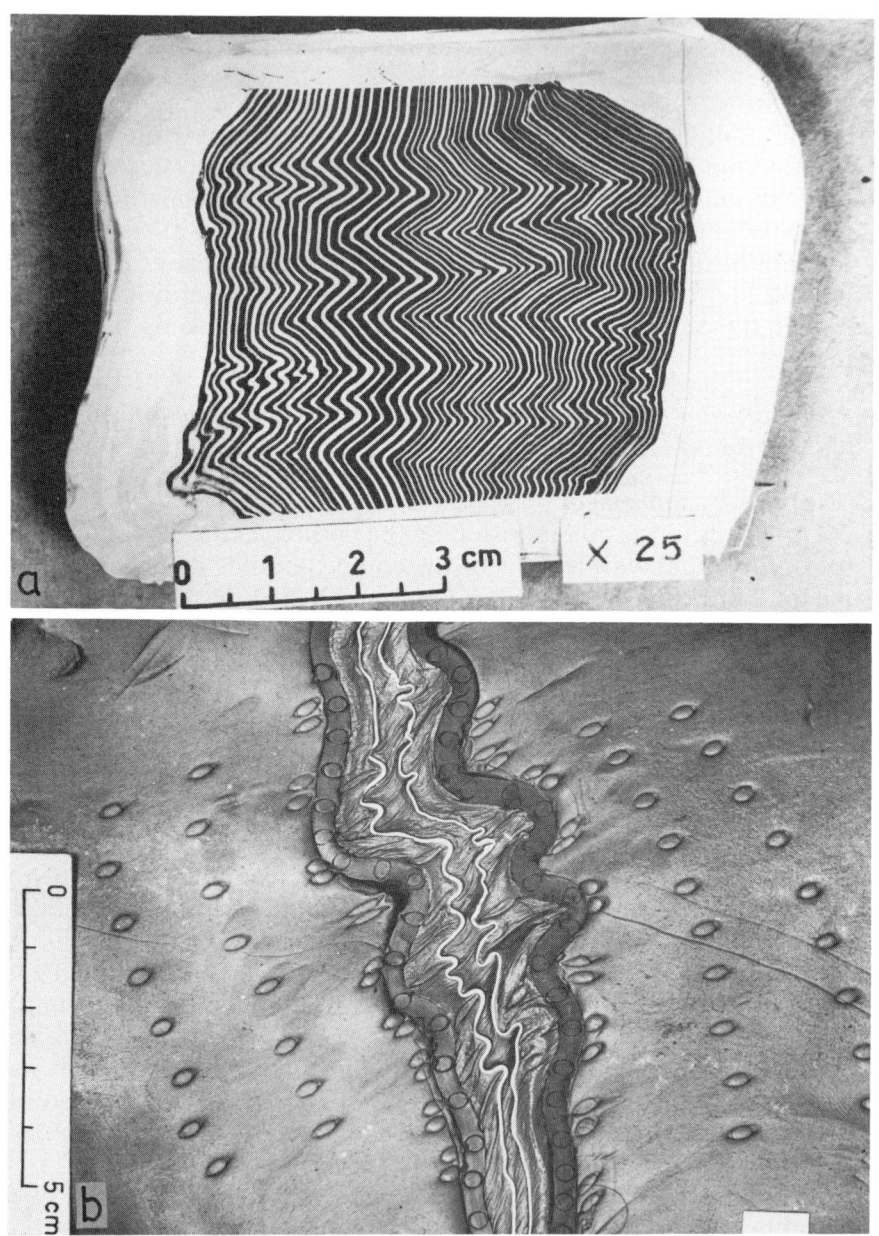

FIG. 13.12.

13.10. Strain in buckle-folded layers

The competent and incompetent layers undergo different types of strain during buckle folding. Although these different types of strain occur together, it is convenient to describe their separate effects under the following headings: (i) *layer-parallel homogeneous strain*, (ii) *tangential longitudinal strain* and (iii) *layer-parallel shear strain* (de Sitter 1958b, Ramberg 1961, 1964, Ramsay 1967).

From experiments of buckle folding with analogue models we find that there is a layer-parallel homogeneous shortening and consequent thickening of the layers in an initial stage of folding. The layer-parallel homogeneous strain is significant as long as the limb dip does not exceed 10–15°. This initial homogeneous strain (Fig. 13.14a) is essentially the same in both the competent and the incompetent layers.

The tangential longitudinal strain, depending upon the curvature of a folded surface, causes a layer-parallel extension on the outer part of a folded competent layer and a layer-parallel compression in its inner arc. In between these two surfaces there is a surface along which there is neither stretching nor compression (Fig. 13.14b). This is known as the *neutral surface*. The effect of tangential longitudinal strain may be seen by the development of radial tension cracks on the outer part and thrust faults towards the inner part of a competent layer. Since the curvature of a fold is maximum at the hinge and decreases to zero at the point of inflection, the absolute value of tangential longitudinal strain decreases from the hinge to the inflection point. It decreases also as it comes closer to the neutral surface.

The magnitude of layer-parallel shear strain is maximum at the middle of a layer and decreases towards the inner and the outer arcs (Fig. 13.14c). On any surface parallel to the layer the shear strain is a maximum at the inflection point and decreases to zero at the hinge.

The total shortening of a layer takes place by a combination of external rotation and layer-parallel homogeneous strain (Fig. 13.15). The ratio of these two strain rates depends on the competence contrast and on the amplitude of the fold. If the competence contrast is very large, the layer-parallel homogeneous strain is negligible and the total shortening takes place entirely by external rotation of the competent

---

Fɪɢ. 13.12 (*opposite*). (a) Internal buckling in a multilayer. The multilayer is composed of layers of modelling clay with greased interfaces. The confining material (white walls) is modelling clay. (b) Buckle-folded multilayer under a dextral simple shear. The dark layers are of soft modelling clay. The two thin white layers of stiff modelling clay have been deformed to two orders of folds. The thin layers are embedded in silicone putty. The entire multilayer is embedded in painter's putty. Note that the long axes of strain ellipses in the folded dark layers converge towards the fold cores. The long axes of the greatly elongated strain ellipses in the incompetent silicon layers converge towards the convex side of the fold.

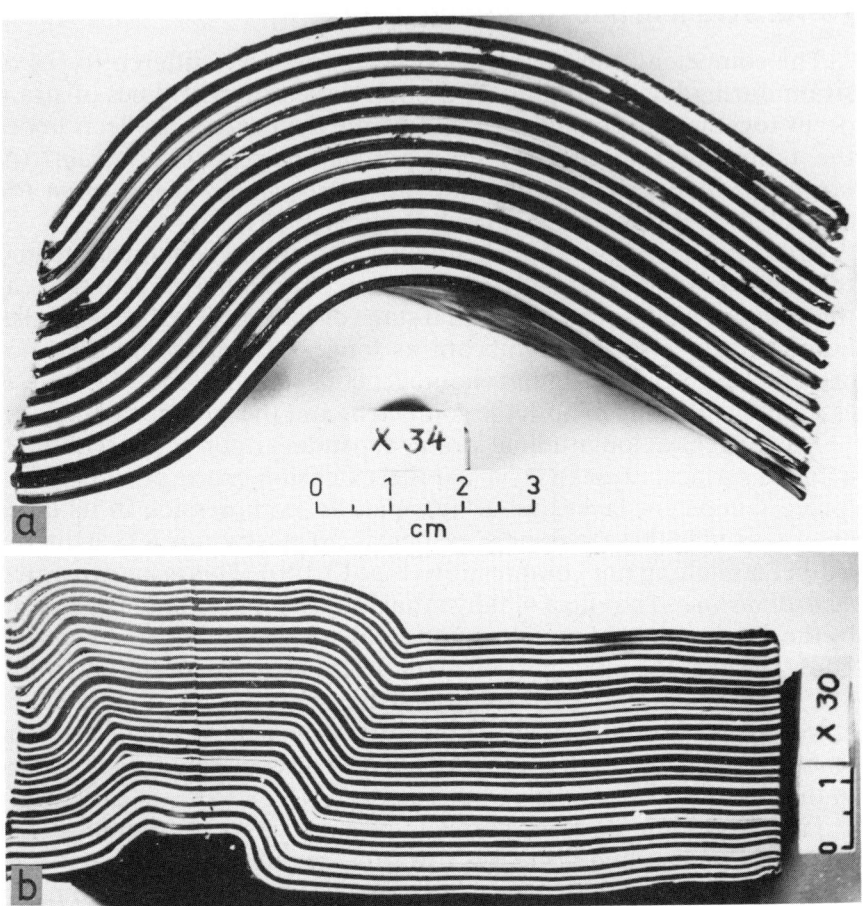

FIG. 13.13. (a) Smooth round-hinged fold in free (unembedded) multilayer with oiled interfaces. The multilayer had a great ease of gliding. (b) Conjugate fold in free multilayer with restricted ease of gliding. The interfaces of the layers of modelling clay were smeared with a viscous grease.

units. At low or moderate competence contrasts there is a significant amount of homogeneous strain before the fold acquires a few degrees of limb dip. Thus, in buckle-folded competent units there is a combination of tangential longitudinal strain and tangential shear strain when the competence contrast is very large. At low or moderate competence contrasts, there is a combination of all three types of strain in the competent units. The orientations of the local strain ellipses along the middle line of a fold profile are shown in Fig. 13.16.

The combined effect of these different types of strain is not the same in the competent and the incompetent units of a multilayer. A commonly

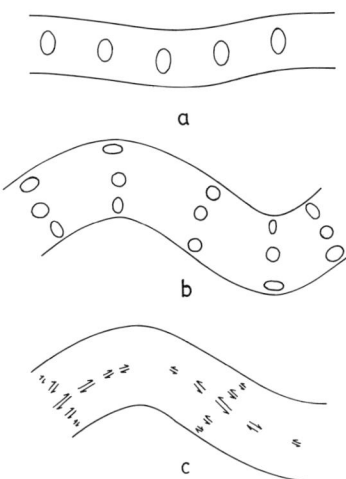

FIG. 13.14. (a) Initial layer-parallel homogeneous strain in a low-amplitude buckling fold. The long axes of strain ellipses are more or less perpendicular to the layering. (b) Tangential longitudinal strain in a buckle-folded layer. (c) Variation of layer-parallel shear strain in a buckle-folded layer.

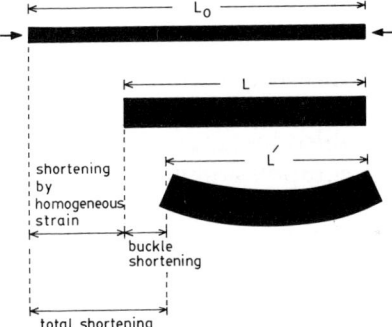

FIG. 13.15. Total shortening of a layer composed of buckle shortening and layer-parallel strain.

observed pattern is shown in Figs. 13.12b and 14.4 in which the long axes of the strain ellipses in the competent units converge towards the core of the fold while those in an incompetent unit converge towards the outer arc of the fold. Because of a larger degree of layer-parallel shear strain at the fold-limbs in incompetent beds, the angle between the bedding and the long axes of the strain ellipse is smaller than in the competent beds (for a detailed discussion of this problem see Ramberg & Ghosh 1968, Treagus & Treagus 1981).

13.11. Three stages of buckle folding

The history of development of a buckling fold can be conveniently divided into three stages:

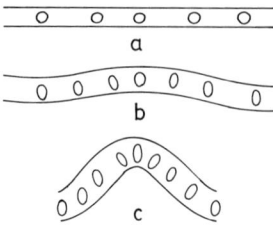

FIG. 13.16. Strain ellipses on the profile plane along the middle surface of a buckle-folded competent unit. (a) Undeformed layer with inscribed circles. (b) and (c) show successive stages of folding. The orientation of the strain ellipses is controlled by both layer-parallel compressive strain and layer-parallel shear strain.

(1) *Initial stage*: The shortening of the layer is essentially by layer-parallel homogeneous strain as long as the limb dip is very low. The direction of maximum extension may be normal to the layer or parallel to the fold axis. In the latter case the competent units may undergo boudinage with boudin axes at a right angle to the fold axis (Fig. 13.17a). If the stretching parallel to the fold axis is very large the layer-parallel shortening may not be accompanied by significant thickening of the layer.

(2) *Middle stage*: At this stage there is a rapid increase in limb dip in comparison with the rate of layer-parallel homogeneous shortening. The folding is accompanied by tangential longitudinal strain. As long as the limbs are in the compression field, there is no boudinage on the profile plane of the fold. Boudinage or layer-parallel homogeneous extension may, however, take place at this stage provided the fold axis is a direction of extension (Fig. 13.17b).

(3) *Late stage*: When the fold has become tight the rate of external rotation becomes rather small. If the competence contrast is not too high the limbs are then thinned and the hinges are thickened. The strain at this stage of folding may be described as late stage strain. If there is an extension perpendicular to the fold axis, the competent unit may show boudinage or pinch and swell structures in the plane of the fold profile (Fig. 13.17c). Depending on the orientations of the bulk strain axes there may or may not be an extension parallel to the fold axis.

13.12. Development of different orders of folds

From the discussions in the previous sections we obtain the following two important principles, which, taken together, explain why the same layer is often affected by different orders of folding.

(1) Depending on the thickness and the mechanical properties, a single layer or a multilayer may develop larger or smaller buckle folds.

274

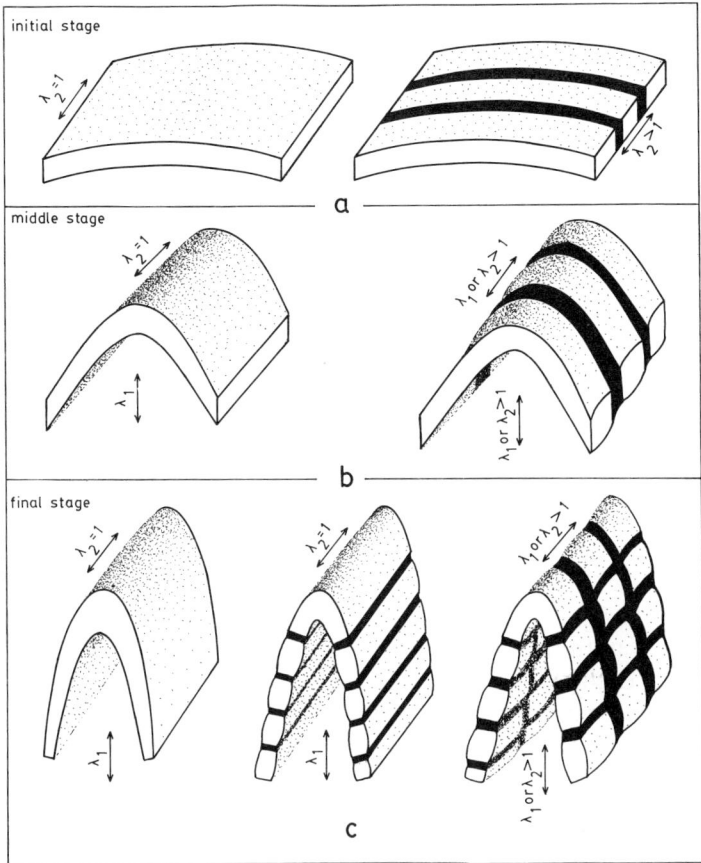

FIG. 13.17. (a) Initial stage of folding. If the fold axis is a direction of no strain, there is no boudinage either in the profile plane or in a section parallel to the fold axis. If the fold axis is a direction of stretching, boudin lines can form at a right angle to the fold axis. (b) Middle stage of folding. On the profile plane the limbs have not undergone a finite extension. There is no boudinage on the profile plane. Boudinage can take place only if there is a stretching parallel to the fold axis. (c) Late stage of folding. On the transverse profile the limbs are being extended and the hinge is being thickened (*left*). If there is no stretching parallel to the fold axis (*centre*) and λ_1 is perpendicular to the fold axis, the limbs are boudinaged with boudin lines parallel to the fold axis. Chocolate tablet boudins may form (*right*) if, in addition, there is a stretching parallel to the fold axis.

Other conditions remaining the same a thicker or a stiffer layer will form larger folds. Similarly, a coarsely bedded thick multilayer will form larger folds than a thinly bedded thin multilayer.

(2) The rate of growth of the fold-amplitude depends largely on the competence contrasts. Other conditions remaining the same a stiffer layer or a multilayer will buckle faster than a less stiff layer or multilayer.

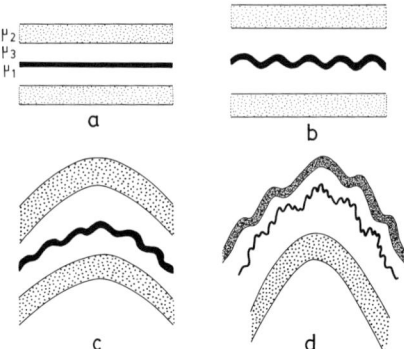

FIG. 13.18. Development of different orders of folds by buckling.

Thus, in natural situations, where layers of dissimilar thicknesses and viscosities are associated, folds of different wavelengths, and some growing at different rates, may interfere with each other. Folds of different orders can develop by such interference. This theory, first proposed by Ramberg (1964) and confirmed by experiments with analogue models (Ramberg 1964, Ghosh 1968), may be clarified by the following example.

Consider a thin layer of viscosity μ_1 and two thicker layers of viscosity μ_2 embedded in a medium of viscosity μ_3 (Fig. 13.18a), where $\mu_1 > \mu_2 \gg \mu_3$. Under layer-parallel shortening the small folds on the thin stiff layer (μ_1) will grow fast while the less stiff layers (μ_2) are mostly undergoing homogeneous thickening and very little external rotation (Fig. 13.18b). At a later stage the larger folds on the thicker layer will grow in amplitude and will distort the enveloping surface of the small-folded unit to folds of larger waves (Fig. 13.18c). In a similar way, a layer may also be affected by waves of more than two orders (Fig. 13.18d).

The smaller folds on the back of a larger fold are variously described as *subsidiary folds* or *drag folds* or *parasitic folds*. Because of flexural flow the parasitic folds (de Sitter 1958a & b) become asymmetrical with opposite senses of asymmetry (S- and Z-folds) on the two limbs of the larger fold; the parasitic folds at the hinge of the larger fold remain symmetrical or M-shaped. With tightening of the larger fold its limbs undergo an extension. As a consequence the parasitic folds on the limbs may be unfolded in certain cases. The parasitic folds on the hinge zones, however, continue to grow in amplitude with progressive deformation (Fig. 13.19). This is one of the reasons why mesoscopic folds are comparatively so prevalent at the hinge zones of large-scale folds. It is noteworthy however that, if the larger fold is asymmetric, then the parasitic folds are mostly concentrated at the short limbs rather than at the hinge only (Fig. 13.20).

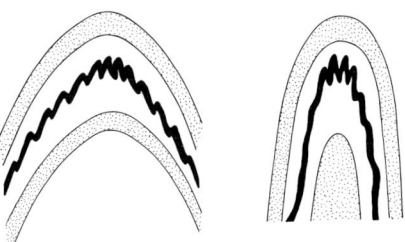

FIG. 13.19. *Left*: development of S- and Z-folds on the limbs and M-folds at the hinge of a larger fold. *Right*: when the limbs of the folds are extended, the smaller folds on the limbs are partially unrolled while the M-folds at the hinge are further tightened.

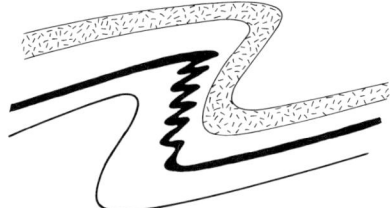

FIG. 13.20. Parasitic folds localized at the short limb of an asymmetric fold.

13.13. Orientation of buckle-fold axis, hinge migration

Consider a single competent layer or a stack of such layers embedded in an incompetent host. Let this complex undergo, prior to buckling, an initial homogeneous strain, with the principal strain axes X, Y, Z. Parallel to layering, a section through the strain ellipsoid will appear as an ellipse, with long axis X'. Only for certain special orientations of the layering can the X'-axis be parallel to the X- or Y-axis. In the general situation X' will be oblique to the principal axes of the strain ellipsoid.

The hinges of the buckle folds will be nucleated parallel to the X'-axis (Ramberg 1959, Flinn 1962, Ghosh 1966, Borradaile 1978, Treagus & Treagus 1981). Consequently, the initial orientation of the fold axis will depend on the initial orientation of the layering with respect to the orientations of the axes of principal strain:

(1) Let the layering be designated as S. If S is parallel to the YZ-plane (Fig. 13.21a), the fold axis will be initially parallel to the Y-axis. (2) If S is parallel to the XZ-plane (Fig. 13.21b), the fold axis will be parallel to the X-axis. (3) The fold axis will be parallel to the Y-axis if S is initially parallel to it but oblique to the X- and Z-axes (Fig. 13.21c). (4) Similarly, the fold axis will be parallel to the X-axis if S is initially parallel to it but is oblique to the other two axes (Fig. 13.21d). Since in all these four cases the fold axis is either parallel to the X- or to the Y-axis, the fold axis lies in the XY-plane. (5) If S is parallel to Z and is oblique to

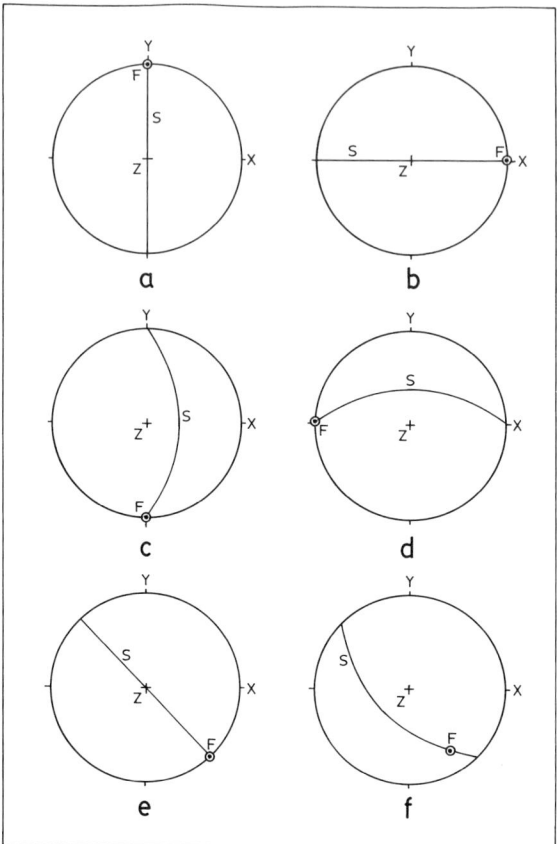

FIG. 13.21. Stereographic projections showing how the initial orientation of the fold axis depends on the orientation of the S-surface. X, Y, Z are the principal axes of finite strain.

the X- and Y-axes, the fold axis will not be parallel to any of the principal axes of strain but will still lie on the XY-plane (Fig. 13.21e). (6) However, if S is oblique to all the principal axes, the fold axis will neither coincide with the X- or Y-axis nor will it lie on the XY-plane (Fig. 13.21f). With progressive deformation the fold axis will gradually rotate towards the X-axis.

The fold axis is parallel to the hinge line of a cylindrical fold. If the fold remains symmetrical at all stages of its growth and the hinge line does not undergo a rotation, then the hinge line will always coincide with a single material line. It may or may not do so in other situations. *Hinge migration*, the process of shifting of the hinge through successive material lines, often takes place by moving the hinge line parallel to itself during the growth of an asymmetrical fold. This is a geometrical consequence of a progressive change in the ratio of the lengths of the

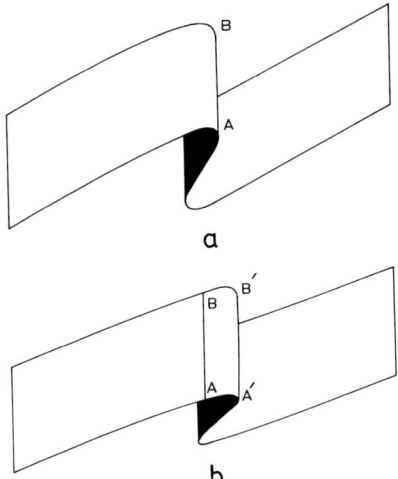

FIG. 13.22. Parallel hinge-line migration during asymmetric folding. (a) An asymmetric fold with *AB* as hinge line. (b) With progressive tightening a new hinge line $A'B'$ has formed.

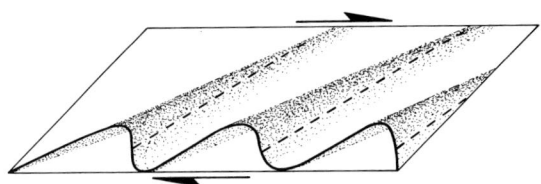

FIG. 13.23. Hinge-line migration in non-coaxial deformation. The newly formed hinge line may make an angle with the material line which was parallel to the earlier hinge.

long and the short limbs of an asymmetrical fold while the arc length of the fold remains unchanged (Fig. 13.22 a,b). In Fig. 13.22b, the material line *AB* was earlier (Fig. 13.22a) positioned along a hinge but does not now coincide with the new hinge $A'B'$, although the two lines are still parallel. The X'-axis, the long axis of the strain ellipse on the enveloping surface, is only a *geometrically defined line*. If the X'-axis rotates, either due to rotational character of strain or due to the change in orientation of an obliquely aligned enveloping surface, new hinge lines may grow parallel to the successive orientations of the X'-axis. In this kind of hinge migration a *material line* which was parallel to an earlier hinge will make an angle with the new hinge (Fig. 13.23).

A folded layer shows the maximum curvature along the hinge. Hinge migration involves a reduction of the curvature at the position of the old hinge and creation of a stronger curvature at the position of the new hinge. If the fold has already acquired a strong curvature at the hinge of a tight or isoclinal fold, a further tightening of the fold along the old hinge may require less energy than shifting the hinge to the position of a new material line. Perhaps, this is the reason why, in experimental deformation of soft (non-elastic) models, hinge migration ceases to be significant beyond a certain stage of tightening of the folds.

The conclusions of this section may be summarized in the following way:

(1) The fold axis will coincide with the X-axis or Y-axis for certain special orientations of the layering.

(2) In layers oblique to the principal axes the fold axis will neither coincide with a principal axis nor will it lie on the XY-plane.

(3) During such oblique folding the fold axis will progressively rotate towards the X-axis and may finally end up being virtually parallel to it.

(4) A fold hinge is nucleated parallel to the X'-axis. In certain circumstances the fold hinge migrates through successive material lines. In other circumstances it may continue to coincide with the same material line. Progressive hinge migration is unlikely beyond a certain stage of tightening of the fold.

13.14. Diagnostic morphological characters of buckle folds

On the basis of the discussions in the foregoing sections buckle folds are expected to show one or more of the following morphological features:

(1) The majority of folds in a single-layer or a multilayer have nearly the same arc length.

(2) In single-layer folds with the same lithology, the arc length/thickness ratio is more or less constant or shows a small range of variation. This implies that a thicker layer will show larger folds.

(3) In multilayers with intercalated competent and incompetent units, the competent units show class 1B or 1C geometry, while the incompetent layers show class 3 geometry. The competent and incompetent layers are distinguished by the fact that the ratio of orthogonal thickness at the hinge and the limb is always smaller in the competent layer than in the incompetent layer (Fig. 13.24a).

(4) Because of tangential longitudinal strain, radial tension cracks may occur on the outer side of a folded competent layer (Fig. 13.24b).

(5) Since the sense of layer-parallel shear strain is opposite in the two limbs of a fold, porphyroblasts of garnet will have, in thin sections, rotations similar to those shown in Fig. 13.24c.

(6) Buckling of certain types of multilayers gives rise to disharmonic or polyharmonic folds or a regular train of parasitic folds with opposite senses of asymmetry in the two limbs as shown in Fig. 13.24d.

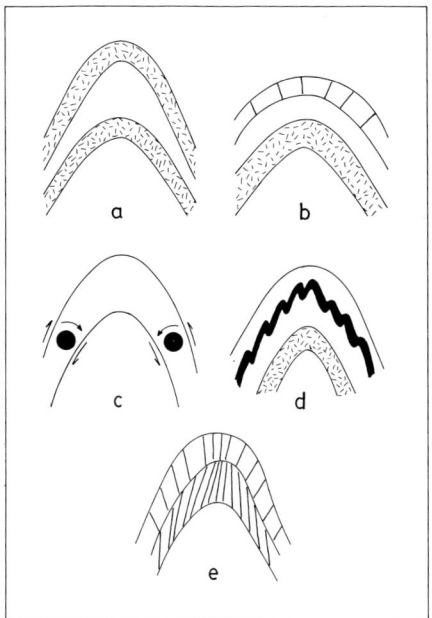

FIG. 13.24. Some diagnostic morphological characters of buckling folds.

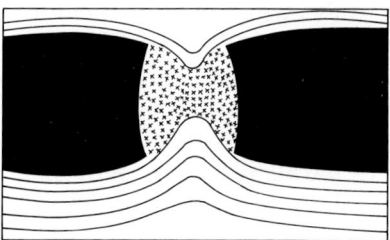

FIG. 13.25. Bending folds near the separation zone of boudins.

(7) Buckle-folded multilayers may show a refraction of cleavage. The angle between the layering and the cleavage at the fold limbs is larger in the competent units than in the incompetent units (Fig. 13.24e).

13.15. Bending folds

Bending folds develop in response to forces acting across the layers. These folds are much less common than the buckle folds. The most common types of bending folds found in mesoscopic scale are those which develop near the separation zones of boudins. During the development of boudinage structures the incompetent host material tends to be pushed inside the progressively opening gaps between the boudins. If the layers in the host rock do not have a strong competence contrast the hinges of the bending folds are generally thickened with respect to the limbs. In buckle folds the layers undergo a shortening when folds have a low amplitude. In contrast, the layers in a bending fold, opposite the separation zone between boudins, extend along their arc lengths at all the stages of folding. The bending folds associated with boudinage structures rapidly die out as we move away from the boudinaged layer (Fig. 13.25).

Salt domes and granite domes are common examples of large-scale bending folds. These are produced by unstable mass distribution in the field of gravity. The process of development of such large-scale bending folds, powered entirely by the force of gravity, has been studied in detail through laboratory experiments (Nettleton 1934, Ramberg 1963d, 1966, 1967). The theory of development of such folds has also been described in detail by Ramberg (1967, 1968).

The development of bending folds on the interface of two gravitationally unstable liquids can be demonstrated by a simple experiment (Fig. 13.26). Into an airtight oblong plexiglass box pour two viscous immiscible liquids of different densities. The pair of liquids can be, for example, gear oil and syrup, or oil and condensed milk. The lid of the box is then stuck at the top with adhesive. Care must be taken to ensure that not a single air bubble remains inside the box. The denser liquid (syrup or condensed milk) will remain at the bottom and the lighter oil will remain at the top (Fig. 13.26a). The box is then put upside down. Now the lighter liquid forms the substratum and the denser liquid forms the overburden. If the liquids are sufficiently viscous the interface between them will slowly form a uniform train of approximately cylindrical folds with axes parallel to the length of the box (Fig. 13.26b). After a certain stage the folds will form evenly spaced axial culminations and depressions on the anticlinal crests (Fig. 13.26c). With increase in amplitude these domes will grow faster and the interface will be deformed to mushroom-shaped folds (Fig. 13.26d). In the final stage the sidewise spreading crests or troughs will coalesce and a stable configuration will be achieved with the lighter liquid at the top and the denser at the bottom (Fig. 13.26e).

In Ramberg's experiments with centrifuged models much stiffer materials could be used, with silicon putty as the substratum and a layered complex of painter's putty as the overburden. The rising of the domes of silicone putty and the corresponding sinking of the overburden produced a series of bending folds remarkably similar to the folds found in salt and granite domes.

The implications of these experimental results go much beyond the problem of development of large-scale bending folds. The combination of an up-and-down movement and the sidewise spreading of the mushrooming crests and troughs in the models gave rise to a variety of

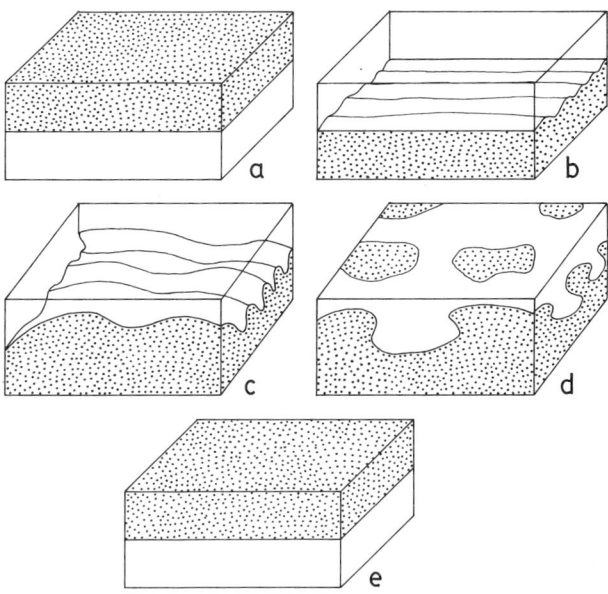

FIG. 13.26. Stages of development of bending folds on the interface of two gravitationally unstable immiscible liquids. Initially, the denser liquid (dotted) is at the top.

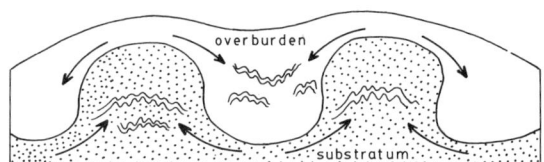

FIG. 13.27. Rising of domes of a buoyant layer (dotted) and sinking of the denser overburden. Although the overall movement is vertical, there are local zones of sidewise movement causing horizontal shortening and horizontal extension.

natural-looking orogenic structures, e.g. upright and recumbent folds, imbrications, recumbent fold nappes cored by basement material and fold interference patterns resulting from interference of upright and recumbent folds. In these models the folding of the layers is powered entirely by the pull of gravity and the total model does not undergo any horizontal shortening. Nevertheless, because of the sidewise movement of the overburden away from the crests of rising domes and the sidewise movement of the substratum from beneath the sinking troughs, zones of compression are created both in the overburden and the substratum (Fig. 13.27). Competent layers lying in such zones undergo buckle folding. The experiments therefore furnish a model for the creation of true local shortening along the width of an orogen while the distance between the edges of the orogen remains unchanged (Ramberg 1967).

283

Although it is acknowledged that Ramberg's model is valid for interpreting salt domes and granite domes, the wider implications of these experiments in the context of evolution of large-scale orogenic structures have not yet been fully assessed.

13.16. Passive folds

Passive folds are produced by accentuation of pre-existing curvatures of layers by more or less homogeneous strain. The homogeneity of strain is not an essential condition. What is required is that the strain in neighbouring points of the successive layers is approximately the same. This implies that passive folding can take place only if there is no significant competence contrasts among the layers.

In passive folds there is neither a competence contrast which can give rise to buckle folds under layer-parallel shortening nor a system of laterally acting forces to produce bending folds. The folds are merely produced by a passive accentuation of pre-existing curvatures (Fig. 13.28). These initial curvatures can be produced in several ways. Hudleston (1976, 1977, 1983) has shown that small-scale passive folds are often produced in almost all glaciers. Because of irregularities in the bedrock the old foliation in the glacier ice may have gentle warps. Under changed conditions (e.g. change in ice thickness, surface slope) the particle paths of the simple shear deformation may no longer be parallel to the old foliation. The gentle warps of the old foliation, under certain situations, can then be passively deformed to tight recumbent folds (Fig. 13.29). Passive folds may develop in rocks in a similar way in shear zones, provided the layers do not have an anisotropy or strong competence contrast. Initial curvatures in the layers may also develop by local distortions around pebbles, porphyroblasts, porphyroclasts, etc. Later deformation causes a passive accentuation of these curvatures.

Passive folds have similar or sub-similar shapes and a uniform orientation of the cleavage in all parts.

A common mode of development of a regular train of passive folds is by deformation of penecontemporaneous structures during later diastrophic movements. Different stages of development of such folds have been identified in the garnet zone mica-schists of the Ghatsila area of Singhbhum, E. India (Ghosh & Lahiri 1983).

These Precambrian rocks have a large variety of remarkably well-preserved sedimentary structures. In the least deformed rocks the penecontemporaneous contortions are only slightly tighter in the profile planes of diastrophic folds (i.e. in sections sub-perpendicular to bedding–cleavage intersections) than in sections parallel to their axial planes (Fig. 13.30). With a larger shortening across the cleavage the

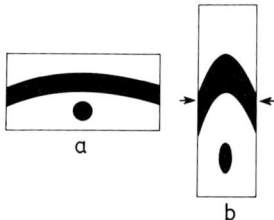

FIG. 13.28. Passive folding of a layer by homogeneous strain. The initial curvature of the layer is accentuated by folding.

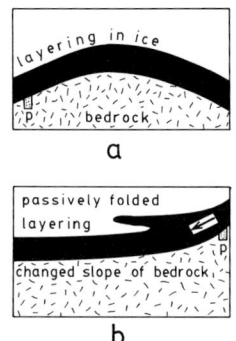

FIG. 13.29. Passive folding in glacier ice over uneven bedrock. After Hudleston 1983.

penecontemporaneous contortions become progressively tighter in sections normal to the bedding–cleavage intersection, while the contortions remain open in sections parallel to the cleavage. The diastrophic mesoscopic folds are cylindrical and have a uniform orientation of fold axes in small areas. In contrast, the passive flattening of the penecontemporaneous contortions have given rise to plane non-cylindrical folds which have closed oval outcrops in sections sub-parallel to the general orientation of bedding. These passive folds are tighter than the buckle folds of neighbouring outcrops, have sub-similar shapes and have sharp anticlinal and rounded synclinal hinges (Fig. 2.9).

13.17. The models of flexure folding and shear folding

Flexure folds

The broad category of flexure folds includes the following types:

1. Pure flexural folds.
2. Flexural slip folds.
3. Flexural flow folds.

In the model of pure flexure folding, a layer undergoes an external rotation to give rise to the folds. The outer surface of the layer (the

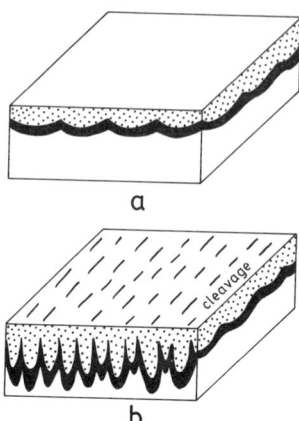

FIG. 13.30. (a) Penecontemporaneous contortions. (b) Deformed contortions. The initial curvatures are accentuated by passive folding under layer-parallel compression.

extrados) is extended while the inner surface (the intrados) is shortened. The middle surface of a flexure fold remains a neutral surface and there is no tangential longitudinal strain along it. The orthogonal thickness of a pure flexure fold remains the same all over it. A pre-existing lineation on the surface of the layer is only externally rotated and the angle between the lineation and the fold axis remains constant all over the fold.

When a stack of competent layers with low cohesion at the interfaces is folded, each layer slips past the other as in a flexed deck of cards. This is the model of flexural slip. In symmetrical folds the flexural slip is zero at the fold hinges and increases to a maximum value at the inflection points. In well-bedded rocks, e.g. sandstones, quartzites and cherts, evidence of flexural slip is furnished by offsets of transverse quartz or carbonate veins at layer boundaries (Fig. 13.31).

The following simplified model gives an insight into the factors which control the magnitude of flexural slip between two layers. Let AB and CD represent traces of the middle surfaces of two layers with thickness h_1 and h_2. EF is the interface between the two layers (Fig. 13.32a). After the layers are folded, the arc FG in Fig. 13.32b represents the magnitude of flexural slip. In this figure the left-hand side of the layers is held fixed as in the hinge of a symmetrical fold, while the layers are folded in the form of circular arcs. Let r be the radius of curvature of the interface EF and let θ be the angle between the two radii at E and F. From Fig. 13.32b, we find

$$AB = (r - h_1/2)\theta$$
$$CD = (r + h_2/2)(\theta - \triangle\theta).$$

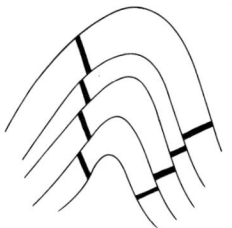

FIG. 13.31. Offset of veins by flexural slip folding.

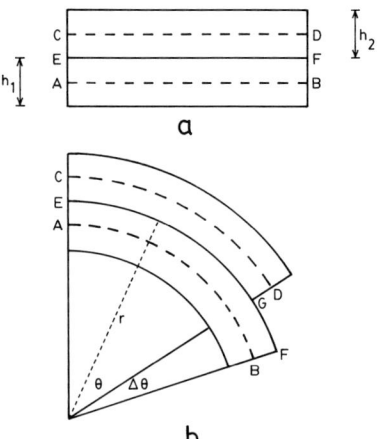

FIG. 13.32. Magnitude of flexural slip when the surfaces are deformed to circular arcs. (a) CD and AB are the lengths of neutral surfaces of two adjoining layers. (b) After folding, the lengths of CD and AB remain the same. FG is the magnitude of flexural slip.

Since $AB = CD$, we have, from these two equations,

$$\triangle\theta = \frac{\frac{1}{2}(h_1 + h_2)\theta}{r + \frac{1}{2}h_2}.$$

The flexural slip, FG, is

$$r \triangle\theta = \frac{\frac{1}{2}(h_1 + h_2)r\theta}{r + \frac{1}{2}h_2}.$$

If the horizontal segment of the layer at A is regarded as the hinge of the fold, θ becomes equal to the dip angle α. The flexural slip is zero at the hinge and increases with the dip angle. It also increases with an increase in the thickness of the layers.

Folding may take place by flexural slip on penetrative cleavage surfaces (Fig. 13.33). Such a fold is called a *bend glide fold* or *flexure glide fold* (Turner & Weiss 1963, pp. 474–478). Although the transverse veins are distorted by the penetrative flexural slip, they may not show discrete offsets.

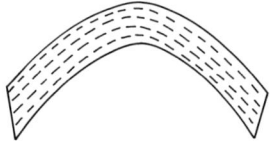

FIG. 13.33. Bend gliding during the folding of a set of penetrative cleavage surfaces.

FIG. 13.34. Flexing of competent layers with layer-parallel shear in intervening incompetent layers.

If competent and incompetent layers are intercalated and folded harmoniously, each competent layer may undergo pure flexure while the intervening incompetent layers undergo a distributed layer-parallel shear without the presence of kinematically active layer-parallel S surfaces (Fig. 13.34). These have been described as *flexural-flow folds* (Donath & Parker 1964). If the thickness of the incompetent layers become vanishingly small the folds become flexural slip folds. The flexural-flow folds in nature do not always show a constant orthogonal thickness of the competent layers; they often show a class 1C type of geometry.

The geometry of natural flexure folds of all the three types can hardly be distinguished from that of buckling folds. However, the ideal model of flexure folding requires parallel folding and the absence of layer-parallel strain. For all practical purposes the flexure folds belong to a special category of buckle folds.

Shear folds

Shear folds or slip folds are those which develop by heterogeneous slip along a set of closely spaced slip planes which cut across the layering (Fig. 13.35). The folding can develop only if the direction of slip is at an angle to the line of intersection of the slip plane and the layering.

Shear folds often show parallel slices of rocks in which the shearing is small or absent. Such slices have been called *gleitbretter*.

In a shear fold the slip planes must be parallel to the axial plane of the fold. The axial plane thickness must remain constant all over the fold. Consequently, shear folds should show the ideal geometry of a similar fold. In the ideal model of shear folding, the fold limbs are lengthened

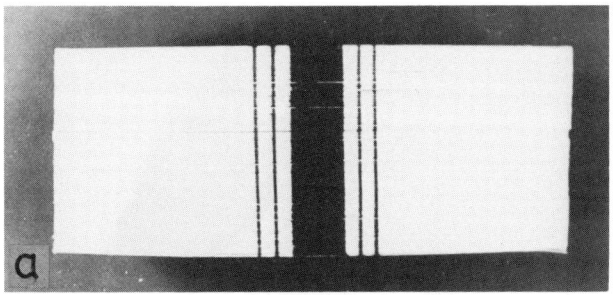

FIG. 13.35. Shear folding in card deck. (a) Parallel marker lines drawn across a deck of cards. (b) Shear folding of the markers by smoothly varying heterogeneous slip on the cards.

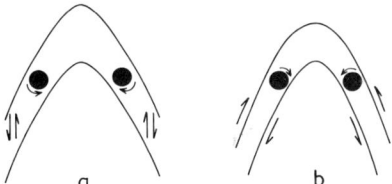

FIG. 13.36. Dissimilar senses of rotation of spherical porphyroblasts in the limbs of (a) shear fold and (b) flexural slip fold.

and thinned but there is no shortening across the axial plane. The sense of shear should be different in the two limbs of a fold. In an antiform the right limb should show a dextral shear and the left limb should show a sinistral shear. As a consequence, a rigid porphyroblast will show a clockwise rotation on the right-hand limb and an anticlockwise rotation on the left limb of the antiform. These senses of rotation are opposite to those of buckling folds or flexural-flow folds (Fig. 13.36).

The model of shear folding is often described in terms of the *kinematic axes*, *a*, *b* and *c*, with *a* as the direction of slip, *b* being at a right angle to *a* and on the slip plane itself and with *c* as a normal to the slip plane (Fig. 13.37a). In a shear fold, the fold axis may or may not be

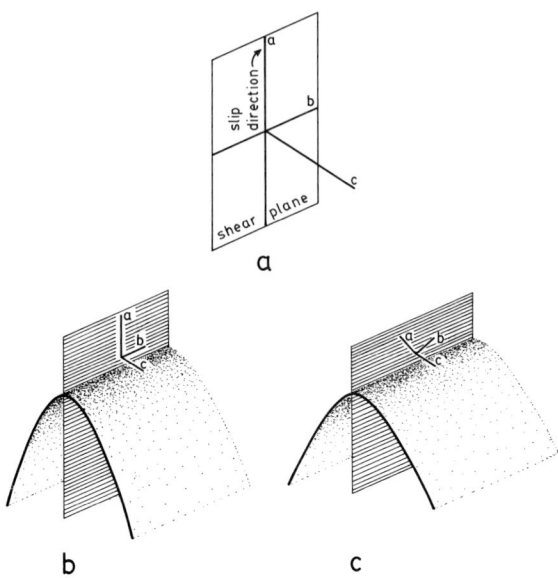

FIG. 13.37. (a) Kinematic axes *a*, *b* and *c*, with slip direction *a*, and with *b* at a right angle to *a*, and lying on the slip plane. *c* is normal to the *ab*-plane. (b) Shear folding, with *b* parallel to the fold axis. (c) Shear folding, with *b* at an angle to the fold axis.

parallel to the *b* kinematic axis (Fig. 13.37b, c). As a consequence, the transverse profile of a shear fold does not always contain the direction of shear.

A line lying parallel to the slip planes neither changes its orientation nor undergoes a strain in the course of slip. Hence, during shear folding there should not be a rotation of the fold axis. Nor is there a stretching parallel to the fold axis.

The model of shear folding can be easily simulated by card-deck slip (Fig. 13.35). If the slip is heterogeneous, parallel marker lines drawn on the edge of the card deck are deformed to shear folds. Within a small area in which the deformation is essentially homogeneous, a circle is deformed to an ellipse. It is evident from this model that the slip planes are parallel to the circular sections of the strain ellipsoid. The *XY*-plane of the strain ellipsoid will not be parallel to the slip plane or the axial plane of the fold. Therefore, unless the deformation is extremely large, we do not expect a preferred shape orientation of the plastically deformed grains parallel to the axial plane cleavage of shear folds.

The ideal model of shear folding cannot explain many of the features of deformed natural rocks. Thus, for example, most natural folds do not show the geometry of similar folds. They often show clear evidence of a shortening across the axial planes. They also show a disharmony of

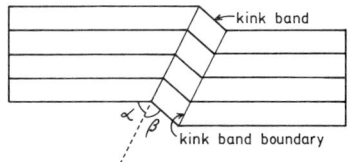

FIG. 13.38. Geometrical elements of kink band.

the folded layers, with smaller folds on thin beds and larger folds on thick beds. Moreover, contrary to what is predicted by the model of shear folding, most natural folds show a preferred dimensional orientation of plastically deformed grains (and pebbles) parallel to the axial planes.

13.18. Kink band and conjugate fold

A *kink band* is a sharply defined band within which the layering is straight but is at an angle with the layering outside the band. The band may be lenticular or bounded by straight parallel walls. The layers are continuous across the kink bands but form sharp hinges at the *kink band boundaries*. A kink band boundary is sometimes described as a *kink plane* (Hills 1963, p. 239). It forms the axial plane of the kink fold.

The layering of the rock is rotated within the kink band. For parallel-walled kink bands the geometry can be described by the orientation of the kink plane, the width of the band and the two angles α and β (Fig. 13.38). α is the angle between the kink plane and the unrotated layer outside the band. β is the angle between the kink plane and the rotated layer inside the band (Anderson 1968).

Kink bands are very common in deformed crystals, especially in chlorite, micas, calcite and kyanite, and are often found in the mesoscopic scale in well-foliated rocks, such as slates, phyllites and schists (e.g. Voll 1960, Ramsay 1962c, 1967, Anderson 1964, Dewey 1966, Clifford 1968, Fyson 1966, Hobson 1973, Hanmer 1979, 1982, Wallace & Clifford 1983, Williams 1987). Kink bands in a scale of decimetres have been described from the Western Sahara by Collomb & Donzeau (1974). In finely foliated rocks the kink bands are sharp-hinged. The hinges become less sharp in coarsely laminated rocks. Large kink folds or *megakinks* in the scale of maps have been described by Rixon *et al.* (1983) and Powell *et al.* (1985) and are likely to be fairly common.

The definition of kink bands as given above includes two different types of kinks. If the layer outside the kink band is kept horizontally, then the rotation of the layer within the kink band may correspond with the sense of movement of either a reverse fault or a normal fault (Fig. 13.39). These two types of kink bands may be described as *reverse* and

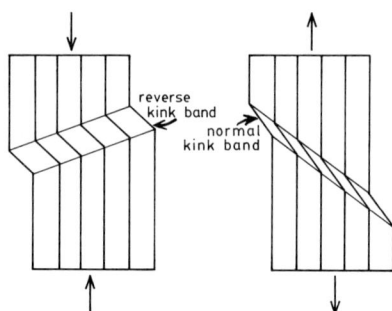

FIG. 13.39. Reverse and normal kink bands. A reverse kink band causes an overall shortening, while a normal kink band causes an overall extension.

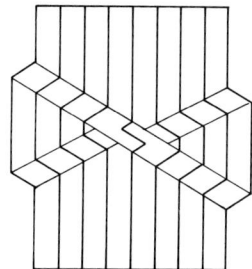

FIG. 13.40. Conjugate kink bands. The band running from upper left to lower right is dextral and the band running from upper right to lower left is sinistral.

normal kink bands (Dewey 1965) or as *contractional* and *extensional kink bands* (Ramsay & Huber 1987, p. 427). In the majority of areas the structures commonly described as kink bands are reverse kink bands and Weiss (1968, p. 361) suggested that the term kink band should be restricted to these. The present discussion will be confined to reverse kink bands which develop as a result of layer-parallel shortening.

The unrotated layer on one side of a kink band shows a side-stepping with reference to the unrotated layer on the other side. By analogy with strike-slip faults, and depending on the sense of side-stepping (Fig. 13.40), a kink band can either be *dextral* (right-handed) or *sinistral* (left-handed). A *conjugate fold* is produced when the layering is affected by a conjugate set of dextral and sinistral kink bands.

Field studies of kink bands show a small range of *kink angles*, α and β. From a large number of measurements in the kink bands of slates and thinly laminated siltstones of the Ards Peninsula, Northern Ireland, Anderson (1968) found that, typically, the angle α is close to 60° and the angle β is close to 78°. It should be noted that the total thickness of the multilayer inside and outside the kink band can remain the same only if α = β (Fig. 13.41a). If β is greater than α (Fig. 13.41b), there should be a dilation across the layering. Evidence of such dilation is sometimes

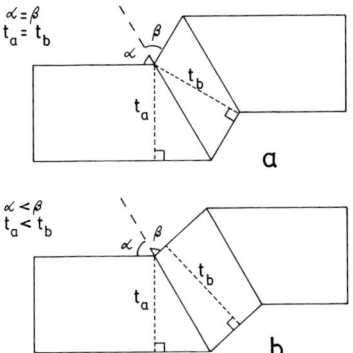

FIG. 13.41. (a) Kink band with α = β. The total thickness of the multilayer inside and outside the kink band remains the same. (b) Kink band with β > α. There is a dilation across the layering within the kink band.

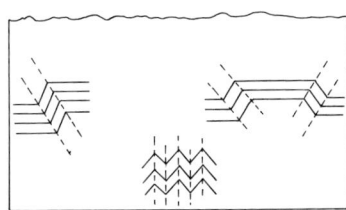

FIG. 13.42. Kink band, conjugate fold and chevron fold of the same generation developed from the same initial orientation of a foliation. The structures show three orientations of axial planes.

found by the occurrence of extensional veins parallel to the layering within the kink bands.

Kink bands have been produced by deformation of phyllites and slates under laboratory conditions (Paterson & Weiss 1966, Donath 1968) and by deformation of various kinds of analogue models (Ghosh 1968, Honea & Johnson 1976, Cobbold et al. 1971, Means & Williams 1972, Latham 1985). These studies considerably enhance our understanding of the mechanism of kink band formation. The investigations by Paterson & Weiss (1966) and of Weiss (1968) are particularly important in this respect. The important results of these experiments are summarized below.

(1) Conjugate kink bands form when the compressive stress σ_3 is parallel or sub-parallel to the layering of the specimen. Only one set of kink band forms when the layering is oblique to σ_3.

(2) σ_3 bisects the obtuse angle between symmetrical conjugate kink bands.

(3) For symmetrical conjugate kinks α is close to 60° and β is somewhat larger than α but is always less than 90°. Throughout the course of deformation lenticular kink bands are nucleated in different parts of the specimen and chevron folds are produced where dextral and sinistral kink bands meet. These chevron folds have axial planes sub-perpendicular to the compression axis. With progressive deformation the areas of chevron folds increase and finally nearly all areas of conjugate folds and unrotated layers are replaced by chevron folds.

The experiments with kink bands are important in several respects. In particular, they enable us to determine the direction of principal compressive stress in certain situations. When the conjugate folds are symmetrical and have parallel fold hinges, σ_3 is parallel to the lamination and is perpendicular to the fold axis. The experiments suggest that σ_2 is parallel to the fold axis and σ_1 is normal to the unrotated layering. From the experiments we further learn that conjugate sets of kink bands may be closely associated with chevron folds. As a consequence of such an association a single generation of folds of an area may show at least three different attitudes (e.g. Naha *et al.* 1987) of the axial planes (Fig. 13.42).

CHAPTER 14

Cleavage

14.1. Nomenclature

Rock cleavage is a set of closely spaced planar parallel secondary fabric elements that impart a mechanical anisotropy to the rock and do not cause an apparent loss of cohesion in the rock (Dennis *in* Bayly *et al.* 1977). The cleaved rocks generally show a domainal structure, with *cleavage domains* and *microlithons*. In the cleavage domain the original fabric of the rock has been strongly altered. In the microlithons the original fabric has undergone little or no alteration. The cleavage domains, sometimes described as films, folia or seams, usually show a strong preferred mineral orientation parallel to the cleavage. In one extremity of the spectrum, the cleavage domains are so thick that microlithons are non-existent or indistinguishable in the microscopic scale. The cleavage is then called a continuous cleavage. On the other extremity the cleavage domains consist of sharp extremely thin discontinuities along which there is little or no preferred orientation of minerals. There are all gradations in between these two extremities. The rock cleavages are thus divided into two basic types, *continuous cleavage* and *spaced cleavage*. The term cleavage has been used irrespective of the type of rock in which the structure occurs (Wilson 1961, p. 457). The term *schistosity* is restricted to the continuous cleavage in somewhat coarser-grained rocks. *Slaty cleavage* is the corresponding term for finer-grained slaty rocks. *Foliation*, in the sense of Hobbs *et al.* (1976), includes all types of cleavage, both continuous and spaced, as well as bedding and metamorphic banding.

Spaced cleavages may be either a *crenulation cleavage* (Knill 1960) or a *disjunctive cleavage* (Powell 1979). In the development of a crenulation cleavage there is microfolding of an earlier cleavage. Disjunctive cleavages are not associated with such microfolding.

The morphological classification as given above is of comparatively recent origin (Dennis 1972, Bayly *et al.* 1977, Powell 1979). Continuous cleavage and spaced cleavage were earlier distinguished by the genetic terms *flow cleavage* and *fracture cleavage* (Leith 1905).

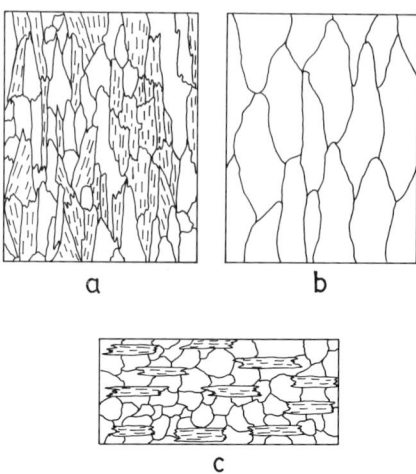

a   b

c

FIG. 14.1. Some microscopic characters of cleavage. (a) Cleavage marked by parallel
flakes of mica and elongate grains of quartz. (b) Cleavage marked by preferred shape
orientation of elongated grains of quartz. (c) Preferred orientation of isolated mica
flakes in a matrix of equiaxed quartz grains.

14.2. Microscopic characters

In layers rich in phyllosilicate the cleavage is usually defined by
sub-parallel alignment of the flakes of these minerals. The grain
boundaries are sub-parallel to the mineral cleavage. Thus, in such
domains the cleavage is defined by parallel orientation of crystallographic planes (Fig. 14.1a). In strongly deformed quartzites, on the
other hand, the cleavage is defined by a sub-parallel array of
deformed flat grains of quartz. In a thin section perpendicular to the
cleavage the elongated grains of quartz have a preferred shape
orientation (Fig. 14.1b). Crystallographic preferred orientation, if
any, is not immediately discernible. In certain rocks, such as quartzose mica schists or micaceous quartzites, the cleavage in certain
domains is marked by isolated flakes of mica which run through a
groundmass of equidimensional quartz grains (Fig. 14.1c). In the
majority of schistose rocks such different types of domains are
closely associated (Fig. 14.2). In slates, phyllites and mica schists, the
cleavage in the mica-rich domains often anastomose around
lenticular quartz-rich domains.

Many of the metamorphic rocks, both schists and gneisses, show a
layering marked by dissimilar mineral assemblages or of dissimilar
grain size; the layering is parallel to the cleavage. This may represent
the remnant of a bedding or may have been produced by metamorphic
differentiation.

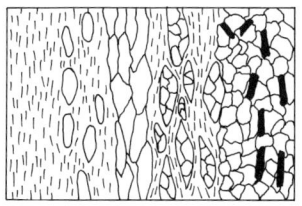

FIG. 14.2. Association of different types of domains in a schistose rock.

14.3. Geometrical relations with folds

Axial plane cleavage

That cleavage surfaces in most deformed terrains are approximately parallel to the axial planes of folds was recognized long ago (Sedgwick 1835, Darwin 1846 and Sorby 1853 quoted by Wilson 1961, p. 465). This common variety of cleavage (Fig. 14.9a) is called *axial plane cleavage*. An axial plane cleavage may be a slaty cleavage, a schistosity or a crenulation cleavage. In areas of superposed folding the successive generations of cleavages, S_1, S_2, S_3, etc., are usually axial planar to the successive generations of folds F_1, F_2, F_3, etc. The axial plane cleavage may show local deviations from the orientation of the axial plane. The deviation may show up in the form of a cleavage fan or as a cleavage refraction (Fig. 14.3). In either case the trace of the cleavage on the form surface is approximately parallel to the fold hinge. Moreover, the cleavage at the hinge zone is generally parallel to the axial plane. When an axial plane cleavage and a bedding fissility are both prominent, the rock has a tendency to break up along elongate fragments sub-parallel to the fold axis. The resulting structure is known as *pencil structure*.

Cleavage refraction

There is often a systematic change in the angle between the bedding and the cleavage in different parts of a folded layer. This causes a fanning of the cleavage. The *cleavage fan* may be *convergent* or *divergent* (Fig. 14.3a), depending on whether the cleavage surfaces converge towards the core or towards the convex side of a fold (Ramsay 1967, p. 405). Where competent and incompetent layers are intercalated, the bedding–cleavage angle is larger in the competent unit than in the incompetent unit. Such a change in orientation of the cleavage from layer to layer is known as *cleavage refraction* (Fig. 14.3b). As we shall see later, rock cleavage is approximately parallel to the XY-plane of the finite strain ellipsoid. The cleavage fan and the cleavage

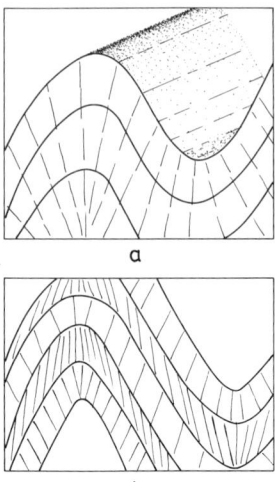

FIG. 14.3. (a) Cleavage fan. (b) Cleavage refraction.

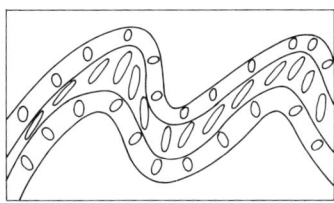

FIG. 14.4. Orientations of strain ellipses on the transverse profile of a fold. Note that the long axes of strain ellipses in the competent units (upper and lower layers) converge towards the core. The long axes in the incompetent layer converge towards the convex side of the folds.

refraction are explained by the change in orientation of the XY-plane in different parts of a fold and in layers of different competences (e.g. Ghosh 1966, Ramsay 1967, p. 404, Ramberg & Ghosh 1968, Treagus & Treagus 1981). The fan-shaped arrangement of strain ellipse axes in competent and incompetent layers in profile plane of experimentally produced buckling folds are similar to those of convergent and divergent cleavage fans found in folded multilayered rocks (Figs. 13.12b, 14.4).

Transected cleavage

In some places, apart from the fanning and refraction, the cleavage occurs at a distinct angle (Fig. 14.5) with the fold hinge. As a result, on the exposed form surface the trace of the cleavage is found to transect

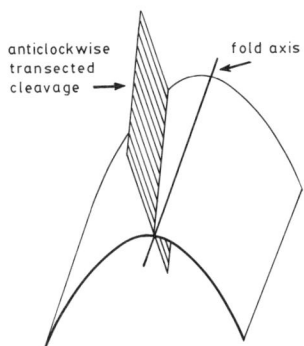

anticlockwise
transected
cleavage ⟶

fold axis

FIG. 14.5. Transected cleavage occurring at an angle to the fold hinge.

the fold hinge. This type of cleavage is called a *non-axial planar cleavage* (Stringer & Treagus 1980) or *transected cleavage* (Ramsay & Huber 1987, p. 334). A transected cleavage is not a later generation of cleavage superimposed on an earlier generation of folds. Like the axial plane cleavage it forms during some stage of folding (Powell 1974, Treagus & Treagus 1981). Thus, for example, a transected cleavage S_1 should be correlated with F_1 folds. In profile sections of the folds the cleavage may fan and refract in sympathy with the fold, indicating thereby that the cleavage development is synchronous with folding (Treagus & Treagus 1981, p. 2). The transection of the cleavage may be in a clockwise or an anticlockwise sense. In some areas there is a systematic change in the sense of transection over a large-scale structure. Thus, Murphy (1985) reports different senses of transection from the two limbs of a large-scale fold, while at the core of the fold the cleavage is axial planar. Similarly, Treagus & Treagus (1981, p. 1) record the observation of Sutton & Watson (1956) that clockwise and anticlockwise transection of two areas occur on either side of a third area where the cleavage is axial planar.

Transected cleavage is not a rare feature. It has been reported from several regions, e.g. the Dalradian of N.E. Scotland (Treagus & Treagus 1981, pp. 1–2, re-interpreting Sutton & Watson 1956), the Ordovician and Silurian rocks of S.W. Scotland (Stringer & Treagus 1980), the English Lake District (Soper & Mosley 1978), the Lower Palaeozoic rocks of Central Ireland (Phillips *et al.* 1980, Sanderson *et al.* 1980, Murphy 1985), northern Norway (Roberts 1971), the Precambrian rocks of N.W. Tasmania (Powell 1974), the Adelaide Geosyncline (Bell 1978b), northern and south-east Virginia (Wickham 1972, Gray 1981) and from different places of Canada (Stringer 1975, Stringer & Lajtai 1979, Borradaile 1978). According to Duncan (1985), the cleavage in Sulphur Creek area of Tasmania, which was earlier identified as a transected cleavage, pre-dates the folds that are transected by the cleavage.

Although axial plane cleavage and transected cleavage are closely related and are often associated in the field, the distinction between the two is essential for a correct analysis of the geometry of the folds. The bedding–cleavage intersection can be used to determine the orientation of the fold axis only if the cleavage is axial planar and not transected. A distinction between the two types of cleavage cannot be made from observation on the profile of a fold only; in addition, a special effort should be made to observe the structures in the neighbourhood of the exposed hinge lines of mesoscopic folds.

Transected cleavages can develop when folding takes place in a layer oblique to all the principal axes of strain (Borradaile 1978, Stringer & Treagus 1980, Ramsay & Huber 1983, p. 229, Ramsay & Huber 1987, p. 334). Cleavage surfaces in a rock are approximately parallel to the XY-plane of the strain ellipsoid while the fold axis is approximately parallel to the long axis of the strain ellipse (Ramberg 1959, Flinn 1962, Ghosh 1966, Treagus & Treagus 1981) on the plane of the layer. When the layer is inclined to the X-, Y- and Z-axes, the long axis of the ellipse may not be parallel to the XY-plane. This is the reason why the cleavage may transect the fold hinge at a low angle. Mawer & Williams (1991) envisaged a continuous process of fold formation and transposition. The cleavage will transect the fold when a new cleavage has formed by transposition at the microscale, but the fold marked by a coarse layering is still unaffected.

Other relations

Not all cleavages are associated with folding. When a ductile shear zone develops in massive rocks like granites or greenstones the cleavage which initially develops from the massive rock (e.g. Berthé et al. 1979) is not associated with a synchronous system of folds.

The axial planar nature of a cleavage is not easy to identify in all cases. In areas of isoclinal folding the cleavage is parallel to the bedding in the fold limbs but cuts across the bedding in the fold hinges. In such terrains the axial planar nature of the cleavage can be identified only by observations at the hinge zones. Where there are several generations of folds the hinges of the earliest folds are difficult to locate. Some typical problems encountered in the field are illustrated in Fig. 14.6. In the major part of each of the three characteristic domains of three different areas represented in the figure, the cleavage S is parallel to the bedding. In the domain represented by Fig. 14.6a, an early fold hinge can be identified. The cleavage is axial planar to this early fold. In Fig. 14.6b we can identify two generations of folds. The cleavage is axial planar

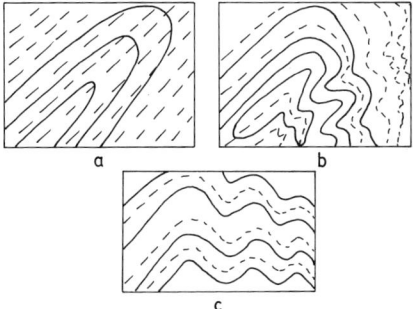

FIG. 14.6. (a) Cleavage (dashed lines) axial planar to fold on bedding. (b) Cleavage axial planar to early folds. The early cleavage wraps around the noses of the later folds. (c) Cleavage parallel to bedding. It is likely, but not quite certain, that the cleavage is axial planar to an earlier generation of folds.

to the earlier generation of these folds. In Fig. 14.6c, the cleavage is everywhere parallel to the bedding. In the first two cases the cleavage, S, can be proved to be an axial plane cleavage. In the third case we are left in doubt. Since the relations such as are shown in Fig. 14.6a and b are found in many areas, it is likely that in the third case too the cleavage is axial planar to an early generation of folds the hinges of which have either been obliterated by transposition and shearing or have not yet been located by a careful search. Nevertheless, we cannot entirely rule out the possibility that the cleavage is a *bedding plane cleavage*, i.e. a cleavage which is initiated parallel to the bedding and is neither transected nor axial planar with respect to any system of folds.

In certain areas a supposed bedding plane cleavage is found to be deformed around the hinges of the earliest recognizable folds. If the cleavage is identified as a partially modified depositional fabric, it is represented by the symbol S_0 while the early folds are designated F_1. If later investigations indicate that the cleavage is not everywhere parallel to the bedding, the cleavage should then be regarded as post-S_0 and pre-F_1. Powell & Rickard (1985) suggested the use of the symbol S^* in referring to this early cleavage.

14.4. Use of axial plane cleavage in geometrical analysis

The use of axial plane cleavage in the geometrical analysis of an area has been emphasized by the early workers and discussed in detail by Leith (1923), Billings (1954) and Wilson (1946, 1961).

(1) Once a cleavage in an area is identified as an axial plane cleavage from observations in a few critical outcrops, the orientation of the

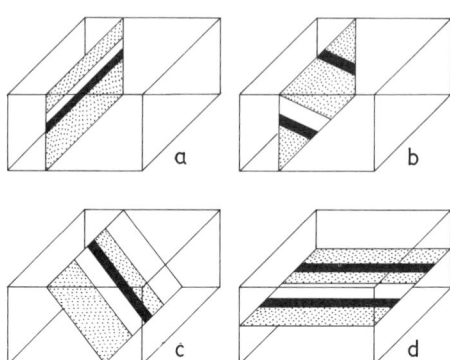

FIG. 14.7. Method of identification of (a) horizontal upright fold, (b) plunging upright fold, (c) reclined fold and (d) recumbent fold from attitudes of the axial plane and the bedding–cleavage intersection.

cleavage can be taken as the orientation of the axial planes throughout the entire area.

(2) As pointed out by Leith (1923), the trace of bedding on the cleavage surface is approximately parallel to the fold axis. This intersection lineation is often utilized to determine the orientation of the fold axis where the hinge lines of mesoscopic folds are unexposed.

(3) The attitude of the cleavage together with the pitch of the intersection lineation on the cleavage surface give a fair idea of the geometry of the fold. Figure 14.7 shows four cases in which a horizontal upright fold, a plunging upright fold, a reclined fold and a recumbent fold can be identified by noting the attitudes of the cleavage and the intersection lineation. The average orientation of the cleavage gives us some information about the geometry of the large-scale folds. Thus, if the cleavage is sub-vertical in the majority of outcrops, we may conclude that the large-scale fold, if any, is upright. Similarly, if from a reconnaissance study we find that the cleavage in an area is often sub-horizontal, we search for the presence of large-scale recumbent folds in that area.

(4) The bedding and the axial plane cleavage remain at a right angle to each other at the hinge of a fold. This geometrical relation can sometimes be used to locate the hinge zone of a large-scale fold.

(5) The relative steepness of bedding and cleavage enables us to distinguish between normal and overturned limbs of antiforms and synforms. In the normal limb the cleavage is steeper than the bedding and in the overturned limb the bedding is steeper than the cleavage (Fig. 14.8). Hence, where the antiforms are anticlines and the synforms are synclines, we can also deduce the stratigraphic succession from the relative steepness of bedding and cleavage.

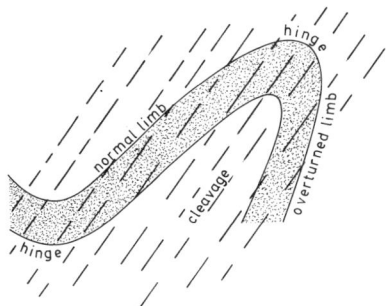

FIG. 14.8. Bedding–cleavage relation in normal limb, overturned limb and hinge. Cleavage is steeper than bedding in normal limb; it is less steep than the bedding in overturned limb and the two are mutually perpendicular at the hinge.

14.5. Axial plane cleavage and strain

As remarked by several early workers, notably Tyndall (1856) and Harker (1886), cleavage and folding are both manifestations of the same deformation. However, cleavage does not appear in the rocks until the deformation has progressed to a certain extent. Apart from folding, the deformation in a cleaved rock is shown in many other ways. In a slaty or schistose rock the fossils are distorted (Phillips 1844, Sharpe 1847, Haughton 1856, Daubrée 1876, Harker 1886, Wettstein 1886, Heim 1919, Breddin 1956, 1957, Ramsay 1967, Tan 1973 and many others), reduction spots are flattened (Sorby 1853, 1856, Tullis & Wood 1975), oolites are deformed to ellipsoidal shapes (Cloos 1947b) and all primary inclusions such as amygdales in volcanic rocks and the lapilli in tuffs (Oertel 1970, Helm & Siddans 1971, Mukhopadhyay 1972) are strongly stretched parallel to the cleavage surfaces. Sharpe (1847) also observed that with increase in the intensity of distortion of the fossil shells the cleavage in the slates become more pronounced. The threshold value of shortening beyond which a cleavage appears in a rock probably depends on the lithology as well as on the prevailing metamorphic conditions. From a very detailed study of deformation of the oolites from the South Mountain area of the Appalachian fold belt, Cloos (1947b) found that a cleavage first appears in the limestones when the shortening is in excess of about 30 per cent. From a similar detailed study of slates from several areas Wood (1974, p. 399) has found that development of a prominent slaty cleavage entails a shortening in excess of 50 per cent.

14.6. Axial plane cleavage in relation to principal axes of finite strain

In the purple Cambrian slates of Wales, Sorby (1855) found green reduction spots which were nearly spherical or were in the shape of uniaxial oblate type of ellipsoids in the undeformed rocks. In the deformed rocks the reduction spots are greatly shortened perpendicular to the cleavage, with the long and intermediate axes of the deformed ellipsoids parallel to the cleavage planes of the slates. Sorby concluded that the slaty cleavage plane is perpendicular to the direction of maximum shortening in the rocks. Later strain measurements from natural strain gauges (e.g. Cloos 1947b, Ramsay & Graham 1970, Siddans 1972, Wood 1974) confirm this finding. On the basis of such evidence the majority of structural geologists have concluded that slaty cleavage is parallel to the XY-plane of finite strain. Indeed, there is hardly any controversy regarding the conclusion that the axial plane foliation, including slaty cleavage, schistosity and crenulation cleavage, is either parallel or is nearly parallel to the XY-plane of the finite strain ellipsoid. In strongly deformed rocks we may find, in addition, on the plane of a slaty cleavage or schistosity, a mineral lineation approximately parallel to the direction of maximum stretching in the rock. Evidently, such a linear structure is not found when the bulk strain ellipsoid is of the uniaxial oblate type. These conclusions are of great importance in regional structural analysis. Thus, the orientation of the axial plane cleavage not only gives us information about the geometry of the folds, it also gives us the orientation of the XY-plane. Where, in addition, the lineation can be identified as the X-axis, the approximate orientations of all the principal axes of finite strain become known.

14.7. Problems concerning the origin of cleavage

The origin of cleavage has been a subject of discussion for many years. There is a vast literature on this topic and excellent reviews (Harker 1885, 1886, Leith 1905, 1923, Swanson 1941, Wilson 1946, 1961, Voll 1960, Siddans 1972, Wood 1974, Williams 1977, Oertel 1983). Even from the earliest researches it was obvious that the development of cleavage is a complex process and involves several factors. Wood (1974) and Ramsay & Huber (1983, p. 185) have pointed out that the broad groups of these factors were identified in the classical studies on cleavage by Sorby (1853, 1856, 1858, 1879b).

A large part of early discussions about cleavage centred on two alternative hypotheses. According to one of these, cleavage develops parallel to the XY-plane of the strain ellipsoid (Sharpe 1847, Sorby 1853, 1856, Tyndall 1856, Haughton 1856, Harker 1885, 1886, Van Hise 1896, Leith 1905, Ramsay 1967, Siddans 1972, Wood 1976, Tullis 1976). The other hypothesis states that cleavage forms parallel to a plane of shear strain (Phillips 1844, Fisher 1884, Becker 1896, 1904, Sander 1930, Hoeppener 1956, Wunderlich 1959, Voll 1960, Williams 1976). The difference between the two hypotheses rests on the fact that the principal axes of the strain ellipsoid forms a unique set of material lines which remain perpendicular to one another before and after deformation. As a consequence there cannot be any finite shear strain on the XY-plane. Initially, the second hypothesis suffered from the incorrect assumption by Becker (1896) that cleavage develops parallel to the circular sections of the strain ellipsoid; these sections were also taken by him to be parallel to the surfaces of no distortion as well as to the surfaces of maximum finite shear strain. As we have seen in Chapter 8 and as initially pointed out by Griggs (1935), the surfaces of no distortion are generally elliptical cones which do not coincide with circular sections. Moreover, neither of these surfaces coincides with the surfaces of maximum shear strain. In its present form the second hypothesis avoids this error and states that at least some foliation is initiated parallel to a plane of shear strain though not necessarily the plane of maximum shear strain (Hobbs *et al.* 1976, Williams 1977).

The advocates of the first hypothesis put emphasis on the fact that whenever the XY-plane is determined by measurement of strain from natural strain indicators, it is essentially parallel to the orientation of the cleavage. On the other hand, we sometimes find evidence of shear strain and displacement of transverse compositional layers (Fig. 14.9b) by slip along cleavage surfaces. The advocates of the second hypothesis emphasize these observations and conclude that, in certain cases at least, the cleavage is initiated parallel to a plane of shear strain.

Slaty cleavage and schistosity are defined by a preferred orientation of platy minerals. The processes which lead towards the development of preferred orientation are often discussed under the headings of (1) rotation, (2) recrystallization, (3) plastic deformation, (4) pressure solution and (5) dewatering. It is possible that some of these processes act together in such way that their separate contributions are hard to distinguish. Moreover, it may happen that the actual mechanism by which a particular fabric is brought about is too complex to be described in terms of factors listed above. In any case, for an understanding of the process of cleavage development a description of each of these processes is a convenient starting point.

14.8. Rotation

Cleavage in slates and schists is largely defined by preferred orientation of phyllosilicate grains. In undeformed pelitic rocks the detrital grains of phyllosilicates are either poorly oriented or show a rough parallelism with the bedding plane. It is likely that in a freshly deposited clay the very fine flakes of phyllosilicates are randomly oriented while the coarse flakes are sub-parallel to the bedding. Under the load of the overburden the soft sediments undergo compaction. As a result the

FIG. 14.9. (a) Cleavage parallel to axial surface traces of plunging folds on bedding in mica schist. Ghatsila, India. (b) Offset of bedding laminae by slip on cleavage. The front face of the specimen is oblique to both the bedding and the cleavage. Aravalli phyllite, Rajasthan, India. Specimen from collection of A. B. Roy.

detrital grains rotate towards the bedding. Before the beginning of tectonic deformation we may thus have a compaction-induced preferred orientation of detrital grains leading to the development of bedding fissility. In slates in which the cleavage is initiated at an angle to the bedding fissility the cleavage-parallel flakes of phyllosilicates can sometimes be identified as detrital grains (Williams 1972). This is a clear proof that the preferred orientation of phyllosilicates in slates is at least partly caused by the purely mechanical process of rigid body rotation. For similar reasons Sorby (1853) considered rotation to be the dominant mechanism for the preferred orientation of phyllosilicates in slates.

Although it is realized that development of cleavage is not entirely a mechanical process, recent researches suggest that rotation is an important mechanism in the process of development of cleavage. Several authors suggested that the rotation of phyllosilicates takes place in accordance with the theoretical model proposed by March (1932). The *March model* assumed that the initial platy grains in a rock are randomly oriented. It was also assumed that the grains deformed in the same manner as the matrix. In other words, the platy grains play only the role of passive markers. March's analysis shows that after homogeneous deformation the largest concentration of the platy grains will be along the XY-plane.

To get an insight into the March model consider a simpler situation in which randomly oriented acicular grains, within a circular area of radius r, are deformed by homogeneous plane strain, such as simple shear. The following simple derivation of a March model for the special case of simple shear is based on the analysis of Fernandez (1987).

Let θ and θ' be the initial and final angles which an acicular grain makes with the direction of shear (Fig. 14.10). We have, then, by eqn. (8.73),

$$\cot\theta' = \cot\theta + \gamma. \tag{14.1}$$

In the undeformed circular area, let there be n passive markers oriented within an infinitesimally small angle $d\theta$. Then $n/d\theta$ can be taken as a measure of the orientation density in the undeformed state. In the deformed state the orientation density in a direction θ' is $n/d\theta'$. The relative orientation density is

$$D = \frac{n}{d\theta'} \bigg/ \frac{n}{d\theta} = \frac{d\theta}{d\theta'}.$$

Within the circular area, the area between the lines θ and $(\theta + d\theta)$ is $\frac{1}{2}r^2 d\theta$, where r is the radius of the circle. After deformation the circle is deformed to an ellipse and r is changed to the length r'. The area contained between the lines θ' and $(\theta' + d\theta')$ is $\frac{1}{2}r'^2 d\theta'$. If the area does not change during deformation

$$\tfrac{1}{2}r^2 d\theta = \tfrac{1}{2}r'^2 d\theta'$$

or,

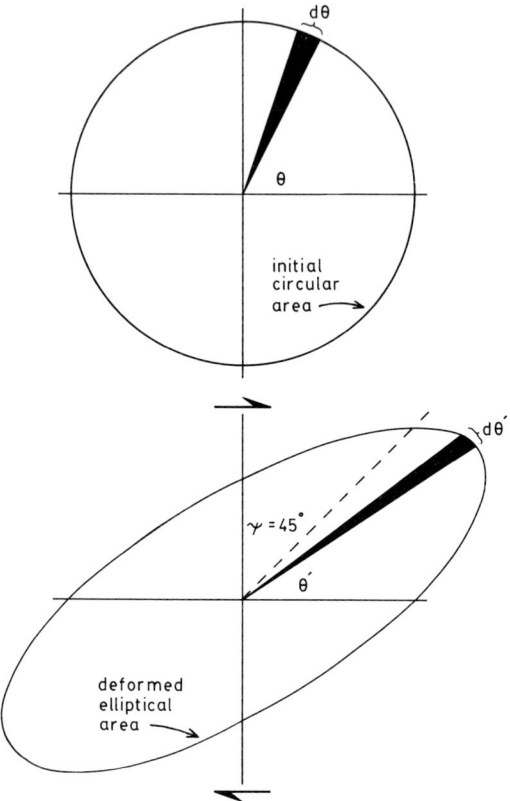

FIG. 14.10. An initial circular area (*upper figure*) is deformed to an elliptical area (*lower figure*) by simple shear ($\gamma = 1$). A small angle $d\theta$ is changed to $d\theta'$.

$$\frac{r'^2}{r^2} = \frac{d\theta}{d\theta'}.$$

Since r'^2/r^2 is the quadratic elongation along the direction θ',

$$D = \frac{d\theta}{d\theta'} = \lambda.$$

Since λ is greatest along the major axis of the strain ellipse, the relative orientation density is greatest in that direction. This simple derivation therefore shows that when randomly oriented passive markers are deformed, the largest concentration of the markers will be along the X-axis.

By differentiation of eqn. (14.1) we find that

$$\operatorname{cosec}^2\theta' = \operatorname{cosec}^2\theta \frac{d\theta}{d\theta'}$$

or,

$$D = \frac{d\theta}{d\theta'} = \frac{\sin^2\theta}{\sin^2\theta'}.$$

To express D in terms of θ' only, we note that

$$\frac{\sin^2\theta'}{\sin^2\theta} = \sin^2\theta' \operatorname{cosec}^2\theta = \sin^2\theta' (\cot^2\theta + 1).$$

Expressing $\cot\theta$ in terms of $\cot\theta'$ by eqn. (14.1), we find after simplification

$$D = \frac{\sin^2\theta}{\sin^2\theta'} = [(\cos\theta' - \gamma \sin\theta')^2 + \sin^2\theta']^{-1},$$

or, in another form,

$$D = (\gamma^2 \sin^2\theta' - 2\gamma \sin\theta' \cos\theta' + 1)^{-1}.$$

It is easy to confirm that D is a maximum when θ' coincides with the orientation of the major axis of the strain ellipse. Thus, D is a maximum or $1/D$ is a minimum when

$$\frac{d}{d\theta}\left(\frac{1}{D}\right) = 2\gamma^2 \cos\theta' \sin\theta' - 2\gamma \cos2\theta' = 0$$

or, as in eqn. (8.69),

$$\tan2\theta' = \frac{2}{\gamma}.$$

The three-dimensional derivation of the March model has been described by Owens (1973), who also generalized it for the situation of an initial non-uniform angular distribution of linear or planar elements.

The mechanism of development of a preferred orientation is much more complex than the March model. When the grains are closely spaced the rotating grains may interfere with each other. The grains of phyllosilicates may rotate by bending or development of kink bands. When the initial fabric is a bedding fissility the rotation of the detrital grains may be partly through development of crenulations (Williams

1972, Gray & Durney 1979a). Moreover, as we shall see in the following sections, mechanical rotation is not the only factor which may cause a preferred orientation.

To test the validity of the March model one may determine the degree of preferred orientation of the phyllosilicates and also independently determine the magnitudes and directions of principal strains from natural strain indicators. The data can then be compared with the degree of preferred orientation as predicted by the March model for that strain. Some of these studies show a close correspondence between the observed and predicted preferred orientations (Wood 1974, Tullis & Wood 1975, Wood *et al.* 1976, Wood & Oertel 1980).

A test for the Marchian behaviour may also be obtained from experimental deformation. An advantage of this method is that both the initial and final fabrics are known. Means *et al.* (1984) performed experiments with layered samples with alternate layers of KCl–mica and NaCl–mica mixtures with a marked layer-parallel preferred orientation of the mica flakes. The experiments showed a remarkable feature. For moderate values of shortening (56 and 68 per cent) the degree of preferred orientation of the mica flakes is 20 and 30 per cent stronger than the concentrations predicted by the March model employing the initial preferred orientation of the samples. However, the observed preferred orientations are consistent with those predicted by the March model with a random initial fabric. For larger values of shortening the observed preferred orientations deviated significantly from those predicted by the March model.

14.9. Syntectonic growth

The development of cleavage is by no means a purely mechanical phenomenon. Among the early workers it was Van Hise (1896) in particular who emphasized this aspect of cleavage development. He pointed out that the fine flakes of mica and chlorite oriented parallel to the slaty cleavage were not detrital but had grown in the rock. Indeed, recent researches show that among the constituents of slates only illite, some detrital muscovite and some of the iron-rich chlorites show no evidence of growth; the major constituents such as phengite, chlorite, quartz and possibly paragonite all show evidence of growth and have formed by metamorphic reactions (Knipe 1981). With further reconstitution slates may grade into phyllites and schists.

There is a general consensus that the growth of phyllosilicates in slates and schists is dominantly syntectonic and that the development of slaty cleavage and schistosity results from an interaction of deformation and metamorphic growth processes. The preferred orientation of the growing phyllosilicates may be brought about by any one or a combination of the following processes: (1) oriented growth in a stress field, (2) oriented growth controlled by strain, (3) oriented growth controlled by pre-existing fabric and (4) unoriented growth synchronous with mechanical rotation.

(1) *Syntectonic growth in a stress field*

It has been suggested that preferred orientation of anisotropic minerals is brought about by oriented growth in a stress field (Kamb 1959). According to this theory the micas grow with (001) perpendicular to the greatest compressive stress. The theory in its present form is difficult to reconcile with the fact that rock cleavage is approximately parallel to the XY-plane of finite strain. The principal axes of stress and strain coincide in case of coaxial deformation and in case of both coaxial and non-coaxial infinitesimal deformations. The principal axes of stress and strain deviate in case of finite non-coaxial deformation.

(2) *Oriented growth controlled by strain*

In low grade metagraywackes and slates we sometimes find a localized microstructure known as *mica beard* (Ransom 1965, referred to by Williams 1977, Powell 1969, Williams 1972, Means 1975, Etheridge & Lee 1975, Roy 1978). These tapering tufts of mica grains are oriented parallel to the megascopically visible rock cleavage and grow outward from the clastic quartz grain boundaries which are at a high angle to the cleavage. The microstructure indicates that the mica beards grow in local sites of extension and are nucleated parallel to the local direction of maximum extension. According to Williams (1977, p. 317), the cleavage in some of these low grade rocks is marked out dominantly by the mica beards. The microstructures of slates may show other evidence of oriented growth. An important conclusion of recent studies is that a slaty cleavage may develop by crenulation of a bedding fissility marked by orientation of detrital phyllosilicates. A detailed description of the development of a slaty cleavage through such a stage of crenulation of bedding foliation has been given by Williams (1972), Knipe & White (1977) and Knipe (1981). In such situations we may find flakes of phyllosilicates aligned parallel to the axial planes of the crenulations and cutting across the bedding foliation in the hinge zones of crenulations (Knipe & White 1977, p. 372, Knipe 1981, fig. 2a). The occurrence of such a microstructure suggests the absence of rotation of the newly grown phyllosilicate.

Oriented growth of phyllosilicates may also take place along kink band boundaries of old grains. Such microkinking is likely to be common when there is a shortening along the bedding foliation. Along the kink band boundaries the bending of the lattice is intense and hence such boundaries serve as preferred sites of recrystallization (Tullis 1971, Hobbs *et al.* 1976, p. 249, Knipe & White 1977, p. 372).

(3) *Mimetic crystallization*

In metamorphic terrains the slaty cleavage may grade into the schistosity of the adjoining phyllites and schists. The passage from slates to schists is accompanied by a coarsening of the grain size of the phyllosilicates. In certain regions at least, it is likely that the higher grade schists had evolved from lower grade rocks with progressive increase in grain size accompanied by some compositional change of the oriented phyllosilicates. In other words, it is likely that in some rocks the phyllosilicates have grown in size, mostly along their length, the growth being controlled by the pre-existing fabric. This process, known as mimetic crystallization, is thought to be fairly common (Sander 1930, p. 207, Turner & Weiss 1963, p. 441). Mimetic crystallization does not explain the initial development of preferred orientation; however, the process may be of importance in accentuating the degree of preferred orientation. Mimetic crystallization has sometimes been regarded as post-tectonic (Spry 1969, p. 257). Yet as Read (1949) has suggested, mimetic crystallization "may conclude a series of related deformations and crystallizations — it may simply outlast deformation".

(4) *Unoriented growth synchronous with rotation*

There are hardly any microstructural criteria by which we can test the efficiency of this process. Some insight into this problem may be obtained from the experiments by Tullis (1976). The starting material in these experiments was a phlogopite oxide mix or powdered and compacted fluorophlogopite, biotite and very fine-grained muscovite. When the experiments were conducted under a hydrostatic pressure and high temperature, a decussate texture was produced with randomly oriented recrystallized flakes of fluorophlogopite. In another experiment a similar recrystallization under hydrostatic heating was followed by a uniaxial shortening of the specimen at a temperature in which no recrystallization was possible. A weak foliation developed in the specimen. The individual flakes showed signs of strong post-crystalline deformation. In the other experiments the specimens were shortened at a temperature in which recrystallization was also possible. The deformed specimens show a well-developed cleavage in which the individual flakes of phyllosilicates were relatively free from undulatory extinction and microkinking. For these latter experiments the degree of preferred orientation, corresponding to the strain, is compatible with March model. This series of experiments led Tullis to conclude that possibly selective nucleation and (or) oriented growth was not significant and that preferred orientation parallel to XY-plane was achieved essentially by rotation of the recrystallizing grains. On the other hand, in similar experiments by Etheridge *et al.* (1974, p. 282) the preferred orientation of syntectonically recrystallized micas was found to be stronger than that predicted by the March model. Tullis (1976, p. 751) therefore keeps open the alternative that selective nucleation or oriented growth may "fortuitously give the same relationship between strain and preferred orientation as does mechanical rotation due to shape".

14.10. Plastic deformation

Cleavage in quartzites or marbles is defined by preferred orientation of elongate grains of quartz or calcite. The most important deformation mechanisms in these rocks are dislocation creep and pressure solution. There are very few quantitative studies (Mitra 1976, 1978) to assess the relative importance of these two mechanisms in natural examples. Mitra's investigations demonstrate that the two mechanisms may operate together; under certain situations pressure solution is the more active mechanism and in other situations dislocation creep is more important. Most geologists believe that textures characteristic of pressure solution are common at low temperatures while crystal plastic flow textures dominate in rocks which have been deformed at high temperatures (Kerrich 1978, Rutter 1976, Mitra 1976, 1978, Ramsay & Huber 1983, pp. 120–124).

Unless the metamorphic temperature is very low, cleavage in mono-mineralic rocks forms dominantly by plastic flattening of grains. The foliation produced by such flattening is essentially parallel to the XY-plane. The flattening of the grains involves intracrystalline slips and twinning. Under laboratory conditions, similar cleavages have been produced in both marbles (Griggs *et al.* 1960) and quartzites (Tullis *et al.* 1973).

Evidence of intracrystalline deformation in foliated quartzites (e.g. Tullis *et al.* 1973, Mitra 1976, White 1976, Bouchez 1977, Behrmann 1985) is furnished by the occurrence of undulatory extinction, deformation bands and deformation lamellae in optical microscopic studies and by the recognition of dislocation sub-structures in transmission electron microscopy. Plastic deformation is also indicated by the development of preferred crystallographic orientations of the minerals. In certain rocks the grains of quartz contain inclusions of rutile needles. When the grains are deformed plastically the rutile needles are boudinaged or folded. As shown by Mitra (1976, 1978), the magnitude of plastic deformation of the individual grains can be estimated from the boudinaged and folded rutile needles. This excellent work provides us with a simple method of estimating the relative importance of the different deformation mechanisms in the total deformation of a rock. The method may also be applied to higher grade rocks in which the quartz grains may show boudinage of included sillimanite or kyanite needles.

With progressive deformation the flattened grains may be partly or wholly replaced by an aggregate of syntectonically recrystallized smaller equant grains (White 1973a, b, Bell & Etheridge 1976). When partially replaced, the elongate remnants of the flattened grains, occurring as porphyroclasts, mark out the foliation in the rock. In some mylonites all the flattened grains may be replaced by a mosaic of small equant grains so that the foliation is no longer defined by a preferred shape orientation of the grains. The rock nevertheless shows a megascopically well-defined fissility sub-parallel to the XY-plane of bulk strain. In thin sections the foliation may also be recognized by subparallel alignment of lenticular areas of slightly different grain size and by orientation of trails of dusty particles which mark out the boundaries of previous elongate grains.

In addition to producing a preferred orientation of elongate grains, plastic flow may produce a banding which enhances the fissility of the rocks. Where elongate grains are present they are invariably parallel to the bands. In contrast with the differentiated layering resulting from chemical redistribution of materials, the banding in plastically deformed rocks is produced by an essentially mechanical process. So far, three types of banding have been recognized. In low temperature quartzite mylonites some of the quartz grains are strongly flattened and attain a ribbon-like shape. The banding is produced by the quartz ribbons alternating with zones of finely recrystallized quartz (Bouchez 1977, fig. 5). The banding produced in mylonites made up of several minerals have been sub-divided into two types (Boullier & Gueguen 1975, pp. 101–102). The first type corresponds to a high stress deformation at a temperature which is much lower (less than half) than the melting point of the minerals. When the rock is deformed there is an initial stage of intragranular gliding. After this stage one of the minerals recrystallizes first (olivine in peridotite or quartz in granite) while other minerals (orthopyroxene in peridotite or feldspar in granite) continue to deform by gliding and remain as long augen. A fine banding is thereby produced by strongly elongated augen of one mineral alternating with recrystallized mosaics of another mineral (Figs. 14.11, 21.2b). Boullier and Gueguen showed that the augen of orthopyroxene are finally boudinaged, with the train of boudins forming a discontinuous band. If the mylonite is produced in a high temperature, low stress regime, the augen begin to recrystallize at a later stage and the cluster of small recrystallized grains is drawn out as a long tapering tail. Such a drawing out of the recrystallized mosaic and the consequent development of a tectonic layering imply deformation by grain boundary sliding. This deformation mechanism does not have any importance at low temperatures, i.e. at temperatures below half the melting point of a mineral. At higher temperatures the grain boundaries become weak

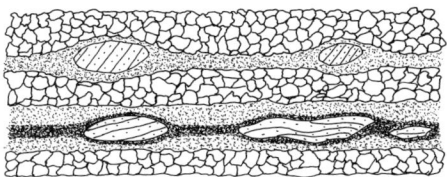

FIG. 14.11. Microscopic banding produced in granite mylonite. Bands of equant recrystallized quartz alternate with bands of fine-grained feldspar mosaic containing elongate feldspar porphyroclasts.

zones and a very fine-grained rock may deform by sliding along grain boundaries much as a volume of loose sand deforms by intergranular movements. Grain boundary sliding is considered to be dominant in superplastic deformation, a type of deformation in polycrystalline materials which allows an unusually large tensile elongation without necking. It has been suggested that when the grain size of quartzite mylonites becomes very small (less than 10 microns), there is a switch of deformation mechanism from plasticity to superplasticity (Behrmann 1985). Behrmann and Mainprice (1987) have suggested that during deformation of some high temperature (amphibolite facies) granite mylonites there is a partitioning of plasticity and superplasticity in alternate bands. The superplastic bands are created by drawing out of a very fine-grained mosaic as long tails of feldspar porphyroclasts.

14.11. Solution transfer

Pressure solution or *solution transfer* refers to a process by which the material from a crystal or an aggregate of crystals is dissolved where it faces the principal compressive stress and is redeposited on the faces which are at a high angle to the principal tensile stress. Its importance in the development of metamorphic textures was recognized long ago. An early evidence of pressure solution came from Sorby's (1863) study of pitted pebbles of limestones. The pitted appearance of the pebbles was explained by him by a process of pressing together of the pebbles; at the pressed contacts the most soluble parts were dissolved and gave rise to the pitted appearance. The importance of pressure solution in the development of cleavage was also emphasized by Van Hise (1896) and Becke (1903).

The theory of pressure solution predicts continuous variation of solubility around differentially stressed grains. There is thus the possibility of diffusion of materials from points of higher to points of lower stress (Durney 1978). Diffusion may take place either through the crystal lattice or along grain boundaries. Since lattice diffusion requires temperatures (Ashby 1972) much higher than what is prevalent in low

315

grade metamorphism, diffusion along grain boundaries is the most likely mechanism of pressure solution in rocks (Durney 1972, Elliott 1973).

Several authors have emphasized the role of pressure solution in the development of a penetrative cleavage. Because of solution-redeposition, the grains of quartz become elongated in a direction normal to the principal compressive stress. Again, solution transfer of quartz from the interstices of mica flakes leads to the formation of long narrow stringers of mica seams. With continued shortening the mica flakes within the seams rotate towards the XY-plane. In the final microstructure, the mica seams anastomose around lenticular aggregates of quartz (Williams 1972, Morris 1981). Plessman (1964) has shown that similar mica-rich seams may form by solution transfer of carbonate in some carbonate-rich slates.

Foliated rocks often show a layering parallel to the cleavage. In the previous section we have seen how such a layering can form by plastic and superplastic deformation. More commonly, however, diffusive mass transfer is likely to be the most dominant mechanism for the development of a cleavage-parallel layering. Differentiated layering parallel to crenulation cleavages is usually explained by such diffusive mass transfer. According to Williams (1972), Cosgrove (1976) and Gray (1979) the normal compressive stress is larger in the limbs than in the hinge zones of crenulations. It is suggested that dissolution of quartz from the limb regions and its redeposition in the hinge zones may produce alternate mica-rich and quartz-rich zones parallel to the axial planes of crenulations. A similar differentiated crenulation cleavage was produced experimentally by Means & Williams (1974) in salt–mica mixtures. Gray & Durney (1979b, p. 79) have pointed out that, since quartz is not stiffer than the micas, the mean stress in quartz cannot be greater than in the micas. Solution transfer from crenulation limbs takes place, not because the stress is greater in quartz, but because quartz is more soluble and therefore more mobile than the micas.

The modification of a texture by solution transfer may be recognized by the following microscopic features. Inert particles in the form of dusty opaque minerals may accumulate at the boundary in which the grain has been dissolved and along the old grain border where the new material has been added. In certain cases the old grain contains dusty inclusions while the newly added material growing outward is free of inclusions. The contact between these two areas marks out the old grain boundary. The old grain and the freshly deposited material may or may not be optically continuous. Pressure solution can be recognized by the truncation of the old grain boundary (Fig. 14.12a, b). In some cases pressure solution may also be recognized by the stylolitic appearance of

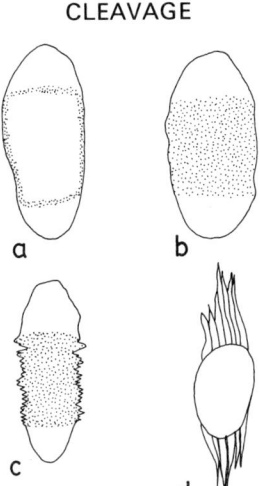

FIG. 14.12. Some microscopic features indicating solution transfer.

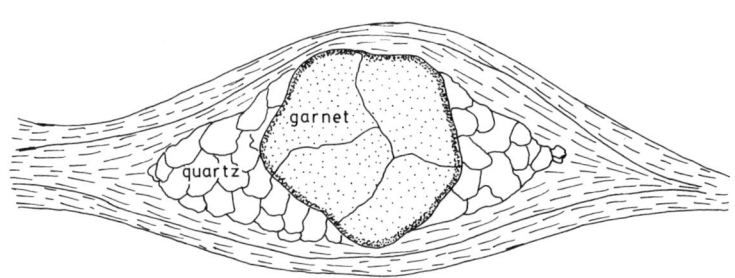

FIG. 14.13. Quartz in pressure shadow zone of garnet porphyroblast.

the grain borders facing the shortening direction (Fig. 14.12c). The freshly deposited material on the tensile side has sometimes a fibrous appearance (Fig. 14.12d), the fibres being parallel to local direction of incremental extension. The structure known as pressure shadow is a result of solution-redeposition process. These tapering zones, growing outward from grain borders facing the extension direction, occur on opposite sides of rigid grains of minerals like garnet (Fig. 14.13), pyrite, magnetite, etc. When visible with the naked eye, the pressure shadow zones form a linear structure on the foliation surface (Fig. 16.2a).

14.12. Dewatering hypothesis

Maxwell (1962) suggested that axial plane cleavage in slates may develop during deformation of water-rich unconsolidated sediments.

According to Maxwell the Martinsburg slate in Pennsylvania and New Jersey, U.S.A., contains sandstone dykes parallel to the cleavage planes of slates. Since sandstone dykes can only form in unconsolidated sediments, he suggested that water escaping from a volume of deformed unconsolidated sediments caused a mechanical rotation of the clastic grains and thereby gave rise to an axial plane cleavage. Later studies have shown that Maxwell's hypothesis is unacceptable as a general mechanism for the development of slaty cleavage. It can neither explain fossil and pebble distortion associated with cleavage formation nor can it explain the development of cleavage in metamorphosed igneous rocks. Maxwell's hypothesis implies that the slaty cleavage is mainly represented by elongate detrital grains. However, as discussed in section 14.9, the major constituents of slates are not detrital; they have formed by metamorphic reactions.

14.13. Crenulation cleavage

When an earlier cleavage is microfolded, a new cleavage may develop sub-parallel to the axial surface. This new cleavage is called *crenulation cleavage* (Knill 1960, Rickard 1961). This structure was earlier variously described as *Ausweichungsclivage* (Heim 1878), strain-slip cleavage (Bonney 1886), *Umfaltungsclivage* (Sander 1911), transposition cleavage (Turner & Weiss 1963, p. 92), etc. Crenulation cleavages generally form in phyllosilicate-rich rocks.

There are two main types of crenulation cleavage, discrete and zonal (Gray 1977, Powell 1979). *Discrete crenulation cleavages* are thin, sharply defined discontinuities, either planar or anastomosing, which truncate the pre-existing cleavage (Fig. 14.14a). *Zonal crenulation cleavages* are laminar domains through which the earlier fabric is continuous (Fig. 14.14b). Crenulation cleavage is a *spaced cleavage* with two different types of domains. In rocks showing a discrete crenulation cleavage the discontinuities may be represented by only a concentration of opaque minerals, without any concentration of phyllosilicates. With progressive deformation, the discontinuities are marked by thin zones of oriented phyllosilicates truncating the earlier fabric. The laminar zones in between these *cleavage domains* are called *microlithons*. When the crenulation cleavage becomes better developed, the widths of the cleavage domains increase at the expense of the widths of microlithons (Fig. 14.15). A larger concentration of phyllosilicates in the cleavage domains than in the microlithons gives the rock a banded appearance.

The embryonic stage of a crenulation cleavage may not be represented by the formation of a discontinuity. It is often seen that microfolding of a pre-existing cleavage causes a solution transfer of quartz from the limbs towards the hinges (e.g. Williams 1972, Cosgrove 1976, Gray & Durney

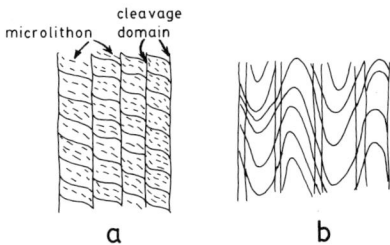

FIG. 14.14. (a) Discrete and (b) zonal crenulation cleavages.

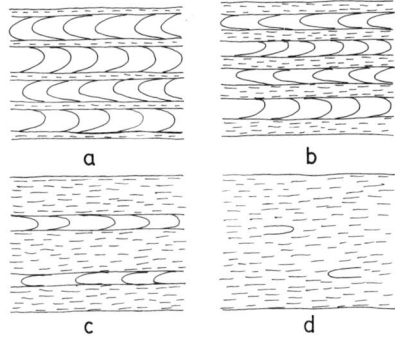

FIG. 14.15. Stages of transformation of crenulation cleavage to a continuous cleavage.

1979a, b). This gives rise to a banding roughly parallel to the axial planes of crenulations. This banding may then represent the embryonic crenulation cleavage. The earlier fabric may be traced continuously from the microlithons to the cleavage domains. With progressive tightening of the microfolds, the cleavage domains increase in width and the flakes of phyllosilicates become sub-parallel to the axial surface of crenulations. In the extreme case the microlithons are completely eliminated and the crenulation cleavage is transformed to a continuous cleavage (Fig. 14.15). Such a transformation of crenulation cleavage to a continuous cleavage is not rare. Evidence of such transformation is found by tracing transitional stages of progressive intensification of the crenulation cleavage to a continuous cleavage (Ghosh 1959, pp. 175–176, 1963, p. 214), by the preservation of relict zones of crenulated microlithons and by preservation of helicitic inclusion trails of crenulations (Fig. 14.16) in porphyroblasts occurring in a matrix which shows no sign of crenulation (e.g. Jacobson 1983a, Bell & Rubenach 1983). Vernon (1987) describes sillimanite-bearing rocks in which the crenulations have been completely destroyed in the main matrix while residual crenulation hinges of sillimanite folia are locally preserved in strain shadows of garnet porphyroblasts.

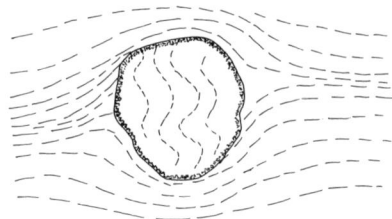

FIG. 14.16. Preservation of helicitic inclusion trails of crenulation in a porphyroblast. The crenulations in the matrix have been replaced by a later continuous cleavage.

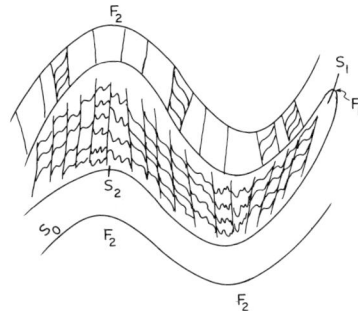

FIG. 14.17. Crenulation cleavage axial planar to later generation (F_2) folds.

Crenulation cleavages are approximately axial planar to mesoscopic and macroscopic folds which develop on an earlier cleavage (Figs. 14.17, 14.18a). Hence, a crenulation cleavage has the same importance in geometrical analysis as a penetrative axial plane cleavage. Recognition of crenulation cleavage is therefore important in the reconstruction of the structural history of a multiply deformed terrain. It should be noted that a crenulation cleavage may itself be folded and give rise to a later generation of crenulation cleavage (e.g. Cosgrove 1980, fig. 8, Price & Cosgrove 1990, p. 444).

The mechanism of crenulation cleavage formation involves buckling of the pre-existing cleavage folia. This is associated with a mechanical rotation of the limbs towards the axial surfaces and solution transfer of the most mobile constituents from limbs towards hinge zones. The flakes of phyllosilicates in the cleavage domain do not always show signs of post-crystalline deformation. Hence crenulation cleavage development, especially in its advanced stages, is likely to be accompanied by recrystallization of the phyllosilicate flakes. Such recrystallization of phyllosilicate flakes sub-parallel to the cleavage

FIG. 14.18. (a) Subvertical crenulation cleavage axial planar to upright folds on an earlier schistosity in mica schist, south of Ghatsila, India. (b) Extensional crenulation cleavage (ecc). The earlier cleavage is from left to right. The ecc runs from upper left to lower right.

domains must also have taken place during widening of the cleavage domains at the expense of the crenulated hinge zones of microlithons.

The crenulation cleavage is approximately parallel to the XY-plane of the bulk strain ellipsoid. However, there is no clear evidence to indicate that it is initiated strictly parallel to the XY-plane or that it continuously tracks the XY-plane of the finite strain ellipsoid. The drag of the earlier fabric at the boundary between the microlithons and the cleavage domains suggests a shear strain parallel to some of the crenulation cleavages. Offsets of lithological layers are sometimes seen along crenulation cleavage surfaces (e.g. Roy 1973, pl. 9). In certain cases the offsets are produced by pressure solution. However, in certain cases the lithological layering and the crenulation cleavage are at a right angle to each other. The offsets of the layers in such cases must have taken place by microfaulting. Such features led Hoeppner (1956), Turner & Weiss (1963, pp. 464–465) and Williams (1976) to suggest that crenulation cleavages are initiated parallel to planes of high shear strain. With continued deformation the cleavage surfaces rotate to become sub-parallel to the XY-plane. However, as yet there is no unequivocal evidence which indicates that crenulation cleavage is initiated parallel to planes of shearing.

14.14. Extensional crenulation cleavage

Platt & Vissers (1980) have suggested that crenulation cleavage may develop by an extension parallel to a pre-existing cleavage. The cleavage may occur in single or conjugate sets. This type of cleavage is known as *extensional crenulation cleavage*. The crenulations associated with this cleavage are always very open microfolds with one set of limbs much thinner than the other. The new cleavage forms in discontinuous segments along the thinner limbs of microfolds (Fig. 14.18b). Along the cleavage domains there is a high deformation intensity and sometimes a displacement of the S-surfaces. The sense of displacement is compatible with an overall extension along the pre-existing cleavage. Extensional crenulation cleavages are rather weakly developed structures which do not impart a fissility to the rocks. The microstructures associated with this type of cleavage suggest that it is initiated as microscale shear bands at a low angle to an original cleavage which was undergoing a cleavage-normal shortening. There is generally no sign of solution transfer in the crenulation cleavage domains. An analogous structure was produced by Means & Williams (1972) with a layer-normal shortening of salt–mica multilayers. The multilayer contained an initial layer-parallel fabric. A structure similar to extensional crenulation cleavage was produced by initiation of microscale conjugate faults or ductile shear zones symmetrically oriented about the maximum shortening direction.

14.15. Disjunctive cleavage

A *disjunctive cleavage* (Powell 1979) is a spaced cleavage in which there is no reorienting of the sedimentary layering or of the earlier fabric in the microlithons. The thin cleavage domains may show a concentration of inert opaque materials or may have some preferred orientation of phyllosilicates. Disjunctive cleavages show a variety of forms and they have been classified into different categories on the basis of spacing, planarity, continuity and thickness of cleavage domains (Alvarez *et al.* 1978, Powell 1979, Borradaile *et al.* 1982, Engelder & Marshak, 1985, Schweitzer & Simpson 1986). The cleavage domains may be planar, sutured, wavy or anastomosing. The structure previously described as fracture cleavage is a type of disjunctive cleavage.

Disjunctive cleavage is sub-parallel to the axial planes of associated folds, often with a fanned arrangement or refraction when passing through one lithology to another. In the majority of cases, disjunctive cleavages form by pressure solution (e.g. Plessmann 1964, Nickelson 1972, Groshong 1975a, b, Alvarez *et al.* 1976, 1978, Beach 1979, 1982, Mitra *et al.* 1984, Engelder & Marshak 1985, Schweitzer & Simpson 1986). This is clearly indicated by the occurrence of partially dissolved grains, oolites and fossils at the border of the cleavage domains. The most soluble material (quartz or calcite in most cases) is dissolved from the cleavage domain leaving a residue of insoluble materials. As a result of such pressure solution the rock undergoes a bulk shortening normal to the cleavage. The shortening is accommodated by a collapse of the solution zone while the microlithons remain essentially undeformed. A preferred orientation of phyllosilicates in the cleavage domains may develop by oriented growth (Mitra 1984) or by rotation of the phyllosilicates when the residual framework collapses under shortening like a "house of cards" (Gray 1981).

Disjunctive cleavages often show offset of bedding along the cleavage domains. There are three theoretical models by which such offsets can be produced (Foster & Hudleston 1986, fig. 10): (i) shearing along the cleavage domain, (ii) removal of material accompanied by shortening across the cleavage domain and (iii) addition of material and consequent extension across the domain. In the case of shearing, the offsets of differently oriented markers will have the same sense (Fig. 14.19d). In the other two cases, the magnitude and the sense of offsets will depend upon the orientation of the marker. The senses of offsets will be different for the second and the third cases (Fig. 14.19b, c). Since such differently oriented markers are rarely available, the exact mechanism which produces an offset may not be identifiable everywhere. However, where the offsets are seen in elongate microfossils, the mechanism of formation of the cleavage domains can easily be identified. If solution transfer has taken place, the two parts of a fossil cannot be matched by putting together the microlithons.

Groshong (1975a) has shown that the variation of dissolution-induced offsets along the inner and the outer arcs of a folded layer will depend upon the geometry of the zones of pressure solution. Removal

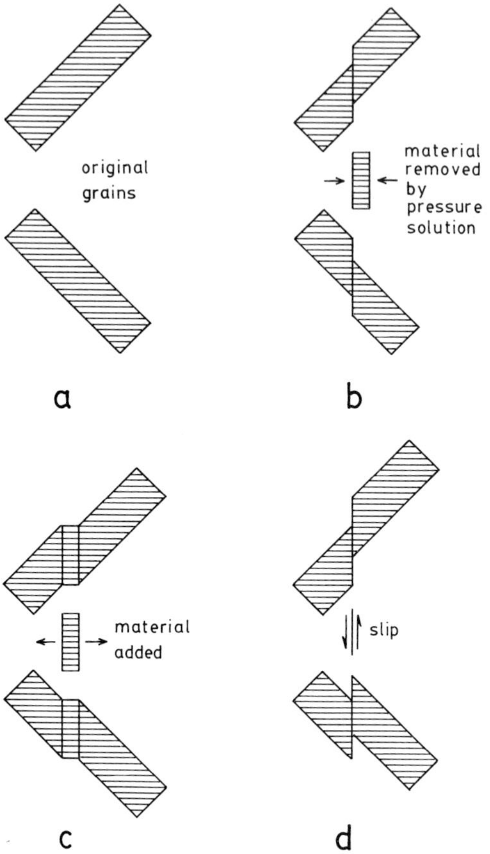

FIG. 14.19. Offset of two differently oriented laths (a) of plagioclase feldspar by (b) pressure solution, (c) addition of material, (d) slip along a shear plane. After Foster & Hudleston 1986.

of materials along zones perpendicular to the bedding does not cause any offsets (Fig. 14.20a). If selective removal takes place from laminar domains parallel to the axial plane of a fold, the offsets at the inner and the outer arcs will be the same (Fig. 14.20b). If there is an increasing amount of removal from the inner arc of a fold, with zones of removal oblique to bedding, the offsets will be greater in the concave side than in the convex side of a fold (Fig. 14.20c). In the fold described by Groshong the offsets are in agreement with the third model.

Not all disjunctive cleavages are produced by solution transfer and not all microscopic offsets along cleavage laminae are produced by this mechanism. There is clearly a need for more detailed studies of different types of disjunctive cleavages. Foster & Hudleston (1986) have demonstrated in an ingenuous way that spaced disjunctive cleavages occurring in the Duluth Complex, Minnesota, U.S.A., were indeed

CLEAVAGE

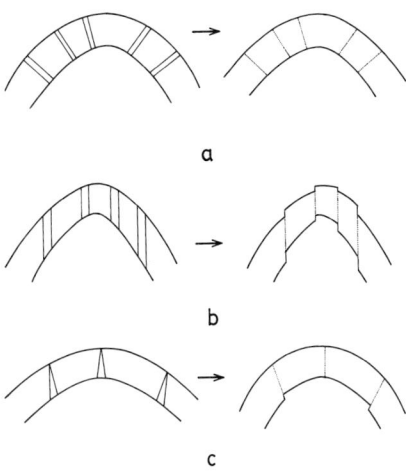

FIG. 14.20. Dissolution-induced offset of inner and outer arcs of a folded layer. The nature of offset depends upon the geometry of the zones of pressure solution.

produced by fracturing and not by pressure solution. The authors have shown that twinned plagioclase laths are offset along the cleavage laminae. The laths oriented perpendicular to the fractures have negligible offsets, thereby ruling out the possibility of shear along the laminae. When the laths are inclined in opposite senses to the cleavage trace, the offsets are in agreement with the model depicted in Fig. 14.19c, indicative of an extensional origin of the fractures. Intragranular fracture may also become important at some stage of cleavage development under greenschist facies conditions in psammitic rocks. We often find in such rocks cleavage seams of mica cutting across grains of quartz and feldspar. It is likely that there has been a gradual lengthening of cleavage seams in these rocks and the process was associated with development of fractures in the quartz and feldspar grains and crystallization of mica along the fractures (e.g. Gregg 1985, fig. 7).

The offset of passive markers across disjunctive cleavages may also result from a combination of pressure solution and slip along the cleavage. Murphy (1990) has described from the Irish Variscides a disjunctive cleavage which rotated out of parallelism with the XY-plane with progressive tightening of the folds, and the total offset of transverse passive markers took place partly by pressure solution and partly by intermicrolithon slip.

14.16. Concept of steady-state foliation

Means (1981) suggested that, under certain situations, rocks may develop a *steady-state foliation*. We have seen that the orientation of the foliation changes continuously in a non-coaxial flow. The intensity of foliation also increases with an increase in the finite strain. On the other hand, a steady-state foliation is conceived as a foliation with constant orientation and constant intensity in a rock undergoing a steady-state flow (flow in constant stress and constant strain rate). Since, in most situations, the intensity of foliation tends to increase with progressive deformation, there must also be concurrent foliation-weakening processes during the development of a steady-state foliation. A steady-state grain-shape foliation has recently been produced experimentally (Ree 1991) by the deformation of octachloropropane (OCP) in simple shear at 80 per cent of the absolute melting temperature of OCP. The history of development of the microstructure was followed in the course of the deformation by using the technique of synkinematic microscopy (Means 1989a). The experiments showed that the foliation intensity and the foliation orientation became steady after a bulk shear strain of about 0.9. The development of this steady-state foliation was associated with both foliation-strengthening and foliation-weakening processes. Foliation-weakening in this experiment took place mainly by migration of grain boundaries so as to reduce the aspect ratios of the grains and by division of an elongated grain into separate, more equiaxed parts by growth of other grains across it.

CHAPTER 15

Interfering Folds

15.1. Introduction

The occurrence of complex folds with curved hinges or curved axial surfaces was recognized long ago (e.g. Clough in Gunn *et al.* 1897, Peach *et al.* 1907, p. 601, Crampton in Peach *et al.* 1913, p. 57, Van Hise & Leith 1911, p. 123, Argand 1912). From these early studies it was clear that in an orogenic belt there might be more than one system of folds which interfere with one another. However, a systematic study of interfering folds started only since the middle of the 1950s (e.g. Sutton & Watson 1955, 1959, Gilmour & McIntyre 1954, Reynolds & Holmes 1954, Weiss *et al.* 1955, Cummins & Shackleton 1955, Scotford 1955, Johnson 1956, King & Rast 1956, Naha 1956, Brown & Engel 1956, Clifford *et al.* 1957, Weiss & McIntyre 1957, Berthelsen 1957, 1960, Ramsay 1958a, b, 1960, Rast 1958, Weiss 1959a, b, Rutland 1959). The major part of this early investigation was from the Scottish Highlands. Very soon, however, it became clear that development of complex structures by the superposition of successive generations of folds was common in all orogenic belts.

Fold interference may take place synchronously when the deformation is of the constriction type and the layers are so oriented that there is a shortening along the layering in all directions. In most cases, however, the folds which interfere with one another are produced successively. Such superposed folding may take place (i) in the course of a single continuous deformation with changing orientations of the stress axes, (ii) in the course of a single orogeny within which there was a superposition of separate deformations with different orientations of the stress or incremental strain axes and (iii) by superposition of deformations belonging to separate orogenies. The identification of each of these cases involves detailed geometrical analysis over large regions. It may not be possible in many cases to conclude with certainty that the superposed folding took place during a progressive deformation or by superposition of discrete deformations, in a single orogeny or in successive orogenies.

15.2. Interfering patterns in a single deformation

As shown by experimental deformations of soft models, even folds produced in a single deformation of non-constriction type may show small localized departures from the cylindrical shape of folds. These experimentally produced folds often show bifurcation of hinges (Ghosh & Ramberg 1968, fig.3A) very similar to the small-scale crenulations in schistose rocks. However, the hinges of these folds meet at low angles and the fold system as a whole remains roughly cylindrical.

Interfering fold patterns may also be produced when the bulk deformation is inhomogeneous. The geometry of the interfering folds will then depend upon the nature of inhomogeneity of the deformation. Ramsay (1962a) and Wood (1974), for instance, showed that the hinge of a large-scale antiformal fold can become curved as the fold grows by different amounts of shortening across the axial plane. The fold shows an axial culmination where the shortening is greatest. Moreover, stratified rocks folded over an irregular basement or stratified rocks distorted by the diapiric rise of the basement may show different types of fold interference. It is interesting to note that the fold arcuations and the axial culminations and depressions of the Pennine nappes were explained by Argand (1912) by the forward movement of the nappes over an irregular Hercynian basement. He emphasized that the folds as seen in transverse profiles, the arcuations or the curving of the fold hinges in plan view, and the axial culminations-depressions as seen in the longitudinal sections, all developed at the same time. The status of Argand's theory is likely to have changed in the course of detailed geometrical analysis (see "Alpine Tectonics" edited by Coward *et al.* 1989) of the Western Alps in the later years. Yet, for some terrains at least, it may be worthwhile to consider the general form of his proposed model in which he envisages the continuous development of folds by tangential forces and the continuous modification of their shapes by the intervention of the infrastructure.

Strictly simultaneous interfering folds may develop in a constriction type (Ramsay 1967) of deformation, with the layers so oriented that there is a shortening in all directions parallel to the layering. Experiments with soft models show that when the rates of shortening in all directions are nearly equal or have small differences, the folds usually show domes and basins of roundish or slightly elongate ground plan (Ghosh & Ramberg 1968). These are associated with strongly arcuate fold hinges without any preferred orientation (Fig. 15.1). The experiments by Ghosh & Ramberg show that, if there is a large difference between the rates of shortening in the two directions, one set of folds grows much faster than the other set. The total shortening in the direction of slower strain rate becomes significant only at an advanced stage of deformation; hence, although the shortening along the two directions is strictly simultaneous, the structures produced at an advanced stage of deformation resemble those produced by superposed folding.

Theoretical studies have shown that the size of domes and basins produced by constrictive deformations would be somewhat larger than the cylindrical folds which would have developed in the same system under uniaxial layer-parallel compression. As shown in Chapter 13, the arc length (L) or the Biot–Ramberg wavelength of cylindrical buckle folds embedded in a viscous medium is given by the relation

$$L = 2\Pi h \sqrt[3]{\frac{1}{6}\frac{\mu_1}{\mu_2}}. \tag{15.1}$$

For domes and basins the corresponding equation is

$$\frac{1}{L_1^2} + \frac{1}{L_2^2} = \frac{1}{L^2} \tag{15.2}$$

(Ghosh 1970, eqn.18) where L_1 and L_2 are the arc lengths along the two principal directions and where L is given by eqn. (15.1). The ratio of L_1 and L_2 will depend upon the ratio of the rates of shortening along the principal directions parallel to the layer. Equation (15.2) shows that the arc lengths in both directions are larger than those of cylindrical folds which would have developed by unidirectional compression of the same set of beds. For equal rates of shortening along all directions parallel to the layer,

$$L_1 = L_2 = \sqrt{2}L. \tag{15.3}$$

15.3. Morphological types of superposed folds

Superposed folding may give rise to a large variety of morphology of a fold surface. These can be classified into three broad categories (Turner & Weiss 1963, p. 108): (1) *plane non-cylindrical* folds, in which the axial surface of the early fold (say, F_1) is planar but the hinge line is curved within that plane, (2) *non-plane non-cylindrical* folds, in which both the hinge line and the axial surface of F_1 are curved and (3) *non-plane cylindrical* folds, in which the hinge of F_1 is straight but the axial surface of F_1 is curved (Fig. 15.2). We shall see later that superposed deformations may cause an early fold to rotate and tighten without causing a curvature of its axial surface or hinge line so that the fold system remains *plane cylindrical*.

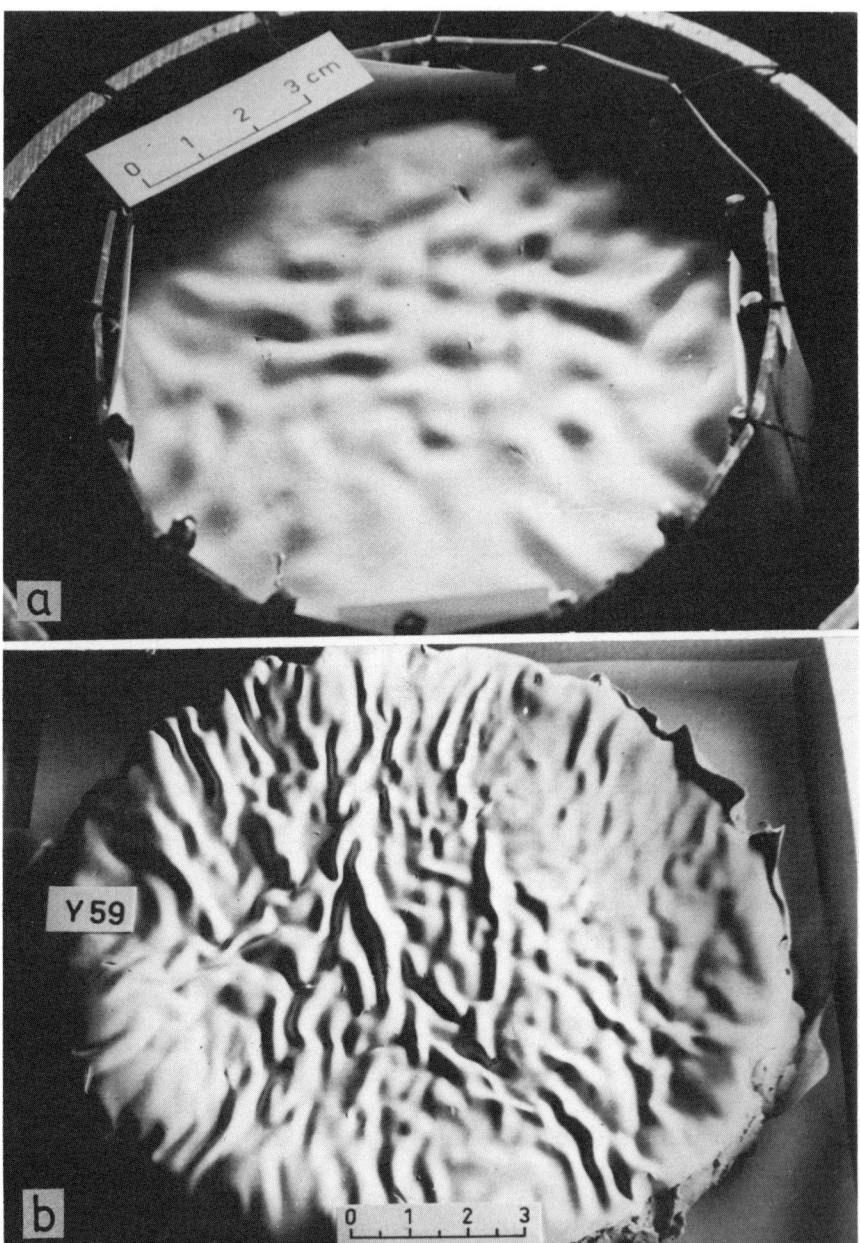

FIG. 15.1. (a) Development of domes and basins in a layer of modelling clay overlying a slab of painter's putty in a constrictive deformation with equal shortening in all directions. (b) Development of diversely oriented buckle folds in test model with unequal shortening in two directions in the plane of the layer. From Ghosh & Ramberg 1968.

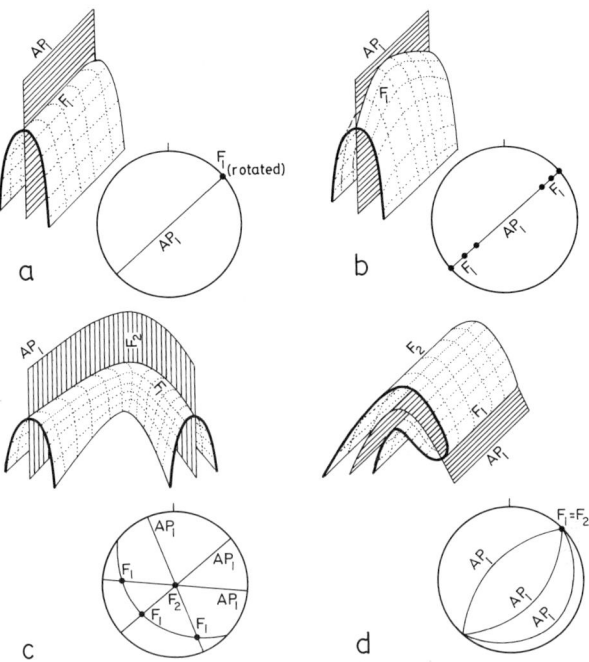

FIG. 15.2. Four types of fold interference and their representation in stereographic projection. (a) Type 0 pattern in which an early fold has been deformed but has remained plane cylindrical. (b) Type 1 interference pattern giving rise to a plane non-cylindrical geometry. In stereographic projection F_1-axes show diverse orientations but lie on the planar axial surface of F_1. (c) Type 2 interference giving rise to a non-plane non-cylindrical geometry. In stereographic projection both the axis and the axial plane of F_1 are diversely oriented. F_1 may or may not lie on a plane. (d) Type 3 interference giving rise to a non-plane cylindrical geometry. AP_1 = axial plane of F_1.

The axial surface of F_1 is cylindrically folded in a non-plane cylindrical fold. In a non-plane non-cylindrical fold the axial surface of F_1 may or may not be cylindrically folded and the hinge line of F_1 may or may not lie in a plane. The most complex forms of non-plane non-cylindrical folds cannot be described in terms of a few geometric parameters. The geometry of such a structure is best represented by sub-dividing it in approximately cylindrical parts and by plotting the structural elements of each part in stereographic or equal area projection. However, for superposed folds in which the F_1 axial surface is either planar or is cylindrically folded, the *axial surface angle* and the *hinge angle* (Fig. 15.3) together give us a good deal of information about the shape of the folded surface produced by superposition of two generations of folds. The axial surface angle is the angle subtended by a folded axial surface at its points of inflection within a plane perpendicular to the axis of curvature of the axial surface. The hinge angle is the

331

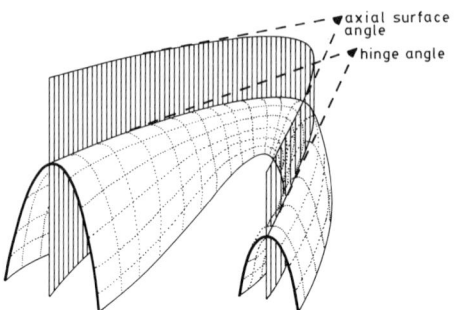

FIG. 15.3. Axial surface angle and hinge angle in a non-plane non-cylindrical fold.

angle subtended by a curved hinge line at its points of inflection (Williams & Chapman 1979). The broad categories of fold interference patterns may be distinguished by the axial surface angle and the hinge angle:

Plane cylindrical folds (Type 0)	axial surface angle = 180°, hinge angle = 180°.
Plane non-cylindrical folds (Type 1)	axial surface angle = 180°, hinge angle ≠ 180°.
Non-plane non-cylindrical folds (Type 2)	axial surface angle ≠ 180°, hinge angle ≠ 180°.
Non-plane cylindrical folds (Type 3)	axial surface angle ≠ 180°, hinge angle = 180°.

These four categories are generally regarded as type 0, type 1, type 2 and type 3 of fold-superposition (Ramsay 1967, Ramsay & Huber 1987). The majority of complex folds are produced by superposed buckle folding. Superposed buckling is a complex process and is as yet poorly understood. Thus, in many cases, and especially for type 2 super-position, it is not yet possible to predict the exact geometry of a folded surface produced by buckling of a given system of folds. In any event the distinction among the four categories of structures represented in Fig. 15.2 can be made entirely on the basis of their morphology and without any assumption about the mechanism.

Let us consider a simplified kinematic model for the development of each of these interference patterns. Let the first fold axial plane and axis, in their initial orientation, be designated as AP_1 and f_1 respectively. Let us now consider an imaginary set of cylindrical second generation folds, with axes b_2, which would have been produced by deflections on a planar surface, with the direction of deflection, a_2, normal to the planar surface. It is evident that a_2 must also be normal to b_2. Ramsay (1967) and

332

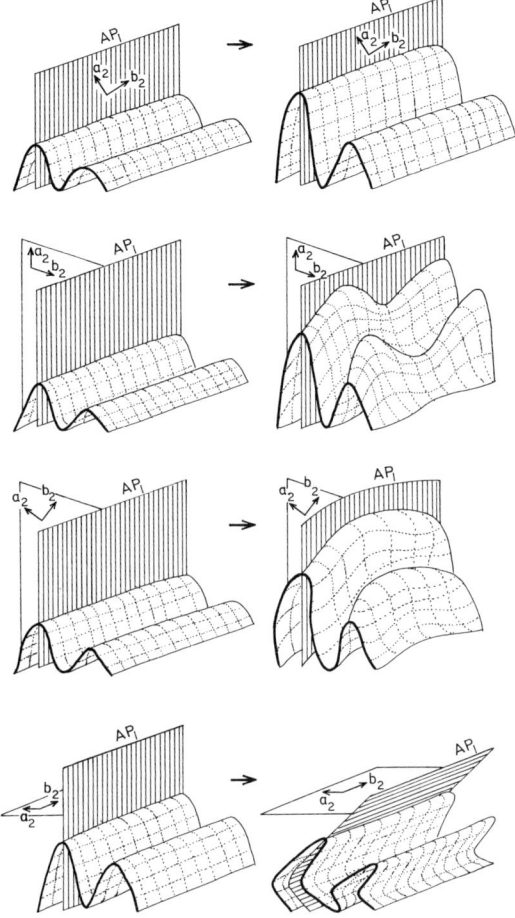

FIG. 15.4. Development of type 0, type 1, type 2, and type 3 fold interference (from top to bottom respectively) according to the shear fold model. On the left-hand side the shear direction a_2 and the shear plane a_2b_2 for the second deformation are shown with respect to a cylindrical F_1. The corresponding deformed fold is shown on its right-hand side.

Ramsay & Huber (1987) derived the four types of fold interference patterns by superimposing these deflections on the geometry of the first generation folds. The type of interference pattern will depend mainly on the angle between f_1 and b_2 and the angle between AP_1 and a_2 (Fig. 15.4). As suggested by Thiessen & Means (1980), we may also generate the four categories of superposition under the following conditions:

	Angle between a_2 and AP_1	Angle between a_2b_2-plane and f_1
Type 0	$= 0$	$= 0$
Type 1	$= 0$	$\neq 0$
Type 2	$\neq 0$	$\neq 0$
Type 3	$\neq 0$	$= 0$

If we assume, as Ramsay (1967) and Ramsay & Huber (1987, p. 493) have done as a first approximation, that the second generation folds are produced by heterogeneous simple shear, then a_2b_2 will be parallel to the shear planes and a_2 will be parallel to the shear direction. In the following discussion, however, the four types of interference patterns will be distinguished only on their morphological basis. Thus, for example, the type 0 interference pattern will refer to cylindrical folds whose final geometry is a product of superposed deformations.

In the experiments of superposed buckle folding by Ghosh & Ramberg (1968, pl. IV), the early folds were tightened and their hinges and axial surfaces were rotated, but no new folds were initiated when the direction of the later shortening was more than 30° with the hinges and axial surfaces of the early folds. However, the geometry of the folds was greatly modified by the later deformation; the final structure must there-fore be regarded as a product of superposed deformations. Since the folds retain their plane cylindrical geometry, it is reasonable to regard the final product as type 0 pattern. Such interference patterns are likely to be fairly common (cf. Odonne & Vialon 1987), though difficult to distinguish, in areas of superposed folding. Thus, for example, from southern Rajasthan of India, Naha & Mohanty (1988, pp. 84, 86) described an area in which an E–W horizontal compression has produced a set of F_2 folds with vertical axial planes striking N–S. The F_2 folds, deforming the hinges and axial surfaces of the F_1 folds, are localized in the sectors with steep S-surfaces striking nearly E–W. They are consistently absent in sectors in which the S-surfaces strike between NW–SE and N–S; the F_1 folds in these sectors are, instead, further flattened. The geometry of the folds in these sectors should be interpreted as a product of type 0 interference rather than by the localized absence of the second deformation.

15.4. Outcrop patterns of superposed folds

Superposed folds may have a large variety of outcrop patterns on horizontal or inclined sections. On the basis of the fold geometry alone, the type 0 pattern cannot be distinguished from plane cylindrical folds produced in a single deformation. However, some of the outcrop patterns of superposed folds enable us to identify the type 1, type 2 and type 3 patterns (Ramsay 1967, fig. 10–13) of fold interference. No doubt, the three-dimensional form of a superposed fold is much better understood when, along with the outcrop pattern itself, the orientations of the fold axes are also known.

The type 1 pattern of interference, or the *dome-and-basin structure*, can usually be identified from oval, somewhat rhombic or lozenge-shaped outcrops (Fig. 15.5a). Such two-dimensional interference patterns emerge when the curved hinge line of an early fold meets the plane of outcrop in at least two points (Fig. 15.6). Evidently, a plane non-cylindrical fold will not give this characteristic outcrop pattern in all outcrop faces. A section plane parallel to the F_2 axial plane will only show sinusoidal traces of the F_1 folds. Similarly, a section plane parallel to the F_1 axial planes will show the sinusoidal traces of the F_2 folds.

The type 2 pattern of fold superposition on a large scale may be identified from crescent-shaped (Fig. 15.7a) or mushroom-shaped out-crop patterns and sometimes from mirror-image-like repetition of fold noses (Fig. 15.7b). The crescent- or mushroom-shaped outcrops emerge when each of the F_1 and F_2 fold-hinges meets a more or less planar outcrop face at more than one point. The mirror-image-like pattern is observed when the outcrop face is parallel to the F_2 hinge but meets the F_1 hinge at more than one point. A variety of other patterns may emerge in other sections (Thiessen & Means 1980, Thiessen 1986) through type 2 fold interference. Thus, a sinusoidal pattern will emerge if the outcrop face is parallel to the F_2 hinge and meets the F_1 hinge at one point only, and a hook-shaped pattern will form if the outcrop face intersects both the F_1 and F_2 hinge lines once only.

The *coaxial fold* or the type 3 fold interference often gives rise to hook-shaped outcrops (Figs. 15.5b, 15.7c). These appear on outcrop faces that intersect both F_1 and F_2 hinges (Fig. 15.8). However, the sole presence of hook-shaped outcrop should not be taken as an evidence of coaxial folding. As shown by Thiessen (1986, p. 567), hook-shaped patterns may also appear on certain sections through non-plane non-cylindrical folds. Moreover, in very long crescent-shaped outcrops one of the crescent noses may be poorly exposed or may be outside the field of observation, so that the outcrop may appear as a hook. However, once the existence of coaxial folding has been established in an area, hook-shaped outcrops, along with the fold style, may be used to identify individual type 3 superpositions.

The folding of the axial surface of a non-isoclinal fold, producing either a type 3 or type 2 fold interference, is generally associated with a shifting of the position of the axial surface trace of the second generation fold on either side of the axial trace of the first generation fold

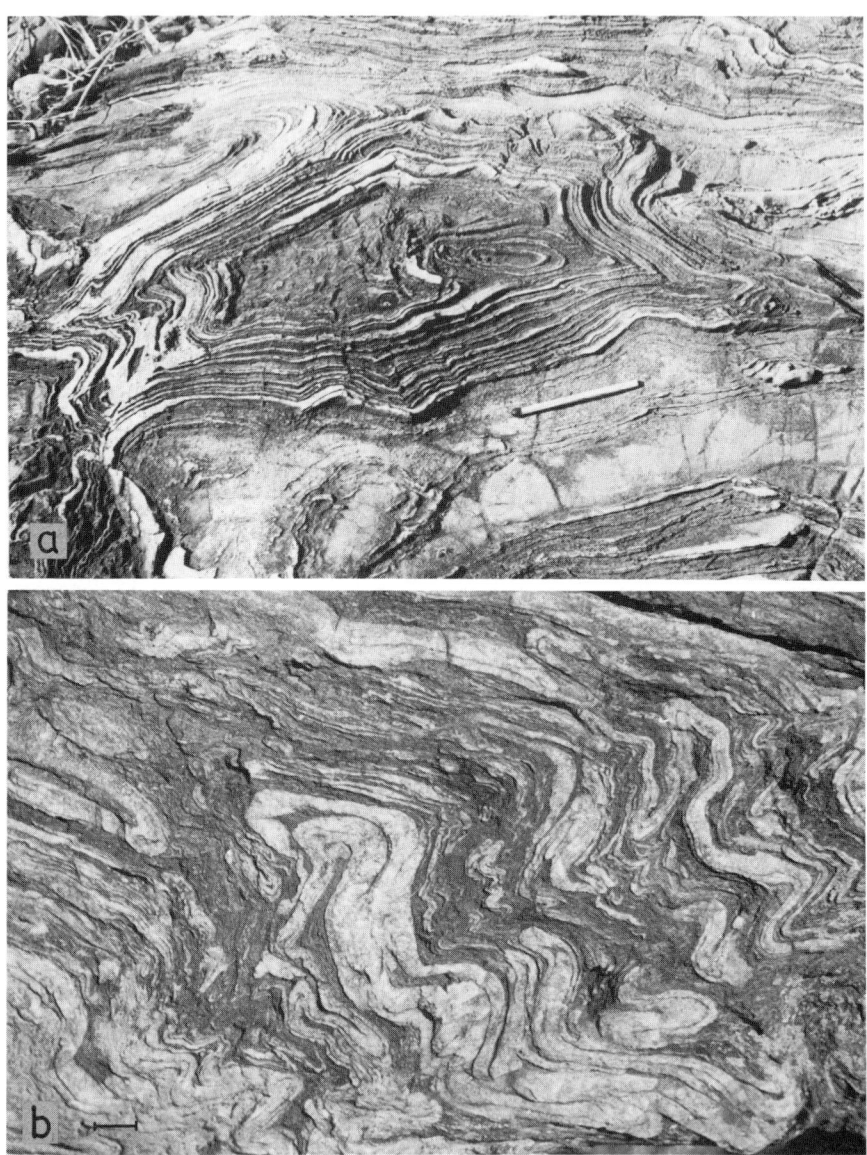

FIG. 15.5. (a) Type 1 interference giving rise to oval outcrops in Precambrian rocks of Jhamarkotra, Rajasthan, India. (b) Hook-shaped outcrops of coaxial folds in banded iron formation. Kolar schist belt, India.

(Ramsay & Huber 1987, p. 477 and fig. 22.2). Such a *shifting of the axial surface trace* should always be noted in the field since it enables us to conclude that there is an earlier generation of fold even where the noses

INTERFERING FOLDS

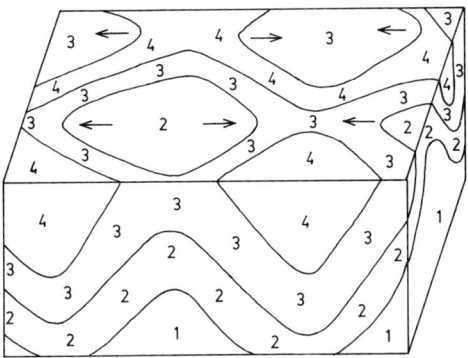

FIG. 15.6. Outcrop pattern for type 1 fold interference. The arrows on the top horizontal surface of the block diagram show the plunge directions of F_1.

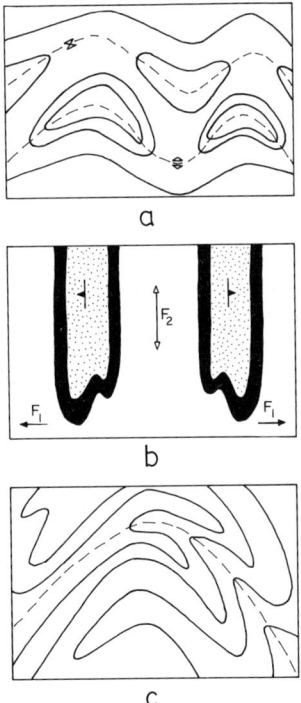

FIG. 15.7. (a) Crescentic and (b) mirror image outcrop patterns of type 2 fold interference. (c) Hook-shaped outcrop pattern of type 3 fold interference.

of the early folds are not exposed in the outcrop. The shifting of the axial surface trace can occur in both shear (Fig. 15.9) and buckle folding.

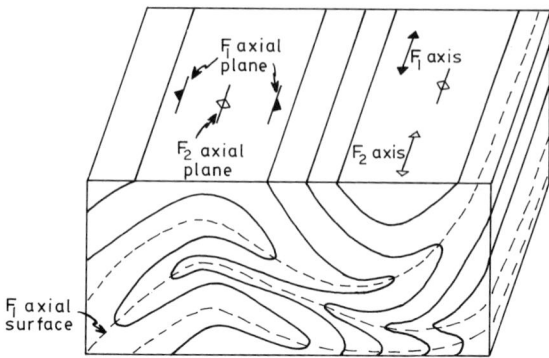

FIG. 15.8. Block diagram showing different outcrop patterns on different section planes through a type 3 fold interference.

Although two-dimensional fold interference patterns give us valuable information concerning the nature of superposed folding, the overall shape of the folded surface and, sometimes, the type of fold interference, cannot be determined unless the orientations of the different generations of lineations and axial plane cleavages are also determined. Superposed folding is often heterogeneous, in the sense that the first folds are not deformed in the same manner by the second folds in all places. Although in areas of superposed folding the earlier folds generally show a larger amount of tightening, there are instances in which an early fold opens out locally in certain places and is tightened elsewhere (Fig. 15.10). The right-hand side of Fig. 15.10 shows a crescentic outcrop appearing on a section through a model of experimentally produced superposed buckling folds of two generations. It should be noted that the trace of axial plane of the early fold passes through only one of the sharp noses of the crescent. This resulted from the fact that during the second deformation the early fold opened out towards the upper and the right portion of what is now the crescentic outcrop; the second folds superimposed on the opened out portion gave rise to the sharp nose of the crescent on the right-hand side.

15.5. Modes of superposed buckling

General considerations

Earlier experiments on superposed buckling (Ghosh & Ramberg 1968, Skjernaa 1975, Watkinson 1981) indicated that the geometry of superposed folds depends to a large extent on the shape of the early folds. A similar dependence of the buckling mode on the shape of the early folds was also noticed in the field. Julivert & Marcos (1973, p. 374) reported from the Cantabrian zone of north-west Spain that the open

FIG. 15.9. Shifting of axial surface trace of F_2 formed by shear folding in a card deck model. (a) Initial fold drawn on edge of card deck. (b) Shear folds (F_2) produced by heterogeneous card deck slip. Note that axial traces of F_2 have different positions on the two limbs of F_1.

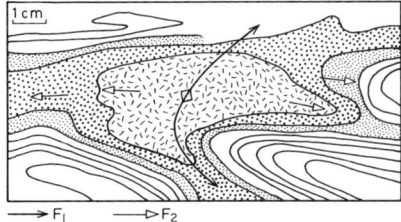

FIG. 15.10. Outcrop pattern of type 2 interference appearing on a section through a model of experimentally produced superposed buckling folds of two generations. The axial surface trace of the early folds passes through only one of the noses of a crescentic outcrop. The axial surface trace of a second generation fold passes through the other sharp nose of the crescent. From Ghosh 1974b, fig. 5.

early folds were refolded to domes and basins but the tight folds were deformed by folding of axial surfaces. Again, from the Douarnenez Bay area in Brittany, Watkinson (1981) showed that tight early folds

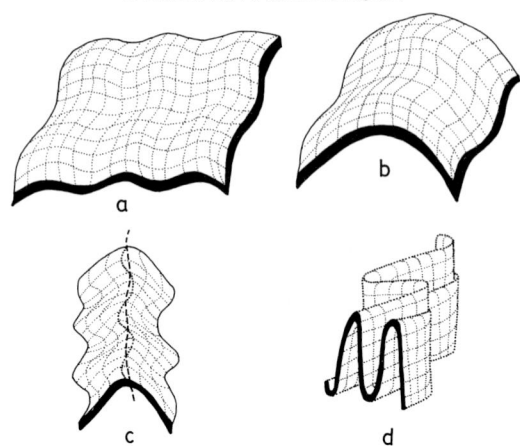

FIG. 15.11. Four standard modes of superposed buckling in single layer.

with narrow hinge zones had been deformed to a type 2 interference pattern. On the other hand, small second generation folds rode over the hinges of early folds when the latter had rounded hinges.

Thus, the mode of superposed buckling depends upon the shapes of early folds; since the profile shapes of folds are more varied in a multilayer than in a single layer, the refolding modes are more varied in multilayered rocks.

Superposed buckling in single layer

Let us first consider the modes of superposed buckling of a single competent layer embedded in an incompetent host (Fig. 15.11). In such a layer, the early folds may be refolded in any one of the following four modes (Ghosh *et al.* 1992).

1. First mode of superposed buckling

This mode of buckle folding is obtained when the early fold (F_1) is very gentle, with a small curvature of the layer at the hinge zone or with a large interlimb angle ($>$ about 135°). The superposed folds (F_2) are roughly of the same size as F_1 and the interference of F_1 and F_2 gives rise to a dome-and-basin pattern (type 1 interference). When the two fold systems are at a right angle to each other and the tightnesses in the two directions are nearly equal, the domes and basins are equant in plan view. On the other hand, when the second generation fold waves are superposed oblique to the first generation fold waves, the F_2 folds always appear as elongate domes and basins (Fig. 15.12). The long direction of their oval outcrops may not be perpendicular to the direction of shortening of either F_1 or F_2 (Fig. 15.13). If the shortening in the second

340

deformation is large, the F_2 folds become tight or isoclinal, with planar axial surfaces and curved hinge lines. With a large amount of stretching normal to the overall orientation of the layer, the F_2 hinge line curvatures of those plane non-cylindrical folds may be greatly accentuated and may give rise to sheath folds.

The dome-and-basin pattern in the first mode of superposed buckling is quite distinct from the interference pattern produced by constrictive deformation with $\lambda_3 < \lambda_2 < 1$ along the layering. In the latter case (Fig. 15.1b), the domes and basins are always associated with diversely oriented arcuate fold hinges.

2. Second mode of superposed buckling

The first mode of superposed buckling is inhibited when the first folds have a moderate tightness (i.e. when the interlimb angle has a range of roughly 135–90°). Instead of a dome-and-basin pattern we then obtain a set of F_2 folds much smaller in size than the F_1 folds (Ghosh & Ramberg 1968). These small folds ride over the hinges of the larger F_1 folds (Fig. 15.14a).

3. Third mode of superposed buckling

This mode of superposed buckling can be seen in the experiments in which the F_1 fold in its initial state is moderately tight (interlimb angle < about 90°) but not very tight or isoclinal. In this mode the early and the late folds have roughly similar arc lengths but they neither form a dome-and-basin pattern nor do the F_2 folds ride over the F_1 hinges. If the gross orientation of the layer is sub-horizontal, then the F_2 folds plunge in opposite directions on the two limbs of an F_1 fold, with an antiformal F_2 on one limb of F_1 passing to a synformal F_2 on the other limb (Ghosh & Ramberg 1968). This mode of superposed buckling (Fig. 15.14b) is characterized by the following two features:

(a) Each F_2 fold occupies a roughly triangular area; an antiform narrows down as it approaches an F_1 synformal hinge zone and a synform narrows down as it approaches an F_1 antiformal hinge zone (Fig. 15.14b).

(b) With progressive tightening of the F_2 folds, the F_2 synforms propagate across the F_1 antiformal hinges and the F_2 antiforms propagate across the F_1 synformal hinges. This process of fold propagation is associated with the replacement of an old F_1 hinge by a sinuous new hinge (Figs. 15.14b, 15.15). At a point where a synformal F_2 propagates across an antiformal F_1 hinge, the initial upward convexity of the layer changes to an upward concavity (Fig 15.15b). The resultant structure is that of a type 2 fold interference (Figs. 15.15a and 15.14b). It should be

FIG. 15.12. (a) Low-angle interference of two sets of open buckle folds giving rise to domes and basins in thinly banded marble–phyllite intercalations. Near Narayani temple on Jaipur-Alwar Road, India. (b) Domes and basins produced in the first mode of superposed buckling. The shortening in the second deformation (parallel to the scale bar) is at an angle of 30° with the first generation fold hinges.

noted, however, that what appears to be a strongly curved F_1 hinge is in reality a feature that is newly created during the second deformation. A material line marking out the position of an old F_1 hinge remains less sinuous than the new hinge and runs across the new hinge. As a result of

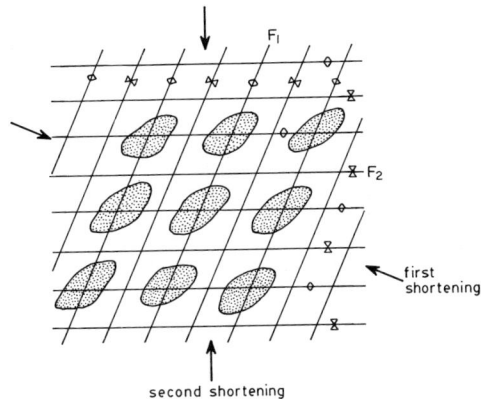

FIG. 15.13. Oval outcrops of elongate domes where F_1 and F_2 antiformal waves interfere and are oblique to each other. Note that the long axes of domes are parallel to neither F_1 nor F_2 hinges.

this process of *hinge replacement*, a lineation parallel to the old F_1 hinge will make an angle with the newly created hinge.

4. Fourth mode of superposed buckling

If the F_1 fold in its initial state is very tight or isoclinal, its hinge and axial surface are folded during the second deformation (Fig. 15.11d) involving a shortening along the F_1 hinge line. The resulting structure is a type 2 fold interference. Unlike the third mode of superposed buckling there is no hinge replacement during the second deformation. If the second deformation involves only a flexure, the angle between the early and the late fold axes remains constant in different parts of the structure. The angle may not remain constant if the later folding involves both flexure and flattening (Fig. 15.16). The mechanism of reorientation of hinges of such isoclinal early folds by combined flexure and flattening is similar to that of reorientation of early lineations as described in Chapter 16.

The experiments clearly show that the mode of buckle folding during the second deformation is greatly dependent on the tightness of the first folds. Thus, for example, a dome-and-basin pattern or a type 1 fold interference never developed in the experiments (Ghosh & Ramberg 1968, Skjernaa 1975, Ghosh *et al.* 1992) when the early fold was tight or isoclinal. The deformation of the tight or isoclinal folds invariably produced a type 2 or type 3 interference pattern. Among natural structures the type 1 interference pattern is most commonly encountered in areas where two generations of open folds interfere with each other and give rise to open domes and basins (e.g. Naha & Chatterjee 1982, p. 40, fig. 7). However, plane non-cylindrical isoclinal folds do occur in some

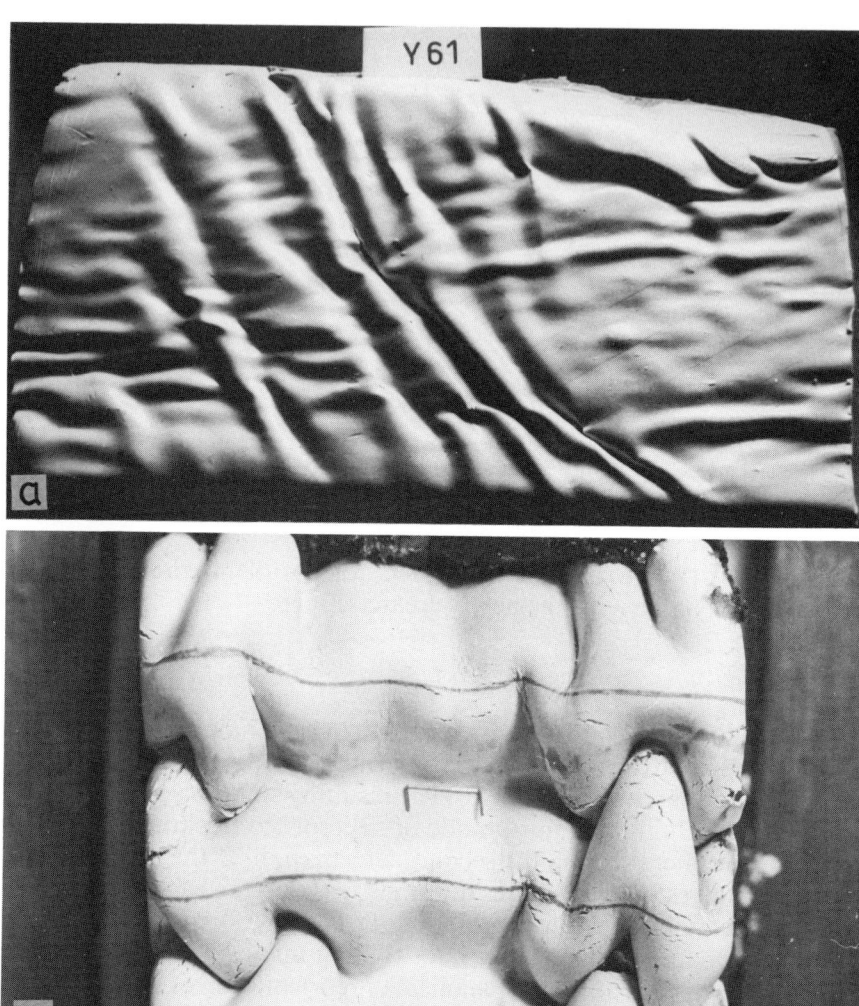

FIG. 15.14. (a) Superposed buckling of the second mode. The second direction of compression in the model was at about 30° to the first generation fold axes. The smaller folds running from left to right are of the second generation. (b) Third mode of superposed buckling in single layer. The first fold trends from left to right and the second fold runs from front to back. The dark lines running from left to right were initially parallel to F_1 antiformal hinges. The strongly sinuous hinge line of the present non-plane fold does not coincide with the early fold hinge line. There has been a hinge replacement of the early fold during the second deformation (Ghosh *et al.* 1992).

areas of superposed folding. In some of these structures the early and the late fold hinges are both tightly or isoclinally folded (Turner & Weiss 1963, p. 143, Tobisch 1966, Mukhopadhyay & Sengupta 1979),

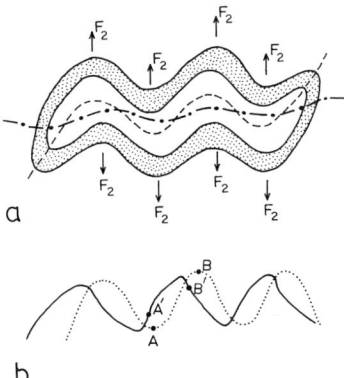

b

FIG. 15.15. (a) Outcrop pattern of third mode of superposed interference. Line with dot and dash represents position of old hinge of F_1 and dashed line represents position of the new hinge formed by replacement of the old F_1 hinge. (b) Section across F_1 axes and along F_2 hinges. A and B are two material points on a synformal and an antiformal hinge of initial F_1. A' and B' are positions of the points after hinge replacement.

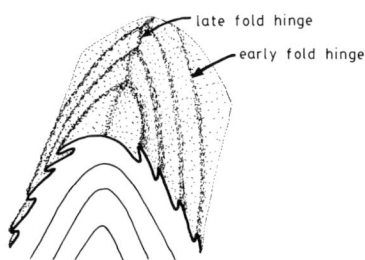

FIG. 15.16. Variable angles between F_1 and F_2 hinges when isoclinal F_1 folds are refolded by combined buckling and flattening.

the three-dimensional form of the structure being that of a tightly appressed test tube with an elliptical cross-section. Such plane non-cylindrical folds with small hinge angles (Williams & Chapman 1979) are described as *sheath folds* (Carreras *et al.* 1977). As discussed in Chapter 21, sheath folds usually develop in ductile shear zones in the course of a single continuous deformation. In certain regions sheath folds have also developed outside shear zones by superposition of two discrete deformations. The mechanism of development of sheath folds by superposed deformation, especially by superposed buckling, is not fully understood as yet. However, as mentioned earlier, sheath folds can develop in the first mode of superposed buckling by a tightening of a set of initially gentle domes and basins during the second deformation, with a very large amount of layer-normal stretching. The long

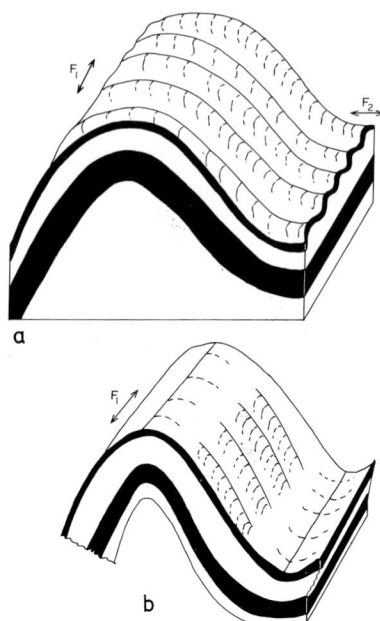

FIG. 15.17. Modified second mode of superposed buckling in multilayer. The small folds ride over the hinges of larger F_1 in (a), while in (b) the small F_2 folds are restricted to the limbs of F_1.

axis of an oval outcrop of such a sheath fold will be parallel to the axial surface trace of F_2.

Modes of superposed buckling in multilayers

A thin multilayer which gives rise to a set of sinusoidal buckling folds is refolded in the same manner as a single layer; it can undergo superposed buckling in any one of the four modes as described above. However, the experimental deformations of multilayered models (Ghosh et al. 1993) show that the four basic modes of superposed buckling may be modified in other types of multilayers. The modifications are essentially concerned with the relative size of the F_1 and F_2 folds in the different modes.

In superposed buckling of a single layer in the first, third and the fourth modes, the F_1 and the F_2 folds are roughly of the same size. Although in the second mode of superposed buckling in a single layer the F_2 folds are distinctly smaller than the F_1 folds, the arc length of F_2 is not smaller than one-fourth the arc length of F_1; in multilayers, on the other hand, we may have a modified second mode or mode 2a in which the arc length of F_2 folds are much smaller (i.e. much less than one-fourth) than those of F_1 (Fig. 15.17a). In another modification of the second mode, i.e. in mode 2b, the small F_2 folds are overprinted on the long limbs of narrow-hinged F_1 folds and the F_2 folds do not ride over their hinges (Fig. 15.17b). In a similar modification of the third mode of superposed buckling in multilayers, the hinge line and the axial surface of a large F_1 fold is deformed in a much smaller scale. This relatively small-scale folding of the axial surface does not distort the overall trend of F_1.

The relative size of F_1 and F_2 may also be modified in the fourth mode of superposed buckling in multilayers. When a stack of layers is deformed to a series of isoclinal folds with long limbs and small hinge zones the effective thickness of the multilayer is increased by repetition of the layering. Under a hinge-parallel shortening, these folds deform in the fourth mode and give rise to a type 2 interference pattern. The size of the F_2 may then be much larger than that of F_1. On the other hand, the effective thickening may cause a kind of "self-confinement" (Biot 1964) and the F_2 folds may then be much smaller in size than the F_1 folds.

The morphology of superposed folds become quite complex when the mechanical property of a layered stack is such that different orders of folds can develop in it. If two orders of folds are produced, with thin competent layers deformed to relatively small folds and thicker and somewhat less competent layers giving rise to disharmonic larger folds, the growth rates of the smaller folds is usually faster, and consequently the smaller folds may become somewhat tighter than the larger folds. As a result, depending upon the contrast in tightness of the two sets of folds, the mode of superposed buckling may be identical or different in the smaller and the larger folds. In either case, the morphology of the superposed folds becomes complex because there is a distortion of the interference pattern of the small folds by that of the disharmonic larger folds (Figs. 15.18, 15.19, 15.20 and 15.21). The outcrop pattern of a small-folded thin competent layer can then be much more complex than the outcrop pattern of disharmonic larger folds in a thicker layer.

The experiments (Ghosh et al. 1993) show that if superposed buckling takes place where two orders of non-isoclinal disharmonic folds are present, the axes of the smaller folds may not always be parallel to those of the neighbouring larger folds of the same generation (Fig. 15.20, for example). An important consequence of this observation is that, in the presence of disharmonic folds, the small folds of a later generation may not always be parallel to the larger folds of the same generation as is required by Pumpelly's rule (Pumpelly et al. 1894). Thus, it may happen that the axial surfaces of the small F_2 folds in a relatively thin layer are bodily rotated around the differently oriented axis of a larger disharmonic F_2 fold. In Fig. 15.22a, for

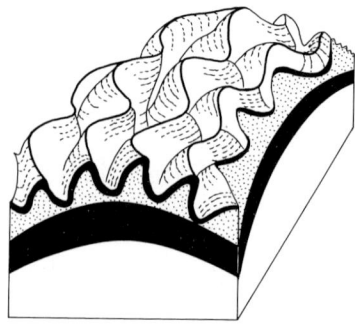

FIG. 15.18. Schematic diagram showing refolding of disharmonic folds. After the first deformation the small F_1 folds (running from front to back) in the thin upper layer were moderately tight, while the larger fold in the thick lower layer was gentle. Because of this difference in tightness, the smaller F_1 folds were deformed in the third mode during the second deformation, while the larger folds were deformed into a dome. The smaller interfering folds on the thinner layer are distorted by the dome (Ghosh *et al.* 1993).

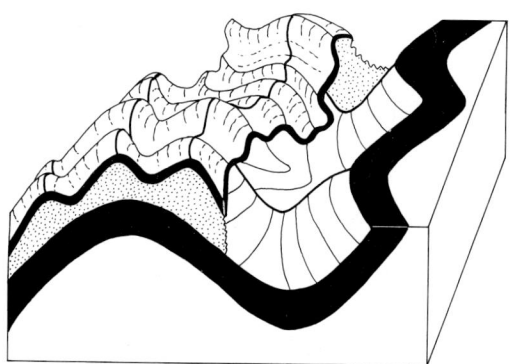

FIG. 15.19. Schematic diagram showing superposed buckling of two orders of folds. Both the smaller and the larger F_1 folds, running from front to back, are refolded in the second mode. The small-folded thin layer is not shown on the right-hand side in order to reveal the morphology of the larger superposed folds in the thicker layer. Note that the hinge lines of the smaller F_2 folds (with left-to-right trend) are disharmonic with the hinge lines of the larger F_2 folds (Ghosh *et al.* 1993).

example, we find that the small isoclinal F_1 folds have been refolded to a type 2 pattern during the second deformation. The axial surfaces and the hinges of the small F_2 folds have been bodily rotated around the sub-horizontal axes of the larger F_2 folds. The structure may give the erroneous impression that there have been three separate deformations instead of two. Figure 15.22b shows the details of the same model in which the type 2 interference pattern of the small folds in the thin layer is distorted over a large dome in a thick layer. Unless the overprinting

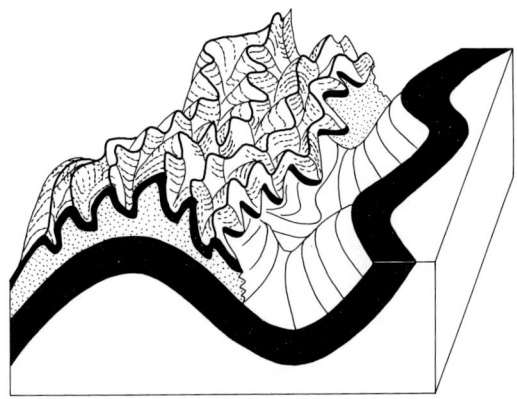

Fɪɢ. 15.20. Schematic diagram showing superposed buckling of two orders of folds. Because of the difference in tightness of the F_1 folds (running from front to back) in the thin and the thick competent layers, the refolding in the thin upper layer is in the third mode and the refolding in the thick lower layer is in the second mode. The thin layer is not shown in the right-hand part in order to reveal the geometry of the larger folds in the thick layer. Note that the attitudes of the hinge lines of the smaller F_2 folds (left to right) are different from that of larger F_2 folds in the same domain (Ghosh *et al.* 1993).

Fɪɢ. 15.21. Superposed buckling of two orders of folds. The smaller folds in the thin competent layer have been refolded in the fourth mode while there has been superposed buckling in the third mode in the thicker layer. The hinges and axial surfaces of the smaller F_2 folds have been distorted by the interference pattern in the thicker layer.

relations of lineations and cleavages are properly studied, such a structure, seen either in the outcrop or in the map scale, may be incorrectly interpreted by assuming that the larger domes and basins have grown in a distinctly later and separate event.

Relative tightness of old and new folds

Although there is no reason for the bulk shortening in the earliest deformation to be the largest in all cases, it is often observed in areas of

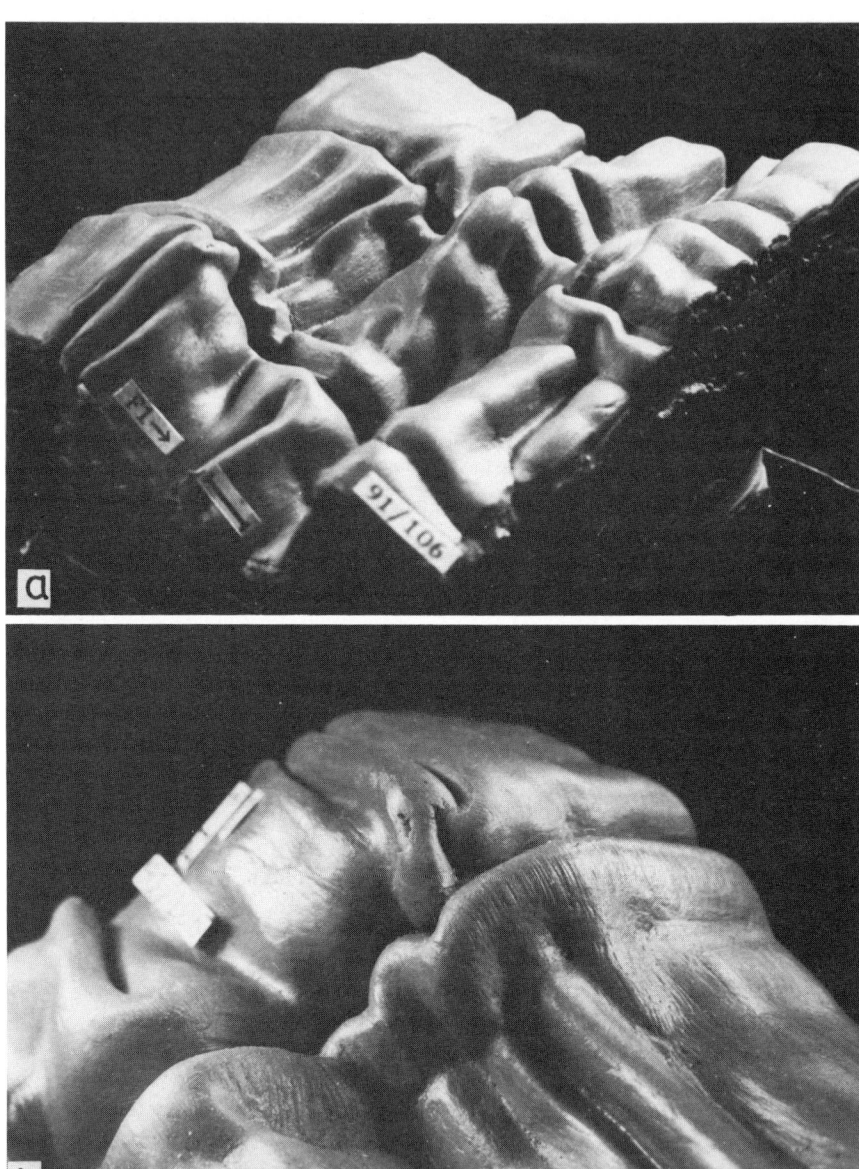

FIG. 15.22. (a) Model showing superposed buckling on the exposed surface of a thin layer of modelling clay. This multilayered model gave rise to two orders of F_1 folds (trend shown by arrow). The smaller F_1 folds were isoclinal in some places and open or moderately tight elsewhere. The larger F_1 folds were gentle or open. During the second shortening (parallel to F_1-axis), the isoclinal F_1 folds were deformed to type 2 interference pattern; the resulting F_2 folds had steep to moderately steep hinge lines. This interference pattern was distorted over the larger F_2 folds which developed on an

superposed folding that the earliest generation of folds is the tightest. In the light of the experimental results (Ghosh *et al.* 1993) this field observation may be explained in the following way:

(a) In certain areas the earliest folds may indeed be the tightest because the shortening in that deformation was the largest.

(b) Experiments on superposed buckle folding indicate that, during deformation of the early folds in either the third or the fourth mode, the initial tightness of the early folds increases as their axial surfaces become folded. Hence the greater tightness of the early folds may have been induced in the course of a later deformation (cf. Ramsay 1967, p. 547).

(c) In certain areas the early fold may indeed be much more open than the later folds. We have seen that, when the early folds are gentle, a later deformation initially gives rise to a dome-and-basin pattern. If the later deformation is sufficiently large, the new folds may become very tight or isoclinal. In that event the separate identity of the hinges of the early folds becomes obscure and their presence is only manifested by the axial culminations and depressions of the later folds (Fig. 15.23). Figure 15.24 shows a striking example from the Aravalli metasediments of W. India where isoclinal F_1 folds are coaxially folded to upright, gentle or open (F_{1A}) folds. The figure shows the development of domes and basins by interference of open F_{1A} and tight F_2 folds. That the tight fold is a later fold is shown by the fact that it deforms the F_1 lineation.

15.6. Geometrical analysis of superposed folds in mesoscopic scale

General

The geometrical analysis of superposed folds of a region is done in conjunction with analyses of the associated cleavages and lineations. The objective of the geometrical analysis is to reconstruct the three-dimensional form of the superposed folds, to determine the sequence of development of cleavages and lineations and to reconstruct, as best as possible, the course of geometrical evolution of the structure as a whole. These objectives are comparatively easy to achieve when the three-dimensional forms of the interfering folds are observable in the

---

Fig. 15.22 (*Continued*)

underlying thick competent layer. The larger F_2 fold had subhorizontal axis. (b) Details of the model shown in (a). Note that the type 2 interference of small folds is distorted over a larger-scale dome formed by the interference of larger F_1 and F_2 folds. Where the small F_1 folds were not well developed, the small F_2 folds (as in the right foreground) rode over the hinges of the gentle and larger F_1 folds (Ghosh *et al.* 1993).

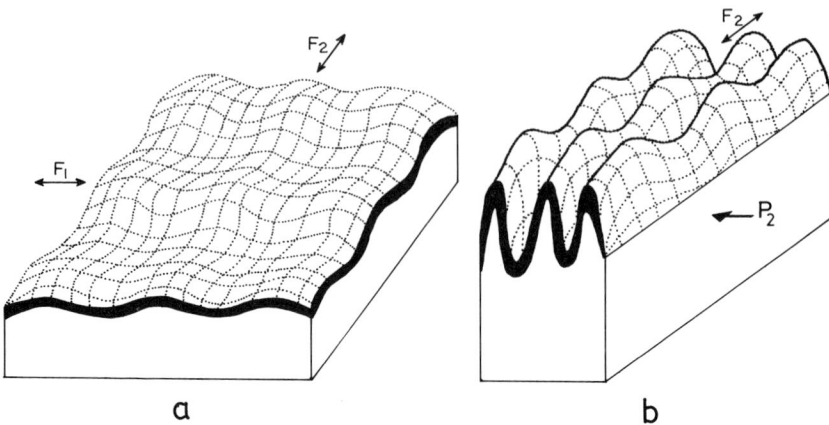

FIG. 15.23. (a) Development of gentle domes and basins in the first mode of superposed buckling. (b) Obliteration of F_1 hinge lines by tightening of second generation folds. The presence of F_1 folds can now be recognized only by the axial culminations and depressions of F_2-axes.

FIG. 15.24. Development of elongate domes and basins in a thin garnet-rich layer in phyllite by interference of a set of open early folds (F_{1A}) and tight late folds (F_2). The F_2 folds, running from left to right in the figure, have a roughly N–S trend and are very tight. A lineation parallel to the early fold axis (parallel to pencil) is deformed over the hinges of F_2. In the light of experimental results it is suggested that the structure was initiated as gentle domes and basins in the first mode of superposed buckling. These were later greatly flattened during the tightening of the F_2 folds. Photograph from the phyllite quarry in Aravalli metasediments, 5 km E of Gogunda near Udaipur, Rajasthan. Visit to this exposure was by courtesy of Prof. B. L. Sharma.

mesoscopic scale. However, the conclusions from such mesoscopic analyses cannot be immediately extended to the interpretation of a large-scale structure. Large areas of superposed folding do not always show a uniform course of geometrical evolution in all domains. In general, geometrical analyses in the mesoscopic and the macroscopic scales supplement each other and both sets of data are utilized to obtain the total picture. The essential aspects of geometrical analyses are described in the following sections. For a detailed description of the methods of geometrical analysis in different scales the reader should consult Turner & Weiss (1963, pp. 76–193).

Overprinting relations

The initial objective of geometrical analysis is to identify the different *generations* of folds and associated cleavages and lineations. There are areas of multiple deformations where there are five or six generations of folds and are designated F_1, F_2, F_3, etc. The fold-related cleavages and axial surfaces are designated S_1, S_2, S_3, etc. Similarly, the successive generations of lineations are called L_1, L_2, L_3, etc. If the bedding can be identified it is usually represented by the symbol S_0. The earliest generation of cleavage in certain areas may not appear as axial planar to a set of folds on bedding, either because such folds are absent or because they are as yet undiscovered. The cleavage may then be referred to as S_1 and the earliest observable fold which deforms S_1 may be called F_2. Some authors follow the convention of labelling the first recognizable fold F_1 and its axial surface S_1; an earlier bedding-transecting cleavage may then be referred to as S^* (Powell & Rickard 1985) or as pre-F_1 cleavage.

Hobbs *et al.* (1976) and Williams (1985) have pointed out that, although between any two successive generations of folds, say F_1 and F_2, F_1 is older than F_2 in each locality, the folds of each generation might have started to grow at different times in different localities. In other words, all the F_1 folds of a region may not be of the same age and F_1 of one locality may be of the same age or may even be somewhat younger than F_2 of another locality.

The distinction between an earlier and a later generation of folds can be made by their *overprinting* relation. The fold which deforms the hinge or the axial surface of another fold is of the later generation. The time relation between the two sets of folds can be easily determined from those outcrops where the hinges of both sets of interfering folds are exposed and the three-dimensional form of the fold-interference can be directly observed. Some examples of the overprinting relations as seen in outcrops and in oriented specimens are shown in Fig. 15.25. It may be noted that in Fig. 15.25b the hinge line and the axial surface of

F_1 are diversely oriented because of the later folding. The hinge lines of F_2 have different orientations on the two limbs of F_1 although the two hinge lines tend to lie on a plane which is the common axial plane of the two F_2 folds. On any one limb of F_1 the angle between F_1 and F_2 axes remains constant at different parts of the F_2 fold. In Fig. 15.25c, on the other hand, the angle between F_1 and F_2 axes is approximately a right angle in the hinge zone of F_2 while the angle sharply decreases in the limbs of F_2. In this specimen F_1 is isoclinal and F_2 is moderately tight. If we place a transparent overlay on the folds, draw on it the hinge lines of F_2 and F_1 and then flatten out the transparent sheet, we find that the hinge line of F_1 remains strongly curved in the shape of a U. In other words, if we straighten out the F_1 axial surface by an external rotation around the F_2 axis, the F_1 fold does not attain a plane cylindrical geometry but becomes plane non-cylindrical. If field studies indicate that the interference pattern was produced by only two phases of folding, F_1 and F_2, we may conclude that during the second generation of folding there was a strong flattening along with an external rotation. This example shows that the so-called geometrical analysis is not always purely geometrical; it may sometimes involve theoretical considerations about the mechanism of folding.

Figure 15.25d is a sketch from a part of a large outcrop which shows two sets of folds developed on a schistosity (S_1) which, elsewhere in the outcrop (Fig. 15.25a), is axial planar to a set of isoclinal folds (F_1) on bedding. The schistosity is deformed by two sets of later open folds, F_2 and F_3, of which F_2 is coaxial with F_1 and has a prominent axial planar crenulation cleavage. The size of the F_3 folds is distinctly smaller than that of F_2. The sketch shows F_3 folds riding over and across the F_2 hinges in conformity with the second mode of superposed buckling.

Where three-dimensional exposures of interfering folds are not easily available, we can often distinguish between an early and a late fold from the geometrical relations between the folds and a cleavage or a lineation. Figure 15.26a shows a typical case in which a prominent and characteristic lineation is consistently parallel to hinge lines of an early generation of folds. A later generation of fold can be identified when this lineation is deformed by another set of folds. Similarly, as Fig. 15.26b shows, the earliest cleavage (S_1) in some areas is axial planar to a set of folds (F_1) on bedding. A later generation of folds can be identified when they are developed by deformation of S_1. Even where the fold hinges are not exposed, the folds and the associated cleavages of different generations can be often distinguished from the overprinting relation observed in two dimensions of an outcrop face. Thus, Fig. 15.27 shows a crescentic and a hook-shaped outcrop in each of which the earlier and the later folds can be distinguished from the outcrop pattern along with the overprinting relation of the two cleavages.

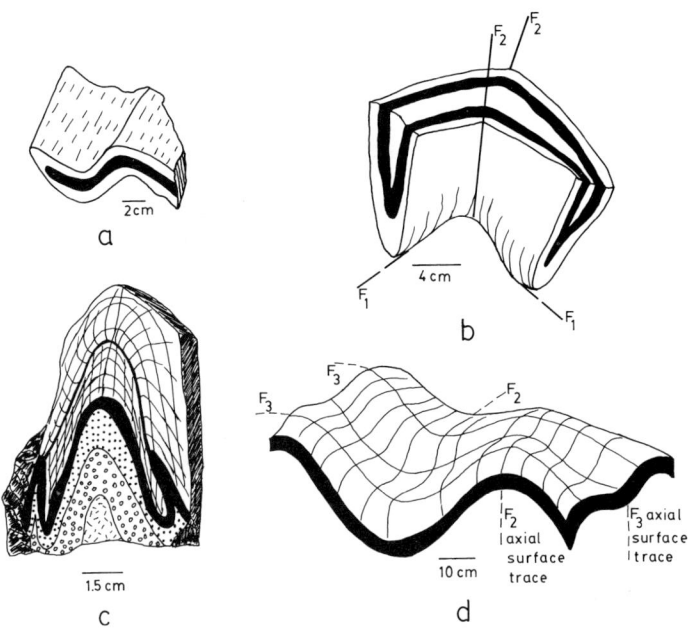

FIG. 15.25. Examples of fold interference in mesoscopic scale. (a) Coaxial folding (type 3 interference). (b) Type 2 interference involving the folding of axial surface and hinge of a tight early fold. (c) An isoclinal early fold refolded by combined flexure and flattening. The angle between F_1 and F_2 varies from place to place. (d) Small folds (F_3) riding over larger folds (F_2). Both these folds have developed on the axial plane cleavage of still earlier folds (F_1).

Style and orientation

Apart from overprinting relations, two other features, *style* and *orientation*, are important in identification of structures of different generations. "The term *tectonic style* introduced by Lugeon", as Turner & Weiss (1963, pp. 78–79) explain, "refers to the total character of a group of mesoscopic structures that distinguishes it from a group of comparable structures of another place or age, in the same way that the total character of a group of buildings or an art object can distinguish it from similar objects of other periods, places or influences." The style of a fold refers to its characteristic morphology as seen in transverse profile and also indicates whether the fold is upright, recumbent, inclined or reclined. The orientation of a fold is generally described by the orientations of its axis and axial plane. It is worth pointing out that if we know the orientations of both these structural elements we can say whether the fold is upright, recumbent, inclined or reclined. However, the statement that a fold is upright or recumbent does not indicate the orientation of its fold axis and the statement that a

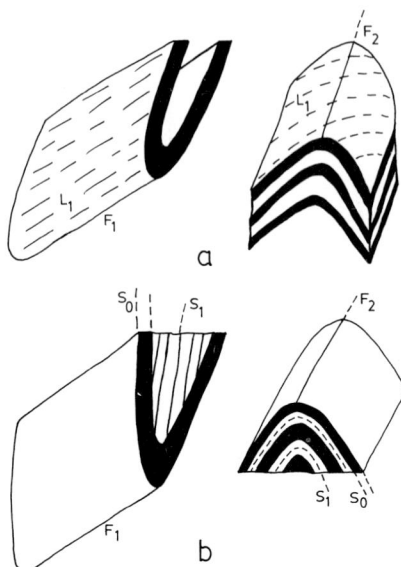

FIG. 15.26. (a) Recognition of later fold by folding of early lineation. (b) Recognition of later fold by folding of early cleavage.

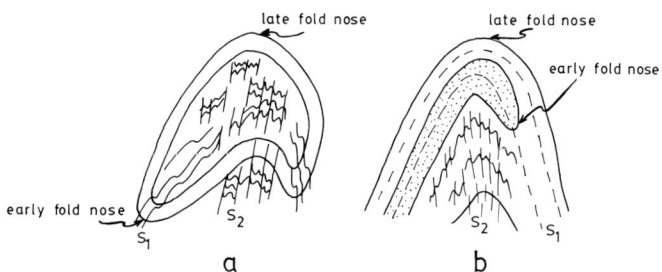

FIG. 15.27. Recognition of different generations of folds from outcrop pattern and overprinting relation of cleavages.

fold is inclined or reclined does not specify the orientations of its axis and axial plane.

For distinction among folds of different generations the style and the orientation should always be used with caution. Folds of different generations may have very similar styles. In certain areas, for instance, the two earliest generations of folds (Fig. 15.28) are isoclinal and show a type 3 interference with a common fold axis. Found separately, the two generations of folds cannot be distinguished from their style and orientation. Moreover, folds of the same generation may have different

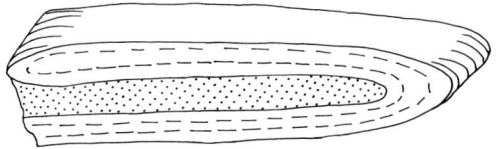

FIG. 15.28. Coaxially folded isoclinal F_1 and F_2. Found separately, the folds cannot be distinguished from style and orientation alone.

styles and orientations in different parts of an area. The tightness of a set of folds, for instance, may strongly vary from place to place. The orientations of a set of early mesoscopic folds may be different in different parts of a later large-scale fold. The orientations of a set of later mesoscopic folds superposed on a large-scale non-isoclinal fold may also be very much different in different parts of the large-scale fold.

In spite of such limitations, the fold style and orientation, either singly or taken together, and used with caution, often enable us to distinguish different generations of folds. As an example let us consider the method of age correlation that has been used by a number of workers over a large terrain in western India (e.g. Naha *et al.* 1966, 1969) where overprinting relations in numerous outcrops show three phases of folding. The earliest folds are invariably isoclinal with a very high amplitude/wavelength ratio. Where least disturbed, these show a reclined geometry with northerly striking axial planes and westerly plunge of the fold axis. In certain localities these are coaxially folded to a set of open upright westerly plunging folds. Over a large part of the terrain these two phases of folds are deformed by a group of upright folds of varying tightnesses with a roughly N–S strike of the axial planes. Since the axial planes of these folds are sub-vertical, the fold axes have a nearly uniform northerly or southerly trend, although the amount of plunge may vary strongly from place to place. Once this age correlation is made from overprinting relations in selected outcrops, the style and orientation of the folds in other outcrops may be used to identify the different generations of folds (Fig. 15.29). Thus, even where no overprinting relations are available, an isoclinal E–W trending reclined fold (Fig. 15.29A) can be identified as a first generation fold and an open westerly plunging upright fold (Fig. 15.29C) can be identified as a second generation structure. In most places the open or moderately tight upright folds with northerly striking steep axial plane (Fig. 15.29F) can be regarded as the third generation folds. However, in some of the sub-areas where these folds are often tight, an upright fold with a N–S strike of axial plane may either be a third generation fold or a reoriented first generation fold. For age correlation of structures the style and orientation have therefore to be used with caution. Some

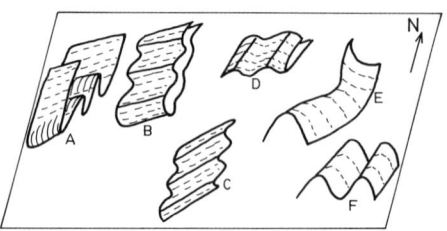

Fig. 15.29. Once overprinting relations are established in critical outcrops, the different generations of folds may sometimes be identified from style and orientation. Overprinting relations are seen in *B, D* and *E*. There is no fold interference in *A, C* and *F*, but from their style and orientation they have been distinguished as first, second and third generation folds.

other limitations in the use of style and orientation in correlation have been discussed by Williams (1985).

15.7. Macroscopic analysis

Mapping

Geometrical analysis of macroscopic superposed folds is difficult and requires considerable skill. The three-dimensional form of the large-scale structure can be reconstructed only if the outcrop pattern of the large-scale folds can be obtained by mapping the lithological units or by form surface mapping. The outcrop pattern can be mapped out with comparative ease where more or less continuous lithological units are present. Elsewhere the lithological boundaries of adjoining isolated outcrops are joined along their strikewise extensions. During mapping, the hinge regions of large-scale plunging folds are often located by the orthogonal relations of the axial plane cleavage and the lithological layering. Certain terrains of monotonous lithology may have a few marker units sufficiently continuous but too thin to show their outcrop thicknesses in the map scale. The mapping of such a marker unit, even as a thin line on the map, is important because in such monotonous terrains the geometry of the large-scale fold can only be determined on the basis of its outcrop pattern. This method is particularly useful to determine the large-scale fold geometry of migmatitic terrains containing a few mappable thin bands of such resistant rocks as quartzite, calc gneiss and amphibolite. Where more or less continuous lithologic units are not available, *form surface mapping* (e.g. Hobbs *et al.* 1976, p. 366) may reveal in an approximate manner the outcrop patterns of large-scale superposed folds. The form surface maps are usually prepared to reveal the closures of large-scale folds by tracing out on the map by continuous or discontinuous lines the trends of thin impersistent layers and the strike of foliation.

The lithological mapping is accompanied by determination of younging directions and measurement of the attitudes of bedding, cleavages, lineations and the axes and the axial planes of mesoscopic folds of different generations. Depending on the availability of outcrops, the attitudes of structural elements are measured at regular intervals as far as possible. The intervals are generally made larger or smaller depending on whether the attitude of a structural element remains uniform or changes sharply within a short distance. In addition, it is often useful to record on the map whether the mesoscopic folds are S-, M- or Z-shaped. The structural data, as much as possible, are put on the map. In addition, the linear structures and the poles to S-planes are plotted in equal area projections.

Plotting of poles of S surfaces in equal area or stereographic projection

An important aspect of the macroscopic analysis is to subdivide the regions into domains or sub-areas in each of which the large-scale fold is approximately cylindrical. The large-scale fold may have formed by deformation of either a lithological layering or a cleavage. In a domain of cylindrical folding the poles to the folded S-surface will lie close to a great circle in equal area or stereographic projection. The S-surface poles are called the π-poles and the great circle is called the π-circle. The normal to the π-circle gives us the β-axis or the fold axis for the particular domain. If the fold axis does not have a uniform orientation throughout a domain the π-poles will not lie on a well-defined great circle. An attempt is then made to further subdivide the domain into smaller units until each unit shows the pole distribution characteristic of a cylindrical fold. Interference of the folds may, however, have taken place in such a manner that a part of a large-scale structure is not divisible into macroscopic cylindrical domains. The large-scale fold geometry of such sub-areas will remain undetermined; however, even in such sub-areas, the pattern of S-pole distribution often gives us an approximate idea about the non-cylindrical geometry of the large-scale structure.

Division into cylindrical domains

The delineation of the sub-areas is usually made by trial and error. The outcrop pattern in the large-scale, along with the structural data plotted in it, give us a rough idea of making a preliminary choice of the sub-area boundaries. If the preliminary plotting of the S-poles does not

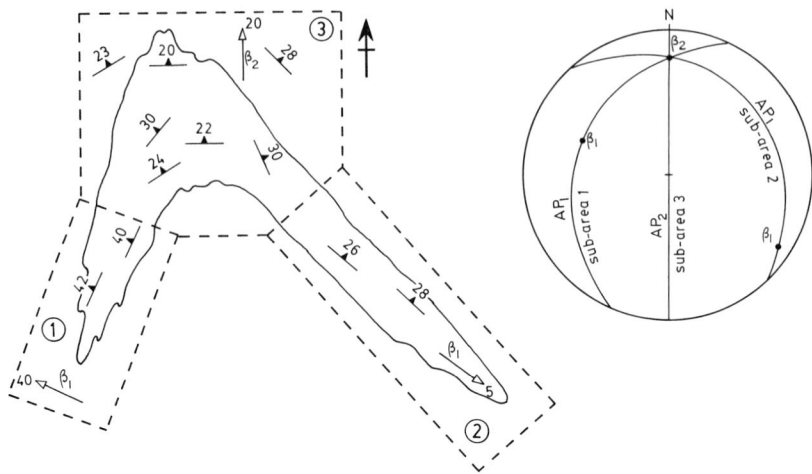

Fig. 15.30. Schematic map showing division into sub-areas. Arrows represent β-axes of bedding poles of each sub-area. The planar structure is of axial planes of early folds. The β-axis and the average orientation of axial plane of each sub-area are plotted in equal area projection (synoptic diagram).

give a well-defined π-circle, the sub-area boundaries are shifted by trial, till it shows cylindrical or sub-cylindrical systems of folds. Ramsay & Huber (1987) suggest that a fold system may be considered as cylindrical if the π-poles occur within 10° of the π-circle; it is sub-cylindrical if the π-poles occur within 20° of the π-circle.

The identification of a β-axis with respect to the folds of a particular generation (e.g. $β_1$ corresponding to F_1, $β_2$ corresponding to F_2) is done both from the large-scale outcrop pattern, as in Fig. 15.30, and from the orientation data of the mesoscopic linear structures. Thus, if F_1 linear structures in equal area projection cluster around the β-axis, the latter may be identified as $β_1$.

Reconstruction of large-scale geometry

For a complete macroscopic analysis the poles of axial planes of a large number of mesoscopic folds should also be plotted in equal area projections of each sub-area. The great circle at 90° with their point of maximum concentration may be taken as the average attitude of the axial surface of the macroscopic fold. Evidently, such a clustering of the points will occur only if the axial surface of the large-scale fold is approximately planar. The outcrop pattern, together with the orientations of the β-axis and of the axial planes, enable us to determine the geometry of the macroscopic fold in a sub-area. In Fig. 15.30, for example, the β-axis in sub-area 1 plunges 40° towards 294°. The fold closure in this sector has been interpreted as that of an isoclinal F_1 fold. Since the trend of the β-axis is nearly at a right angle to the strike of the axial plane of F_1, the F_1 fold in this sector is reclined. In sub-area 2, however, β_1 trends at a low angle to the F_1 axial trace and the F_1 fold here is an inclined fold. In sub-area 3 the β axis corresponds with the F_2 fold axis. Since the trend of β_2 is essentially parallel to the axial trace of F_2, the large-scale fold in this sector must be upright. It is often easy to visualize the gross geometry of the large-scale structure from a synoptic diagram (Fig. 15.30) in which the β-axes and the axial planes of the macroscopic folds of all the sub-areas are plotted together in an equal area projection. Thus, the reclined and inclined forms of F_1 in sub-areas 1 and 2 and the upright character of F_2 in sub-area 3 become immediately apparent from the synoptic diagram of Fig. 15.30. AP_1 in this figure represents the mean axial plane of F_1 in sub-areas 1 and 2. Evidently, the axial surface of F_1 has also been folded around the F_2 axis. We can thus see from the synoptic diagram that the entire region, comprising of all the three sub-areas, represents a single large second-generation antiform on the axial surface of an isoclinal first-generation fold. We further see that, as expected in flexural slip folding, the angle between F_1 and F_2 (i.e. between β_1 and β_2) is the same on the two limbs of this antiform, so that if the limbs are unrolled by rotation around β_2, the β_1-axes on the two limbs will have the same orientation.

The macroscopic analysis of many regions is complicated by the interference of more than two generations of folds as well as by the heterogeneity of deformations. A particular generation of folds may not be equally tightened or equally distorted by later folds in all the parts of a region. This is illustrated by an idealized map (Fig. 15.31). This map has assembled in a very schematic manner the characteristic forms of certain large-scale outcrop patterns from a large terrain in south-central Rajasthan in western India (Naha 1983, Naha *et al.* 1966, 1969, 1973). The actual outcrop patterns have been considerably modified in Fig. 15.31. Its purpose is to illustrate how the geometry of the large-scale folds can be determined with the S-pole diagrams (Fig. 15.32) and the outcrop patterns, along with certain crucial field observations of the structures in the mesoscopic scale.

To interpret the large-scale structure we need the following information about superposed folding in the mesoscopic scale. The rocks of this terrain have undergone three main phases of deformation. The earliest folds (F_1) are everywhere isoclinal. Where unaffected by later folding, the axial planes of the folds strike N–S with low plunges towards E or W. These have been deformed by a set of coaxial upright folds with E–W striking steep axial planes. These second generation folds (F_2) are weakly developed in some places and are well developed but fairly open elsewhere. The third generation folds (F_3) are more or

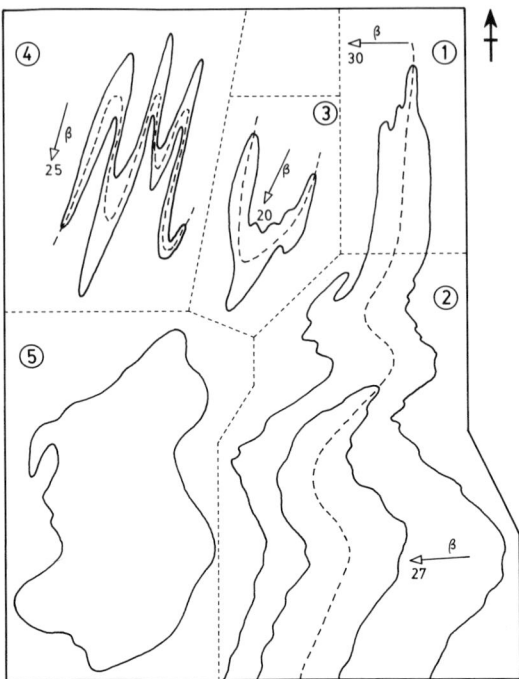

Fig. 15.31. Schematic map of some characteristic outcrop patterns of metamorphic rocks in a gneissic complex. In sub-area 5 the S-poles do not show a clear π-circle or a clearly defined β-axis.

less upright, with N–S to NNE–SSW striking axial planes, and range in tightness from open to isoclinal. The variety of outcrop patterns and of the S-pole diagrams in the different sub-areas has resulted mainly from the relative prominence and the varying tightnesses of the large-scale F_2 and F_3 folds.

In sub-area 1 we have the northward closure of a fold with N–S striking axial trace. Figure 15.32a is a contour diagram of the poles of lithological layering of this sub-area. The S-poles occur along a girdle or π-circle with the β-axis plunging 30° towards west. Since the trend of the fold axis is perpendicular to the axial trace, the large-scale fold in this sub-area must have a reclined geometry. The diagram shows that an overwhelming majority of the S-planes have a very small range of orientation. In other words, they are nearly parallel. The outcrop pattern, together with the point maximum of the S-poles, indicate that the large-scale fold is isoclinal with a comparatively small areal extent of the hinge zone. The axes of the mesoscopic F_1 folds (not shown in the diagram) cluster around the β-axis. This large-scale isoclinal reclined fold can thus be identified as F_1.

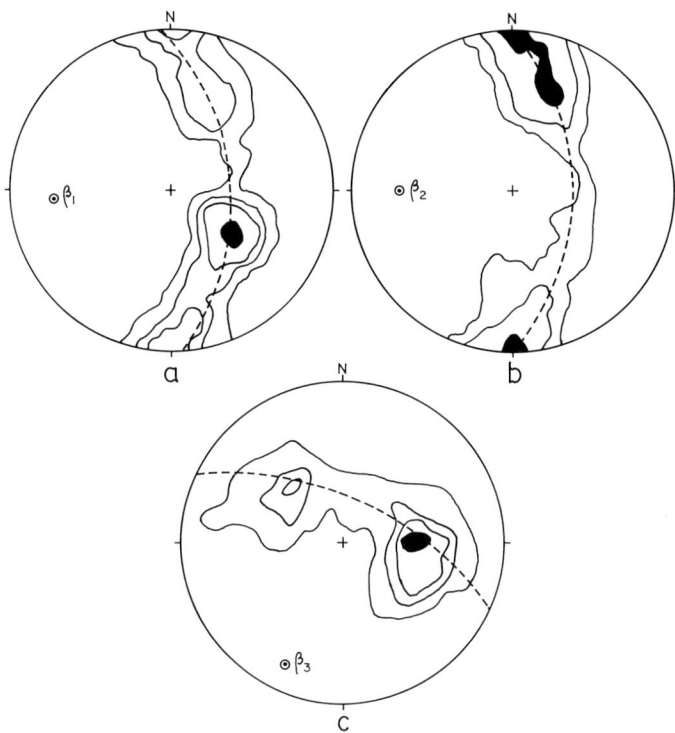

Fɪɢ. 15.32. Contoured diagrams of poles of axial planes in three sub-areas of Fig. 15.31. (a) Sub-area 1, (b) Sub-area 2 and (c) Sub-area 3.

The axial trace of the F_1 fold is strongly curved in sub-area 2. The diagram of the poles of axial planes of mesoscopic F_1 folds of this sub-area (Fig. 15.32b) shows a weakly defined girdle, with β-axis plunging 27° towards 267°. This orientation is very close to that of $β_1$ of sub-area 1. We can conclude, therefore, that the axial surface of F_1 has been folded about an axis subparallel to that of F_1. The axes of mesoscopic F_1 and F_2 folds both cluster around the β-axis. The mean orientation of the axial planes of the mesoscopic F_2 folds is vertical and has a strike of 267°. Since the easterly and westerly closing folds of the sub-area have developed by deformation of the axial planes of F_1 folds, they must be later than F_1, and the β-axis of this sub-area can be correlated with the F_2 axis ($= β_2$). The large-scale F_2 folds in this sub-area are thus upright and are coaxial with F_1. Although northerly trending F_3 folds do occur in the mesoscopic scale in sub-areas 1 and 2, from the π-pole diagrams it is evident that the F_3 folds have not affected the large-scale geometry of the earlier folds in these sub-areas.

The poles of axial planes of mesoscopic F_1 folds in sub-area 3 (Fig. 15.32c) occur along an incomplete girdle, with β-axis plunging 20°

towards 206°. Unlike sub-areas 1 and 2, SSW plunging mesoscopic F_3 folds are abundant in sub-area 3. The mean orientation of their axial planes is sub-vertical and has a NNE strike. The maximum concentration of the F_3 axes coincides with β-axis. The axes of F_1 folds (not shown in figure) are widely scattered in equal area projection. From these observations we can conclude that the crescent-shaped outcrop of sub-area 3 was produced by an interference of large-scale F_1 and F_3 folds. The SSW plunging β-axis should be correlated with the F_3 axis. The two maxima of Fig. 15.32c should correspond with the limbs of F_3. The F_3 folds here are fairly open, with an interlimb angle of about 90°. The large-scale F_3 fold in this sub-area has distorted both the hinges and the axial surfaces of the F_1 fold. The sub-cylindrical form of the large-scale F_3 fold (as indicated by the π-diagram) suggests that the pre-F_3 structure in this sub-area had a large areal extent of more or less planar S-surfaces. In other words, the F_1 hinge region must be very weak in this sub-area; otherwise the π-diagram would not have shown a clearly defined π-circle.

Sub-area 4 shows a geometry very similar to that of sub-area 3. However, the F_3 folds here are very tight, so that in the major part of the sub-area the F_1 and F_3 axial surfaces are nearly parallel. The poles of axial planes of mesoscopic F_1 folds of sub-area 5 are widely scattered and it is not possible to draw a π-circle from the π-diagram. The structure of this sub-area is thus non-cylindrical. A cylindrical domain could not be found by further sub-division of such a sector. This part of the area shows an abundance of isoclinal F_1 folds the axial surfaces of which have been deformed into domes and basins. The mesoscopic domes and basins have formed by interference of open F_2 and F_3 folds. On the basis of observation of mesoscopic structures we can suggest that the scattering of π-poles in sub-area 5 has resulted essentially from the interference of two sets of open upright folds, F_2 and F_3 in the large scale.

15.8. The problem of coaxial folding

In spite of differences in the detailed geometry of superposed fold systems from region to region, we are often struck by certain broad similarities in the history of their morphological evolution in many widely separated terrains. In this respect, one of the most striking features is the prevalence of coaxial folds. Coaxial folding commonly involves refolding of tight or isoclinal folds (say, F_1) by a set of later folds (say, F_2); the F_2 folds may vary in tightness from gentle to isoclinal. For most large terrains of coaxial folding the openness of at least some of the F_2 folds indicates that the coaxiality was not brought about by rotation of the fold hinges as in many ductile shear zones. In that event, the development of coaxial folds requires that the direction of shortening during the formation of F_2 is approximately perpendicular to the F_1 axis and is nearly parallel or at a low angle to the F_1 axial surface. Such a special situation may develop merely by chance in certain localities only, and the resulting coaxial folds are then extremely localized.

In certain terrains the sub-vertical axial surfaces of tight or isoclinal folds with sub-horizontal axis are deformed to recumbent folds with approximately parallel early and late fold hinges. It has been suggested that this kind of coaxial folding is produced by a sagging and flattening of the rock mass under its own weight at a late stage of structural evolution. The coaxial folds which are most commonly found in nature show a different type of geometry. Thus, for example, over a large terrain the Delhi Group of Proterozoic rocks in Western India were first deformed to isoclinal, recumbent or gently plunging reclined folds with NE trending axes. Their axial surfaces were coaxially folded to upright folds of various tightnesses (e.g. Naha *et al.* 1984, 1988). The parallelism of F_1 and F_2 axes over such a large terrain strongly suggests that the coaxial folding took place in the course of a single continuous deformation. Naha *et al.* (1988) suggested that the recumbent folds were produced in a rotational or non-coaxial deformation which, due to resistance to forward movement, later changed to a horizontal shortening and refolded the recumbent structure to an upright pattern. The horizontal shortening must have been by an irrotational or coaxial deformation. This is indicated by the fact that throughout the entire region the F_2 axial surfaces maintain the NE striking sub-vertical attitude in spite of large variations in the fold-tightness. A more or less similar sequence of style is shown by the coaxial early and the late folds in many other terrains (see Fyson 1971 for detailed references).

Fyson (1971, fig.1) suggested that the recumbent folds initiate early and continue to grow in the lower and inner part (the infrastructure) of an orogenic belt in response to a horizontal shear along with a vertical flattening. The upright folds grow at a later stage in relatively superficial rocks in which vertical relief is easiest. The interference of the early recumbent folds and the late coaxial upright folds takes place in an intermediate zone. However, as Fyson noted, it is not obvious why in many areas the upright folds were superposed on the recumbently folded rocks of the infrastructure. Evidently, the problem of prevalence of such coaxial folds is as yet unsolved. This problem is rendered more difficult by the fact that in many areas the early recumbent folds show a lineation which appears to be a stretching lineation parallel to the fold axis and to the trend of the orogenic belt (see section 16.5).

CHAPTER 16

Lineations

16.1. Types of lineations

Linear structures of different types (Cloos 1946) are almost always found in deformed rocks. Although the term *lineation* is generally restricted to penetrative linear structures, the distinction between penetrative and non-penetrative linear structures is not always sharp. Moreover, a non-penetrative linear structure can be often used in geometrical and kinematic analyses in the same way as a lineation *sensu stricto*. Other than the axes of mesoscopic folds, the following are the most common types of linear structures (Fig. 16.1).

1. *Mineral lineation*. This is a penetrative structure marked by sub-parallel alignment of elongate minerals. A mineral lineation may develop by parallel arrangement of crystals of elongate habit, e.g. long hornblende crystals in some amphibolites and sillimanite needles in khondalites. A common type of mineral lineation is found in phyllites and mica schists with sub-parallel arrangement of leaf-shaped mica flakes on the foliation surface.

2. *Lineation marked by stretched grains*. The grains of quartz, feldspar, calcite and other minerals are often strongly elongated by deformation. Even if the grains are very small the lineation marked by them can be easily identified in the field.

3. *Lineation marked by deformed pebble, ooid, reduction spot, etc.*

4. *Lineation marked by elongate clusters of grains*. Within each cluster the individual minerals may or may not show a preferred shape orientation.

5. *Intersection of bedding and axial plane cleavage*.

6. *Intersection of two cleavages*. The trace of bedding on a cleavage surface appears as colour stripes. A similar colour striping may also develop by the intersection of two cleavages when the earlier cleavage had a colour banding parallel to it. In certain rocks a prominent lineation is marked by a colour striping on the cleavage even when the bedding or an earlier colour banding is not identifiable in the rocks. The

structure is best described as a *striping lineation*. In all likelihood such striping lineations are initiated as an intersection lineation.

7. *Axes of crenulations.*

8. *Pressure-shadow lineation.* These are parallel spindle-shaped pressure-shadow zones on the foliation surface. The spindle-shaped bodies, composed of quartz, biotite, chlorite or other minerals, occur as tails on either side of very competent grains of garnet, magnetite or other minerals (Fig. 16.2a).

9. *Mullions and rods.*

10. *Axes of boudins.*

11. *Striations on slickensides.*

The last three are not penetrative structures. *Mullions* are parallel, pillar-like cylindrical structures which develop on discrete surfaces. The surface on which a set of mullions develops may be a bedding or a cleavage. Wilson (1953) has described a type of mullion in which the structures develop on a surface which is not everywhere parallel to the bedding or the cleavage. Wilson (1953, 1961) has recognized three types of mullions, *fold mullions, cleavage mullions* and *irregular mullions*. The fold mullions (Fig. 16.2b) are nothing but a set of exposed fold hinges. In cross-section the folds are generally cuspate in shape with broad smoothly curved convex surfaces separated by narrow, sharp, inward-closing hinges. The cleavage mullions have a similar shape and the mullion surface is accordant with the cleavage in some parts but may intersect the cleavage elsewhere. The surface of a cleavage mullion has a polished and striated appearance and has often a mica veneer. The irregular mullions are also cylindrical bodies with a grooved or fluted surface covered with a mica veneer and with an irregular cross-section.

Rods (Wilson 1953, 1961) are long cylindrical bodies of vein quartz or other segregated materials in a metamorphic rock. Similar cylindrical structures may develop by the breaking up of sheet-like veins; however, such structures derived from the breaking up of a continuous body are not considered as rods.

16.2. Usefulness of lineations in structural analysis

Geometrical relation with folds

There are generally different types of lineations in a metamorphic terrain. Their orientations should be separately recorded during structural mapping. Even the same rock may have different types of lineations. Thus, for example, the cleavage surface of most slates shows the bedding–cleavage intersection and a lineation marked by stretched grains at a high angle to the intersection lineation.

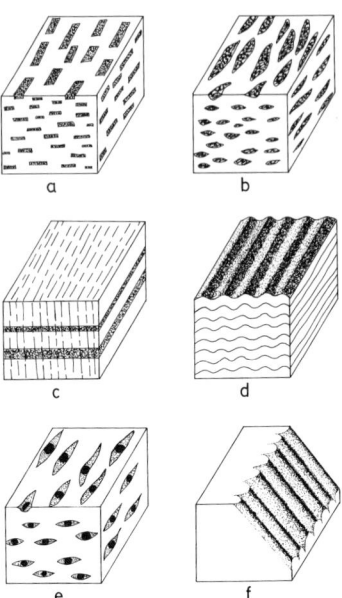

FIG. 16.1. Some common types of linear structures: (a) Lineation marked by parallel orientation of mineral grains. (b) Preferred orientation of elongated cluster of grains. (c) Bedding–cleavage intersection. (d) Axes of crenulations. (e) Pressure-shadow lineation. (f) Mullions.

The intersection of bedding and axial plane cleavage is always parallel to the axis of folds which develop on the bedding. Where the cleavage surface is deformed to large-scale folds, the crenulation axes or the intersections of the crenulation cleavage and the earlier cleavage are parallel to the axes of these folds. Mullions, especially fold mullions, are also parallel to the axes of large-scale folds. Mineral lineations and lineations marked by stretched grains or grain aggregates may or may not be parallel to the fold axis. These linear structures can be utilized to determine the orientation of the fold axis only if their geometrical relations with the folds can be clearly established from observations in critical outcrops.

Geometrical relation with strain

The pressure-shadow lineation is parallel to the direction of maximum stretching in the rocks. Mineral lineation and lineations marked by deformed grains, grain aggregates and other objects are commonly parallel to the X-axis of the strain ellipsoid. Evidently, the mapping of these lineations is of great importance since the lineation map immediately gives us the gross pattern of variation of the orientation of the X-axis throughout the area.

FIG. 16.2. (a) Pressure–shadow lineation (parallel to matchstick) in mica schist. The pressure shadow has developed at the border of garnet porphyroblasts. Ghatsila, India. (b) Steeply plunging mullions, Zawar, Rajasthan, India.

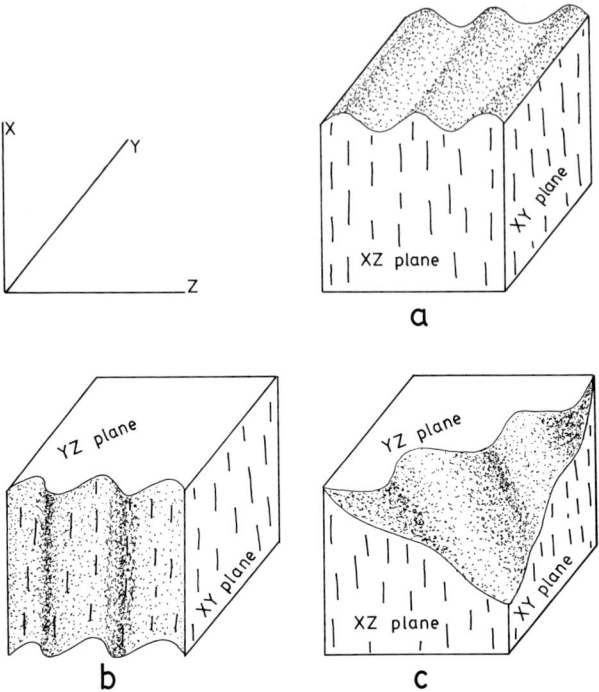

FIG. 16.3. Stretching lineation (dashed lines on XY-planes) perpendicular to fold axis (a), parallel to fold axis (b) and oblique to fold axis (c).

Whether a mineral lineation or a grain-elongation lineation is parallel to the fold axis or not depends upon the relation between the geometry of the fold and the orientation of the principal strain axes. The fold axis initiates parallel to the long axis of the strain ellipse on the plane of the layer. This ellipse axis may or may not coincide with a principal axis (either X or Y) of the strain ellipsoid. We may have any one of the three situations:

1. The layer is initially parallel to the YZ-plane of the strain ellipsoid or somewhat inclined to it but parallel to the Y-axis. The fold axis initiates parallel to the Y-axis and the stretching lineation on the cleavage surface develops perpendicular to the fold axis (Fig. 16.3a).

2. The layer is initially parallel to the XZ-plane of the strain ellipsoid or is somewhat inclined to it but parallel to the X-axis. The fold axis is then parallel to the X-axis and the stretching lineation (Fig. 16.3b).

3. The layer is initially inclined to the X- and the Y-axes. The fold axis initiates neither parallel to X nor parallel to Y and is oblique to the stretching lineation (Fig. 16.3c).

With progressive deformation a marker line oblique to the principal axes of the strain ellipsoid rotates towards the X-axis. Hence, if the stretching along the X-direction is extremely large, the fold axis may finally become nearly parallel to the stretching lineation even if it was oblique to it in the initial stage.

Linear structures indicating slip direction

The striations on slickensides are parallel to the direction of fault movement. The movement on a fault may take place over a long period of time and the direction of movement may change during this period. The striations on slickensides are generally parallel to the last phase of the fault movement.

Lineation in superposed deformation

Structural analyses of deformed terrains almost always involves measurement of attitudes of lineations and representing them in the structural map and in equal area projections. It is clear from many of the earlier studies that lineations play an essential and sometimes a key role in unravelling the structural history of an area (Kvale 1948, Sutton & Watson 1955, Weiss & McIntyre 1957, Clifford *et al.* 1957, Ramsay 1958a & b, 1960, Clifford 1960, Fleuty 1961, Baird 1962, Naha & Halyburton 1977 and many others).

When an early lineation (say, L_1) is deformed by later folds (say, F_2), the pattern of the deformed lineation gives us valuable information regarding the nature of superposed deformation. The lineation pattern also gives us some information about the mechanism of folding. The significance of different types of lineation patterns is described in the following section.

16.3. Patterns of deformed early lineation

Lineations deformed by flexural slip folds

When an early lineation (L_1) is deformed by a later flexural slip fold (F_2), the angle between the lineation and the fold axis remains constant in all parts of the fold (Fig. 16.4a, b). If we place a transparent overlay on the folded surface, draw a line along the fold-hinge and trace out the lineation, we find that the lineation becomes straight when the transparent sheet is flattened out. In other words, the lineation is straightened out when the fold is unrolled. Where the folds are somewhat larger so that the attitudes of the lineation and of the folded layer can be measured in the different parts of the folds, the lineations and the fold axis can be plotted in equal area projection. If the lineation is unrollable the lineation plots will lie on a small circle around the fold axis

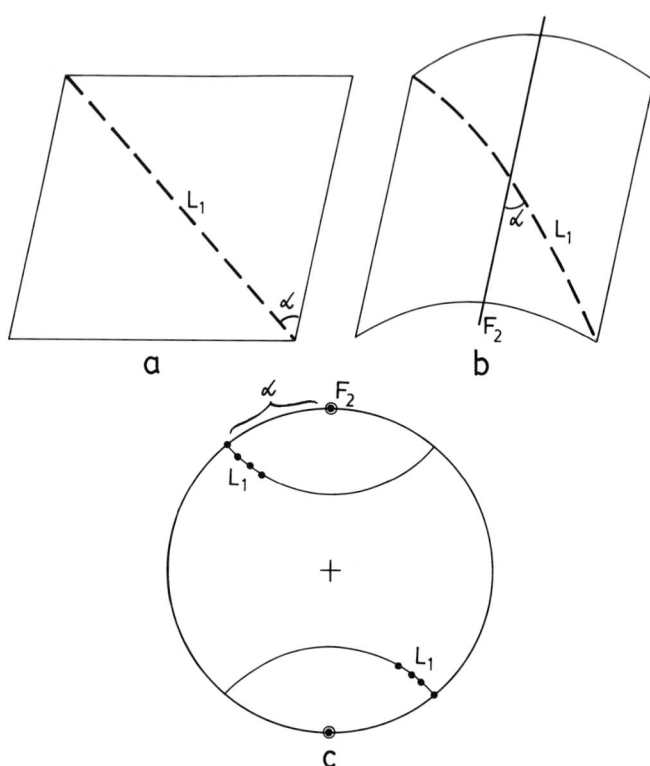

FIG. 16.4. (a) Initial orientation of early lineation L_1. (b) Deformation of L_1 by flexural slip folding. The angle (α) between L_1 and F_2 remains constant all over the fold. (c) Equal area projection of L_1 of (b) shows a small circle distribution of L_1 around F_2.

(Fig. 16.4c). If L_1 had initially a uniform orientation but the F_2 axis was differently oriented in different places, each of these places will show a pattern of small circle distribution of L_1 around the local F_2. The initial orientation of the lineation can be determined from the intersection of these small circles when they are plotted in a single diagram (Ramsay 1967, pp. 549–550).

Lineations deformed by shear folds

The orientation of an early lineation changes during shear folding. The deformed lineation (Fig. 16.5a) always lies on a plane defined by the initial orientation of the lineation and the direction of slip of the later shear fold (e.g. Weiss 1955). In equal area projection the lineations are distributed along a great circle (Fig. 16.5b). The orientation of the slip direction can be determined from the intersection of the lineation

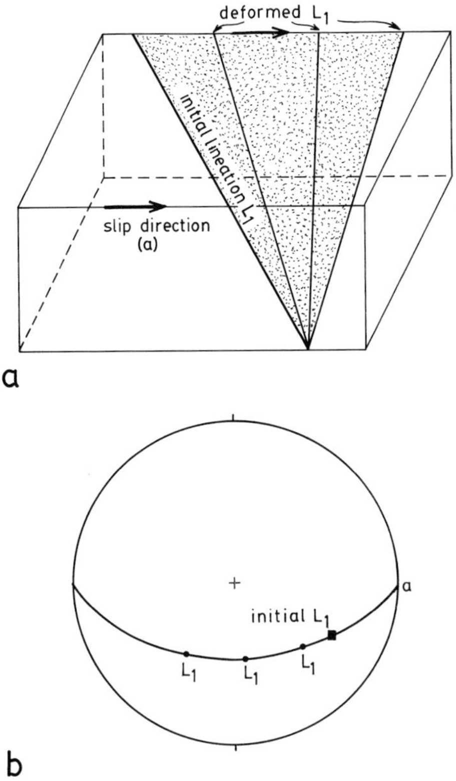

LINEATIONS

deformed L$_1$

initial lineation L$_1$

slip direction
(a)

a

initial L$_1$

a

L$_1$ L$_1$ L$_1$

b

FIG. 16.5. (a) During shear folding an early lineation L$_1$ changes its orientation but always lies on a plane defined by direction of shear and the initial attitude of L$_1$. (b) Equal area projection of L$_1$ deformed by shear folding. The differently oriented lineations lie on a great circle.

plane and the axial plane of the fold (Ramsay 1967, pp. 474–475). It should be noted, however, that the occurrence of a great circle pattern does not prove that the lineations have been deformed by shear folding. Indeed, in many cases the great circle pattern develops by other mechanisms.

Lineation patterns in other types of folding

A variety of lineation patterns may develop when the later folds form by a combination of buckling and flattening (Ramsay 1960, 1967, Hudleston 1973c, Mukhopadhyay & Ghosh 1980, Ghosh & Chatterjee 1985). When folding takes place by a shortening parallel to the layers, the total shortening is generally composed of two parts. While a part of the shortening is achieved by external rotation, i.e. a body rotation of the layers, another part is essentially a layer-parallel homogeneous

373

strain (Ramberg 1964) which causes a change in the orthogonal thickness. The ratio of the rates of external rotation (or buckle shortening) and layer-parallel strain depends mainly on two factors, (i) the competence contrast of the layer and its environment and (ii) the stage of folding as indicated by the inter-limb angle.

If a layer is very competent relative to its host, the layer-parallel strain in it is negligible at all stages of evolution of the fold. The folding then takes place almost entirely by external rotation and the deformed early lineation shows a simple small circle pattern.

If the competence contrast is not so large the lineation changes its orientation both by an external rotation of the layer and by layer-parallel strain. The effect of layer-parallel strain is relatively large in the initial stages when the inter-limb angle is very large. In the middle stage of folding the rate of buckle shortening becomes large compared to the rate of layer-parallel strain. In the late stage of folding, when the fold has already become tight, there is not much scope of shortening by external rotation of the limbs. If the deformation continues the fold must change its shape mainly by layer-parallel strain. The lineation must also rotate so that there is a decrease in its angle with the long axis of the strain ellipse on the layer.

The final pattern of deformed lineations is controlled by (1) the competence contrast, (2) the initial angle between L_1 and F_2, (3) the initial orientation of the fold axis with respect to the orientations of the principal axes of bulk strain and (4) the nature and intensity of deformation. Depending upon these factors we may have a variety of lineation patterns (Fig. 16.6). In most of these patterns the angle between L_1 and F_2 varies from place to place in different parts of the fold. If the fold is unrolled the lineation remains curvilinear. The lineations do not necessarily lie on a great circle in equal area projection. However, if the deformation is very large the lineations may show a great circle pattern in certain situations.

The recognition of "non-unrollable" lineations not only enables us to identify folds formed by simultaneous buckling and flattening, it may also be possible in certain instances to determine the rotation history of folds from a study of the lineation patterns (Ghosh & Sengupta 1987a & b). A V- or U-shaped pattern of L_1 on the unrolled form surface of F_2 is particularly useful in this respect. The sharply curved segment of the lineation is located either at the hinge zone of F_2 or very close to it (Figs. 16.7, 16.6f & g). This type of pattern develops only when the undeformed L_1 was initially at a large angle to F_2. Its presence further indicates rotation of the fold axis towards the stretching direction through a large angle (Fig. 16.8). In general, the curvature of this lineation pattern is opposite in antiforms and synforms (Hudleston 1973c, Ghosh & Chatterjee 1985). In another type of

LINEATIONS

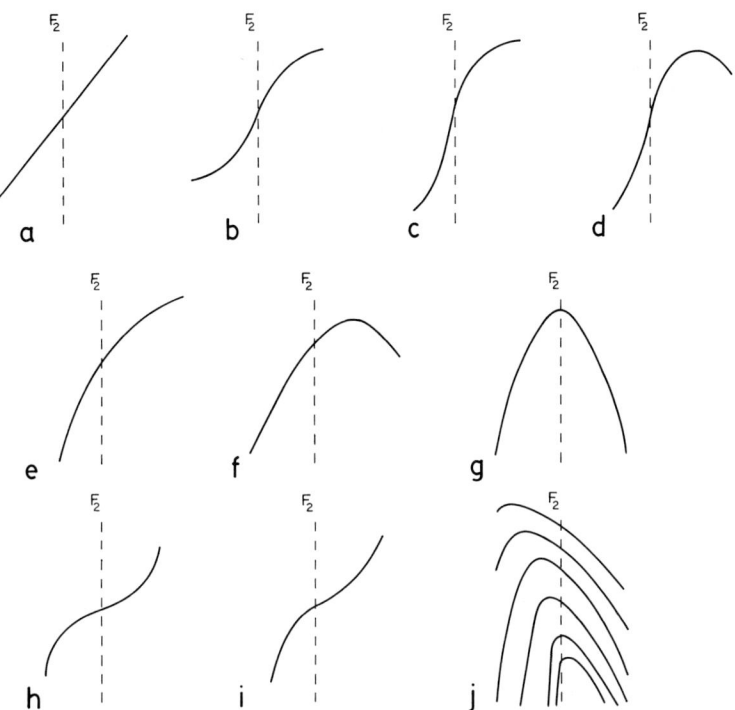

FIG. 16.6. Diverse patterns of L_1 when the form surface of F_2 is unrolled around its axis. The curved lineation patterns have developed because of a combination of buckling and flattening. After Ghosh & Chatterjee 1985.

pattern on the unrolled form surface (Fig. 16.6b, c) the acute angle between L_1 and F_2 is variable, with a minimum at the hinge zone, and opens in the same sense in all parts of the fold; if the angle is small we may conclude that the fold hinge was initiated at a low angle to the lineation (Ghosh & Chatterjee 1985, Ghosh & Sengupta 1987a, p. 278) and there was no significant rotation of the fold axis.

Whatever be the mechanism of folding, an early lineation often has different orientations on the two limbs of a later fold. Hence, in areas of isoclinal folding we may get differently oriented lineations on parallel surfaces of foliation (Fig. 16.9). The narrow hinge zones of the iso-clinal folds may not be identifiable in all such outcrops. However, from observations in a few critical outcrops it is usually possible to locate the fold hinges and to confirm that the different orientations of the lineation on parallel foliation surfaces are indeed a result of isoclinal folding.

In certain places a curved lineation may occur on more or less planar foliation surfaces (Fig. 16.10). In all likelihood this pattern may develop in different ways. It may, for example, develop in a single deformation by

FIG. 16.7. Early lineations deformed by later fold. The lineations remain curved when the folds are unrolled. On the unrolled form surface the maximum curvature of the U-shaped lineation pattern is in the neighbourhood of the fold hinge. In (b) the lineation is on an isoclinal fold. The lineation on the backside of the fold is seen on mirror reflection. The black band is the edge of the mirror. Quartzite mylonite in Singhbhum Shear Zone.

a mechanism similar to that of sheath folding (Fig. 16.11). It may also form by superposed deformations (e.g. Ramsay 1967, p. 471, Ghosh & Chatterjee 1985, p. 661).

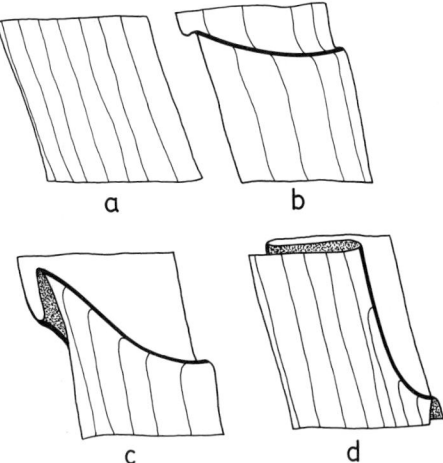

FIG. 16.8. Mechanism of development of U-shaped lineation pattern (on unrolled form surface) by rotation of fold hinges in shear zone. (a) Foliation with down-dip stretching lineation. (b) Development of sub-horizontal fold. (c) and (d) Rotation of fold hinges towards the down-dip stretching direction. In (d) the fold has become reclined. After Ghosh & Sengupta 1990.

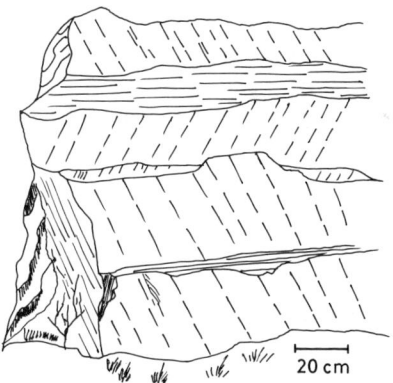

FIG. 16.9. Different orientations of mineral lineation on parallel foliation surfaces. Singhbhum Shear Zone, India.

16.4. Slickenside, slickenfibre, slickolite

The *slickenside* lineation which we often see in fault zones indicates the direction of slip. In some cases the sense of slip can be determined from a step-like arrangement along the slickenside lineation. Slickenside lineations may be of different types (e.g. Means 1987). The grooves or scratches on the slickensides form in the same way as groove casts in soft sediments, by the ploughing of a protuberance on a fault face.

377

FIG. 16.10. Strongly curved lineation on planar mylonitic foliation. Specimen from ductile shear zone at Phulad, Rajasthan, India. The area also shows abundant sheath folds.

Slickensides may also form when the powdered rock flour or gouge is drawn out in the form of streaks or accumulate in front or behind protuberances. Long ridges may develop by erosional sheltering (Means 1987). Slickenside lineations may also develop by growth of sheets of fibres of vein material with the fibres parallel to the direction of displacement (Durney & Ramsay 1973, Ramsay & Huber 1983). The fibres are slightly oblique to the general surface of movement. This gives a step-like arrangement of the fibre sheets. The acute angle between the fibre sheets and the slickenside surface is sympathetic to the sense of shear. The fibres, often described as *shear fibres* or *slickenfibres*, are generally composed of quartz or calcite. *Slickolites* (slickenside + stylolite) are stylolitic teeth (Bretz 1940) at a low angle to the fault surface. The slickolite lineation points towards the displacement direction. Means (1987) has recorded a type of slickenside with both grooves and ridges, each ridge fitting into a groove in the opposite wall of the fault. The origin of this type of slickenside is not yet understood.

16.5. Relation between stretching lineation and the fold axis

As mentioned in section 16.2, the fold geometry may show diverse relations with the direction of maximum stretching. In many terrains,

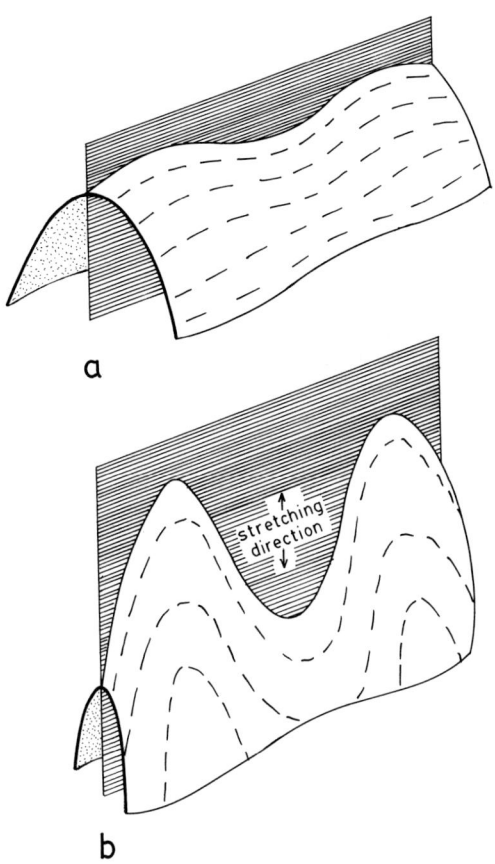

FIG. 16.11. Mechanism of development of curved lineation on more or less planar limbs of isoclinal folds with strongly curved hinge line.

especially in regions of upright folding in relatively superficial rocks, a steep lineation which can be clearly identified as the direction of maximum stretching occurs on the axial plane cleavage and at a high angle to the fold axis (Fig. 16.12a). The occurrence of such a steep stretching lineation is easy to understand, since, under a horizontal shortening, the vertical direction is expected to be the direction of easiest relief. The occurrence of a lineation sub-parallel to the fold axis is much more difficult to explain. There are two different cases to be considered for such a geometrical relationship. In one of these the lineation is approximately parallel to the axes of mesoscopic folds in ductile shear zones (Fig. 16.12b). Recent studies in ductile shear zones have clearly demonstrated that this parallelism is brought about by a rotation (Fig. 16.8) of the fold hinges (see Chapter 21). Thus, in a ductile shear zone with a

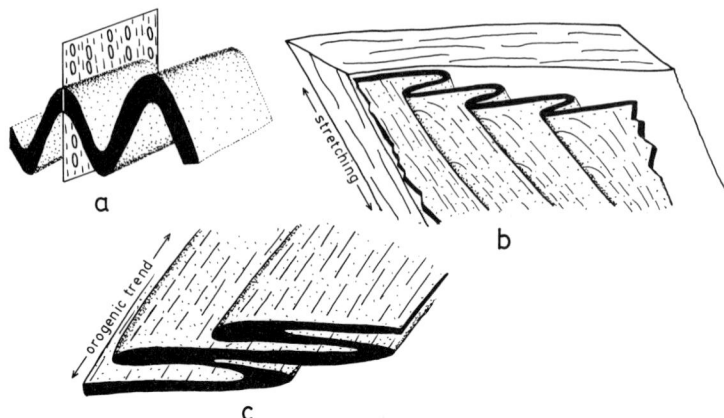

Fig. 16.12. Different relations between stretching lineation and fold axis.

thrusting sense of movement the stretching lineation is nearly down-dip at all stages of deformation. The folds initiate at a high angle to this lineation, rotate towards the direction of stretching in the course of progressive deformation and finally become nearly reclined. In that event the fold axis becomes sub-parallel to the stretching lineation in most segments. A careful examination of the outcrops may reveal the rotation history of the folds (Ghosh & Sengupta 1987a, 1990) through the pattern of the bent lineations (Figs. 16.6g, 16.7b) over the hinges of the reclined folds.

In many fold belts, especially in highly metamorphosed and gneissic terrains, a pervasive lineation appears parallel to the fold axis and sub-parallel to the trend of the orogenic belt (Fig. 16.12c). The lineation may appear as a pebble elongation and as a mineral lineation. The occurrence of such lineations is difficult to explain because a large stretching parallel to the orogenic belt creates a room problem. So far no geotectonic model has been proposed which explains a large extension parallel to the orogenic trend. The reason for the occurrence of "stretching lineation" parallel to the fold axis in such terrains is not yet fully understood. Before a solution to this problem is attempted there is clearly a necessity of detailed strain analyses to confirm that the mineral lineation in such fold systems is indeed sub-parallel to the direction of maximum extension.

CHAPTER 17

Boudinage

17.1. Introduction

The term boudinage was coined by Lohest *et al.* (1909, also Lohest 1909) to describe the process of formation of a certain type of structure by fragmentation of a sheet of brittle rock sandwiched between layers of ductile rocks (Fig. 17.1). The structure was first described from the sandstone beds in the Bastogne region of Belgium. In plan view, i.e. on a surface parallel to the stratification plane, the sandstone beds are separated into "enormous cylinders or boudins aligned side by side". The fragments are separated by pods of vein quartz. In transverse sections the fragments are barrel-shaped and are aligned in a row. The individual fragments are called boudins and the process itself is called boudinage. The term boudin means sausage. In making this analogy Lohest emphasized the plan view of the cylindrical boudins. However, to some people the analogy meant the similarity of the cross-sectional view of a row of boudins to a chain of sausages tied end to end.

Pinch-and-swell (Fig. 17.2a) structures are often found in association with boudinage structures. They show localized zones of thinning similar to the necked zones of stretched metal rods without the formation of fractures. Boudinage and pinch-and-swell structures are very common and occur in all terrains which have undergone large penetrative deformations. Similar structures were described by Ramsay (1881) and Harker (1889) even before the term boudinage was coined. The early studies on such structures (Holmquist 1930, 1931, Corin 1932, Wegmann 1932) clearly indicated that boudinage and pinch-and-swell structures were produced by extension of a brittle or more competent layer encased within a ductile or less competent host.

In a boudinage structure the fragments occur in a row so that there is no doubt about the continuity of a layer before boudinage occurred (Fig. 17.1b). In certain areas we find irregular isolated fragments of a competent rock within an incompetent host. The fragments are occasionally folded. Although they are derived from one or more layers, the general continuity of a layer cannot be deciphered with certainty by joining the fragments. These should be called *tectonic inclusions* (Rast 1956) or *fish* (McIntyre 1951) rather than boudins.

The classical type of boudinage structures, such as those described by Lohest (1909), Corin (1932), Wegmann (1932) and Cloos (1947a), developed by extension fracture of a competent layer and separation of the fragments. This is also the commonest variety of boudinage. Boudinage may also develop in a competent layer by shear fracture and offset of fragments resulting from a layer-parallel extension. Further, a structure similar to the boudinage structure may form within foliated rocks without an apparent lithological or competence contrast between a boudinaged zone and its host. These are called foliation boudins or internal boudins. These three processes (Fig. 17.3), e.g. (1) *extension fracture boudinage*, (2) *shear fracture boudinage* and (3) *foliation boudinage*, will be described separately. Unless otherwise mentioned, the term boudinage in the following discussion will refer to the classical type.

17.2. Geometry of boudinage structure

The geometry of boudinage structures (Fig. 17.4a) may be described in terms of the following structural elements (Wilson 1961): (i) The *length* of a boudin is measured along its longest direction. This direction is the *boudin axis.* (ii) The *width* of a boudin is measured in a section transverse to the boudin axis and along a direction parallel to the layering. (iii) The thickness of a boudin is the thickness of the layer which has suffered boudinage. (iv) The distance between two adjacent boudins in a transverse section is called *separation.* In a general way the gap between two boudins may be described as the node or the separation zone.

If boudinage takes place in two directions (*chocolate tablet boudinage* of Wegmann 1932), the individual boudins in plan view may either be equidimensional or, more commonly, inequidimensional (Fig. 17.2b). In the latter case, the longer direction may be called the boudin axis. If necessary, we may distinguish between the *longitudinal boudin axis* and the *transverse boudin axis*. In a similar way we may distinguish between *longitudinal separation* and *transverse separation* (Fig. 17.4b). In case of pinch-and-swell structures, the line along which the layer has undergone the maximum thinning is called the *neck line* (Fig. 17.4c). The boudin axis and the neck line have the same tectonic significance in the sense that both the lines are perpendicular to a direction of principal extension in the plane of the layer.

The layer which undergoes boudinage is more brittle or less ductile than the host rock. The layers of the ductile host flows towards the zones of separation and are deformed to localized bending folds. The folds, described as *scar folds* (Figs. 17.1a, 17.4a) by Whitten (1966), rapidly die out as we move away from the boudinaged layer.

FIG. 17.1. (a) Boudinage of a quartz vein in a gneissic host. (b) Boudinage of quartzite bands in banded iron formation. Ramagiri, Andhra Pradesh, India.

The plan view of boudins is not exposed in most cases. This is the reason why orientation data of boudin axes are so seldom given in the geological literature. However, where boudinage structures are plentiful, a careful search generally reveals some features from which the orientations of the boudin axis can be measured. (i) When the plan view of a boudinaged layer is exposed (Fig. 17.5a), the long direction of

FIG. 17.2. (a) Pinch-and-swell structure in an amphibolite band in granite gneiss. The layer-normal shortening of the amphibolite band is shown by folding of the transverse quartzofeldspathic bands. Chotanagpur Gneiss, Jasidih, India. (b) Casts of chocolate tablet boudins in phyllite. The two sets of boudin axes are parallel to the two pencils. The rock sample was broken along the interface of a boudinaged quartz vein and its host phyllite. Ramagiri, Andhra Pradesh, India.

cylindrical boudins can be directly measured. (ii) In some cases, a transverse or oblique section of the boudins is exposed while the plan view is covered by a thin layer of host rock in which the hinges of the scar folds

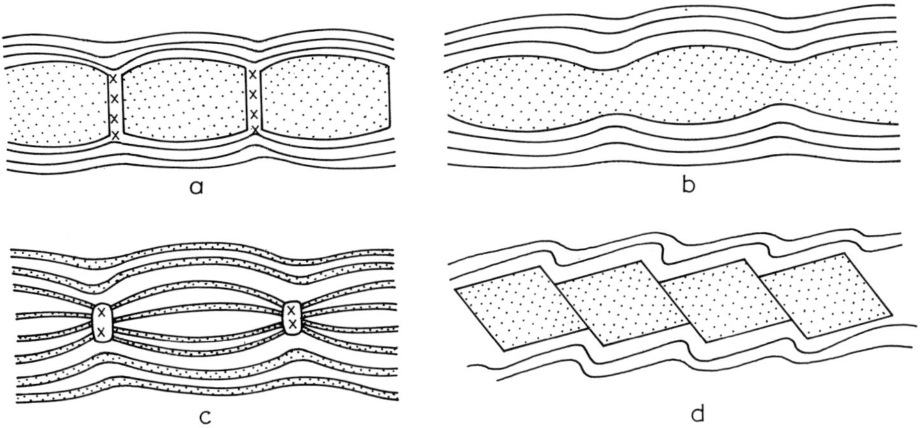

FIG. 17.3. (a) Extension fracture boudinage, (b) pinching-and-swelling, (c) foliation boudinage and (d) shear fracture boudinage.

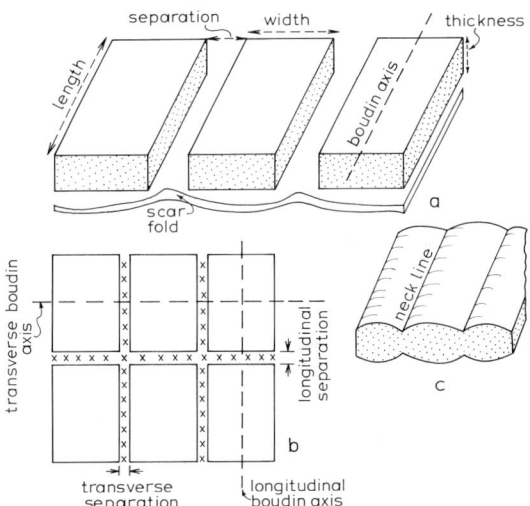

FIG. 17.4. Structural elements for describing the geometry of boudinage and pinch-and-swell structures.

can be seen (Fig. 17.5b). The orientation of the hinges gives the orientation of the boudin axis. (iii) While following a boudinaged layer we may come across an exposure where the boudins have been partially or completely eroded away but the casts of boudins are exposed in plan view (Figs. 17.5c, 17.2b) on the enveloping foliation surface. The boudin axis can be measured from such casts. (iv) The traces of the nodal pegmatite or vein material on the enveloping bedding or foliation surface can also give us the attitudes of the boudin axes (Fig. 17.5d).

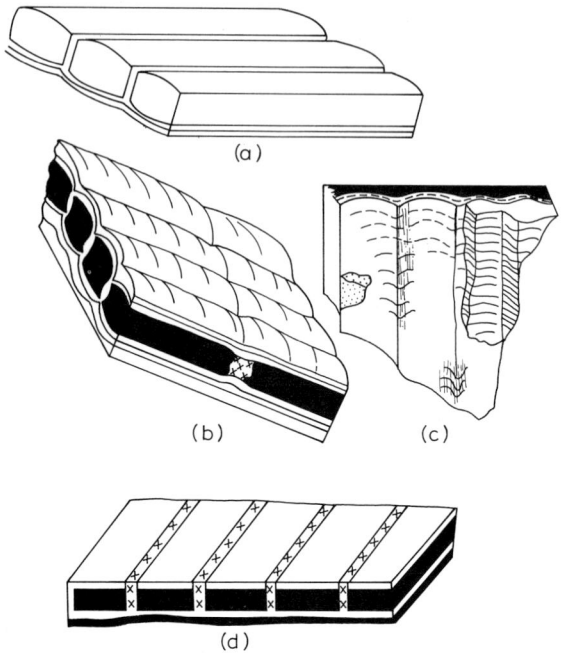

Fig. 17.5. Methods of determining orientation of boudin axis in the field. (a) Boudins with exposed plan view. (b) Scar fold hinges exposed on the enveloping host rock. (c) Casts of boudins in the host rock. (d) Traces of nodal veins on the foliation of the host rock.

17.3. Significance of boudin shape in transverse section

In transverse sections the boudins may have varied shapes (Fig. 17.6): rectangular, barrel-shaped, lenticular, fish-head and rhombic. Such dissimilar shapes give us some information concerning the competence contrast between a boudinaged layer and its host and also about the extent of *pre-boudinage* and *post-boudinage plastic deformation.*

The presence of *rectangular boudins* with edges at a right angle to the general layering indicates that throughout the course of development of boudinage the layer behaved in a brittle manner (Fig. 17.6a). At the corners of the boudins there is a relatively high shear stress which tends to deform the boudins. Preservation of the sharp right-angled corners of rectangular boudins therefore indicates that the competence contrast between the boudins and the host rock is very large. *Barrel-shaped boudins* with straight edges (Fig. 17.8b) are produced when some amount of necking is followed by extension fracture at the necked zone.

The shapes of the boudins may be further modified by post-boudinage plastic deformation. Since boudins are more competent

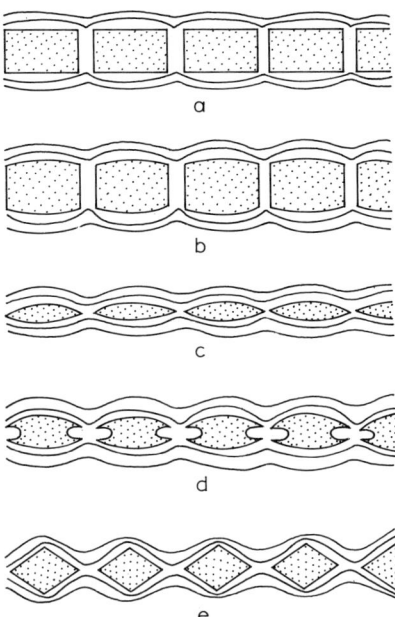

FIG. 17.6. Diverse shapes of boudins in transverse section. (a) Rectangular, (b) barrel-shaped, (c) lenticular, (d) fish-head and (e) rhombic boudins.

than the host rock, they deform more slowly than the surroundings and a high shear strain develops at the longer edges of the boudins (Fig. 17.7). Since the sense of shear is opposite in the two halves of a boudin, the shear is associated with an effective lengthening of that part of the boudin which is close to the contact. The shear strain and the associated lengthening decreases towards the mid-level of the boudinaged layer. As a result, the lateral walls of the boudin curve inward (Fig. 17.7). This zone of curving also undergoes a larger layer-normal compression so that the barrel-like shape is exaggerated. A very large amount of such post-boudinage plastic deformation leads to the formation of *fish-head boudins* (Fig. 17.8) similar to those described by Wegmann (1932, fig. 1) and Ramberg (1955, fig. 1B).

Experimental deformation with soft models suggests that with decreasing competence contrast between a layer and its embedding medium, extension fracture is preceded by a greater layer-parallel homogeneous elongation as well as a greater localized deformation by necking. If the competence contrast is rather small the necking continues till the pinched zones taper off and *lenticular boudins* (Fig. 17.6c) are separated without the formation of a clearly defined layer-normal extension fracture (Ramberg 1955, p. 514).

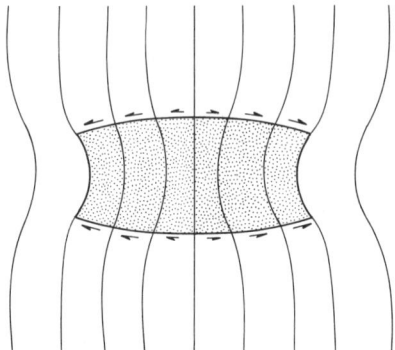

FIG. 17.7. Development of barrel-shaped boudins with inward curving of the lateral edges.

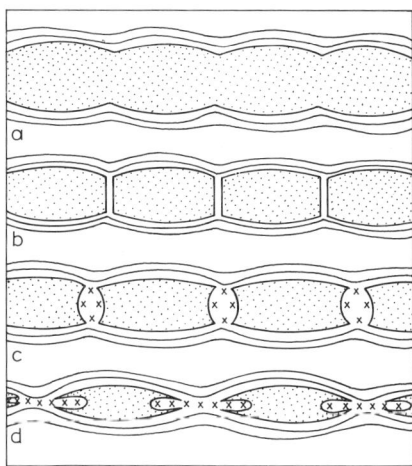

FIG. 17.8. Stages of development of barrel-shaped and fish-head boudins.

17.4. Aspect ratios of boudins in cross-section

The aspect ratios of boudins are measured in sections perpendicular to a boudin axis. For unidirectional boudinage the aspect ratio is the ratio of width to thickness. For two-dimensional boudinage we have also to measure the length-to-thickness ratio. The width-to-thickness ratio of boudins may vary within a wide range. However, in most areas there is a general tendency of thicker beds to form boudins with larger width (Ramberg 1955, Uemura 1965). Ekström (1975) and Troëng (1975) report a range of 2–4 of width-to-thickness ratio, while Talbot (1970) reports a much greater range of 2–20. The measurements by Uemura (1965, pp. 105–106) show a smaller range, 1.4–3.3, with an average of 2.1.

The theory of boudinage predicts that, in any one competent bed, boudins with a restricted range of aspect ratio should be more frequent than others. To take a simple situation, assume that a brittle layer sandwiched between two ductile layers is subjected to a layer-normal compression. As the ductile layers are thinned they tend to flow laterally. This lateral flow causes the development of layer-parallel tensile stresses in the brittle layer. When the stress exceeds the strength, an extension fracture develops perpendicular to the layering and the layer is thus broken into two pieces (Fig. 17.9a). With continuation of lateral flow, the separation between the two pieces increases while a fresh mid-point fracture forms in each of the pieces (Fig. 17.9b). The stress within a competent fragment increases with increasing width (i.e. length at a right angle to the boudin axis) and decreases with increasing thickness. This means that it becomes increasingly difficult for a bed to fracture when the boudin width becomes shorter. For each bed there is a critical aspect ratio (say l_c) below which further mid-point fracturing is not possible. This is the reason why it is expected that the boudins of a bed would show a small range of width-to-thickness ratios.

As Lloyd *et al.* (1982) point out, the final aspect ratio which is obtained after such sequential mid-point fracturing is dependent on its initial length. This becomes clear if we consider a few examples. Let the critical aspect ratio be 2. If we start from a layer segment of aspect ratio 20, by successive mid-point fracturing we arrive at a value of 2.5. Since this is larger than l_c, the segment will suffer extension fracture to yield stable boudins with aspect ratio 1.25. If, instead, the starting layer segment has an aspect ratio 30, we get a final aspect ratio 1.875 of stable boudins. Thus, depending upon the initial length of the layer segment in a transverse section, the final boudin aspect ratio may range between l_c and $l_c/2$. In nature, this range is likely to be modified by other factors such as the occurrence of flaws in the competent bed, variation of strength and microstructure in different parts of the bed and post-fracture elongation of the boudins.

17.5. Geometrical relation with folds

The boudin axis may be parallel, perpendicular or oblique to the fold axis. In the majority of areas the boudin axes are roughly parallel to the fold axes (Quirke 1923, Cloos 1946, 1947a, McIntyre 1950b, Wilson 1951, 1961, Sanderson 1974). In many areas there are two sets of boudin axes or neck lines (Fig. 17.10a), parallel and perpendicular to the fold axis (Jones 1959, Coe 1959, Fyson 1962, Sengupta 1985). Thus, in Devonian rocks near Plymouth, England, the most prominent neck lines are oriented down-dip, at a right angle to the sub-horizontal fold axis; less well-developed neck lines are approximately parallel to the fold axis (Fyson 1962, p. 214). The relations are similar to those described by Coe (1959) from West Cork, Ireland. The boudin axes or the lengths of

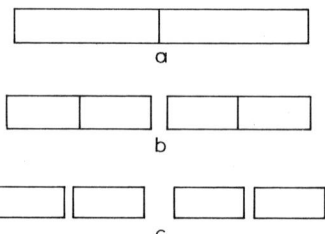

FIG. 17.9. Boudinage by sequential development of mid-point fractures in a brittle layer. Note that the separation between adjoining boudins is not the same everywhere. The separation will be larger across an earlier fracture.

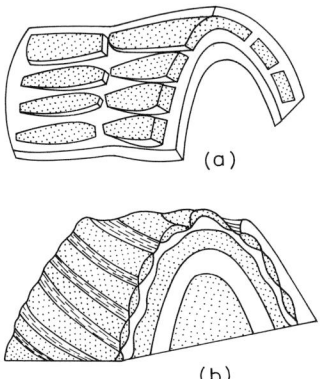

FIG. 17.10. (a) Boudin axes parallel and perpendicular to the fold axis. (b) An early generation of boudins deformed by a fold. The boudin lines are folded and the individual boudins at the hinge zone are folded into half-waves.

boudins need not be either parallel or perpendicular to the fold axis. Wilson (1961) refers to an observation of D. J. Shearman from N. Devon, England, where the lengths of boudins are oblique to the axes of folds. Such oblique boudins may develop if a bed is oriented oblique to the principal axes of stress.

The geometrical relation of boudinage structures with folds should never be taken for granted, especially in rocks which are known to have undergone repeated deformations. The occurrence of boudinage structure in a transverse section of a fold does not imply that the boudin axis is parallel to the fold axis. The reason for emphasizing this obvious point is that the parallelism of boudin axis and fold axis has so often been emphasized that the students tend to make this assumption when they observe boudinage structures on a fold profile. It is true that boudin axis and fold axis are often approximately parallel (Fig. 17.10a), but in making this assumption without actual observation or measurement there is a possibility that we are losing a vital piece of information in the reconstruction of the history of superposed deformations. In areas of

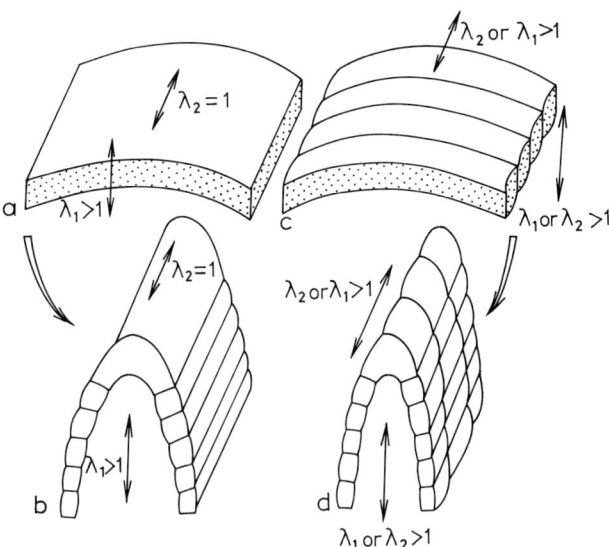

FIG. 17.11. In gentle buckle folds with λ_1 parallel to axial plane and perpendicular to fold hinge, as in (a), the limbs of the folds are in the shortening field and there cannot be any boudinage with boudin axes parallel to the fold axis. However, such boudins can develop, as in (b), when the limbs have rotated to the extension field. Gentle folds may have undergone boudinage (c) if the fold axis is a direction of stretching. With tightening of the fold (d), another set of boudins may develop by a stretching perpendicular to the fold hinge.

superposed deformations the earlier boudin axes may be oblique to the later fold axis (Fig. 17.10b). Since the plan view of boudins is rarely exposed, such evidence, unless a careful search is made, is likely to be overlooked.

During flexure folding or buckle folding, there is initially a layer-parallel shortening perpendicular to the fold axis. Pinch-and-swell structures and extension fracture boudinage, with boudin axes parallel to the fold axis, cannot form at this initial stage (Fig. 17.11a); the boudins may form at a later stage of folding when the limbs of the folds have entered the extension field (Fig. 17.11b). Even at such a mature stage of folding, boudinage will not take place at the hinge zones of folds. The occurrence of small extension fracture boudinage structures at the broad hinge zone of a symmetrical buckling fold should therefore imply that the boudinaged layer was folded subsequently. If there is an extension parallel to the fold axis, then boudinage may develop even at an early stage when the amplitude of the fold is small; the axes of these boudins will be at a right angle to the fold axis (Fig. 17.11c). Thus, among extension fracture boudinage structures which have grown during flexure or buckle folding, those which have their axis parallel to the fold axis must have been initiated when the folds were fairly tight;

on the other hand, the boudins which are perpendicular to the fold axis could have been initiated either at an early or at a late stage of folding (cf. Schwerdtner & Clark 1967).

17.6. Boudinage in superposed deformations

The folding of a boudinaged layer, in certain instances, can be easily distinguished from boudinage structures which are initiated during folding. Thus, when extension fracture boudinage structures occur on the transverse profile of a gentle fold wave, with the layering in both the hinge and the limbs in the compression field, we can conclude that the boudinaged layer has been subsequently folded. We can arrive at the same conclusion when small boudins are situated at the broad hinge zone of a fold (Fig. 17.12a) towards its core. In addition, the shortening of a boudinaged layer may give rise to a variety of characteristic patterns.

Experiments on folding of boudinaged layers (Sengupta 1983) show that, unlike folding, boudinage is not recoverable. If a boudinaged layer is shortened at a right angle to the boudin axis, it cannot go back to its original shape; the displacement pattern is greatly influenced by the discontinuities of the layer. During layer-parallel shortening of a boudinaged layer, the nature of deformation of the layer (Fig. 17.13) is mainly controlled by (1) the competence of the boudins relative to the host rock and (2) the width-to-thickness ratio of the boudins. Short or medium-sized very stiff boudins can only undergo a body rotation. They generally form a herring-bone pattern (Fig. 17.13a). If such stiff boudins have large width-to-thickness ratios, they may show a tile-like piling of one boudin over another (Fig. 17.13b). The flexible boudins are folded (Figs. 17.12b, 17.13c) and if the width-to-thickness ratio is not large the individual boudins are folded into half-waves, the neighbouring folds often closing in the same sense (Fig. 17.13d).

Rectangular boudins are produced in brittle and very competent beds. At some later stage the competence contrast may be reduced. Thus, for example, the competence of amphibolite boudins in a granitized terrain may be reduced by biotitization. During folding of a row of such softened boudins, the thickness of the boudins increases and the width decreases. However, because of tangential longitudinal strain associated with folding, the boudin width along the inner arc of the fold becomes smaller than that along the outer arc. As a result, the cross-sections of the boudins become trapezoidal (Fig. 17.14). Moreover, because of a stress concentration along the lateral edges of the boudins, the thickness along the edges become greater than in the central part of the boudins. For this reason the trapezoidal boudins may have flame-like projections along their lateral edges (Fig. 17.15). Deformed boudins of a similar shape have been recorded from several areas

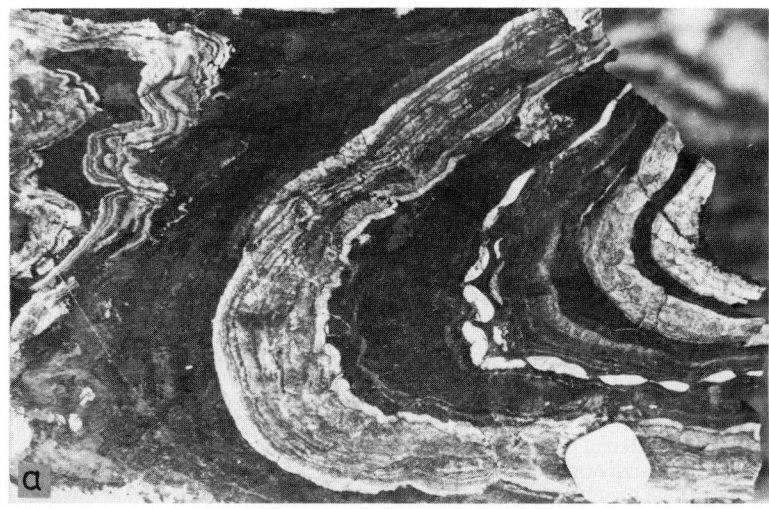

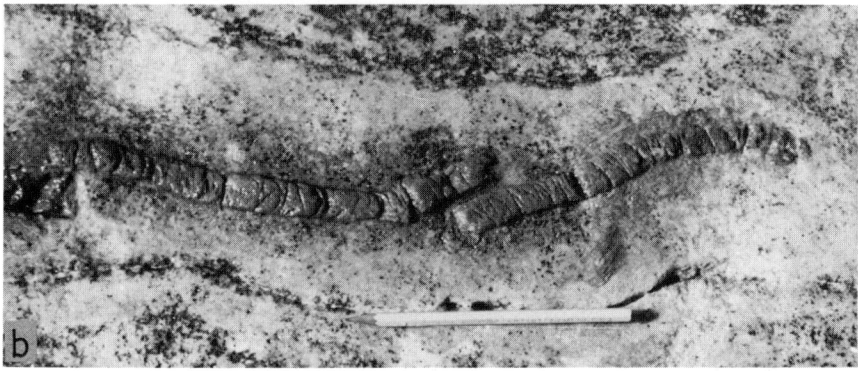

FIG. 17.12. (a) Folded boudins of a thin quartzite layer in a haematite band. The boudinage is earlier than the fold. This is indicated by the occurrence of the boudins at the hinge zone and also by folding of the individual boudins into half-waves. Banded iron formation, Ramagiri, Andhra Pradesh, India. (b) Folding of flexible boudins with riding of one boudin over another. The structure indicates a shortening of the boudinaged layer. The boudins are in a siliceous layer in a carbonate rock from Väddö, Sweden.

(Ramberg 1952, Wegmann 1965, Sengupta 1983, 1985). This characteristic shape of the boudins indicates that: (i) the folding is later than the boudinage and (ii) the competence contrast between boudins and the host rock has been low during or before the folding (Sengupta 1983).

In areas of superposed deformations, boudinage structures may form during both early (F_1) and later (F_2) folding. The early boudins may be recognized by their characteristic deformation patterns (Fig. 17.13) over the F_2 folds. In certain cases the later boudins can be distinguished where the axial surfaces of early folds are distorted by the

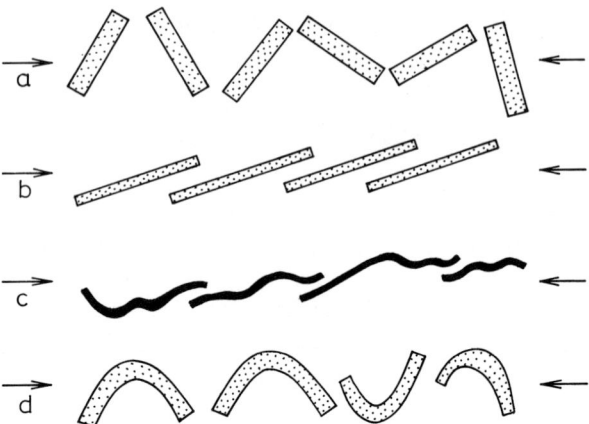

FIG. 17.13. Different patterns obtained in experiments when a row of boudins is subjected to a layer-parallel compression. (a) Herringbone pattern of rigid boudins. (b) Tiled-up long rigid boudins. (c) Folding and tiling of flexible boudins. (d) Folding of boudins into half-waves. From Sengupta 1983.

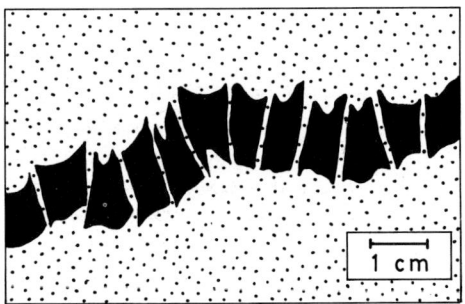

FIG. 17.14. Experimental deformation of a row of deformable "boudins" by a layer-parallel shortening. During deformation the boudins show flame-like projections perpendicular to the layering. At some stage the boudinaged layer as a whole is folded, and over the fold hinges the boudins acquire a trapezoidal shape. Sketched from photograph of model. From Sengupta 1983.

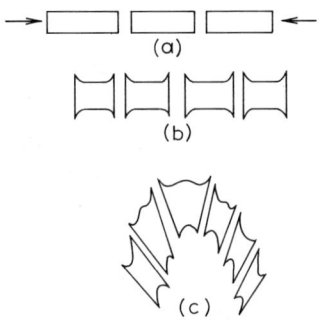

FIG. 17.15. Stages of development of trapezoidal boudins with flame-like projections.

scar folds (Fig. 17.16) which develop near the separation zones between the boudins.

17.7. Origin of extension fracture boudins

The mechanics of origin of extension fracture boudins of the classical type was first enumerated by Ramberg (1955) who considered the model of an elastic-brittle layer embedded in a viscous matrix compressed by a layer-normal compression between two rigid plates (Fig. 17.17). He postulated that the layer-normal compression in the matrix would induce viscous drag or shear stresses on either face of the brittle layer. An extension fracture forms where the resulting tensile stress in the elastic layer exceeds the tensile strength.

Let us first consider the initiation of boudinage by one-dimensional extension in the plane of the layer. Consider a brittle layer of thickness H and length $2a$ sandwiched between two viscous slabs each of thickness h and bounded on the outside by a rigid plate. Let the x-axis be chosen along the upper surface of the elastic-brittle plate, with z-axis vertical and with the origin at the middle of the plate (Fig. 17.17). The upper plate moves down with a velocity $-\dot{w}_0$ and the lower plate moves up with a velocity $\dot{w}_0$. The viscous material adheres to the surfaces of the elastic layer and to the inner surfaces of the two confining plates. As the viscous material is laterally squeezed out during layer-normal compression, marker lines, which were initially vertical in each of the viscous slabs, tend to bow outward, as illustrated in Fig. 17.17. As the figure shows, excepting at the level $h/2$, these marker lines in the upper slab are no longer perpendicular to the elastic plate. Therefore, excepting at this level, there will be shear strain at all points within the upper viscous slab. It has been shown by Ramberg (1955) that, at the contact with the embedded elastic plate, the rate of shear strain of the viscous material is

$$\dot{\gamma}_{xz} = \frac{6\,\dot{w}_0\,x}{h^2},\tag{17.1}$$

so that the shear stress is

$$\tau = \frac{6\eta\dot{w}_0\,x}{h^2},\tag{17.2}$$

where η is the coefficient of viscosity of the embedding slabs. This is the viscous drag imparted to the upper surface of the elastic plate. Since the flow of the lower viscous slab is similar to that of the upper slab, there is an equal viscous drag on the lower face of the elastic plate. These are the external forces imparted to the elastic plate. Under their

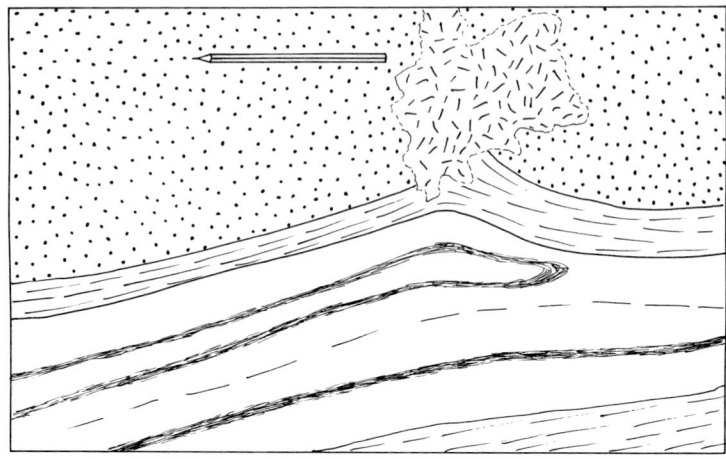

FIG. 17.16. Deformation of axial surface of an isoclinal early fold near the separation zone between two boudins of amphibolite in granite gneiss. From Jasidih Bihar, India.

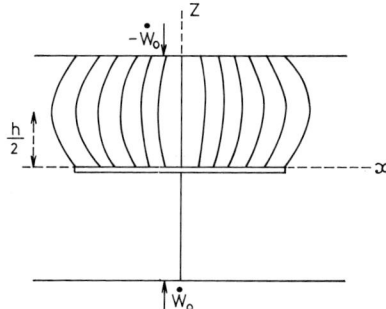

FIG. 17.17. A brittle layer of thickness H is sandwiched between two viscous slabs, each of thickness h. The viscous slabs are confined between rigid plates. The bounding plates approach towards the centre at a velocity of $\dot{w}_o$. The negative velocity of the upper plate indicates that it is moving towards the negative direction of the Z-axis. The viscous slabs are shortened and are squeezed out. Hence vertical marker lines bow outward and there is a shear strain near the surface of the brittle plate.

action the elastic plate tends to extend in the x-direction, the tensile force per unit volume in the plate (Fig. 17.18) being

$$R_x = \frac{2\tau}{H}. \tag{17.3}$$

From eqns. (17.2) and (17.3), we then have

$$R_x = Ax \tag{17.4}$$

where

$$A = \frac{12\eta\dot{w}_0}{h^2 H}.$$

It should be noted that R_x is the external force acting on unit volume of the elastic plate. The tensile stress (σ_x) in the elastic plate is obtained by integrating R_x between the limits x and a:

$$\sigma_x = \int_x^a \frac{12\eta \dot{w}_0 x}{h^2 H} dx$$

or,

$$\sigma_x = \frac{6\eta \dot{w}_0}{h^2 H}(a^2 - x^2). \tag{17.5}$$

The tensile stress is maximum in the middle of the plate, at $x = 0$:

$$\sigma_{x(\text{max})} = \frac{6\eta \dot{w}_0 a^2}{h^2 H}. \tag{17.6}$$

A mid-point fracture forms when the stress exceeds the tensile strength of the plate. The layer is thus broken into two boudins. If the flow of the matrix continues, each of these fragments suffers an extension and a mid-point fracture forms in each of them. Such sequential mid-point fracturing (Fig. 17.9) of the boudins continues till the boudin width becomes so small that the maximum tensile stress at its middle drops below the tensile strength and further sub-division of the boudins is not possible. This is the reason why a boudinaged bed shows a small range of width-to-thickness ratios of the individual boudins.

As discussed in section 17.4, depending on the initial length of the layer segment in a transverse section, the final aspect ratio of the boudins may range between l_c and $l_c/2$, where l_c is the critical aspect ratio. If the boudin aspect ratios are uniformly distributed within this range, the mean aspect ratio will be $\frac{3}{4} l_c$.

In the model proposed by Ramberg (1955), it was assumed that the embedding viscous slabs were confined between rigid plates. We may also consider a model in which the embedding medium is of infinite extent. This corresponds to the situation where the thick embedding ductile layers undergo a homogeneous strain in the major part, while the flow field is altered only in the neighbourhood of the elastic layer. Even in such a situation, the tensile stress will be maximum in the middle of the elastic layer and boudinage will take place by successive halving of the layer. In this case, according to Smith (1975, p. 1607), the maximum tensile stress in the middle of the layer will be

$$\sigma_x = \frac{c \, \dot{\varepsilon}_x \, \eta \, a}{H} \tag{17.7}$$

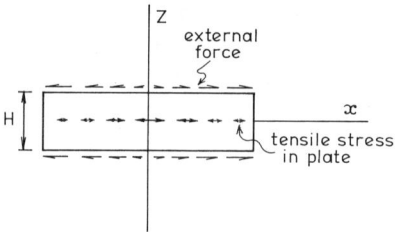

FIG. 17.18. Brittle plate of thickness H as in Fig. 17.17. The upper and lower faces of the plates are subjected to shear stresses. As shown by the curved marker lines in Fig. 17.17, the shear stresses increase from the middle towards the extremities of the plate. Under their action, tensile stresses develop inside the plate. The tensile stress is largest in the middle of the plate. If the stress exceeds the tensile strength, the plate breaks in the middle.

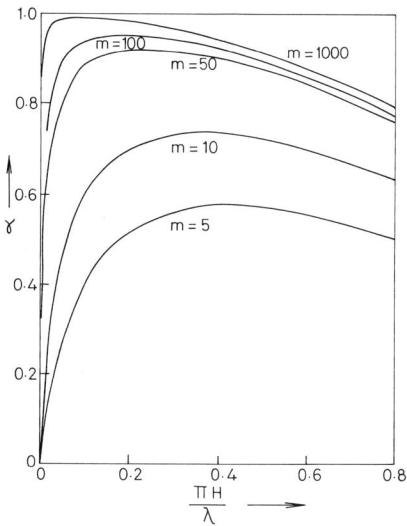

FIG. 17.19. Plot of normalized growth rate (γ) of amplitude against $\pi H/\lambda$ for different viscosity ratios (m). Each curve shows a maximum of γ ($= \gamma_d$) for a certain value of λ/H. The figure shows that $\gamma_d < 1$. From Smith 1975.

where ε_x° is the rate of pure shear in the viscous matrix at a large distance from the boudinaged layer and where c is a constant of order 1.

A theory similar to that of Ramberg (1955) has been proposed by Lloyd *et al.* (1982). In this theory the authors base their analysis on the mechanical behaviour of fibre-reinforced elastic composite materials (Cox 1952, Kelly 1973, Tewari 1978, Stowell & Liu 1961). When such a composite is extended parallel to the fibres, a fibre relieves the surrounding matrix of load by carrying an extra load itself. When the tensile stress which is transferred to the fibre exceeds its breaking strength, the fibre breaks in the middle.

The model of fibre-reinforced composites is relevant for boudinage of cylindrical or prismatic materials such as belemnites (Heim 1919), andalusite crystals (Sanderson & Meneilly 1981) and fibrolite or rutile needles included in quartz grains. The mathematical model of the fibre-matrix system has been modified by Lloyd *et al.* (1982) to incorporate an elastic layer-matrix system.

The theory of development of stable boudins by successive halving predicts that the inter-boudin gaps will not be of equal lengths; the separation will be larger across earlier fractures (Fig. 17.9).

17.8. Origin of pinch-and-swell structures

The mechanics of development of pinch-and-swell structures has been analysed by Smith (1975, 1977, 1979) for both Newtonian and non-Newtonian materials. A similar solution has also been given by Fullagar (1980). Consider the two-dimensional flow of a layer and its embedding medium under a layer-normal compression. The liquids are first considered to be Newtonian. If the layer is infinitely long, it will be strained at the same rate as the embedding medium. Let $\dot{e}_{11}$ be the uniform strain rate along the x-direction parallel to the layer. Let us now specify that in addition to this basic state of flow the interfaces of the layer and the embedding medium have small irregularities of a sinusoidal nature. When the perturbations are superposed on the basic flow, the irregularities may decay or grow in amplitude. A growth in amplitude means that there is an instability and the pinch-and-swell structure can grow.

Let the amplitude be a and the rate of growth of amplitude be B, so that

$$B = \frac{1}{a}\frac{da}{dt}.$$

Let the ratio of the rate of amplitude growth to the strain rate of the basic flow be

$$\gamma = \frac{B}{\dot{e}_{11}}.$$

This ratio is dependent on two factors: (1) the viscosity ratio between the layer and the embedding medium and (2) the ratio of wavelength to thickness of the incipient pinch-and-swell. Smith (1975) has shown that this dependence can be expressed by the equation

$$\gamma = \frac{B}{\dot{e}_{11}} = \frac{2(m-1)}{R}, \tag{17.8}$$

where $m = \mu_2/\mu_1$, the ratio of viscosities of the layer (μ_2) and the embedding medium (μ_1). R is a function of m and the wavelength-to-thickness ratio λ/H. γ is called the normalized growth rate, that is, the growth rate of amplitude divided by the background rate of homogeneous strain. When γ is plotted against $\pi H/\lambda$, for any value of the viscosity ratio, the curve shows a maximum (Fig. 17.19). In other words, for each viscosity ratio there is a certain value of λ/H for which the amplitude has the largest growth rate. These fastest growing waves dominate over other incipient waves and mask them. These wavelengths are called the *dominant wavelengths*. The corresponding value of γ is designated as γ_d. Smith's equations and Fig. 17.19 show that $\gamma_d < 1$. It should be noted that, because of the layer-normal compression, the embedded layer undergoes a continuous thinning. Hence the fact that $\gamma_d < 1$ means that the growth of the disturbance amplitude cannot overcome the opposing effect of overall thinning of the layer. *Smith's analysis, therefore, shows that well-developed pinch-and-swell structures cannot form in Newtonian materials.* This conclusion is supported by the observation that the process of necking, which is so often seen in metals, does not occur in rods and plates of linear viscous materials (Fletcher 1974, p. 1031).

Fletcher's and Smith's studies emphasize the importance of considering the non-Newtonian behaviour of rocks in the theoretical modelling of rock structures. In a linear viscous material or Newtonian material, the stress is proportional to the strain rate, and the ratio of the stress and the strain rate is defined as the coefficient of viscosity. In a non-linear or non-Newtonian material, the plot of the stress as a function of the strain rate is not a straight line (Fig. 17.20). For any strain rate, the ratio between the stress and the strain rate may be defined as the viscosity, μ, while the ratio between the incremental stress and the incremental strain rate may be called the effective viscosity or differential viscosity coefficient η (Biot 1965a, p. 389). Smith (1977) considers a material in which $\mu > \eta$. Such a material is called a strain-rate softening material.

Consider a competent layer of thickness H embedded in an incompetent material of infinite extent. The layer and the embedding medium are undergoing a homogeneous pure shear with $\dot{e}^{\circ}_{11}$ as the uniform strain rate along the length of the layer. At this basic strain rate $\dot{e}^{\circ}_{11}$, each material will have some value of μ and η. Let μ_1 and η_1 be the values for the incompetent material and μ_2 and η_2 be the corresponding entities for the competent material (Fig. 17.20). Let

$$m = \frac{\mu_2}{\mu_1}, n_1 = \frac{\mu_1}{\eta_1} \text{ and } n_2 = \frac{\mu_2}{\eta_2}.$$

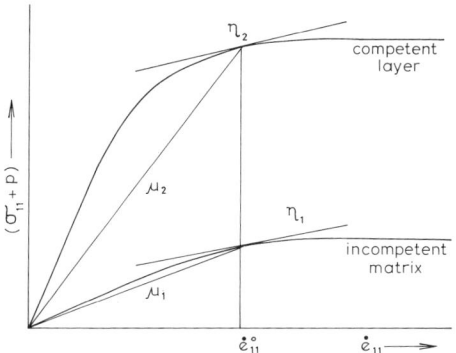

FIG. 17.20. Distinction between μ and η in a non-Newtonian material. From Smith 1979.

n is here a measure of the non-linearity of the materials. In Newtonian materials, $n = 1$. For strain-rate softening materials, $n > 1$. Let us also specify, as in the previous case, that in addition to the basic state of flow the interfaces of the layer and the embedding medium have small sinusoidal irregularities or perturbations. The theory indicates that the rate of growth of amplitudes of these irregularities depends on m, n_1 and n_2. The theory further shows that for a particular combination of m, n_1 and n_2 the amplitude of a particular wavelength has the fastest growth rate. This is the dominant wavelength λ_d. Waves smaller and larger than λ_d will be masked by the growth of λ_d. A pinch-and-swell structure can form only if the rate of growth of its amplitude is significantly larger than the rate of uniform extension of the layer. The ratio between these two rates is called the *normalized growth rate*, γ_d. In other words, a pinch-and-swell structure can be well formed only if γ_d is larger than 1. Smith's theory shows that when $n_1 = n_2 = 1$, i.e. when both the layer and the matrix are Newtonian, $\gamma_d < 1$ and, hence, pinch-and-swell structures cannot grow. If the layer is Newtonian ($n_2 = 1$) but the matrix is non-Newtonian ($n_1 > 1$), even then $\gamma_d < 1$. However, γ_d is always greater than 1 (Fig. 17.21) when the layer is non-Newtonian (i.e. when $n_1 = 1$ or $n_1 > 1$, and $n_2 > 1$). *Thus, a pinch-and-swell structure can only grow when the competent layer is non-Newtonian.* Indeed, the very presence of pinch-and-swell structures indicates a non-Newtonian behaviour of the competent layer. In such cases the maximum growth rate is increased by increase of both n_1 and n_2. However, for large values of the viscosity ratio, the effect of n_1 is negligible. Smith's (1977, p. 317) theory shows the following simple and elegant relationship

$$\gamma_d = n_2,$$

when m is very large. This can also be clearly seen from Fig. 17.21. For a strongly non-Newtonian layer, the theory also gives another relationship of remarkable simplicity. It predicts that if n_2 is very large, the ratio of dominant wavelength to thickness of the layer is 4, i.e.

$$\frac{\lambda_d}{H} = 4.$$

On the whole, Smith's theory suggests that the aspect ratios of a large fraction of natural pinch-and-swell structures should range between 4 and 6.

17.9. Origin of chocolate tablet boudinage structures

Boudinage by a layer-parallel two-dimensional extension is common in many areas. Development of such structures bounded on all sides by layer-normal extension fractures was called chocolate tablet boudinage by Wegmann (1932). In plan view, the chocolate tablet structures may be rectangular, rhombic, polygonal or irregular. A two-dimensional extension may also produce pinch-and-swell structures with neck lines in the shape of a closed curve or with two sets of neck lines at a high angle to each other. In some strongly lineated rocks, a two-dimensional layer-parallel extension may produce boudinage in one direction and pinch-and-swell structures in another (Fig. 17.22). When chocolate tablet structures have a rectangular plan view, we often find that one set of fractures or neck lines is more prominent than the other. Thus Coe (1959) and Fyson (1962) report the occurrence of two sets of boudin axes or neck lines at a high angle to each other, the axes parallel to the fold axis being more prominent than those perpendicular to the fold axis.

There is a general consensus that chocolate tablet structures develop in a flattening type of deformation. When there are equal extensions in all directions parallel to the plane of the layer, the resulting boudins should not have any linear directions. Although the plan view in some natural examples of boudinage do show such a lack of linear orientation of the extension fractures separating the boudins, the majority of chocolate tablet structures show two sets of boudin axes or two sets of extension fractures at a high angle to each other. A flattening type of bulk deformation cannot cause the simultaneous development of these two sets of fractures (Ramsay 1967, p. 112, Sanderson 1974, p. 658, Ghosh 1988); if the tensile stresses in the plane of the layer are unequal, it seems reasonable to expect that only one set of fractures will form

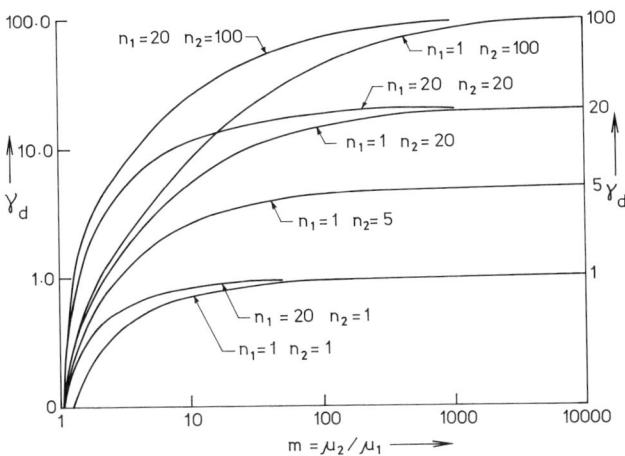

FIG. 17.21. Plot of γ_d against m (μ_2/μ_1) for different combinations of n_1 (μ_1/η_1) and n_2 (μ_2/η_2). Pinch-and-swell structures can form if $\gamma_d > 1$. The figure shows that $\gamma_d > 1$ when $n_2 > 1$. It further shows that if m is very large, $\gamma_d \approx n_2$. After Smith 1977.

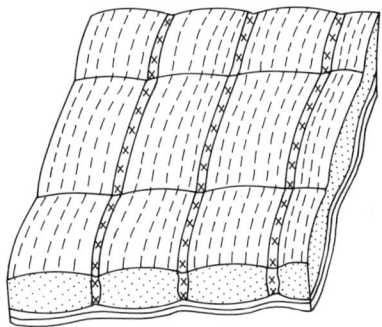

FIG. 17.22. Chocolate tablet structure with pinching-and-swelling in one direction and boudinage in another, with boudin axes parallel to a lineation (dashed lines).

perpendicular to the largest tensile stress. Why then should two prominent directions of boudin axes prevail in the large number of chocolate tablet structures? The answer is that the two sets of fractures are not initiated strictly simultaneously. Both the theory and the experiments (Ghosh 1988) indicate that in a flattening type of bulk deformation, the initiation of the first set of boudins alters the stress distribution in the competent layer, even though the bulk stress field in the incompetent matrix remains unaltered. The development of the second set of fractures at a high angle to the first set is then controlled to a great extent by the geometry of the first set of boudins (Fig. 17.24a).

Chocolate tablet boudinage structures may develop in different ways (Fig. 17.23). When a competent and brittle rock is embedded in an incompetent and ductile matrix, chocolate tablet boudinage may

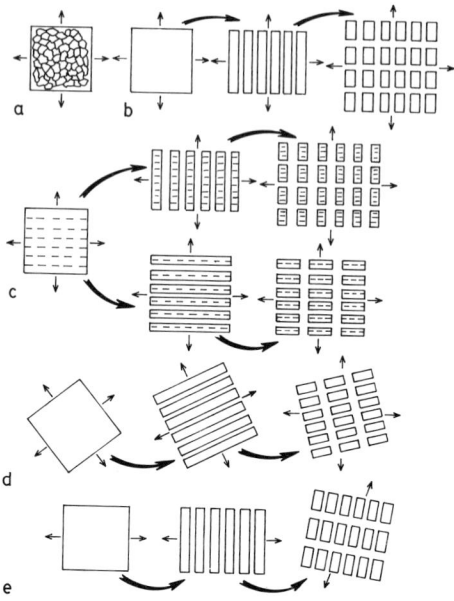

Fɪɢ. 17.23. Different ways of development of chocolate tablet structures. (a) Boudinage by equal layer-parallel extension in all directions. (b) Layer under unequal extensions of matrix. The first set of boudins develops at a right angle to the largest stress. When the boudins become long and narrow, the extension fractures form perpendicular to the length. (c) Boudinage in strongly lineated rock. Upper figure shows the same course of development as (b). The lower figure shows the case when the tensile strength perpendicular to lineation is much smaller than in a direction parallel to lineation. The first set of boudins may then form parallel to lineation, even if this direction is perpendicular to λ_2 and not λ_1. (d) Development of chocolate tablet boudins by superposed deformations of the flattening type. Note that the second set of fractures is perpendicular to the first set of boudins and not necessarily perpendicular to the maximum stretching. (e) Development of chocolate tablet boudins by two superposed uniaxial extensions. Again, the development of the second set of fractures is controlled by the geometry of the first set of boudins (Ghosh 1988).

develop if there is an extension in all directions in the plane of the layer. If the extensions are equal in all directions, the plan view of the resulting boudins do not have any uniform linear orientation (Fig. 17.24b). If the extensions are unequal, two sets of extension fractures may develop sequentially and at a right angle to each other (Fig. 17.25); the plan view of the resulting boudins is often rectangular. In this case the two sets of boudin axes will be parallel to the strain-rate axes of the bulk deformation parallel to the plane of the layer.

In lineated rocks in which the tensile strength in a transverse direction is much smaller than in a longitudinal direction, two sets of extension fractures, parallel and perpendicular to the lineation, are likely to form sequentially when the bulk deformation is of the flattening type.

The orientations of the transverse and longitudinal axes of the boudins are then controlled by the orientation of the lineation and may not be parallel to the principal strain-rate axes in the matrix (Fig. 17.26).

17.10. Sequential versus simultaneous development of boudinage along a layer

The theoretical models indicate that when elastic deformation of a competent layer is followed directly by the formation of an extension fracture, the layer is divided into two equal fragments. If the deformation continues, each of the boudins undergoes successive halving till the boudin widths drop below a critical value. On the other hand, from Smith's theory we know that pinch-and-swell structures with a dominant wavelength can simultaneously grow all over the layer. The necks of the pinch-and-swell structures are preferred sites for extension fracture. There is as yet no theoretical model for the simultaneous development of fractures at all the necked zones of a layer. However, experiments with some model materials suggest that if there is a significant amount of permanent deformation before fracture, then boudinage may be initiated in a layer more or less simultaneously at different places (Ghosh 1988).

17.11. Rotated boudins and their asymmetry

Rotation of rigid boudins

In certain areas boudins initiated by extension fracture are found to be rotated and arranged *en echelon*. The principle of development of this step-like arrangement can be understood from the general principle of rotation of rigid inclusions as described in Chapter 11.

The step-like arrangement is produced if the individual boudins rotate at a rate different from the rate of rotation of a line which joins the boudin centres. The rotation of this line is the same as that of a similarly oriented passive marker line in the matrix.

The rotation may take place in both coaxial and non-coaxial bulk deformations. In case of coaxial bulk deformation, rotation can take place only if the bed is oblique to the principal axes of strain. Consider a bed at an angle of θ with the X-axis and consisting of a row of rigid boudins with long axes parallel to the bed. At that orientation the rate of rotation of the line joining the boudin centres is given by the equation

$$\dot{\theta}_m = \dot{\varepsilon}_x \sin 2\theta$$

where $\dot{\varepsilon}_x$ is the pure shear strain rate along the X-direction. The rate of rotation of the individual boudins of aspect ratio R is approximately given by the equation

FIG. 17.24.

$$\dot{\theta}_R = \dot{\varepsilon}_x \cdot \frac{R^2 - 1}{R^2 + 1} \sin 2\theta$$

(Ghosh & Ramberg 1976, eqn. 2).

Since $R > 1$, the two equations show that

$$\dot{\theta}_m > \dot{\theta}_R.$$

In pure shear the rotation of the boudins is always smaller than the rotation of the bed as a whole. This is true for all values of θ other than 0 and 90°; therefore, the *en echelon* pattern will be similar to that shown in Fig. 17.27. Since at large values of R, $\dot{\theta}_m \approx \dot{\theta}_R$, very long boudins will rotate but will not show a stepwise arrangement.

Let us consider the case of bulk simple shear with the boudin axes and the bed oriented at an angle θ with the direction of simple shear. In this case the two rates of rotation are:

$$\dot{\theta}_m = \dot{\gamma} \sin^2\theta,$$

$$\dot{\theta}_R = \frac{\dot{\gamma}(R^2 \sin^2\theta + \cos^2\theta)}{R^2 + 1}$$

where $\dot{\gamma}$ is the rate of simple shear (Ghosh & Ramberg 1976, eqn. 1, p. 9). Here, too, for very large R, $\dot{\theta}_m \approx \dot{\theta}_R$. From the two equations given above we find that

$$\frac{\dot{\theta}_m}{\dot{\theta}_R} = \frac{R^2 + 1}{R^2 + \cot^2\theta}. \tag{17.9}$$

The last equation shows that:

$$\dot{\theta}_m > \dot{\theta}_R, \text{ when } \theta > 45°,$$
$$\dot{\theta}_m < \dot{\theta}_R, \text{ when } \theta < 45°.$$

Thus, depending on the orientation, a rotating boudin may lag behind or move ahead of the bed as a whole. In both cases there will be *en echelon* arrangement of the boudins (Fig. 17.28a & b), although the

---

FIG. 17.24 (*opposite*). (a) Development of extension fracture in a strip of plaster of Paris resting on a slab of pitch. The pitch flowed unequally with the maximum stretching from right to left. The fracture in the brittle strip formed perpendicular to the length of the strip and not perpendicular to the maximum stretching in the pitch. The experiment indicates that, once a set of long narrow boudins is formed, the boudins may be split by fractures perpendicular to their length and not perpendicular to λ_1 in the matrix. (b) Fractures in square plate of plaster of Paris resting on a slab of pitch. These diversely oriented fractures formed when the pitch flowed outward and the plate was extended equally in all directions. From Ghosh 1988.

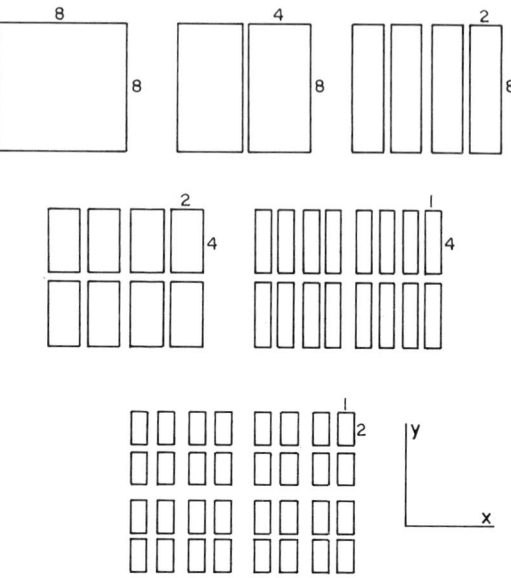

FIG. 17.25. Development of chocolate tablet structures from a square plate by the process of successive halving. If a boudin splits into two, the orientation of the extension fracture will depend on its length-to-width ratio and on the ratio of the far-field strain rates $\dot{\varepsilon}_y^\circ/\dot{\varepsilon}_x^\circ$ in the matrix, with $\dot{\varepsilon}_x^\circ > \dot{\varepsilon}_{y'}^\circ$. Depending on the length-to-width ratio, a plate may be split by fractures either perpendicular to the length or perpendicular to the width. The sequential fracturing continues till the boudin attains a stable shape. This figure shows the case when $\dot{\varepsilon}_y^\circ/\dot{\varepsilon}_x^\circ = 1/10$. From fig. 6 of Ghosh 1988.

sense of offset will be opposite in the two cases. Since the orientation of the boudin axis and of the marker line both come progressively closer to the direction of simple shear, a marker which was rotating ahead of the boudins (with $\theta > 45°$) may at a later stage (when $\theta \ll 45°$) lag behind the boudins. The history of offsetting of boudins can therefore be quite complex.

In simple shear the rotation of boudins is always synthetic, i.e. in the same sense as the sense of rotation of the bulk simple shear. However, as discussed in Chapter 11, in zones of transpression, i.e. when there is a shortening across the shear zones, the rotation may be either synthetic or antithetic.

Asymmetry of deformable boudins

If the boudins are deformable and yet more competent than the matrix, then the rotation of the long edges may be different from that of a marker line in the same direction. The difference in the two rates of rotation is smaller than that of rigid boudins; this difference increases with increasing competence contrast between the boudins and their host. This is borne out by experiments with soft models. In addition to being rotated with respect to the matrix, the boudins are elongated and the right angles at the corners are changed by shear strain so that the transverse sections have the shape of a parallelogram (Fig. 17.29). With boudins having a moderately high competence contrast with the matrix, there is in addition a barrelling of the longer (bedding-parallel) faces along with asymmetric inward curving of the other two faces (Figs. 17.28c, 17.30). Note from Figs. 17.29 and 17.30 that the sense of shear strain along the longer faces depends upon their orientations with respect to the direction of simple shear; the sense of shear along the long faces will be reversed in progressive deformation after these faces make angles of less than 45° with the simple shear direction.

17.12. Origin of asymmetric boudins

Types of asymmetric boudins

Asymmetric boudins or *asymmetric pull-aparts* have been described by many authors (e.g. Cloos 1947a, Edelman 1949, Ramberg 1955, Rast 1956, Beloussov 1962, Uemura 1965, Stromgård 1973, Hanmer 1986, Gaudemer & Taponnier 1987, Malavieille 1987, Goldstein 1988, Jordan *et al.* 1990, Jordan 1991). These are rhomboidal, lozenge-shaped or asymmetric barrel-shaped in transverse sections and may develop by one of the following mechanisms:

(1) Rhombic boudins may form by development of shear fractures oblique to a bed (Fig. 17.31a). These will be described in detail in the next section.

(2) The original shape of extension fracture boudins may change (Fig. 17.31b) by post-fracture plastic deformation. In all such cases there must be a shearing parallel to the layering. The shearing may cause (i) a change in the right angle at the corners of initially rectangular boudins, (ii) an asymmetric barrelling (Ghosh & Ramberg 1976, Hanmer 1986) and (iii) the development of asymmetric lenses by modification of the shape of pinch-and-swell structures (Hanmer 1986). The last two types have been described by Hanmer as type 1 and type 2 asymmetric pull-aparts. Hanmer has described two sub-types of asymmetric pinch-and-swells. In one of these the "pinched" portions are dissected by discrete or zonal oblique shears (Fig. 17.31c), as in an extensional crenulation cleavage. There are no such shear fractures or shear bands in the other sub-type of asymmetric pinch-and-swells (Fig. 17.31d). In both cases the "swells" are antithetically rotated. This means that the sense of rotation of the "swells" is opposite to that of the layer-parallel shear in the host rock.

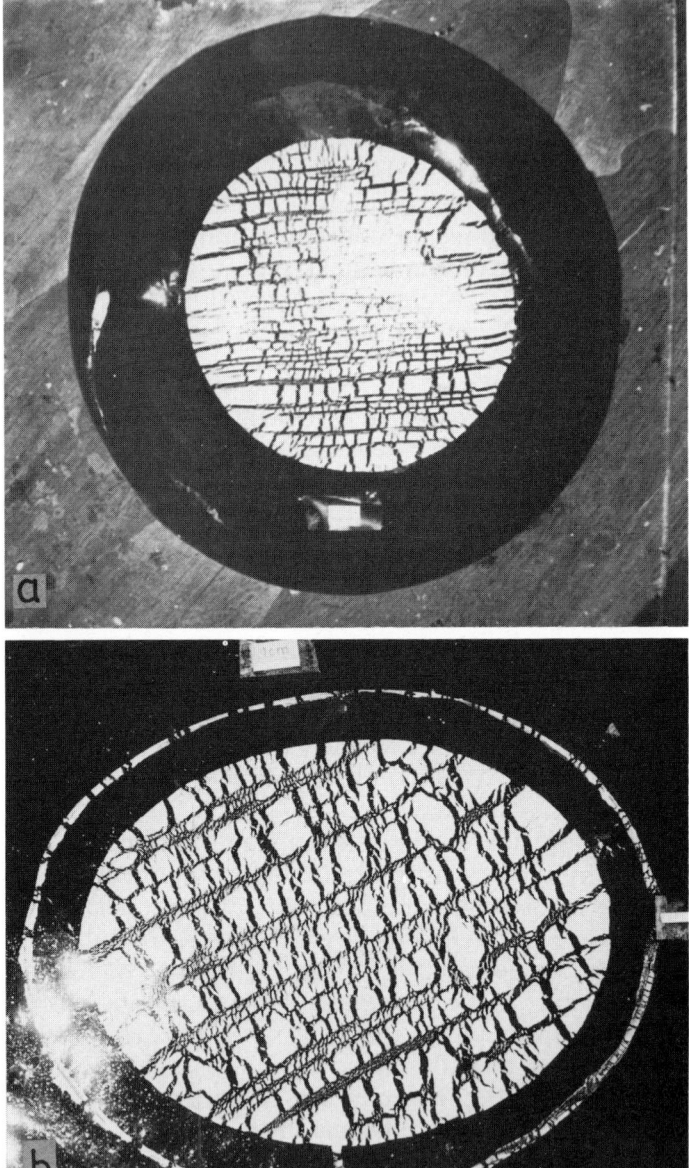

Fig. 17.26.

(3) Stromgård (1973, pp. 28–29, figs. 8 and 9) suggested that rhomboidal boudins may also develop if the bed is oriented oblique to the principal axes of stress and extension fractures are initiated oblique to the bed. Further extension of the bed leads to a separation of the rhombohedral boudins (Fig. 17.31e). A similar mechanism has also been suggested by Goldstein (1988).

Shear fracture boudinage

The majority of asymmetric boudins develop by offset of a bed along shear fractures (Fig. 17.32a). Cloos (1947, fig. 17) has illustrated the formation of boudins by such displacements along fractures oblique to the bedding in a thin sandy bed between shaly layers. The different stages of development of boudinage structures by shear fracturing have been described by Rast (1956, pp. 404–405) in quartzose rocks intercalated with schistose layers in the Perthshire Highlands. As in the case of the classical type of boudinage, there is an extension parallel to the layer and the sense of displacement along shear fractures is such that there is an overall extension of the bed. In the initial stages, with small offsets of the bedding, the array of displaced rhombohedral fragments has the general appearance of a pinch-and-swell structure (Fig. 17.32a). When the fragments are completely separated they look like ordinary boudins. In the initial stages of development the shear fractures are not generally filled with a vein material. When completely separated the boudins are often connected by quartz veins.

Shear fracture boudinage has been described in detail by Uemura (1965) from the Muro Group in Kii peninsula, Japan. Uemura found that, while extension fracture boudinage structures are concentrated at the limbs of a large fold, the shear fracture boudinage structures are localized at the hinge region. Further, the shear fracture boudins are larger and have developed in thicker beds than the extension fracture boudins.

The shear-initiated structure is associated with a body rotation of the boudins. Slip along the fractures and rotation of the fracture surfaces take place simultaneously, so that the boudins are arranged in an *en*

Fɪɢ. 17.26 (*opposite*). Development of chocolate tablet boudins in a lineated plate of plaster of Paris resting on a pitch slab. The lineated plate was produced by brushing a plaster of Paris slurry on the pitch surface. The brush marks gave rise to a mechanical anisotropy when the plaster of Paris became dry. In (a) there was equal extension of pitch slab in all directions. In (b) the pitch flowed out unequally. The lineation was at an angle to λ_1 in the matrix. In both cases, two sets of fractures were produced, parallel and perpendicular to the lineation. The orientation of the boudin axes was controlled by the lineation (Ghosh 1988).

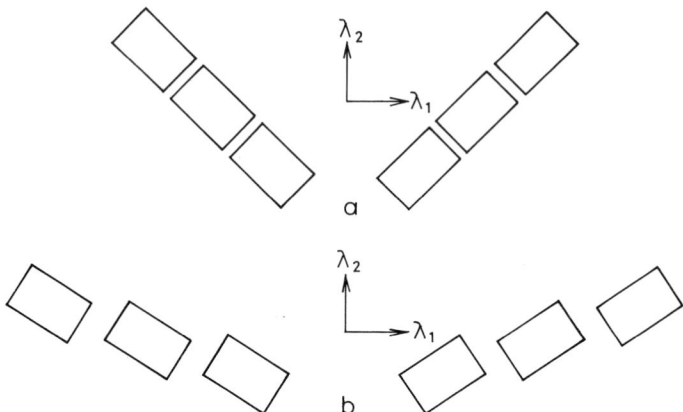

FIG. 17.27. (a) Two boudinaged bands inclined on either side of the compression direction (λ_2). (b) With progressive deformation the bands have rotated towards the λ_1-direction, with clockwise rotation of the band in the right and anticlockwise rotation of the band in the left. The individual boudins have rotated through smaller angles than the line joining the boudin centres. Because of the different senses of rotation, the senses of *en echelon* arrangement are different in the two bands.

echelon pattern. If the shape of the boudins is not modified by ductile deformation, the bedding-parallel elongation can be determined by direct measurement of the initial and final lengths of the bedding-parallel line segment. Thus, in Fig. 17.33 the bedding-parallel quadratic elongation is

$$\sqrt{\lambda} = P'S'/(\tfrac{1}{2}A'B' + B''C' + C''D' + \tfrac{1}{2}D''E').$$

From geometrical construction of shear-initiated boudins we see that when the spacing between fractures is constant, the offsets between adjacent blocks are equal. For unequal spacing, consider two adjoining blocks a and b (Fig. 17.34), with centres P and Q in the unrotated blocks, and let the distance between the bounding fracture surfaces in a and b be δ_a and δ_b respectively. After rotation, the corresponding centres of a and b shift to A and B. Along the general direction of the bed, PQ represents an initial or undeformed line segment while AB represents the stretched line segment. The internal bedding in the blocks is along AC and BD. Let β be the angle between a fracture surface and the general orientation of the bed and α be the angle between the internal bedding and the general orientation of the bed. From Fig. 17.34 it can be seen that

$$\alpha + \beta = \theta. \tag{17.10}$$

Since we have assumed that the shapes of boudins are not changed by ductile deformation, θ remains constant. During rotation of the

BOUDINAGE

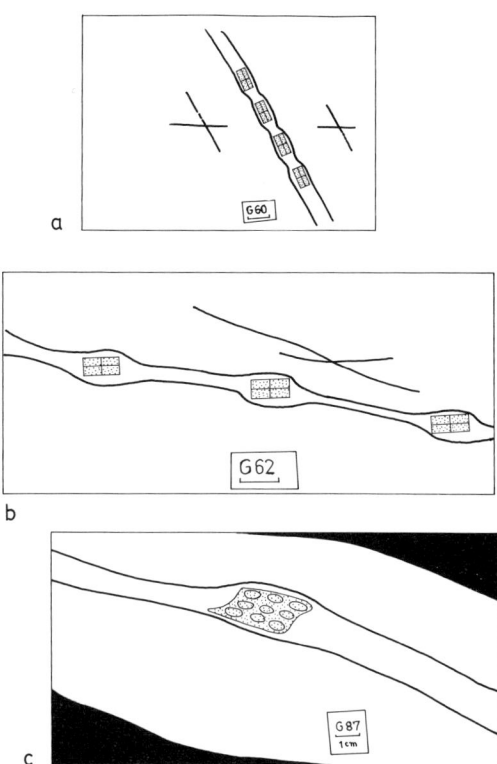

Fig. 17.28. (a) The row of rigid "boudin" blocks was initially perpendicular to the direction of sinistral simple shear. During simple shear the "boudins" have rotated through a smaller angle than a line joining the centres of "boudins". The blocks therefore show a left-stepping arrangement. (b) The row of rigid "boudins" was initially at an angle of 45° with the direction of sinistral simple shear and the blocks have rotated through a larger angle than a marker line with the same initial orientation. The blocks therefore show a right-stepping arrangement. (c) Deformable "boudin" embedded in silicone putty and subjected to sinistral simple shear. The "boudin" initially had a square section with faces parallel and perpendicular to the shear direction. During simple shear it was deformed and was rotated with respect to the matrix. Corresponds to stage (b) of Fig. 17.30. Sketched from photographs of models. After Ghosh & Ramberg 1976.

boudins, α, the magnitude of rigid rotation, increases and β decreases while the sum of the two angles remains constant. From similar triangles ACE and BDE we find that

$$\frac{AE}{AC} = \frac{\sin\theta}{\sin\beta}.$$

Therefore, the quadratic elongation along the general orientation of the bed is

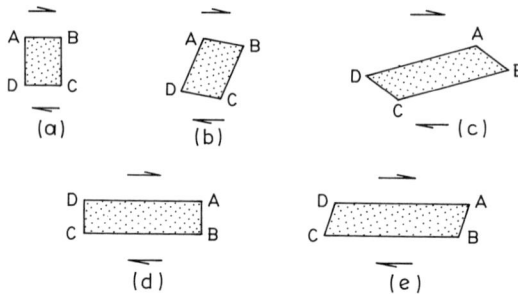

FIG. 17.29. Stages of deformation and rotation of a boudin under dextral simple shear with the long direction of the boudin (faces AD and BC) initially perpendicular to the shear direction. In the initial stage (stage b) there is dextral shear on the shorter faces AB and DC and the angles at B and D are reduced. When, after rotation, the longer faces make a low angle with the shear direction (stage c), there is dextral shear on the longer faces DA and CB. With progressive deformation (stage d), the angles B and D start to increase. If the boudin becomes sufficiently long (stage e), it finally attains a stable position with longer faces parallel to the direction of simple shear.

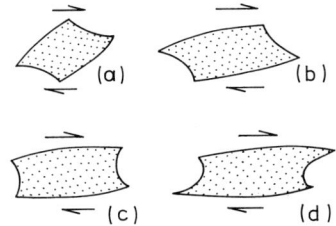

FIG. 17.30. Changes in shape of a deformable boudin with initial square section, under dextral simple shear. The stages (a), (b), (c) and (d) correspond to stages (b), (c), (d) and (e) of Fig. 17.29. In the present case the rotation and change in shape is associated with an asymmetric barrelling. The stage (b) corresponds to the deformation of the model in Fig. 17.28c. In the latter, however, the bulk simple shear is sinistral.

$$\lambda = (\sin\theta/\sin\beta)^2. \qquad (17.11)$$

In a similar way, it can be shown that the offset between the blocks a and b is

$$\gamma_{ab} = \tfrac{1}{2}(\delta_a + \delta_b)(\cot\beta - \cot\theta) \qquad (17.12)$$

or eliminating β with the help of eqn. (17.11),

$$\gamma_{ab} = \tfrac{1}{2}(\delta_a + \delta_b)(\sqrt{\lambda\,\mathrm{cosec}^2\theta - 1} - \cot\theta). \qquad (17.13)$$

The equation shows that the offset between two adjoining blocks is proportional to the average distance between the shear fractures bounding the blocks. The offset increases with increasing rotation of the blocks or increasing layer-parallel extension. The interrelations among the different geometric features described above are valid only if the shear fractures are parallel to one another.

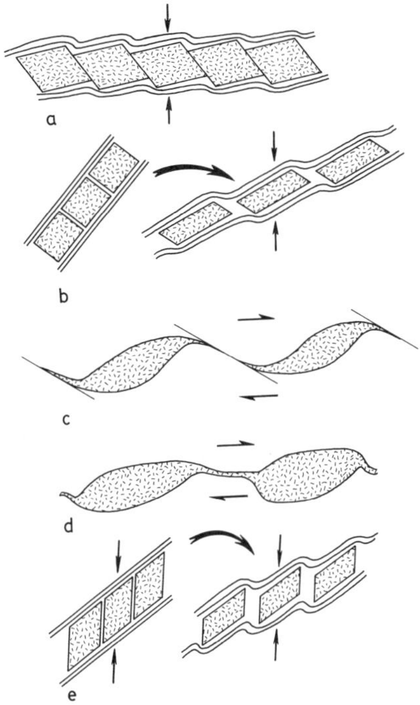

BOUDINAGE

FIG. 17.31. Different ways of development of asymmetric boudins.

In the analysis given above, the total extension of the bed is entirely due to rotation of the boudin blocks lying in contact with each other, with concurrent slip along the shear fractures. However, as shown by Jordan (1991) from the evaporite shear zones of the Swiss Jura, the total layer-parallel elongation may take place partly by slip along the shear fractures and partly by separation of the boudin blocks across the fractures (Fig. 17.35). The separation zones between the offset lozenge-shaped blocks of shale or dolomitic marl, embedded in ductile anhydrite or gypsum, are then filled by sulphate veins. The lozenge-shaped boudins described by Jordan (1991), as well as by Gaudemer & Tapponnier (1987), have undergone antithetic rotations. Jordan has suggested that the antithetic rotation is essentially a result of transpression, i.e. a deformation involving a layer-parallel shearing and a shortening across the shear zone. It should be noted, however, that boudin rotation in a shear zone may not always be antithetic. As explained in section 17.11, whether the boudins will rotate antithetically or synthetically will depend upon several factors, such as the axial ratio of the boudins, their initial orientation with respect to the shear direction and the nature of the bulk deformation (as indicated by the kinematical vorticity number).

FIG. 17.32. (a) Shear fracture boudinage in a band of pyroxene granulite occurring in granite gneiss. The shear fractures in the figure are inclined towards the left. Because of rotation of the boudins, the foliation in the boudins is at an angle to the foliation in the country rock. Schirmacher Range, E. Antarctica. Courtesy Sudipta Sengupta. (b) Foliation boudinage in gneissic rock. Åreskutan, Sweden.

Boudinage associated with shear fractures was obtained in rock deformation experiments by Griggs & Handin (1960, pp. 355–358) and by Paterson & Weiss (1968, pp. 802–803). The importance of shear

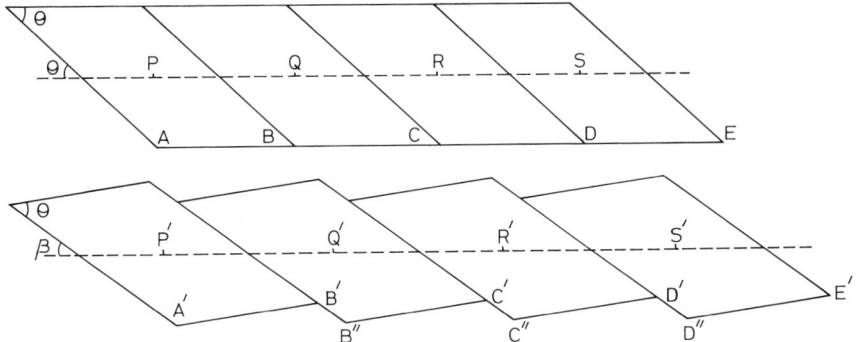

FIG. 17.33. Upper diagram shows a bed dissected by shear fractures. θ is the angle between the bedding and the shear fracture. The lower figure shows slip along the shear fractures with an overall extension of the bed. The line *PS* is lengthened to the line *P' S'*.

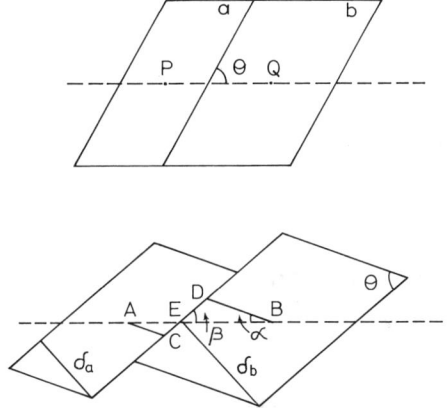

FIG. 17.34. Slip and rotation of shear fracture boudins with unequal spacing between the fractures. Because of the slip and rotation, the line length *PQ* has changed to the length *AB*. In unrotated blocks *PQ* is parallel to the general orientation of bed. After rotation, *AB* is parallel to the general orientation of the bed while *DB* and *AC* are parallel to the bedding inside the boudins. α is the angle between these two directions. β is the angle between *AB* and the shear fracture. The angle between the shear fracture and the bedding inside the boudins is θ. The figure shows that α + β = θ.

fracturing is also evident from the experiments of Gay & Jaeger (1975). In the latter experiments, cylinders of different rocks were embedded in a matrix of crushed rock and were deformed cataclastically by uniaxial compressive loads at a right angle to the cylinder axis. The initial effect of compression was to flatten and elongate the cylinders by homogeneous strain. In the next stage, a necked region formed due to collapse of material along a pair of conjugate normal faults. In the experiments, a shear fracture may cut across the cylinder or shear fractures may join to produce boudins with rounded ends.

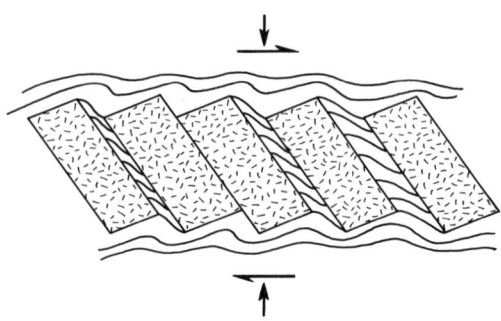

Fig. 17.35. Boudinage partly by slip along shear fractures and partly by separation across the fractures.

In some rocks the boudins may be separated by shear zones instead of shear fractures. The experiments of Cobbold *et al.* (1971) suggest that structures resembling boudinage may be initiated by growth of kink bands. The bands, which have a higher strain along them, may develop into shear zones. The role of small-scale shear zones in the formation of boudinage and pinch-and-swell structure has not been investigated in detail. It is likely that under certain rheological conditions the process is significant since such shear zones are known to develop in association with necking during experimental deformation of cylinders or bars of rocks and metals (Griggs & Handin 1960, fig. 1; Tvergaard *et al.* 1981). During tensile tests of ductile metal cylinders there is an initial stage of homogeneous deformation. Localized deformation in the form of a diffuse neck begins at the maximum load. After this stage there may be either a cup-and-cone type of fracture or the development of one or more bands of high localized shear. The shear bands may be initiated even when the neck is rather shallow. Continued deformation within the shear bands emphasizes the necking of the cylinder and eventually fracture may take place along one of the shear bands. Finite element analysis by Tvergaard *et al.* (1981) show that a variety of band patterns may develop in the necked zone. The particular pattern obtained will depend on the nature of initial thickness inhomogeneity in the necked zone as well as on the constitutive law which characterizes the material behaviour.

17.13. Foliation boudinage

Structures similar to boudins and pinch-and-swells may occur in certain zones of homogeneous strongly foliated rocks. These differ from the classical type of boudins in having no apparent lithological contrast between the boudins and the host rocks (Fig. 17.32b). The structures were first described by Coe (1959) from West Cork, Ireland, where long lens-shaped boudins separated by pods of vein quartz had developed in a sequence of siltstones. Similar structures have also been described by Platt & Vissers (1980) in homogeneous strongly foliated quartzofeldspathic schists from near Agnew in Western Australia. They are also frequently observed in migmatitic terrains. The structures have been described as *foliation boudinage* structures (Hambrey & Milnes 1975, Platt & Vissers 1980) or as *interlocking pinch-and-swell* (Cobbold *et al.* 1971, Cosgrove 1976). Compared to the classical type of boudinage in competent bands, the localization of foliation boudinage may be sporadic; the boudins may or may not occur in a train.

Foliation boudins also occur in foliated glacier ice (Hambrey & Milnes 1975). This clearly shows that a compositional difference is not necessary for their development. Hambrey & Milnes have described two types of boudins from glacier ice. In one of these, boudinage has occurred by extension, necking and rupture of fine-grained ice layers embedded in coarse-grained ice. The development of this type of boudins might have been controlled by competence contrasts. In the other type, the boudins are composed of bundles of laminae whose bulk mechanical property is apparently identical to that of the laminated host. This latter type is similar to the common type of foliation boudins in rocks. In certain cases the boudins in the glacier ice are separated by extension fractures. The fractures filled with water and subsequently frozen have produced a concentration of large radiating crystals of clear ice.

The term foliation boudin should be restricted to the structures in which the individual boudins are separated by fractures. The latter may or may not be filled with vein materials. Where discrete fractures or nodal veins are absent we may use the term interlocking or internal pinch-and-swell structure. Both foliation boudins and interlocking pinch-and-swells may be symmetric or asymmetric.

In *symmetric foliation boudins*, the initial fractures are roughly at a right angle to the foliation. In transverse sections, the fracture-filling veins may be diamond- or lozenge-shaped. From either side, the foliation curves inward towards the fracture tips and imparts a barrel-like shape to a boudin situated between two fractures. In the classical type of boudinage, the barrel-like shape develops because the fractures are initiated after a certain amount of necking. On the other hand, according to Platt & Vissers (1980), the barrel-like shape of symmetric foliation boudins is caused by plastic deformation which follows the initiation of fractures. According to this model, the presence of a prominent foliation restricts the rate of ductile extension parallel to it. Extensional fractures form normal to the foliation. When the rock is shortened normal to the layering, the stress field is strongly disturbed in the neighbourhood of the fractures. This is the reason for the curving of the foliation near the fractures. For small deformations, the shapes of the distorted laminae can be determined from a theoretical analysis of the stress field or displacement field in an infinite elastic medium with an elliptical hole (Muskhelishvili 1953, Lekhnitskii 1981, Amenzade 1979). Essentially similar shapes are also obtained experimentally by a layer-normal compression of a multilayer of soft materials with transverse elongate cuts or holes (Fig. 17.36a). In the experiments, the general shape of a symmetrical foliation boudin is obtained with or without a mechanical anisotropy of the multilayer. Thus, once the transverse fractures are initiated, the foliation merely plays the role of passive markers.

Although both theoretical and experimental studies (Mandal & Karmakar 1989) show that this mode of development of foliation boudins is physically realistic, we cannot entirely rule out the possibility of other modes of symmetric foliation boudinage. It is conceivable that during layer-normal compression of a thick sequence of foliated rocks, certain zones are strongly strain-softened (cf. Price & Cosgrove 1990, p. 409). The zones in which strain-softening is not so prominent may then behave effectively as competent units and undergo pinching-and-swelling followed by extension fracturing. This mode of foliation boudinage differs from the mode of development of the classical type of boudins in the sense that there is no intrinsic difference in competence between the boudinaged zone and its host, but the difference is *induced* in the course of deformation.

The occurrence of *asymmetric foliation boudins* has been recorded by Coe (1959), Hambrey & Milnes (1975) and Platt & Vissers (1980). The asymmetric shape generally develops when either the foliation or the fractures separating the boudins or both are oriented at an angle to the principal stress axes. This may result in different ways. (a) If the bulk deformation is coaxial but a pre-existing foliation is oblique to the principal axes (Fig. 17.31e), extension fractures may develop along certain zones. As in the case of fracture-controlled symmetric boudinage, the foliation swerves inward towards the fracture tips and gives rise to barrel-shaped forms. However, owing to the obliquity of the foliation with reference to the principal strain axes, the structures become asymmetric. (b) After extension fractures are initiated perpendicular to the foliation in bulk non-coaxial deformation, both the foliation and the fracture surfaces will rotate, but at different rates in the course of progressive deformation. Hence the fractures will no longer remain perpendicular to the foliation or to the XY-plane of finite deformation. (c) Asymmetric foliation boudins may also develop in association with shear fractures (Hambrey & Milnes 1975, Platt & Vissers 1980). These will always show displacement of the laminae

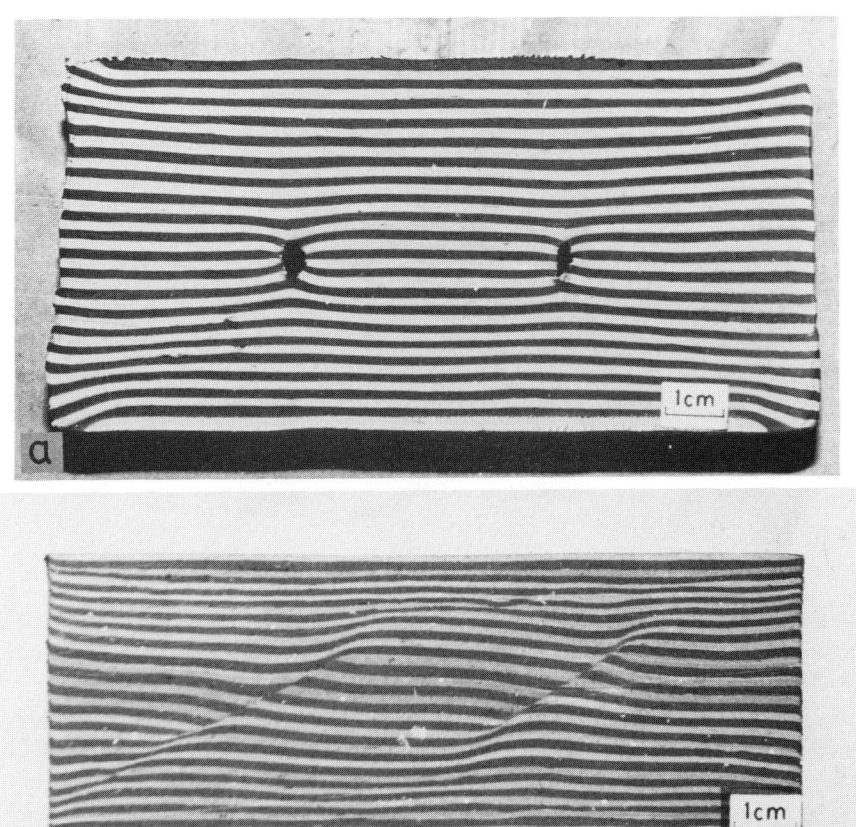

Fig. 17.36. Structures similar to foliation boudins produced in multilayered models. Each model consists of alternate black and white layers of modelling clay with greased interfaces. In (a) the central layers were cut perpendicular to the layering at two places. In (b) the central eight layers were cut at an angle of 45° at two places. Each model was deformed by a layer-normal shortening. In (a) the structure was similar to that of symmetrical foliation boudins. During deformation there was a sidewise opening of the initial cuts. In (b) a structure similar to asymmetric foliation boudin was produced.

along the fractures. The sense of displacement is such that there is an extension along the foliation. Unlike asymmetric boudins associated with oblique extension fractures, those which form after shear fracturing do not generally have vein-filled voids. The progressive deformation of the structures can be studied by deformation of multilayer models. Figure 17.36b shows a multilayer of alternate black and white layers of modelling clay with greased interfaces. The central eight layers were dissected by two parallel cuts at an angle of 45° with the layering. During layer-normal compression of the model, the layers affected by

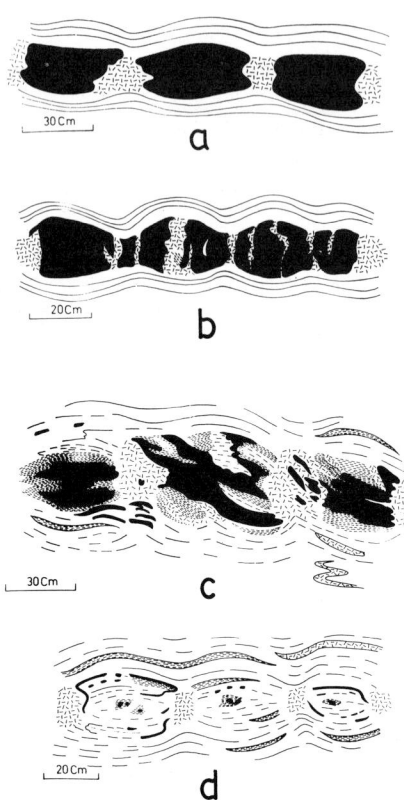

FIG. 17.37. Stages of development of ghost boudins in a granitized terrain. (a) Boudins of amphibolite in granite gneiss. (b) The boudins are partially replaced by pegmatitic clots. (c) A mixed rock or migmatite forms within the boudins. The amphibolite remnants can still be recognized. (d) The boudins are almost completely migmatized and their ghostly remnants can be recognized by the nodal pegmatites and oppositely directed scar folds. These different stages were seen in separate outcrops. Based on field sketches from Chotanagpur Gneissic Complex in Jasidih, India. From Sengupta 1985.

the cuts were displaced with a relative downward movement on the hanging wall side. The displacements were maximum near the "faults" and decreased away from them, so that short segments of initially horizontal layers dipped toward each fault on its hanging wall side (and away from fault on the foot wall side) and gave rise to reverse drags. Similar reverse drags associated with asymmetric foliation boudins have been reported by Platt & Vissers (1980) from metasediments and by Hambrey & Milnes (1975) from glacier ice. The experiments show that, with progressive deformation, the shear fractures rotate away from the compression axis and the rate of shear displacement (compared to the compressive strain rate) decreases with progressive deformation. Because of the displacements of the dissected laminae along the fractures, the enveloping undissected laminae are deformed into kink bands so that the discrete fractures are replaced by shear zones up to a certain distance on either side. (d) Lastly, in areas of superposed deformation, early extension fractures may be oriented oblique to the principal stresses of a later deformation. This situation may also give rise to asymmetric foliation boudins associated with offset of laminae along fractures.

Structures similar to interlocking pinch-and-swell were obtained experimentally by Cobbold *et al.* (1971). The model in this experiment consisted of a multilayer of identical plasticine layers with low cohesion between the layers. Interlocking pinch-and-swell structures developed when the model was compressed perpendicular to the layering. Kink bands, rather than pinch-and-swell structures, were produced if similar experiments were carried out with multilayers that had a high degree of mechanical anisotropy. Cobbold *et al.* (1971) and Cosgrove (1976) consider interlocking pinch-and-swell structures as a type of internal instability. The development of internal instability in a layered or anisotropic medium was investigated in detail by Biot (1964, 1965). Internal instability may occur in a medium of infinite extent or in a finite region bounded by rigid walls so that deflections and shear stresses are absent at the boundaries. Biot's analysis is concerned with the development of internal buckling when such a medium undergoes a shortening parallel to the planar fabric. Cobbold *et al.* (1971) and Cosgrove (1976), using Biot's theory, argue that pinch-and-swell structures will develop when the confined medium is compressed normal to the planar fabric.

17.14. Boudinage in relation to granitization and metamorphism

The enclaves of amphibolite or other metamorphic rocks in gneisses or migmatites often show excellent boudinage structures. Boudinage in such rocks have a special importance, since, as Wegmann (1932, pp. 486–489, 1965, p. 52) has so elegantly emphasized, the boudins may trace out the history of a complex interplay of granitization and deformation. Thus, the separation zones between boudins are often partially or completely filled with pegmatite, indicating thereby that pegmatite emplacement was synkinematic. In addition, in many cases, the boudins themselves become progressively granitized during post-fracture plastic deformation. If the granitization is far advanced, the lithologic contrast between the boudins and the host rock becomes blurred. The boudins, or rather their ghostly remnants, can still be recognized by the oppositely directed scar folds and by the occurrence of nodal pegmatites (Fig. 17.37).

Since the minerals which grow in the gaps between boudins are strictly synkinematic, the boudinage structures may be utilized in certain cases (Misch 1969, 1970) to determine the time relation between deformation and crystallization of minerals in a metamorphic rock. One type of evidence found by Misch (1969) was from boudinaged crystals of sodic amphiboles in blue-schist layers of the Shuksan Greenschist of the Northern Cascades of Washington, U.S.A. The sodic amphiboles here are zoned. A stretching parallel to the schistosity caused rupture across the prism axis of the amphibole crystals. Continued separation of the boudins was compensated by filling of the developing gaps while maintaining the continuity of the amphibole structure. Thus, successive amphibole zones grew, not only on the outer surface of the crystal, but also normal to the walls of the opening rupture. From a regionally consistent zoning sequence, Misch could determine the stages of initial rupture and the history of stretching. The patterns prove that the metamorphism and the preferred orientation of the amphiboles were synkinematic.

In the above example, the mineral which is boudinaged is itself synkinematic. It may also happen that the boudinage is post-crystalline with respect to the mineral which undergoes boudinage but is paracrystalline with respect to other minerals in the host rock or to minerals which newly grow in the separation zone. As an example, Misch (1969, p. 32, fig. 7) describes boudinage of a pistacite-rich band in a crossite-schist. During boudinage, the epidote grains are fractured. The amphibole-rich sandstone matrix flows into the separation zone, while non-aligned flakes of muscovite and chlorite fill the central part of the separation zone. Along the scar-folds of the matrix, crossite shows an association of bent crystals and polygonal arches of undeformed grains. The microstructure indicates that the boudinage was post-crystalline with reference to pistacite but paracrystalline with respect to crossite of the matrix, as well as to neocrystallized muscovite and chlorite of the separation zone.

In certain circumstances, it may be possible to correlate successive episodes of the metamorphic history of a rock with different stages of boudinage. Misch (1970) cites an example from a metamorphosed ultra-mafic rock containing relicts of primary clinopyroxene. The texture of the rock shows the metamorphic reaction sequence clinopyroxene→ tremolite→ Mg chlorite. Many of the clinopyroxene crystals have been boudinaged and the fragments pulled apart by elongation parallel to the schistosity. Large tremolite grains form a rim around pyroxene and fill the expanding fractures. By continued stretching, the large grains of tremolite of the separation zones are themselves fractured, with crystallization of fine-grained tremolite and chlorite in the newly formed gaps. Finally, only chlorite formed where the gaps continued to expand.

The progressive deformation leading to boudinage was throughout paracrystalline, not merely with reference to a single metamorphic episode but with respect to the entire metamorphic history.

.

CHAPTER 18

Faults

18.1. Fault terminology

A fault is a fracture discontinuity along which the rocks on either side have moved past each other. The attitude of the fault plane is expressed by its dip and strike or by its dip and dip direction. The trace of the fault on the earth's surface is the *fault line*. If a fault is not vertical the side above the fault plane is called the *hanging wall* and the side below the fault plane is called the *foot wall*. A fault surface may be curved or planar. Curved faults are known as *listric faults* (Fig. 18.1).

The relative displacement of two adjoining points on either side of the fault plane is known as the *net slip* (Fig. 18.2). Net slip is a vector which is expressed by the attitude of the line of displacement, the sense of displacement and the magnitude of displacement. The components of the net slip along the strike and the dip of the fault plane are known as the *strike-slip* and *dip-slip* components. Similarly, we may speak about the horizontal and the vertical components of the net slip.

Faults may be translational or rotational. In *translational faults* parallel lines on either side remain parallel after faulting. In *rotational faults* they are not. Rotational faults, sometimes described as *hinge faults* and *scissor faults*, have variable net slip in different parts (Fig. 18.3). The movement on rotational faults is best described by the angle of rotation of one block relative to the other, the axis of rotation being the normal to the fault plane. As we shall see in the case of certain normal faults, the angle of rotation and sometimes the sense of rotation may vary from place to place on the same fault.

The continuity of a marker horizon is disrupted by faulting unless the marker horizon is parallel to the fault plane or unless the net slip is parallel to the line of intersection of the marker horizon and the fault plane. In the general situation a marker plane will be separated on either side of the fault plane. The separation measured along the fault dip is known as the *dip separation* and the separation measured along the strike of the fault is known as the *strike separation* (Fig. 18.4). The

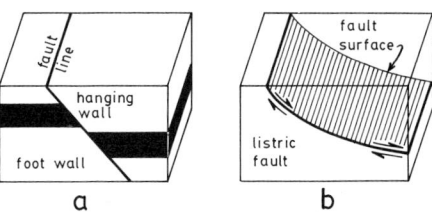

FIG. 18.1. (a) Fault line, hanging wall and foot wall. (b) A listric normal fault.

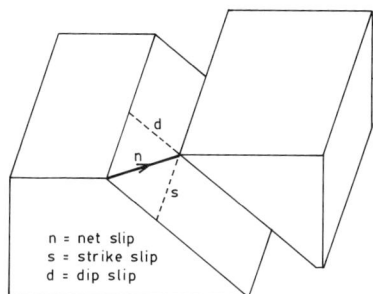

n = net slip
s = strike slip
d = dip slip

FIG. 18.2. Net slip, strike slip and dip slip in an oblique-slip fault.

dip separation is commonly measured in vertical sections perpendicular to the fault plane and the strike separation is commonly measured on a horizontal outcrop. The separation of a marker plane (say, the interface between two beds) along the trace of a fault on any outcrop face is usually called an *offset*.

18.2. Classification of faults based on relative movement between walls

(1) *Dip-slip faults* are those which have a dominant dip-slip component. Depending on the sense of movement, two types of dip-slip faults are recognized. In *normal faults* the hanging wall moves down relative to the foot wall (Fig. 18.5a). In *reverse faults* the hanging wall moves up relative to the foot wall (Fig. 18.5b). A reverse fault in which dip is less than 45° (Hill 1947) is called a *thrust fault*.

(2) *Strike-slip faults* are those which have a dominant strike-slip component. If looking across the fault plane the opposite wall appears to have moved towards the right, the strike-slip fault is called *right lateral* or *right-handed* or *dextral* (Fig. 18.5c). If looking across the fault plane the opposite wall appears to have moved towards the left, the strike-slip fault is called *left lateral* or *left-handed* or *sinistral* (Fig. 18.5d).

(3) In *oblique-slip faults* the strike-slip and the dip-slip components are comparable. Depending on whether the opposite wall moves

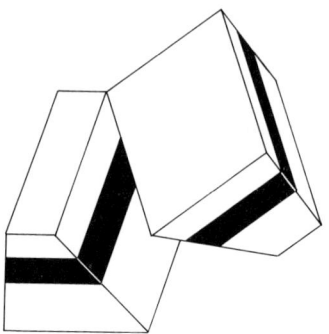

FIG. 18.3. Rotational fault.

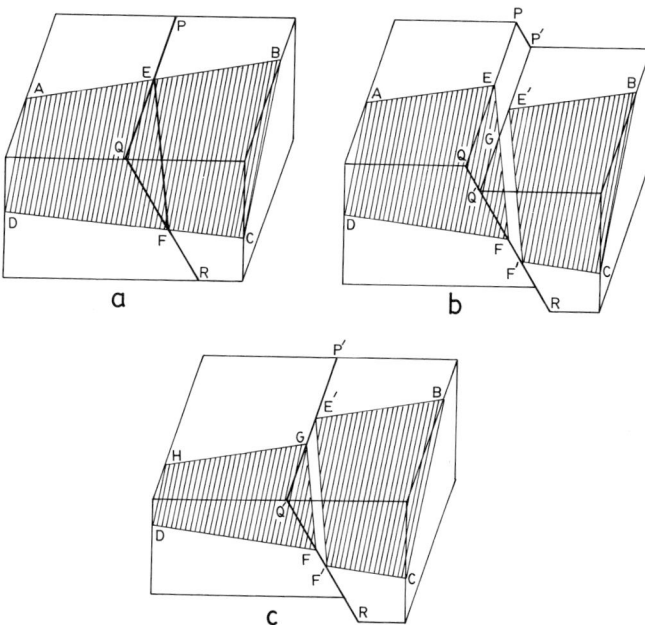

FIG. 18.4. (a) *ABCD* is a bed before faulting. *PQR* is the fault plane. (b) The plane *ABCD* is offset by normal faulting with *QQ'* as net slip. (c) Same as (b), after the fault scarp is eroded away to form a horizontal plane. *GE'* is the strike separation and *FF'* is the dip separation.

towards right or left and whether the hanging wall moves relatively up or down, oblique-slip faults can be further sub-divided into four types (Gill 1971): *right normal, right reverse, left normal* and *left reverse* (Fig. 18.6).

The classification given above is for *translational faults*. It should be recalled that translation of a rigid body is such a motion in which any straight line through the body remains parallel to itself. It should not be

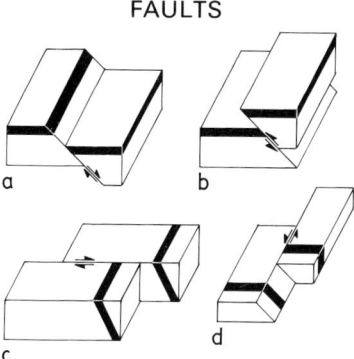

FIG. 18.5. (a) Normal fault, (b) reverse fault, (c) dextral strike-slip fault and (d) sinistral strike-slip fault.

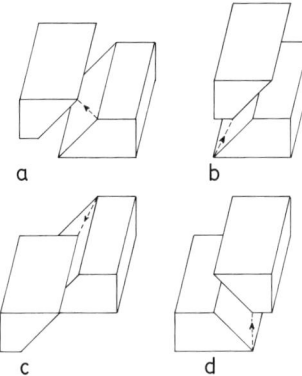

FIG. 18.6. (a) Right normal, (b) right reverse, (c) left normal and (d) left reverse faults.

confused with rectilinear motion. In translation a particle in a body may move in a curved path. Indeed, a listric fault may be translational.

In *rotational faults* the fault surface may be planar or curved. In planar faults the axis of rotation is normal to the fault plane. In curved faults the axis of rotation may be inclined or parallel to the fault surface. Rotational faults may be further sub-divided on the basis of whether the movement of one block relative to the other is clockwise or anticlockwise (Gill 1971).

In *horizontal faults* there is no up and down movement and dextral or sinistral movement is not meaningful. The fault movement is described by specifying the direction of movement of the hanging wall. However, most horizontal faults are parts of larger *inclined faults*, either normal faults or thrust faults. In *vertical faults*, on the other hand, the hanging wall and the foot wall are not defined. The fault movement should be indicated by specifying for one wall the relative up or down movement and the sense of the strike-slip component. Thus, a N–S striking vertical fault may be described (Gill 1971) by saying that the east side has moved down and southward.

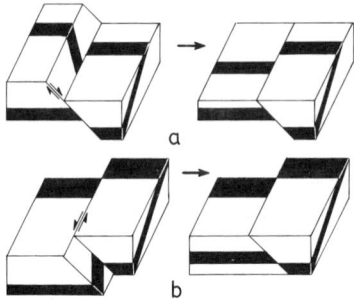

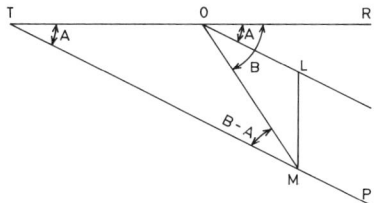

FIG. 18.7. (a) Dip-slip normal fault (with zero strike slip) showing both a strike separation and a dip separation. (b) Strike-slip fault (with zero dip slip) showing both strike separation and dip separation.

FIG. 18.8. View of the foot wall of a fault plane, with TR as its strike line. OL is the trace of a bed of the foot-wall side. TP is the corresponding trace for the hanging-wall side. OM is net slip. Angle $ROL = A$ and angle $ROM = B$. OT and LM are strike and dip separations.

18.3. Relation between slip and separation

Separation and slip are quite different entities. The strike slip or the dip slip evidently depends on the net slip. The strike separation and the dip separation depend both on the net slip and on the orientation of the bed. A bed may show a strike separation even if the strike slip is zero (Fig. 18.7a), or it may show a dip separation even if the dip slip is zero (Fig. 18.7b). A fault may have a large net slip and yet a bed affected by the fault may not have an offset on any section (Fig. 18.9).

If the net slip and the pitch of a bedding trace on the fault plane are known, the magnitude and the sense of strike and dip separations can be easily calculated. Figure 18.8 is a view of a fault plane looking from the hanging-wall towards the foot-wall side. TR is the strike of the fault plane. OL, the trace of a bed on the fault plane on the foot-wall side, is inclined at an angle A with the strike of the fault. TP is the trace of the bed on the fault plane on the hanging-wall side. OM is the magnitude of net slip. B is the angle between OM and the strike of the fault. Positive angles A and B are measured clockwise from 0 to 180° from the right-hand side of the strike line. The magnitude of net slip, n, is positive or negative depending on whether the hanging wall moves down or up relative to the foot wall.

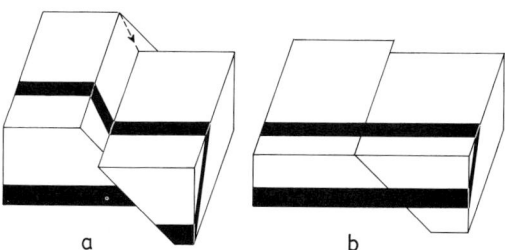

FIG. 18.9. Trace-slip fault with net slip along intersection of bed and fault. There is neither a strike separation nor a dip separation.

In triangle OTM, of Fig. 18.8,

$$\frac{OT}{OM} = \frac{s}{n} = \frac{\sin(B - A)}{\sin A}$$

and in triangle OLM,

$$\frac{LM}{OM} = \frac{d}{n} = \frac{\sin(B - A)}{\sin(90° + A)} = \frac{\sin(B - A)}{\cos A}$$

where OT and LM are the strike and the dip separations s and d, respectively.

Or,

$$s = \frac{n \sin(B - A)}{\sin A}, \qquad (18.1a)$$

$$d = \frac{n \sin(B - A)}{\cos A}. \qquad (18.1b)$$

These simple equations give us immediately the magnitudes and sense of the strike and the dip separations.

(1) If the net slip is parallel to the bedding trace along the fault plane, i.e. if $A = B$, we find that $s = d = 0$. In other words, there is neither a strike separation nor a dip separation. This type of fault is known as a *trace-slip fault* (Fig. 18.9).

(2) If the net slip is parallel to the strike of the fault, i.e. if $B = 0$, eqn. (18.1a) shows that the strike separation equals the net slip. It further shows that, although the dip-slip component is zero, the absolute value of the dip separation is equal to $|n| \tan A$. If the bedding trace pitches towards the right ($A < 90°$), a dextral fault movement causes an apparently normal (positive) dip separation and a sinistral strike slip causes an apparently reverse (negative) dip separation. If the bedding trace pitches towards the left ($A > 90°$), dextral and sinistral fault movements will cause negative and positive dip separations respectively.

(3) If the net slip is parallel to the dip of the fault ($B = 90°$), the dip separation is equal to the net slip and the absolute value of the strike separation equals $n \cot A$. Again, note that there may be a large strike separation even though the strike slip component is zero. For dip-slip normal faults the strike separation will be dextral or sinistral depending on whether the bedding trace pitches towards right or left. In reverse dip-slip faults (n negative) the strike separation will be dextral if bedding trace pitches towards the left (A negative); if the bedding trace pitches towards the right (A positive), the strike separation will be sinistral.

(4) If the strike of the bed is parallel to the strike of the fault, i.e. if $A = 0$, the strike separation as given by eqn. (18.1a) assumes an infinitely large value. This means that the outcrops of beds on a horizontal surface do not meet the fault line. Because of the dip separation, there will be repetition or omission of uniformly dipping beds on the horizontal outcrop (Fig. 18.10). Such faults are called *strike faults*.

(5) In other situations there will be both a strike separation and a dip separation. In general, the magnitudes and senses will not coincide with the strike-slip and dip-slip components. The magnitude and sense of separation can be predicted from eqns. (18.1a) and (18.1b). The sense of strike separation (s) and dip separation (d) for the four types of oblique slip faults are shown in Fig. 18.11. Note that s and d have the same sign (according to the sign convention used here) only when the angle A is less than 90°, i.e. when the bedding trace pitches to the right. This is a consequence of the fact that the ratio $d/s = \tan A$ [from eqns. (18.1a) and (18.1b)]. It is also noteworthy that the absolute value of this ratio is independent of the net slip.

18.4. Determination of net slip

The net slip of a fault can be determined only under special situations. Fault surfaces often have slickensides. These are polished surfaces produced by fault movement and contain parallel striae indicating the direction of fault displacement. These striae should be used with some caution because the slickensides may develop at a late stage of faulting and may not be parallel to the direction of net slip. Where the slickenside striae truly give the net-slip direction, the magnitude and sense of net slip can be determined if the beds are offset along the fault line. Figure 18.12 shows the fault LM striking 85° and dipping 60° southerly. A bed running along CD and EF shows left-handed strike separation of 250 m. In both sides of the fault the bed strikes 35° and dips 50° easterly. The slickenside striae have a trend of 165°. To determine the net slip we have to find from stereographic projection the trend of the line of intersection of the fault and the bed. The lines CG and EH are drawn parallel to this trend. The intercept PR drawn parallel to the net slip gives us the

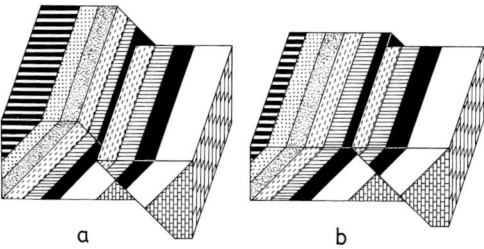

FIG. 18.10. Repetition and omission of uniformly dipping beds in strike faulting.

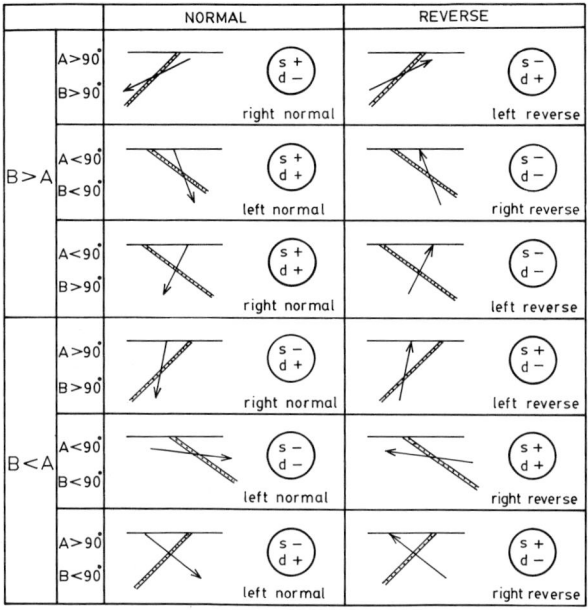

FIG. 18.11. Sense of strike and dip separations in oblique slip faults, depending on whether $A > B$ or $B > A$ and whether A or B is greater or less than 90°. For each type the strike line of the foot-wall face is shown with the trace of the bed and the direction of net slip.

horizontal component of the net slip. On the same diagram, or in a separate diagram, draw the angle RPQ equal to the plunge of slicken-side striae (or apparent dip of fault plane along the direction of PR), with RQ perpendicular to PR. QP gives us the magnitude of net slip, with upward movement of the hanging wall. The net slip can be determined much more quickly by reading off the angles A and B from the stereographic projection and substituting them in eqn. (18.1a). Since the strike separation s in the above example is -250 m, the net slip is

$$n = \frac{-250 \sin A}{\sin (B - A)} = -595 \text{ m.}$$

433

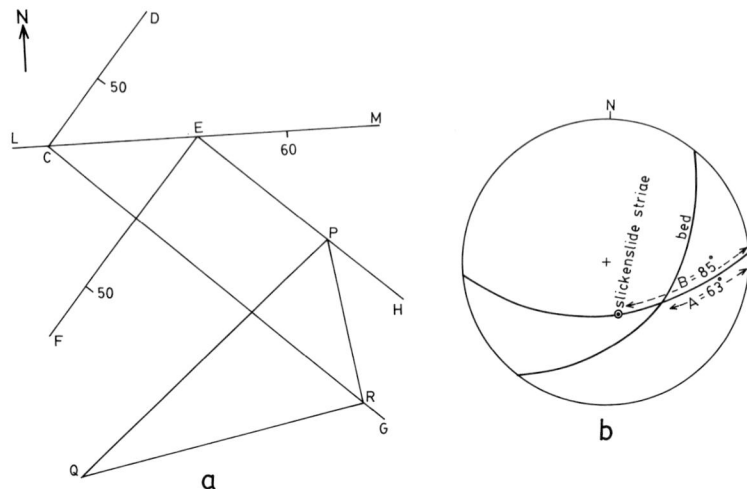

Fig. 18.12. Graphical method of determination of net slip when the attitude of slickenside lineation and the offset of a bed along the fault line are given.

The negative value indicates an upward movement of the hanging wall.

Even if slickenside striae are absent, the net slip can be determined in certain cases. Thus, for example, net slip can be determined if two non-parallel layers are offset along the fault line on a horizontal surface. The non-parallel layers could be two veins or a bed and a dyke or the two limbs of a fold. Figure 18.13 shows the offsets of two non-parallel veins, 1 and 2. The two veins and the fault plane are plotted in stereographic projection. The trend of the line of intersection of each vein and the fault plane is measured. Lines parallel to these trends are drawn as LQ, CP, DP and MQ. The point P in the diagram represents the projection on the horizontal plane of the point where the traces of vein 1 and vein 2 on the fault plane of the hanging-wall side intersect. Q is a similar point on the foot-wall side. PQ is thus the horizontal component of the net slip. The apparent dip of the fault along PQ is determined from the stereographic projection. In a separate diagram, take PQ as a horizontal line, draw the apparent dip angle at P and draw the vertical line QR. PR is the amount of net slip and the sense of slip is such that the hanging-wall side has moved down. The graphical construction can be made somewhat simpler if the pitches of the fault–vein intersections are read off directly from stereographic projection and a graphical construction is made on a plane parallel to the fault plane (Fig. 18.14).

18.5. Recognition of faults

There are several criteria for recognition of faults. Even when a fault line is mapped, the dip of the fault plane may not be measurable. The

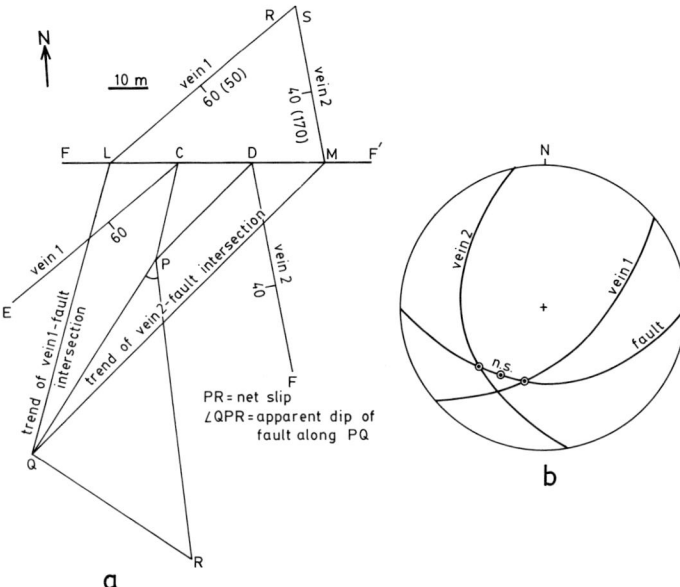

FIG. 18.13. Graphical method of determination of net slip when the attitudes of the fault and of two intersecting veins, along with strike separations of the veins, are known.

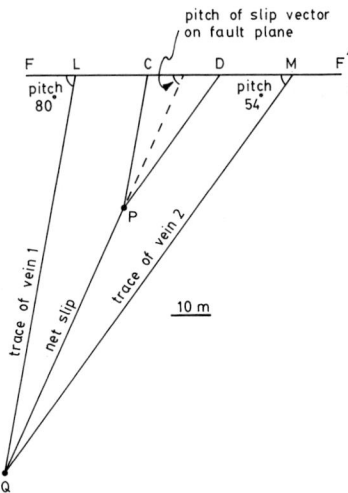

FIG. 18.14. Graphical solution of the same problem as in Fig 18.13, but here the construction is made on the fault plane itself.

435

dip can be measured in certain circumstances, e.g. when the fault trace is exposed in areas of high relief, the fault surface is exposed in scarps or road cuts or when the same fault surface is located at different levels by drilling or mining operations. The following are the most commonly used criteria for recognition of faults.

(1) In the course of geological mapping, faults are sometimes recognized by abrupt termination of beds against a line on the map. If the presence of an angular unconformity can be ruled out (say, from the occurrence of the same stratigraphic horizon on either side), the line of discontinuity can be regarded as a fault line. Sometimes, while tracing a few marker horizons through a monotonous country rock (e.g. thin beds of quartzite in mica schists), the beds appear to be terminated but reappear with an offset. Even if no other evidence of faulting is present, a fault line can be drawn on the map by joining the cut-off points. The orientation of the fault line drawn by this method may not be accurate.

(2) Abrupt termination of beds along a line or a sharply defined structural discontinuity can often be inferred from a study of aerial photographs. Indeed, this is the most rapid method of locating large-scale faults. However, aerial photographs only enable us to suspect the presence of a fault. The occurrence of the fault can be confirmed only after ground checking.

(3) Apart from identifying abrupt termination of beds, aerial photographs are used in various other ways to locate a fault line and sometimes to determine the dip of the fault. Aerial photographs of most areas show some physiographically controlled lines or lineaments. Lineaments may develop due to various reasons. However, from the geological information of an area some of the lineaments may be suspected to be faults. Again, the presence of a fault along a lineament can only be confirmed by ground checking. In any case, a careful study of the aerial photograph may save a lot of time in structural mapping in general and in locating faults in particular.

(4) Faults can sometimes be identified by repetition and omission of beds. Repetition may also be a result of folding; however, in this case the repetition is symmetrical. The repetition of beds caused by faulting is characteristically asymmetrical.

(5) Faults can often be located by the occurrence of *fault breccias* along a continuous or discontinuous line. Fault breccias are crushed rocks with angular fragments within a finely pulverized matrix. The matrix material is often silicified.

(6) Some fault planes contain a pulverized clay-like material known as *gouge*. Others may contain *slickensides* or polished striated surfaces sub-parallel to the fault plane. Some fault planes are characterized by intense silicification. Hot springs are sometimes located along a fault line. However, in practice, a fault can hardly be mapped by the

occurrence of such features. These features are mostly used as confirmatory evidence when the location of a fault has already been inferred from other evidence.

(7) On the other hand, the occurrence of a more or less continuous thin zone of mylonite is a clear evidence of faulting. Mylonites will be described in detail in connection with ductile shear zones.

(8) In certain places a fault plane is directly exposed as a *fault scarp*, a cliff with a more or less planar slope. Evidently, fault scarps are preserved only if the faults are of recent origin or if an old fault is rejuvenated by recent movements.

(9) Thrust faults in orogenic belts often show large horizontal displacements of the order of a few kilometres or even tens of kilometres. Such large displacements may bring together rocks of quite dissimilar sedimentary or metamorphic facies or rocks of very different ages. These features can be taken as criteria of faulting provided other geological features indicate that the discontinuity surface cannot be interpreted as an unconformity. Large-scale thrusting may, for example, bring older rocks over younger, a slice of granite over unmetamorphosed or weakly metamorphosed rocks or may put in juxtaposition rocks of widely different structural histories.

(10) Lastly, deep-seated faults are often located or inferred by interpretation of seismic data.

18.6. Reverse faults

Terminology

Reverse faults are those which have dominant dip-slip components with a relative upward movement of the hanging wall. Reverse faults with a low angle of dip (less than 45°) are called *thrusts* or *thrust faults*. The term thrust fault is usually restricted to large-scale structures. In this restricted sense a thrust fault is defined as a map-scale contraction fault (McClay 1981), a fault which shortens an arbitrary datum line. Thrust faults may show large horizontal displacements in the scale of a few kilometres or tens of kilometres. If the foot wall stays in position and the hanging wall is transported, the fault is called an *overthrust*. If the foot wall moves instead of the hanging wall, the thrust is called an *underthrust*. The hanging-wall sheet of rocks which travelled to some distance over a thrust fault is described as a *thrust sheet*.

If a tectonic unit has moved far over rocks in front of it, it is called an *allochthonous* unit. A rock mass which has not moved over other rocks is described as *autochthonous*. *Parautochthonous* rocks are those which have moved over other rocks to a small extent.

A *thrust nappe* is an allochthonous tectonic sheet which has moved over a thrust fault (Fig. 18.15a). A *fold nappe* is an allochthonous

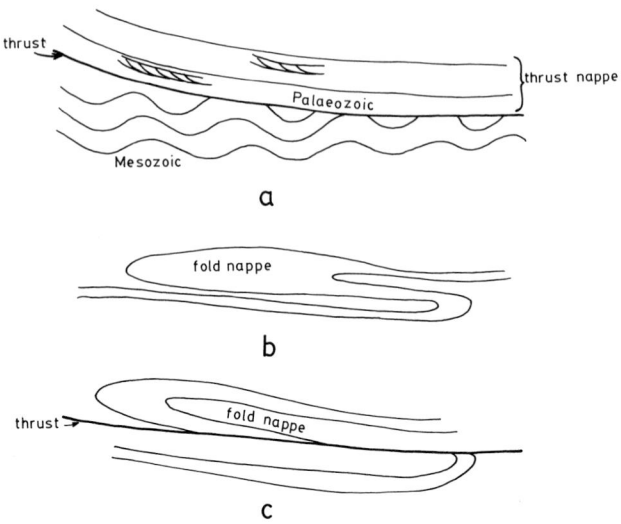

FIG. 18.15. (a) Thrust nappe. (b) Fold nappe. (c) Fold nappe with middle limb sheared out into a thrust fault.

tectonic unit which exhibits large-scale stratigraphic inversion and may have initiated from large recumbent folds. The underlying limbs of these folds may be sheared out into thrust faults (McClay 1981). Thus, a fold nappe might or might not have a fault at its base (Fig. 18.15b, c).

A closed outcrop of a thrust sheet isolated from the main mass by erosion is called a *klippe* (Fig. 18.16). A closed outcrop of the sub-stratum of a thrust sheet, framed on all sides by the trace of the thrust surface, is known as a *window* or *tectonic window* (Fig. 18.17).

Many thrust faults have an upward-curving shape. These *listric thrust faults* start with a gentle dip at the base and have increasing dips at higher levels.

A thrust sheet, emerging on the surface, may have travelled on the erosion surface, sometimes riding over its own debris deposited in front. This is then called an *erosion thrust* or *relief thrust*. The wedge making up the frontal portion of the thrust sheet is known as the *toe*.

Overthrusts and nappes

The discovery of far-travelled overthrusts and fold nappes was an exciting event in the history of geology. A lively description of the early discoveries is given by Edward Bailey in his book *Tectonic essays, mainly Alpine*. According to Bailey (1935), Arnold Escher von der Linth first described large-scale overthrusting from the Glarus area of Switzerland. By the middle of the nineteenth century the presence of overthrusts was

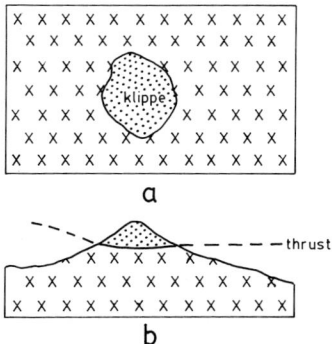

FIG. 18.16. (a) Map and (b) section of a klippe.

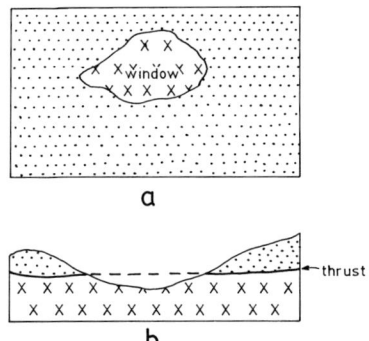

FIG. 18.17. (a) Map and (b) section of a tectonic window.

recorded from the Alps, the Northwest Highlands of Scotland and the Appalachians. The phenomenon of overthrusting was still a controversial topic. Eventually Callaway (1883a, b) and Lapworth (1885) demonstrated that in the Northwest Highlands of Scotland the metamorphosed basement had ridden over a low angle thrust and had travelled over unmetamorphosed Cambrian sediments. The presence of this thrust, known as the Moine Thrust Zone, was later confirmed by the detailed studies by Peach et al. (1907). In the last two decades of the nineteenth and the first two decades of the twentieth centuries the prevalence of nappe structures was well established in the western Alps. In 1884 Bertrand reinterpreted the Glarus structure as a single huge recumbent fold which had travelled northward for more than 35 km. Thrusts and associated overturned folds were also described by him from the Provençal Alps (Bertrand 1899). The Pre-Alps lying north of the Alps was interpreted by Schardt (1893) as huge thrust pieces or klippen and the extremely complex geometry of the superposed alpine nappes was brought to light by the studies of Lugeon (1901, 1902),

Argand (1911, 1916) and Heim (1919). Good descriptions of these studies of the Alps can be obtained in the English language from the books by Collet (1936), Heritsch (1929), Bailey (1935) and Coward *et al.* (1989).

It is now recognized that thrusts are common tectonic features of fold mountains. Our understanding of thrust tectonics in the shallow crust has been considerably enhanced in the last few decades by geological and geophysical studies and interpretation of drilling data obtained in connection with oil exploration in different parts of North America, especially in the Canadian Rockies (Douglas 1950, Bally *et al.* 1966, Dahlstrom 1969, 1970). There has been a resurgence of interest in the tectonics of shallow level thrust systems in recent years. Some of the recent developments are contained in Boyer & Elliott (1982), Suppe (1983), the collections of papers edited by McClay & Price (1981) and by Coward *et al.* (1989) and the special issue of the *Journal of Structural Geology* (Vol. 8, No. 3/4, 1986) on *Thrusting and Deformation*.

Thrust geometry

When a fault is nucleated within a volume of rocks it forms a penny-shaped crack with a closed boundary known as the *tip line* (Elliott 1976a). The blocks on either side of it slip past each other, but the slip must decrease to zero at the tip line. Beyond the tip line the material may undergo some ductile deformation. As the thrust propagates, the tip line also migrates. As long as the tip line of a thrust fault does not reach the ground surface, it remains a *blind thrust*. Most thrusts start as blind thrusts and propagate to reach the surface. Two faults meet along a *branch line*. A tip line and a branch line meet the ground surface at a *tip point* and a *branch point*. For ancient faults the ground surface that we see at present has formed by erosion after the fault movements had ceased. In contrast, the ground surface at the time of development of the fault is the *synorogenic ground surface*.

A subsidiary thrust may branch out as a *splay* from the main thrust. Figure 18.18 shows the outcrops of four different types of splay (Boyer & Elliott 1982). In an *isolated splay* the two tip points of the splay are exposed while the trace of the fault is isolated from the main fault trace. In a *diverging splay* a single tip point of the splay and a single branch point occur on the erosion surface. In a *rejoining splay* the tip line is not exposed and the trace of the splay on the ground surface meets the main fault at two branch points. A *connecting splay* connects two different faults. The shapes of the faults in three dimensions are shown in Fig. 18.19. A slice of rocks bounded by two or more thrust slices is known as a *horse*. A partially eroded horse may be undistinguishable from a

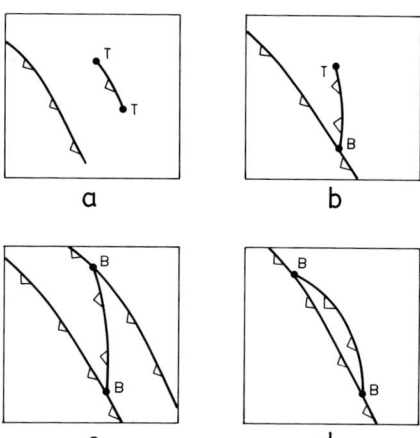

FIG. 18.18. Outcrops of four types of splay: (a) isolated splay, (b) diverging splay, (c) connecting splay and (d) rejoining splay.

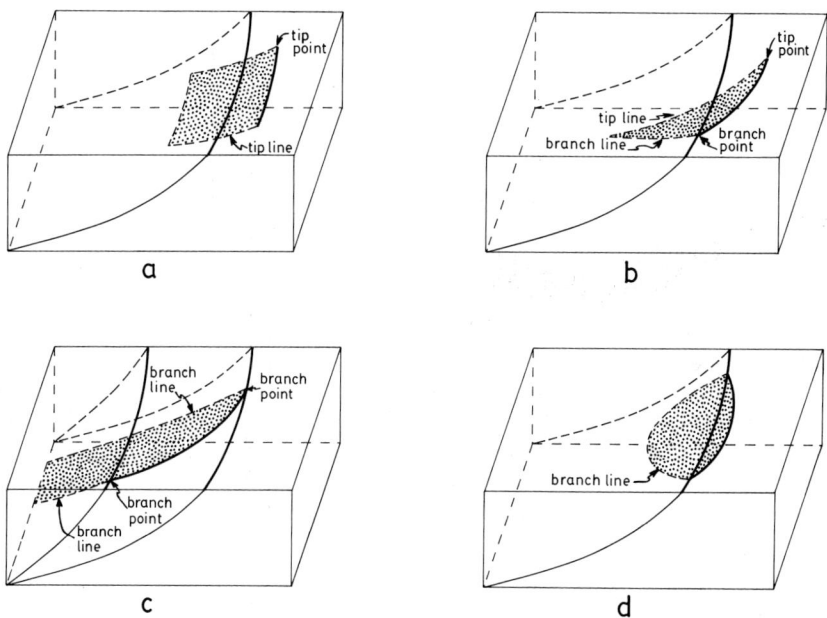

FIG. 18.19. Four types of splay in three dimensions: (a) isolated splay, (b) diverging splay, (c) connecting splay, (d) rejoining splay.

rejoining splay. For a more elaborate treatment of thrust geometry the reader may consult Boyer & Elliott (1982).

Thrusts in some regions follow a ramp-and-flat or a staircase-like trajectory (Douglas 1950). The thrust cuts up-section along a *ramp* and then follows a horizontal zone or *flat* (Fig. 18.20a). When the thrust

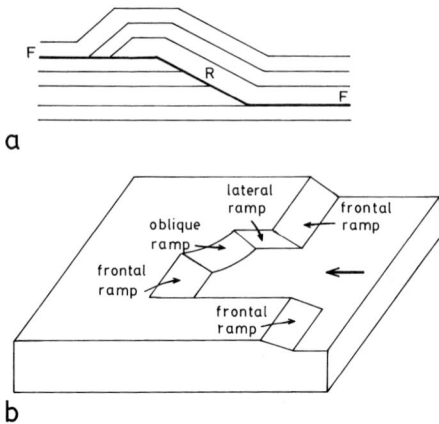

FIG. 18.20. (a) Ramp (*R*) and flats (*F*) of a thrust fault in vertical section. (b) Block diagram of frontal ramp, lateral ramp and oblique ramp.

develops in a previously undeformed sedimentary sequence the flat is parallel to the bedding. If the staircase trajectory develops in a crystalline basement or in a previously folded sedimentary sequence, the flats need not be bedding-parallel (Butler 1982). Ramps are classified (Fig. 18.20b) according to their orientation with respect to the transport direction of thrust sheet (Dahlstrom 1970). A *frontal ramp* strikes roughly perpendicular to the transport direction and has a dominantly reverse dip slip. A *lateral ramp* strikes more or less parallel to the transport direction and hence has a dominant strike slip. An *oblique ramp* strikes oblique to the transport direction and has both strike slip and reverse dip slip (Butler 1982).

The riding of the thrust sheet over a ramp causes the development of a flat-topped anticline on the hanging wall over and in front of the ramp. The development of such folds over the fault steps (fault-bend folds) was recognized by Rich (1934) in the Pine Mountain thrust sheet in the central Appalachians. The concept of a staircase trajectory and the consequent development of a hanging-wall anticline with a kink-like geometry were later incorporated in a generalized geometric–kinematic model of thrust faulting (Royse *et al.* 1975, Suppe & Namson 1979, Elliott & Johnson 1980, Butler 1982, Boyer & Elliott 1982). According to this model the foot wall of the thrust remains essentially undeformed. The position of the hanging-wall anticline remains more or less fixed over the ramp; however, as the thrust sheet is carried forward over the higher level flat and the beds move away from the ramp, their anticlinal form is flattened out while a fresh anticline continually forms over the ramp. A staircase trajectory does not develop on all thrust surfaces. Smooth-trajectory thrusts are also quite common (Cooper & Trayner 1986), especially where thrusting is preceded by some ductile deformation. Evidence of such ductile deformation may be preserved in folds of both hanging walls and foot walls.

Thrust-related folds

Asymmetric folds are often associated with thrust faults (Willis 1893, Heim 1919, Gallup 1951, Badgley 1965, Dahlstrom 1970, Thompson 1981, Elliott & Johnson 1980, Williams & Chapman 1983). However, folds associated with thrusting may develop in different ways. The beds may undergo a considerable amount of ductile deformation preceding the initiation of a thrust; if there are strong competence contrasts among the beds, there is a period of buckle folding *before* thrusting occurs. Buckle folding within the volume of a thrust sheet may also take place *during* the emplacement of a thrust nappe. In other cases the major part of a faulted sequence may not undergo significant folding while an asymmetrical antiform–synform structure develops in the immediate neighbourhood of the thrust due to its drag effect.

A thrust cutting across the gentler backlimb of an asymmetric fold is a *backlimb thrust* (Fig. 18.21). A *forelimb thrust* cuts across its steeper forelimb. Forelimb thrusts are by far the most common and may develop in different ways. A forelimb thrust may develop when the fold is not yet overturned or tightened; the thrust then cuts the forelimb at a large angle (Fig. 18.22a). These are the *break thrusts* of Willis (1893). It may happen that, while individual beds are being displaced by the break thrust at lower levels, the horizontal shortening at higher levels is being accommodated by folding. The beds in these higher level folds are eventually offset when the thrust propagates upward. This is the reason why individual bed displacements across a break thrust often decrease at higher levels (Williams & Chapman 1983). Similar upward diminishing bed displacement was noted by Dahlstrom (1970) in the Rocky Mountain Turner Valley structure. This is a clear evidence of broad simultaneity of folding and thrust propagation. *Stretch thrusts* (Fig. 18.22b) develop at an advanced stage of folding by stretching and shearing out of overturned limbs of tight or isoclinal folds (Heim 1919). Badgley (1965, fig. 6–12) points out that there are faults intermediate between break thrusts and stretch thrusts. In his examples a break thrust is initiated at the core of a moderately tight fold; with continued deformation the fold is further tightened while the upward propagating thrust cuts through the stretched overturned limb at a low angle.

Thrust systems

The large-scale displacements of allochthonous bodies in a mountain belt take place on a comparatively small number of low angle major thrusts. There are subsidiary thrusts of smaller displacements within a

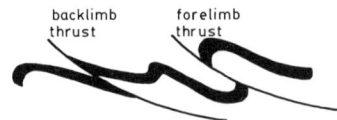

FIG. 18.21. Backlimb thrust and forelimb thrust.

a   b

FIG. 18.22. (a) Break thrust. (b) Stretch thrust.

major thrust sheet. A major thrust may initiate within the basement and may transport large slices of the basement over the cover rocks. Thus, the Main Central Thrust sheet (Auden 1937, Heim & Gansser 1939), with a thickness of more than 10 km of crystalline basement rocks in the Higher Himalaya (Gansser 1964, p. 251), has travelled on the Main Central Thrust and rests on unmetamorphosed or weakly metamorphosed cover rocks. While the main mass of this huge nappe occurs in the Higher Himalaya, its frontal parts are preserved as klippen far to the south in the Lesser Himalaya (Thakur 1981, p. 386). Similar far-travelled thrust sheets of the crystalline basement occur in other mountain chains, e.g. the Alps, the Norwegian Caledonides and the Blue Ridge of the Appalachians.

A major thrust in a mountain chain may develop as a *décollement fault* between the basement and the cover. It should be noted that the term *décollement* refers to a process, not a structure. *Décollement*, or unsticking, generally takes place along a weak horizon above the basement (Fig. 12.21e). The cover detached from the basement undergoes folding (superficial folds of Argand 1922) and subsidiary faulting while the basement remains relatively inert. The classic example of *décollement* is given in the celebrated section of the inner Jura (Buxtorf 1907). The section shows that the entire cover of Mesozoic sediments has slipped over the basement on an intervening zone of evaporite. The nappe lying over the plane of detachment was displaced up to 30 km by the northward push of the Alpine nappes. The existence of a surface of *décollement* below the Jura was deduced by Buxtorf from indirect evidence. The correctness of this interpretation was confirmed later by boreholes sunk in the western Jura (Michel *et al.* 1953, Aubouin 1965, p. 165). The idea of *décollement* was further elaborated by a number of workers, notably by Glangeaud (1944), Aubert (1945) and Laubscher (1972, 1977, 1979, 1981) from the Jura and by Lugeon & Schneegans (1940) and Lugeon & Gagnebin (1941) from the Alps. *Décollement* faults have since been described from other mountain belts. In all such cases the detachment develops along a layer of an exceptionally weak rock ("soap layer" of Lutaud 1924) such as evaporite, salt-bearing clays and shale.

Thrusts in mountain chains generally occur in groups. The geometrical interrelations among the members of such a thrust system have been intensively studied in recent years. A prominent low angle thrust occurs beneath some thrust systems. This is called a *floor thrust* or a *sole thrust*. Butler (1982) has suggested that the term sole thrust (Peach *et al.* 1907, p. 472) should be reserved for the lowest regional thrust surface. In some cases subsidiary faults may splay upward from a floor thrust and may cause a tile-like piling of the subsidiary thrust sheets. This arrangement of the subsidiary thrusts is known as an *imbricate structure* or *schuppen structure*. The imbricate faults meet the floor thrust asymptotically and curve upward with increasing dips at higher levels. An *imbricate fan* (Boyer & Elliott 1982) is a system of imbricate faults spreading out from the sole thrust like an open fan (Fig. 18.23). In a *leading imbricate fan* the thrust in front has the maximum slip. In a *trailing imbricate fan* the thrust at the rear shows the maximum slip. An *imbricate zone* or an *imbricate stack* may be confined between a floor thrust and a *roof thrust*. The imbricate faults usually meet both the floor and the roof thrusts asymptotically. Such a system of thrust faults is known as a *duplex* (Dahlstrom 1970) (Fig. 18.24a). The thrust slice bounded by two imbricate faults within the duplex forms a horse. A roof thrust and a floor thrust meeting at the front forms the *leading edge* of the duplex. They may also meet at the rear or at the *trailing edge* of the duplex (Dahlstrom 1970).

Most thrust sheets travel from the interior or hinterland side towards the external or foreland side of a mountain belt. Those which travel in the opposite sense are known as *backthrusts*. In the majority of duplexes the sense of movement of the floor and the roof thrusts and the sense of movement of the individual imbricate thrusts is such that the hanging wall moves towards the foreland. Thus, commonly the imbricate faults in a duplex dip towards the hinterland. The structure is then described as a *hinterland dipping duplex* (Fig. 18.24a). If the displacements on the imbricate faults are large enough, the horses are bunched up in the form of an *antiformal stack* (Fig. 18.24b). If the displacements on the imbricate faults are still larger, a horse passes beyond an underlying horse and gives rise to a *foreland dipping duplex* (Fig. 18.24c). If the initial spacing between the imbricate faults measured parallel to the floor thrust is s and the displacement on each imbricate fault is u, then

for $u < s$, the duplex is hinterland dipping,

for $u \approx s$, the duplex forms an antiformal stack and

for $u > s$, the duplex is foreland dipping

(Mitra & Boyer 1986).

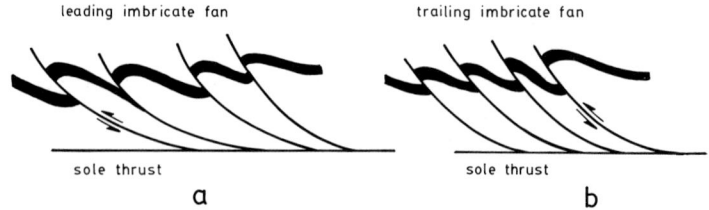

leading imbricate fan   trailing imbricate fan

sole thrust   sole thrust

a   b

FIG. 18.23. (a) Leading imbricate fan with maximum slip in the leading thrust. (b) Trailing imbricate fan with maximum slip in the thrust at the rear.

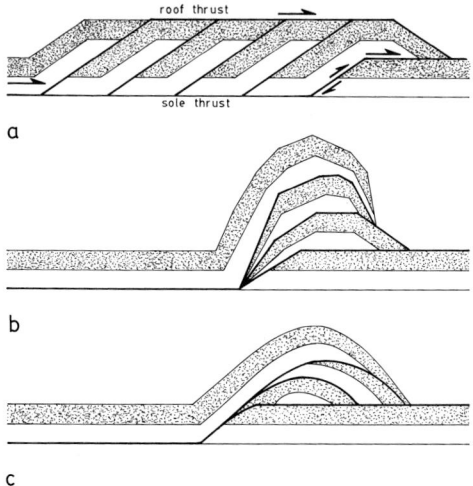

roof thrust

sole thrust

a

b

c

FIG. 18.24. (a) Hinterland dipping duplex, (b) antiformal stack and (c) foreland dipping duplex.

Duplex structures are common in many mountain belts. Although the term duplex is rather recent, similar structures were described long ago, e.g. the Moine Thrust System of the Scottish Highlands (Peach *et al.* 1907) or the Lewis Thrust System of the North American Cordillera (Douglas 1952).

In some instances, as in the Kirthar and Sulaiman mountain ranges of Pakistan, the roof thrust is a backthrust while the horses beneath the roof thrust are hinterland dipping and with the transport directed towards the foreland. The rocks overlying the roof thrust have not undergone significant horizontal translation. This type of structure, described as a *passive roof duplex* (Banks & Warburton 1986), has also been identified in the Alberta foothills of the Rocky Mountains and in the western frontal ranges of the Taiwan thrust belt (Suppe & Namson 1979, Suppe 1980, Davis *et al.* 1983).

It has also been suggested (Shanmugam *et al.* 1988) that duplex-like structures of non-tectonic origin can develop by deformation of soft sediments under the action of sediment gravity flow in submarine fan channels.

There are two models of thrust propagation for the sequence of thrusting.

(1) In an *overstep* sequence a new thrust surface develops behind or towards the hinterland side of an earlier thrust (Elliott & Johnson 1980, Butler 1982). According to this model the higher thrust sheets form in later movements.

(2) In the *piggy-back* model new thrust surfaces are initiated in front or towards the foreland side of an old thrust surface (Dahlstrom 1970); the higher thrusts have formed earlier (Fig. 18.25).

Recent studies indicate that most thrust systems have developed according to the piggy-back model. If, in such a sequence, a new thrust develops on the hanging-wall side it is regarded as an *out-of-sequence* thrust. The occurrence of a piggy-back sequence has been demonstrated in such areas as the southern Canadian foreland belt of the Rocky Mountain (Dahlstrom 1970, Price 1981), the Utah–Idaho–Wyoming belt (Royse *et al.* 1975, Dixon 1982, Wiltschko & Dorr 1983), the Moine Thrust Zone (Elliott & Johnson 1980) and the Helvetic thrusts (Trumpy 1973, Ramsay 1981).

Displacements in hinterland dipping duplexes

Boyer & Elliott (1982) have given an idealized geometric–kinematic model of hinterland dipping duplexes. According to this model, a major thrust with slip U_0 is first initiated (Fig. 18.25). The thrust climbs over a ramp from a lower to an upper flat. A second ramp then forms in front of the first. The thrust surface over the first ramp is then deactivated and the fault movement is transferred over the second ramp. If the slip over the second ramp is U_1, the total slip on the major thrust is now $U_0 + U_1$. A third ramp, connecting the upper and the lower glide horizons, then forms in front of the second. An additional slip U_2 takes place over this new ramp and the horse between the first and the second ramps is carried forward passively. The total slip on the major thrust at this stage is $U_0 + U_1 + U_2$. The characteristic geometry of the duplex is created after the development of a number of such imbricate thrusts. In this idealized model of Boyer & Elliott (1982) the beds are deformed in a kink-like shape while passing over a ramp, the horses are sandwiched between a floor thrust and a roof thrust, and the same stratigraphic unit occurs over the roof thrust for a great distance.

An interesting consequence of this idealized model is that, although the slip on the major thrust increases with time, the shortening (change in length per unit length) within the duplex may remain constant with progressive imbrication provided the geometry of the successive horses

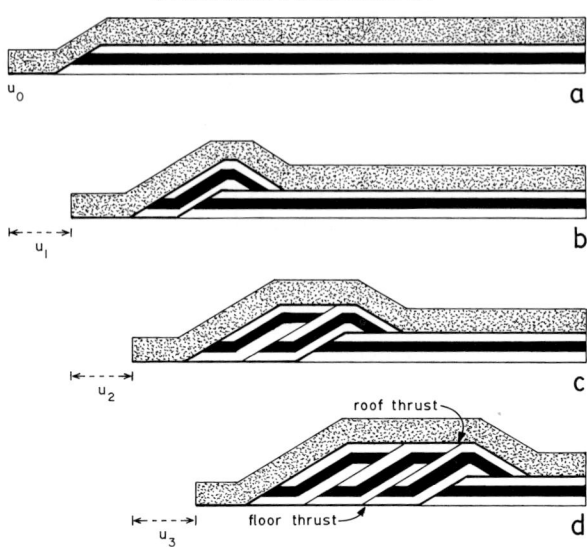

FIG. 18.25. Piggy-back model of development of a duplex. The initial displacement u_0 of a roof thrust is shown in (a). In the final stage (d) the total displacement on the roof thrust is $u_0 + u_1 + u_2 + u_3$.

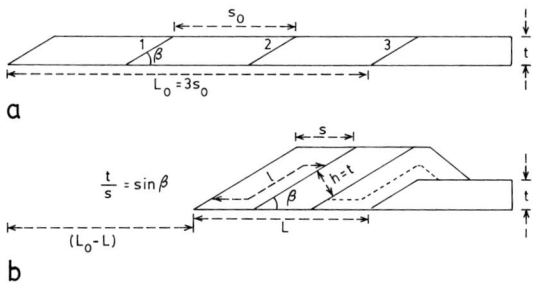

FIG. 18.26. The structural elements of a duplex.

remains the same. Consider the following elements of a duplex (Fig. 18.26):

t = initial thickness of beds affected by subsidiary thrusts,
h = thickness of beds within a horse,
L_0 = initial length of beds affected by imbrication,
L = length of duplex,
n = number of horses,
s_0 = initial spacing between successive ramps,
s = linear dimension of a horse measured parallel to the floor thrust,
l = length of bedding in a horse
β = angle between floor thrust and central segment of a subsidiary thrust.

To make the following analysis simple let us assume that all the horses have the same size and shape. If there is no ductile deformation during the development of the duplex

$$
\begin{aligned}
s_0 &= l \\
t &= h \\
L_0 &= nl \\
L &= ns.
\end{aligned}
\tag{18.2}
$$

The total slip recorded by the duplex is

$$
L_0 - L = n(l - s)
\tag{18.3}
$$

and the shortening is

$$
\frac{L - L_0}{L_0} = \frac{L}{L_0} - 1 = \frac{n(s - l)}{nl}
$$

or,

$$
\frac{L - L_0}{L_0} = \frac{s - l}{l}.
\tag{18.4}
$$

Shortening is usually represented as a percentage by multiplying this value by 100. By comparing eqns. (18.3) and (18.4) we find that total slip increases with an increase in the number of horses but the shortening is independent of n.

From Fig. 18.26 we find that

$$
s = \frac{t}{\sin\beta}.
\tag{18.5}
$$

Therefore, the shortening

$$
\frac{L - L_0}{L_0} = \frac{s}{l} - 1 = \frac{t}{l} \cdot \frac{1}{\sin\beta} - 1.
\tag{18.6}
$$

The equation shows that, for the same value of t/l in a duplex, the absolute value of shortening will be smaller if β is smaller (Boyer & Elliott 1982). The equation also shows that for any value of t/l, there must always be a lower limit of β. The limiting value is obtained when the shortening is zero, or

$$
\beta = \sin^{-1}\left(\frac{t}{l}\right).
\tag{18.7}
$$

Thus, if $t/l = \frac{1}{4}$, β cannot be less than $14.5°$.

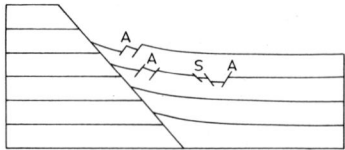

Fɪɢ. 18.27. Antithetic (*A*) and synthetic (*S*) faults associated with a major normal fault.

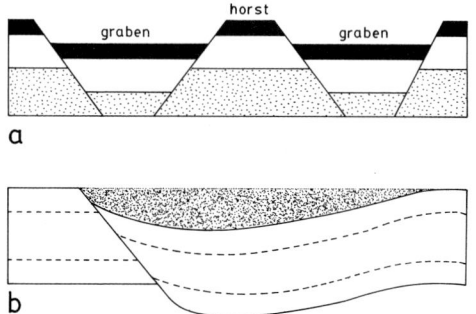

Fɪɢ. 18.28. (a) Horst-and-graben structure. (b) A half graben.

18.7. Normal faults

Normal faults have dominant dip-slip components, with the hanging wall moving down relative to the foot wall. The sense of movement of the fault blocks is such that there is an overall horizontal extension perpendicular to the fault strike. For this reason normal faults may also be described as *extension faults*. The hanging wall of a major normal fault often shows in its neighbourhood subsidiary normal faults either dipping in the same direction or in the opposite direction (Fig. 18.27). These are described as *synthetic* and *antithetic faults* (Cloos 1928).

In areas of extension the normal faults may occur as a set of parallel faults with the same sense of throw in all the fault blocks or they may occur in conjugate sets with opposed dips but with parallel strikes. The block which goes down between two oppositely dipping normal faults forms a fault trough or *graben*. A linear uplifted block between normal faults constitutes a *horst* (Fig. 18.28a). Fault troughs may also form because of the downthrow on a single fault or a single set of faults. Such troughs are known as *half grabens* (Fig. 18.28b).

Sediment-filled ancient grabens and half grabens have been identified in many areas. The large modern grabens, such as the Rhine graben, the East African rift valleys along the string of lakes from Lake Albert to Lake Nyasa and the graben system of the Red Sea, are associated with volcanism and have a complex tectonothermal history. Conjugate normal faulting on a grand scale has given rise to a long submarine graben system along the top of the Mid-Atlantic ridge.

Normal faults in soft materials can be easily produced in the laboratory. Such experiments give us a considerable understanding about the mode of development of structures in areas dominated by normal faulting. Many of the features associated with Rhine graben were reproduced by Hans Cloos (1930, 1936) in soft clay models subjected to a unidirectional extension at the base. Similar experiments were also performed by Ernst Cloos (1968) to produce the fault patterns of the Gulf Coast region of the U.S.A. A convenient way of making such experiments is to mix fine sand with kaolin powder (in the ratio of about 2:1), make a thick paste with water and prepare a square cake within a mould, say 20 cm × 20 cm, and with 6 cm height. The cake is prepared on two partly overlapping metal plates. The mould, open at the top and the bottom, is lifted up when the model is still wet but is strong enough to stand without collapsing under its own weight. Horizontal marker lines may be scratched on two opposite vertical faces of the model by means of a comb. When the basal metal plates are pulled apart the model is initially thinned in the middle and the upper surface sags down. This stage is followed by the development of a pair of conjugate normal faults dipping at about 60° towards each other (Fig. 18.29a). We may get a single set of normal faults if a similar experiment is performed by keeping one basal plate fixed and by slowly pulling away the other plate. While performing such an experiment we can see that in order to maintain the continuity of the model, i.e. to prevent formation of voids, the segment of the hanging wall in the neighbourhood of the fault sinks down with reference to the segments further away from the fault. In the neighbourhood of the fault the horizontal markers therefore dip towards the fault plane, thereby giving rise to the typical *reverse drag* or *roll-over* structure so often seen in areas of normal faulting (Fig. 18.30). The subsidiary synthetic and antithetic faults and the roll-over geometry in front of the master fault are accommodation structures which inhibit the development of gaps in the rocks (Hamblin 1965).

The plan view of the models is also of considerable interest. The fault scarps initially produced on the surface may not run along the entire width of the model and may develop in short segments, with maximum throw in the middle and with zero throw at either end. Except along the central part, the hanging wall of such a fault must have undergone a rotation, with the rotation axis normal to the fault plane. Such rotational faults, hinged at one or both ends, are known as *hinge faults* and are common in many intracratonic basins. The plan view of the models further shows that the total extension may be brought about by the relaying of the displacements through a series of hinge faults with the downthrow varying along each individual fault but with the throw variations on successive faults related in such a manner that, going

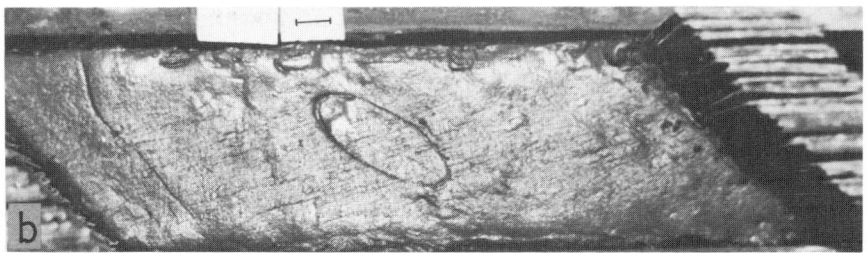

FIG. 18.29. (a) Development of a graben by extension of a clay – sand cake. (b) *R* shears in a clay block undergoing sinistral simple shear.

FIG. 18.30. Reverse drag or roll-over structure associated with normal faulting.

along a line perpendicular to the fault strike, the larger throw on one fault is somewhat compensated by a smaller throw on the fault lying in front of it. As a result, the hanging-wall surface of the model at a distance from the zone of faulting is brought down to a uniform level and the lengthening of the model remains the same along each transverse section. Occurrence of such *relay faults* (Goguel 1962, p. 128) has been recognized in many areas of normal faulting (see, for example, Bristol 1975, Rosendahl & Livingstone 1983, Gabrielsen & Robinson 1984, Ghosh & Mukhopadhyay 1985b, Larsen 1988).

Normal faults may be either planar or listric. The development of a roll-over structure is especially favoured by displacements along listric normal faults (Wernicke & Burchfiel 1982, Gibbs 1983). Because of the roll-over geometry, the bedding dip in the hanging wall is steeper than that of the foot wall. If there is a series of listric faults it is expected that the bedding dips of successive fault blocks will be steeper as we move in the direction of downthrow. Such fault blocks with successively steeper bedding dip were mapped by Anderson (1971) from the Basin and Range province of south-eastern Nevada.

Tilted fault blocks separated by sub-parallel planar faults occur in many areas. It has been suggested that the tilted fault blocks form during crustal extension by a process similar to the simultaneous tilting and sliding of a row of dominoes or a row of books in a bookshelf. This model of faulting, known as the domino model, involves a rigid-body rotation of both the fault plane and the bedding, while the angle between the two remains constant within the rotated fault blocks. Consider a series of parallel planar faults which initially developed in a sequence of horizontal beds. Let the bedding lengths in the tilted fault blocks be l_1, l_2, l_3, etc., and let the lengths in the horizontal direction be b_1, b_2, b_3, etc. Let the angle between the fault plane and the bedding within the tilted block be θ and the angle between the fault plane and the horizontal plane be β (Fig. 18.31). Then the stretch is

$$\frac{b_1 + b_2 + b_3 + \ldots}{l_1 + l_2 + l_3 + \ldots}.$$

Moreover, since the triangles with sides $\frac{1}{2}l_1$ and $\frac{1}{2}b_1$, $\frac{1}{2}l_2$ and $\frac{1}{2}b_2$, $\frac{1}{2}l_3$ and $\frac{1}{2}b_3$ etc., are similar (Fig. 18.31),

$$\frac{b_1 + b_2 + b_3 + \ldots}{l_1 + l_2 + l_3 + \ldots} = \frac{b_1}{l_1} = \frac{b_2}{l_2} = \frac{b_3}{l_3} = \ldots = \frac{\sin\theta}{\sin\beta}$$

and therefore

$$\text{per cent extension} = \left(\frac{\sin\theta}{\sin\beta} - 1\right)100$$

(cf. Thompson 1960, Wernicke & Burchfiel 1982). Since natural examples are more complex, this equation can give only an approximate estimate of the extension.

Steeply dipping normal faults are common in many areas. Indeed, these are the most common type of normal faults. The dips of these faults range between 55 and 70°. Lower dips of normal faults may result because of rotation of the fault planes as predicted by the domino model. Recent studies, however, indicate that many of the large-scale *low-angle normal faults* have a listric form; the faults flatten out at depth where they become sub-parallel to the bedding (Anderson 1971, Armstrong 1972, Wernicke & Burchfiel 1982, Gibbs 1983, 1984, White *et al.* 1986, Davison 1986, Jackson *et al.* 1988, Moretti & Coletta 1988). The flat-lying fault segment may develop along a weak zone of shale or salt above which the upper strata, detached from the basal rocks, are displaced horizontally. If the displacement along these low-angle faults is large, the upper unit may be regarded as an allochthonous body. Some of these structures were earlier misinterpreted as thrust faults. As opposed to thrust faults, however, the low-angle listric faults are characterized by the occurrence of roll-over structures in the hanging wall

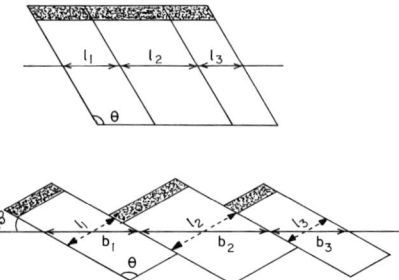

FIG. 18.31. Domino model of development of tilted fault blocks.

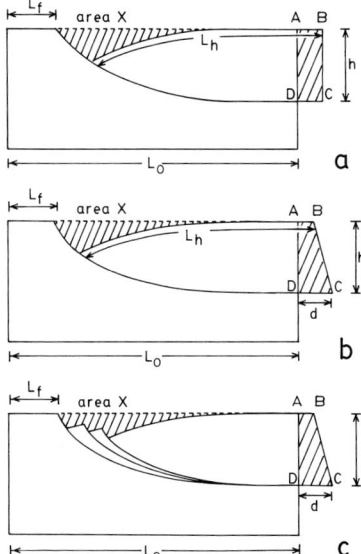

FIG. 18.32. Area balance techniques of calculating depth of detachment of a listric normal fault.

and by the juxtaposition of younger rocks over older rocks. Areas of thin-skinned extensional tectonics often show a system of imbricate normal faults merging into an underlying major low-angle normal fault. Such a low-angle normal fault with large displacement linked to an overlying system of imbricate normal faults is known as a *detachment fault* (Davis *et al.* 1980). Similar structures have been reported from many areas and are well documented from the Basin and Range province of U.S.A. (e.g. Anderson 1971).

There are several graphical methods to reconstruct the listric fault shape and to calculate the depth at which the listric fault flattens out along a horizontal detachment surface. Williams & Vann (1987) have made an excellent review of the different methods. The methods can be applied where the near-surface dip of the fault and the roll-over geometry of the hanging wall are known.

The *area balance technique* (Dahlstrom 1969, Hossack 1979) is applied to calculate the depth of a *décollement* thrust. The same technique can also be used to calculate the depth to detachment of a listric normal fault (Bosworth 1985, Williams & Vann 1987). For plane iso-volumetric strain, the shaded area X in Fig. 18.32a is equal to the shaded rectangular area $ABCD$ at the right-hand end of the section. Since the area X is known from the roll-over geometry, the depth h of the detachment fault can be calculated from the following relations:

$$\text{area } X = \text{area } ABCD = (L - L_0)h$$

where L is the sum of the foot wall and the hanging-wall bed lengths:

$$L = L_f + L_h,$$

and thus,

$$h = \frac{\text{area } X}{L - L_0}.$$

Gibbs (1984) suggested that the development of roll-over geometry may be associated with some bedding-parallel slip in the hanging wall. In such a case the area $ABCD$ will not be a rectangle; it will be approximately a trapezium (Fig. 18.32b) with the side

$$AB = L - L_0$$

and the side

$$CD = d$$

where d is the displacement along the fault. Since the area of the trapezium is the product of half the height (t) and the sum of the parallel sides,

$$t = \frac{\text{area } X}{\frac{1}{2}[d + (L - L_0)]}.$$

The same equation can be used when there is a system of imbricate normal faults splaying out of an extensional sole fault if, as shown in Fig. 18.32c,

$$L_0 = L_a + L_b + L_c + \dots$$

and

$$d = d_a + d_b + d_c + \dots$$

(Chapman & Williams 1984, Bosworth 1985).

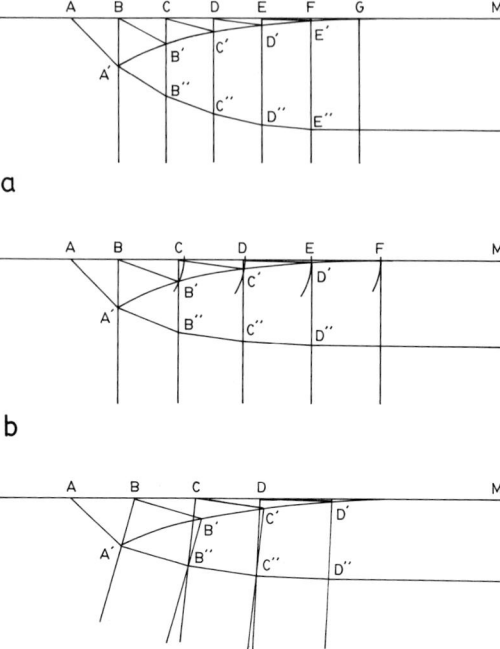

a

b

FIG. 18.33. Determination of the shape of a listric fault from the roll-over geometry by (a) chevron construction, (b) modified chevron construction and (c) slip-line method. Williams & Vann 1987.

To determine the shape of the listric fault from the roll-over geometry we may use any of the following three methods of construction (Williams & Vann 1987) as depicted in Fig. 18.33:

(1) The *Chevron construction* method (Verrall 1982) assumes that the heave (or the horizontal component of dip slip) remains constant all along the length of the fault. In Fig. 18.33a, AA' is the known displacement of a marker bed and the line $A'M$ shows the roll-over structure. AB is the heave of the fault. Vertical lines through points B, C, D, etc., are drawn at intervals equal to the heave. These intersect the curved line $A'M$ at points B', C', D' etc. Draw $A'B''$ parallel to and equal to BB', draw $B''C''$ parallel and equal to CC' and so on. Join A', B'', C'', etc., in a smooth curve to obtain the listric fault geometry.

(2) In the *modified Chevron construction* (Fig. 18.33b) (Williams & Vann 1987), the displacement AA', and not the heave or throw (vertical component of net slip), is assumed to be constant along the length of the fault. The point B here is located in the same way as in the previous method. From B draw an arc of radius AA' to intersects the curved line $A'M$ at B'. The vertical line through B' intersects the horizontal line AM at C. In a similar way locate the points C', D, D', etc. After this

stage of construction the listric fault geometry is determined in the same way as in the previous method.

(3) In the *slip-line method* (Williams & Vann 1987) also, the displacement is kept constant but the hanging-wall deformation is considered in terms of segments perpendicular to the fault rather than in terms of vertical heave segments. Cut out a rectangular strip of paper of width AA'. Place the left edge of the strip at A' and move it in such a way that the short edge ($= AA'$) has two of its corners on the lines AM and $A'M$. The positions of the corners locate the points B and B'. The locations of the points C, C', etc., are found in a similar way. The method of drawing the fault surface is shown in Fig. 18.33c. Wheeler (1987) proposed some modification of the modified chevron model and the slip-line model to ensure that the area of the section is conserved.

The fault geometries obtained by the three methods will not be the same. Moreover, the strains associated with fault propagation are not taken into consideration in any of these construction methods. The methods can therefore give only a rough estimate of the fault geometry.

The foregoing discussion clearly shows that normal faults may be of different types. There are both high- and low-angle faults, planar and listric faults, non-rotational and rotational faults. The rotations may also be of various types. In hinge faults the hanging wall rotates and the rotation axis is normal to the fault plane. In others, especially in listric normal faults, the hanging wall is deformed to give rise to a roll-over structure; the axis of rotation of beds in the hanging wall is parallel to the fault strike. Normal faulting in the domino style is associated with rotation of both the faults and the beds (Fig. 18.34a, b), the axis of rotation being parallel to the strike of the fault.

Although the domino style faulting and listric faulting linked to a flat zone of detachment (Fig. 18.34c) are both known to occur in areas which have suffered large horizontal extensions, there may remain some scope of alternative interpretations (Wernicke & Burchfiel 1982, Jackson & McKenzie 1983, Chenet *et al.* 1983, Le Pichon *et al.* 1983, Wernicke 1985, Barr 1987) regarding the deeper structures of half grabens which affect the entire upper crust (the brittle part of the crust). The large-scale fault-block tilting of the Gebel Zeit area of the Gulf of Suez has been interpreted by Moretti & Coletta (1988) by rotation of the hanging-wall block on a strongly curved listric fault which becomes horizontal at depth in the brittle–ductile transition zone; the fault surface itself does not rotate. In sharp opposition to this model, Jackson *et al.* (1988) argue that the seismogenic normal faults which break through the base of the upper crust almost always have a dip between 30 and 60° and are nearly planar. In the Suez area the large normal faults still dip at an angle of about 35° at a depth of 10 km, i.e. at the brittle–ductile transition zone in the crust. Jackson *et al.* have proposed

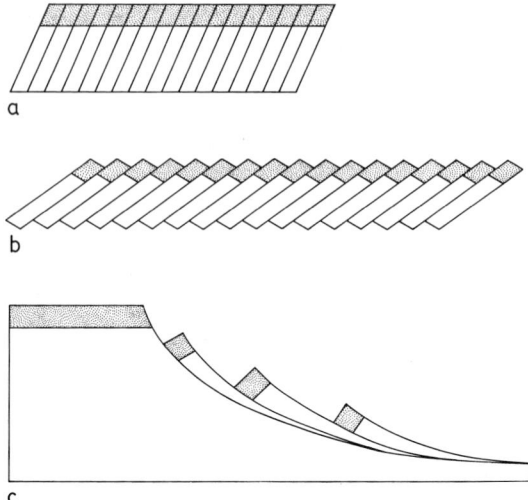

FIG. 18.34. (a) and (b) Horizontal extension by domino style faulting. (c) A series of listric faults linked to a flat zone of detachment.

the theory that the large-scale normal faulting in the Gulf of Suez has occurred in accordance with the domino fault model. The authors have argued that there is no need of a sub-horizontal detachment fault at the base of the tilted blocks.

Large-scale normal faulting is often associated with synchronous sedimentation. Such *syn-sedimentary* faults or *contemporaneous faults* (Hardin & Hardin 1961) are usually recognized by an abrupt thickening of syntectonically deposited beds on the downthrown side. The hanging-wall thickness of these beds decreases away from the fault. The beds which were deposited prior to faulting do not show such thickness variations. Again, if sedimentation and normal faulting are broadly synchronous, the displacements of the beds decrease from older to younger beds. Where the syn-sedimentary faults are rotated during the filling up of half grabens, the older beds undergo a larger amount of rotation and hence have larger dips than the overlying younger beds.

Although, strictly speaking, syn-sedimentary faults and growth faults are synonymous terms, there has been a tendency in recent years to restrict the term growth fault to a special type of syn-sedimentary fault which develops by the collapse of a rapidly deposited pile of sediments, especially muds (Collins & Thompson 1982, Elliott 1986, pp.151–152). These faults typically develop in deposits of large deltas where there is a high proportion of fine-grained sediments. The rapid deposition of a thick pile of sediments leads to its lateral flowage towards the delta front either as a slide or on a listric fault. Similar faults have been extensively described in the large deltas, such as the Gulf of Mexico, the Niger delta, the McKenzie delta and the Mississippi delta (Ocamb 1961, Weber 1971, Coleman & Garrison 1977, Evamy *et al.* 1978). In the prodelta of the Mississippi delta such normal faults form scarps up to 30 m high on the sea floor and show evidence of growth during sedimentation. The deposition on the downthrown side is mainly by mud flow and to a smaller extent by slumping. The fault throw increases from 5–10 m near the surface to 70–80 m at depth.

18.8. Strike-slip faults

Strike-slip faults are those in which the displacement of the blocks is approximately horizontal. In a *dextral* or *right-lateral* or *right-hand* strike-slip fault the relative movement of the block on the far side is towards the right. In a *sinistral* or *left-lateral* or *left-hand* strike-slip fault the relative movement of the far-side block is towards the left (Fig. 18.5). In areas of strike-slip faulting the mean stress axis is usually sub-vertical; consequently, most strike-slip faults are very steep. A *transform fault* (Wilson 1965) is a special type of strike-slip fault. The terms *transcurrent fault* and *wrench fault* are sometimes used to distinguish those strike-slip faults which are not transform faults (Freund 1974, p. 94).

Unlike thrust faults and normal faults, the strike-slip faults do not usually cause a major change in the topography. Active strike-slip faults are, however, often recognized from earthquake-induced offsets of roads, railways, fences and other landmarks. Small strike slips on active faults are also recognized by analysis of seismic data. In the major strike-slip faults such small displacements are compounded over a long period of time to give very large magnitudes of horizontal movement. Thus, according to Kennedy (1946), the Strontian and the Foyers granites which formed at one time a single massif are now separated by a distance of 105 km by horizontal movement along the Great Glen Fault of Scotland. Horizontal movements of several hundreds of kilometres took place along the Alpine Fault of New Zealand (Wellman 1955) and the San Andreas Fault of California.

The *maximum* displacement that can take place along an ordinary strike-slip fault cannot be unlimited. For example, displacement–length ratio of the Marlborough Fault of New Zealand is less than 10 per cent and the ratio for the strike-slip faults in Sistan, south-east Iran, is 13 per cent. It has been suggested that the displacement on an ordinary strike-slip fault is unlikely to exceed one-third the length of the fault (Freund 1970a, b, 1974).

Strike-slip faults may terminate in zones of ductile deformation. Alternatively, near the termination the fault may branch out along several splays along which the total displacement is distributed through a number of smaller displacements. In some cases, to avoid incompatibility with the boundary of the faulted area, the terminations of the faults may bend by ductile deformation (Freund 1974).

A perfectly planar strike-slip fault causes neither an extension nor a shortening; consequently, there is neither a subsidence nor an uplift. However, large strike-slip faults are often curved or are made up of zones of *en echelon* faults connected by short fault segments; movement on such fault systems can cause extension or shortening in the neighbourhood of the bent region of the fault line. Whether an extension or a shortening will occur will depend on whether the fault is right-lateral or left-lateral and whether, on proceeding along a fault line, we have to step towards right or towards left to find the continuity of the fault. A right-lateral right-stepping fault or a left-lateral left-stepping fault will produce an extension in the bent zone (*releasing bend*, Crowell 1974) of the fault (Fig. 18.35a, b). Normal faults may then develop in that zone. As a result a wedge-shaped or rhomb-shaped area near the bend of the fault subsides and forms a small trough described as a *sagpond* or a large-scale *pull-apart basin* (Aydin & Nur 1982). On the other hand, a right-lateral movement on a left-stepping fault or a left-lateral movement on a right-stepping fault (Fig. 18.35c, d) causes a shortening and consequently an uplift of a wedge-shaped area in the bent zone of the fault (*restraining bend*, Crowell 1974). Folds or thrust faults may develop in the shortened zone. In certain cases the uplift or *push-up* is associated with development of a *flower structure* or *palm tree structure* (Lowell 1972, Sylvester & Smith 1976, Harding & Lowell 1979), as seen on a vertical section across the fault line. This is a symmetrical or somewhat asymmetrical arrangement of convex-upward thrust faults (Fig. 18.36a) with combination of reverse and strike-slip displacements. Flower structures have been recognized in many places, for example, in California, from the San Gabriel fault zone (Wilcox *et al.* 1973), the Liebre fault zone of Ridge Basin (Crowell 1975) and the Mecca Hills (Sylvester & Smith 1976).

Pull-aparts may also be underlain by flower structures. The faults associated with these structures show a combination of normal faulting and strike-slip faulting. These are sometimes referred to as *negative flower structures* to distinguish them from reverse faulted *positive flower structures* which underlie push-up zones.

These zones of uplift and subsidence, associated with sub-vertical strike-slip faults, undergo a special type of deformation. The rocks on either side of the deformed zone have negligible vertical displacements while the deformation within the zone involves a shearing parallel to the zone combined with either a shortening or an extension across the zone. These two types of deformation are denoted, respectively, by the

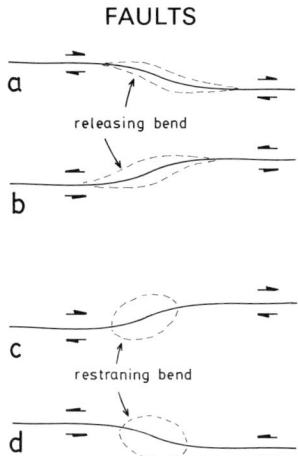

FIG. 18.35. Development of zones of extension at releasing bend at (a) right lateral, right-stepping and (b) left lateral, left-stepping faults and development of zones of compression at restraining bends in (c) right lateral left-stepping fault and (d) left lateral right-stepping strike-slip fault.

FIG. 18.36. Vertical sections across strike-slip faults may show a curving of the faults, sometimes giving rise to (a) a flower structure or (b) a tulip structure.

terms *transpression* and *transtension* (Harland 1971, Sanderson & Marchini 1984).

The flower structure shows a system of convex-upward faults. The sandbox experiments of Naylor *et al.* (1986) show that a pure strike-slip displacement in a basement can cause the development of a zone of strike-slip faults in the cover. These faults are concave upward (Fig. 18.36b) and are not associated with transpression. Such an arrangement of concave-upward strike-slip faults, described by Naylor *et al.* as *tulip structure*, occurs, for example, in the Long Beach oilfield in California and the Moray Firth of Scotland.

Strike-slip fault systems sometimes show imbricate fault arrays (Kingma 1958, Lensen 1958) and *strike-slip duplexes*. These may develop by different kinematic processes (Woodcock & Fischer 1986).

(1) Duplexing may occur at bends of strike-slip faults by propagation of the straight segments of the fault beyond the initial bend and by

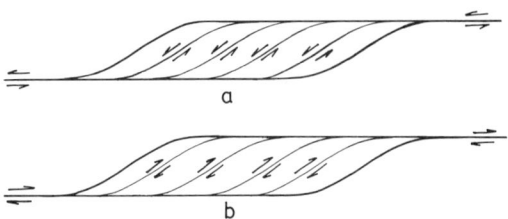

a

b

FIG. 18.37. Development of strike-slip duplexes. (a) Extensional duplex at releasing bend. (b) Contractional duplex at restraining bend.

development of new imbricate faults connecting the two straight segments (Fig. 18.37). The sense of movement of the imbricate faults is such as to cause, in a releasing bend, an extension of a line parallel to the straight segments of the fault. At a restraining bend the imbricate fault movement causes a shortening of the line. In both the cases, however, the sense of movement on the imbricates is the same as on the main fault. Thus, if the main fault is dextral, the imbricates will also be dextral.

(2) If the straight segment of the main faults show an *en echelon* pattern, two approaching fault tips may bend towards each other, connect up and enclose a lozenge-shaped area which represents a horse. A duplex will result by the sequential development of a number of such horses occurring between the overlapping straight segments of the main fault.

(3) It has been shown by Tchalenko (1970) that different sets of shear fractures develop in clay cakes subjected to simple shear. Such different sets of faults and ductile shear zones have been recognized in natural rocks also. It has been suggested that strike-slip duplexes may develop by the linking up of such different sets of shear planes.

There may be a link between the strike-slip duplexes and either the flower structure or the tulip structures. The same three dimensional array of faults may show strike-slip duplexes in plan view and either flower structures or tulip structures in transverse vertical sections (Woodcock & Fischer 1986).

In regions of strike-slip tectonics the faults are often parallel. Such a parallel group of faults is called a *fault set*. The fault sets may occur either in a narrow zone as a part of a large strike-slip system or may be distributed through large regions as in the Mojave Desert of California or the Atlas Mountains in North Africa.

In many distributed strike-slip systems the fault sets occur in separate domains; in some of the *domains of fault sets* all the faults are right-lateral and in others they are all left-lateral. The deformation of the boundary of the entire region is controlled by faulting within an array of these *conjugate domains* (Nur & Ron 1987).

An important finding of recent years is that in many domains of strike-slip fault sets the fault blocks along with the bounding faults have rotated during slipping (Freund 1970a, b, 1974, Garfunkel 1974, Ron *et al.* 1984, 1986, Nur *et al.* 1986, Nur & Ron 1987). The block rotations take place in accordance with the domino model as described in connection with normal faulting. In rotated blocks associated with strike-slip faulting the axis of rotation is sub-vertical. Hence the rotation must cause a change in the palaeomagnetic declination. Such rotated blocks have been identified in several areas from measurements of palaeomagnetic declinations.

We are essentially concerned here with ordinary strike-slip faults. A detailed description of transform faults is outside the scope of this book. It may be mentioned, however, that, although transform faults show strike-parallel displacements, they differ in several respects from ordinary strike-slip faults (Freund 1974, pp. 94–95):

(1) Unlike ordinary faults, a transform fault is located at a lithospheric plate contact.

(2) It usually shows offset of an oceanic ridge opposite to the sense of displacement.

(3) In an ordinary strike-slip fault the displacement is variable along the length of the fault and is smaller than about one-third the length. In a transform fault the displacement is unlimited and is uniform along the length of the fault.

(4) A transform fault terminates abruptly at structural features which accommodate the entire displacement. An ordinary strike-slip fault terminates in zones of ductile deformation or by distributing the displacement along splays or by bending of the fault towards the receding side.

(5) Ordinary parallel strike-slip faults usually show the same sense of displacement. On the other hand, adjacent parallel transform faults often show opposite senses of displacement.

18.9. Balanced cross-sections

The large-scale structures of folded and faulted terrains are often represented by cross-sections. The arbitrariness of structural interpretation involved in the drawing of such sections can be removed to a considerable extent by applying the principles of constructing balanced cross-sections. The method, developed initially by geologists working in the oil and gas fields of the Alberta Foothills of the Rocky Mountains (Douglas 1950, Hunt 1957, Carey 1962, Bally *et al.* 1966), has been described by Dahlstrom (1969) and Hossack (1979) and has now become a standard tool of structural geology. The method enables us to avoid gross errors in section drawing and has been widely applied to check the validity of existing sections, to determine the depth of *décollement* and to measure the minimum contraction or extension across a deformed terrain.

Apart from using the data put into the geological map and the drilling and seismic data if they are available, we should also take into account during section construction the styles of folding and faulting in the mesoscopic scale or exposed in a larger scale on the mountain slope. The large structure inferred in the section should be compatible with structures actually seen in the field.

Balanced sections are usually constructed in regions which have undergone an overall plane strain. For this purpose it is convenient to choose the line of section in a direction along which the contraction or extension of the beds is a maximum. In faulted terrains the plane of the section should include the slip direction. The section line is chosen at a right angle to the fold trend and the strike of dip-slip faults where the two are parallel. If the fold trend or the fault trace is arcuate, the section line is chosen across the central segment of the arcuation.

Balanced sections can be constructed under two situations:

(1) The length of beds in the original and the deformed sections are the same. Balancing a section under this geometrical constraint is known as *line-length balancing*. (2) The total area of the cross-section remains unchanged during deformation. A section constructed under such constraints is said to be *area balanced*.

The first step in constructing a balanced section is to choose vertical reference lines or *pin lines* at each end of the section. A pin line should be chosen at a point where there is no interbed slip, say along the axial plane of a fold or the normal to the dip of undisturbed beds. A preliminary section is then drawn by incorporating the surface and sub-surface data. The next step is to measure the bed lengths of selected horizons between the pin lines. The bed lengths can be measured with a piece of string or by bending a thin strip of graph paper along short segments of the curved bed in the section. The bed lengths between the pin lines should remain the same for different horizons. If the bed lengths do not remain the same, the preliminary section is modified accordingly.

The length-balance and the area-balance methods can be used independently. As suggested by Cooper *et al.* (1983), both methods should be used as a mutual cross-check on the admissibility of the section. The area-balance technique can, however, be used only if the true stratigraphic thickness between an upper and a lower horizon is known from an undeformed terrain. The final check of a balanced cross-section is whether or not the section can be restored to its depositional position without introducing line-length anomalies. A balanced section must be restorable and the usual practice is to prepare simultaneously both the deformed and the *restored sections*. From a series of restored sections

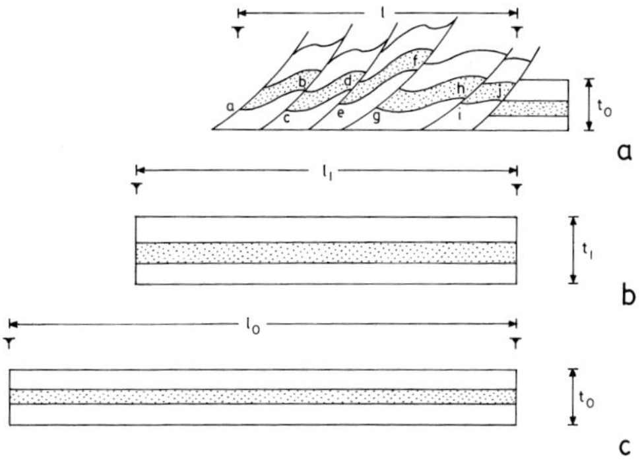

FIG. 18.38. (a) Deformed section with length l between pin lines. The original thickness t_0 is preserved at the right-hand end. (b) Partially restored section between pin lines with length l_1 and thickness t_1. The area of the section is $A = l_1 t_1$. (c) Completely restored section after area balancing, with length l_0 and thickness t_0, with $l_0 = A_0/t_0$.

it may be possible in certain circumstances to prepare a restored or *palinspastic map* (Dennison & Woodward 1963) which shows the beds in their depositional position.

Figure 18.38a is an idealized representation of a section showing a family of imbricate thrusts splaying upward from a floor thrust. The distance between the pin lines is l. The original stratigraphic thickness is known to be t_0. To restore the section the sum of the bed lengths from a to b, c to d, etc., is determined as l_1. After line-length balancing the restored section has a length l_1. Let the total area of this restored section be A_1. This area, divided by t_0, should give us the original line length l_0. In the absence of any homogeneous thickening of the beds, l_0 should be equal to l_1. However, in the present example we find that l_0 is greater than l_1. The section in Fig. 18.38b is therefore a *partially restored section*. In the fully restored section (Fig. 18.38c) the bed length is l_0 and the thickness is t_0. The per cent shortening is $100\ (l - l_0)/l_0$, of which a part is due to the development of the imbricate thrusts and a part due to homogeneous strain before imbrication. It is convenient in such cases to represent the shortenings as natural strain per cents. Thus,

$$\ln\left(\frac{l}{l_0}\right) = \ln\left(\frac{l_1}{l_0}\right) + \ln\left(\frac{l}{l_1}\right).$$

To take an actual example, Cooper *et al.* (1983, p. 148) measured the deformed length of a duplex as 61 m. After line-length balancing the partially restored section had a length of 76 m. The original length, as found by the area-balancing method, was 99.5 m. In this example

$$100 \times \ln\left(\frac{61}{99.5}\right) = 100 \ln\left(\frac{76}{99.5}\right) + 100 \ln\left(\frac{61}{76}\right)$$

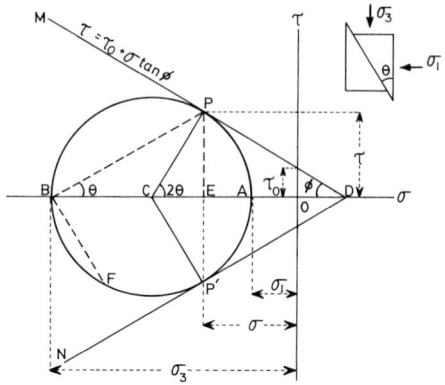

FIG. 18.39. Mohr diagram of stress at shear failure.

or,

$$-48.9 \text{ per cent} = -26.9 \text{ per cent} - 22 \text{ per cent.}$$

The total contraction of -49 per cent was achieved by -27 per cent of contraction by layer-parallel strain and -22 per cent of contraction by thrusting.

18.10. Mechanical aspects of faulting

Criteria of brittle failure

There is no rigorous theory to explain faulting. It is generally assumed that a fault develops when the shear stress acting along a plane is large enough to overcome the cohesive strength as well as the frictional resistance to movement along the fracture plane. A first approximation to the stress condition during initiation of shear fracture was proposed by the French physicist Coulomb. He proposed the following *criterion of shear failure*:

$$|\tau| = \tau_0 + \sigma \tan\varphi \tag{18.8}$$

where τ is the shear stress along the fracture plane, σ is the normal stress across the plane, $\tan\varphi$ $(= \mu_i)$ is the coefficient of internal friction and τ_0 is the cohesive strength of the material. The significance of this equation can be understood from a Mohr diagram (Fig. 18.39).

On a σ–τ diagram, let DM and DN be the two straight lines represented by eqn. (18.8). The lines are at angles of $\pm \varphi$ with the σ-axis and intersect the τ-axis at $\pm \tau_0$. A point on either of these lines specifies the stress condition at which failure should occur. Let P be one such point on DM. The Mohr circle for the condition of failure must have its centre C on the σ-axis and must touch the line DM at point P. If the *normal* to the fracture plane makes an angle θ with the σ_1-axis, then in the Mohr diagram the angle $ACP = 2\theta$ and the angle $ABP = \theta$. BF, the perpendicular to BP, is parallel to the fracture plane. The normal and shear stresses, σ and τ, on the plane are given, respectively, by OE and PE, where PE is perpendicular to the σ-axis. From the triangle DCP, we find that

$$2\theta = 90° - \varphi,$$

or,

$$\theta = 45° - \varphi/2. \tag{18.9}$$

φ is known as the angle of internal friction. It should be noted that θ is the angle between the σ_1-axis and the normal to the shear fracture. Therefore, the fracture surface itself makes an angle of θ with the σ_3-axis, the direction of maximum compressive stress. For non-zero φ, i.e. for cohesive materials, $\theta < 45°$. According to the Coulomb criterion, two sets of shear fractures should form with an angle 2θ between them; this angle is bisected by the direction of maximum compressive stress. Coulomb's criterion suggests that the initiation of shear fractures depends on two properties of the material, τ_0, the cohesive strength, and μ_i ($= \tan\varphi$), the coefficient of internal friction. For a particular combination of τ_0 and μ_i we get according to eqn. (18.8) two straight lines DM and DN on the σ-τ diagram. The angle between the two lines opens up towards the negative side of the σ-axis, since, according to our convention, compressive stress is taken as negative. Regarding the initiation of fracture, the region between the two lines is stable and the region outside these lines is unstable and the shear fractures initiate when the Mohr circle touches these lines which form the border between the stable and unstable regions.

Although the Coulomb criterion is not fully satisfactory, it is in agreement with experimental data of brittle deformation of several rock types at moderate confining pressure (Handin & Hager 1957, Handin 1969).

The lines DM and DN form a *Mohr envelope* which separates fields of stable and unstable stresses. Mohr suggested that τ and σ may not have a linear relation. He suggested that the envelope represented by the general equation $\tau = f(\sigma)$ is curved and is different for different materials. This suggestion is in agreement with triaxial compression tests of brittle failure of rocks. In a triaxial test a jacketed cylinder of rock is put under a confining pressure of a surrounding fluid contained in a pressure vessel. The cylinder of rock is then axially shortened. The axial compressive stress σ_3 gradually increases in absolute value while the confining pressure is kept constant at $\sigma_2 = \sigma_3$. A fracture develops in the brittle rock cylinder when the stress difference $(\sigma_1 - \sigma_3)$ reaches the brittle strength of the rock. The angle θ between the fracture plane and the compression axis is measured and the point of fracture on the Mohr circle is plotted. A number of such experiments are performed at different values of confining pressure.

The strength of an ideal brittle solid is about one-tenth the Young's modulus (see Price 1966, p. 29). However, the observed tensile strengths of materials are much less than this theoretical value. This led Griffith to suggest that most materials have fine flaws in the form of cracks; at a critical stress the cracks propagate and cause brittle failure. A confirmatory evidence of *Griffith's theory* is found from the fact that specially prepared flaw-free fibres have a tensile strength much closer to the theoretical strength. A simplified description of Griffith's theory (Griffith 1920) is given by Odé (1960). From this analysis we find that, at the moment of rupture, the normal and shear stresses are related by the equation

$$\tau^2 = 4T_0^2 - 4T_0\sigma \tag{18.10}$$

where T_0 is the uniaxial tensile strength. This is the equation of the Mohr envelope as predicted by Griffith's theory.

We may write eqn. (18.10) as

$$\tau^2 = -4T_0(\sigma - T_0)$$

or, say, as

$$Y^2 = -4T_0X$$

where $Y = \tau$, $X = \sigma - T_0$.

This is the standard equation of a parabola in which the focus is to the left of the directrix (Fig. 18.40). Although the parabolic form of the Mohr envelope agrees with some experimental data, there is often some discrepancy for fractures caused by compression. It has been argued that in the compressive stress field the Griffith cracks will tend to close. Using this concept, McClintock & Walsh (1962) have proposed a modified Griffith theory according to which τ and σ are linearly related:

$$|\tau| = 2T_0 + \mu_s\sigma, \tag{18.11}$$

where μ_s is the coefficient of sliding friction. This equation is similar to the Coulomb criterion if we replace τ_0 by $2T_0$ and the coefficient of internal friction μ_i by the coefficient of sliding friction μ_s. This modified Griffith theory agrees fairly well with the experimental data for fractures in the compressive stress field.

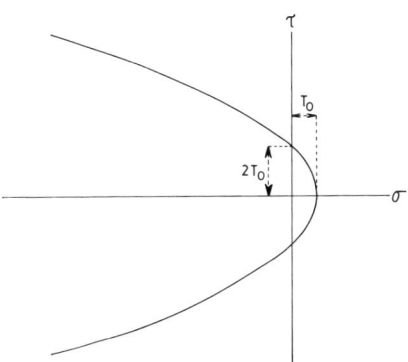

Fig. 18.40. Parabolic form of Mohr envelope.

Anderson's theory of faulting

According to the Coulomb and Mohr theories, one or two sets of shear fracture should form parallel to the intermediate stress axis σ_2 and inclined to the compressive stress axis σ_3 at an angle of less than 45°. Laboratory testing of rock specimens indicate that the coefficients of internal friction generally lie between 0.5 and 1.0 (Jaeger & Cook 1979, p. 155). These values correspond to an angle of 22–32° between the shear fracture and the compressive stress axis. A value of 30° is often considered to be a good overall approximation. Thus, if the orientations of the stress axes remain more or less uniform over an area, the faults should develop in well-defined sets. Occurrence of faults in such parallel sets is common in many parts of the world. It is because of this that a simple dynamic analysis of faults of an area is possible.

With a few simple assumptions, Anderson (1951) has proposed a theory which explains the dynamics of a large number of faults in the shallow crust. Anderson argues that, since there cannot be any shear stress on the free surface of the earth, one of the principal stress axes should be vertical; consequently, the other two stress axes would be horizontal. We get three types of faults depending on whether the tensile stress axis (σ_1), the compressive stress axis (σ_3) or the intermediate stress axis (σ_2) is vertical.

The rocks at depth are under a lithostatic or uniform pressure. In the case when additional stresses are superimposed on this standard state, Anderson considers three possible combinations. (1) There may be an increase of pressure in all horizontal directions. (2) There may be a relief of pressure along all horizontal directions. (3) There may be an increase of pressure along a horizontal direction with a relief of pressure in the horizontal direction at a right angle to it.

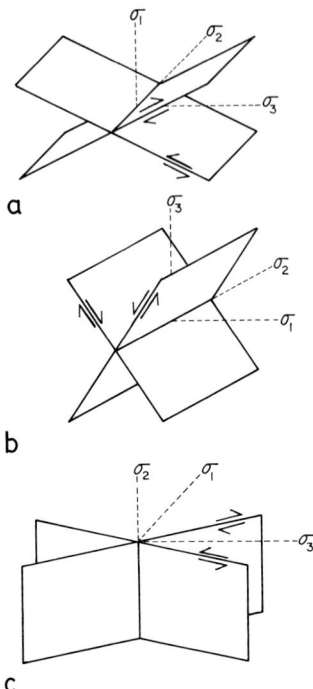

Fɪɢ. 18.41. The orientations of the three principal classes of faults with respect to the orientations of the axes of principal stresses. Note that σ_1 is tensile and σ_3 is compressive. (a) Reverse fault, (b) normal fault, (c) strike-slip fault.

(1) *Reverse faults.* In the first case, in which σ_1 is vertical, it is more probable that the stresses σ_2 and σ_3 in the two horizontal directions will be unequal. The shear fractures will strike parallel to σ_2 and will be inclined to the horizontal σ_3-axis. Moreover, the movement on the fractures will be such that the upper block will move up the slope in a direction perpendicular to σ_2. Thus, thrust faults dipping at an angle of about 20–30° will be produced (Fig. 18.41a).

(2) *Normal faults.* In the second case, the compressive stress σ_3, due to gravity, will be vertical and hence the fault will be at an angle of less than 45° with the vertical. Thus, the fault will strike parallel to the σ_2-axis and will dip at angle of 60°–70°. The upper block will move down along the dip direction and will give rise to dip-slip normal faults (Fig. 18.41b).

(3) *Strike-slip faults.* Since in the third case the σ_2-axis is vertical, the faults will also be vertical. Again, since the movement along the shear fracture will be at a right angle to the σ_2-axis, the faults will be strike-slip faults (Fig. 18.41c).

470

The general conclusion of Anderson's theory is that, among three types of faults in the shallow crust, the reverse faults will be low dipping ($< 45°$), the normal faults will be moderately dipping ($> 45°$) and strike-slip faults will be vertical. Anderson's theory also gives us a mechanical basis of classifying the faults. Anderson's theory is in broad agreement with observations on faulting at shallow crustal levels. Sax (1946, referred to by Hubbert 1951) measured the angle of dip of a large number of normal and reverse faults from the Coal Measures of the Netherlands. Frequency curves for 5°-intervals of dip of each class of faults showed sharp peaks of 63° for normal faults and 22° for reverse faults. Similar observations have been made by other workers from different parts of the world. In Hubbert's (1951) sandbox experiments of faulting, the normal faults had an average dip of 61° and the reverse faults had an average dip of 25°.

Hafner's analysis

Although Anderson's theory explains the common occurrence of low-dipping thrust faults, moderately steep normal faults and sub-vertical strike-slip faults, it is inadequate to explain the fault geometry of many regions. The curved shape of fault surfaces, the occurrence of high-angle reverse faults and the occurrence of either very low-dipping or very steep-dipping normal faults are unexplained by Anderson's theory. Hafner's (1951) analysis explains these features. It shows also that different fault types can form in the same region during a single deformation event.

For specific geologic situations, Hafner assumes a geologically realistic stress condition at the boundary of a region. Starting with these boundary conditions, and with some simplifying assumptions, Hafner derives the orientations of the principal stress axes at different points and then plots the orientations of the potential shear fracture surfaces at an angle of 30° with principal compressive stress. In some thrust-faulted terrains, as in the Rocky Mountain foothills, the intensity of deformation decreases from the interior towards the foreland. This suggests that the horizontal compressive stress is larger in the interior part than in the external part of the mountain. For the sake of simplicity let us consider only the two-dimensional case and take a vertical section with rectangular boundary (Fig. 18.42a). At the bottom of this block there will be vertical stresses to balance the gravitational stresses. The horizontal compressive stresses on the vertical face on the left-hand side (interior) must be greater than on the right-hand vertical face. However, these normal stresses alone cannot keep the rectangular block in static equilibrium. To bring the system into equilibrium we must have shear stresses along the bottom and the vertical faces.

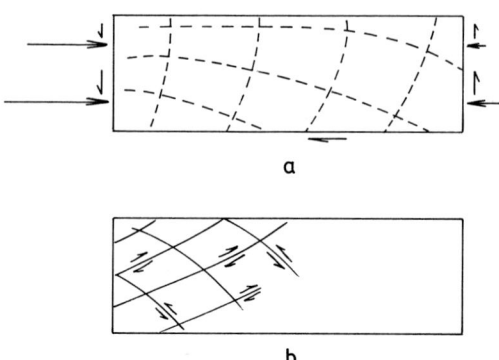

a

b

FIG. 18.42. (a) Schematic representation of stress trajectories inside a rectangular block subjected to horizontal compression from the left-hand side. (b) Curved pattern of faults expected in (a). After Hafner 1951.

Evidently, there cannot be any shear stress on the free upper face of the block. It is also noteworthy that the shear stresses in the vertical and horizontal directions must be equal at each point. On the vertical faces the shear stress is zero at the surface, increases towards the bottom and is equal to the horizontal shear stress at the lower left and lower right corners. A qualitative idea about the orientation of the stress trajectory in the interior of the block can be obtained from this combination of normal and shear stresses at the boundary (Hubbert 1951, p. 370). Since there is no shear stress at the top surface, one of the stress trajectories must end vertically against the top surface; the other stress trajectory will be tangential to the top surface as it reaches this surface. The tilts of each of the stress trajectories will change from surface to depth and as a result the faults, if they develop at all, will be curved (Fig. 18.42b). Faulting in several other geological situations was analysed by Hafner by using the same technique.

Faulting in sedimentary cover in response to vertical movement in basement

Hafner's work was followed by several other investigations with the use of similar mathematical techniques (Sanford 1959, Couples 1977, Gangi *et al.* 1977, Withjack 1979). These analytical solutions can be tested by sandbox experiments in the laboratory (Sanford 1959, Stearns *et al.* 1978, 1981). The analytical solutions are valid only up to the moment of fracture. They are not concerned with displacements along the faults; the sandbox experiments are more informative in this respect. The analytical solutions along with the experiments enable us

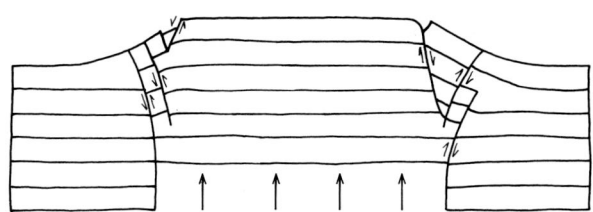

Fig. 18.43. Faults produced in sandbox by upward vertical movement of a piston. After Sanford 1959.

to explain the orientations and the types of faults in a number of terrains.

The investigations by Hafner, Sanford and others are particularly relevant for the development of faults in sediments in response to movement of the basement. The problem is of considerable importance. Thus, for example, high-angle reverse faults (Noakes 1957, Hills 1959, 1963, pp. 194–195) in certain areas are suspected to develop due to basement uplift or introduction of magma at the base of supracrustal rocks.

Figure 18.43 illustrates one of the models in Sanford's experiments in which the sand was uplifted by a vertical movement of a piston from below. The movement of the piston simulates the uplift of a basement underneath a pile of sediments. With the upward displacement of the piston, steep and upward convex reverse faults developed in the overlying sand. As the reverse faults flattened upward they caused a local horizontal shortening. This was compensated by development of normal faults elsewhere. Such theoretical and experimental studies make us realize that the two-fold classification of faults into contraction faults and extension faults (McClay 1981) is inadequate, since it takes into account only the horizontal component of the movement. There is evidently the need of a finer grouping which can distinguish between faults in which vertical changes are either greater or less than the horizontal changes (Stearns *et al.* 1981).

Faulting in simple shear tectonics

Cloos (1955) and Riedel (1929) investigated the development of shear fractures in thin slabs of wet clay subjected to a shearing motion in the boundaries. The experiments produced sets of *en echelon* shear fractures at an angle to the shearing boundaries. The development of such *en echelon* shear fractures or faults is now well established by several experimental and field studies (Morgenstern & Tchalenko 1967, Tchalenko 1968, 1970, Mandl *et al.* 1977, Bartlett *et al.* 1981, Mandl 1988). The different sets of shear fractures resulting from simple shear

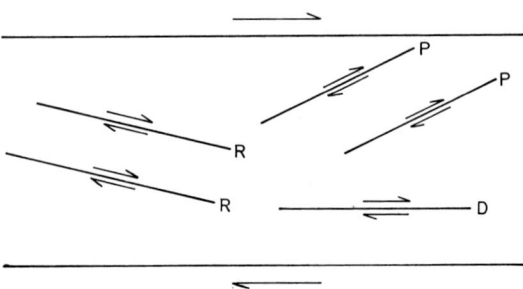

FIG. 18.44. Orientation of R, P and D shears in sheared clay.

are illustrated in Fig. 18.44. Among these the *R shears* or *Riedel shears* form at a low angle to the shearing boundaries. The acute angle between the R shears and the shear zone wall points opposite to the sense of shearing in the boundary (Fig. 18.29b). The sense of shear on the Riedel shears is the same as that of the simple shear in the boundary. The Riedel shears may, therefore, be called synthetic. Another set of *en echelon* synthetic shears, the *P shears*, also develops at a low angle to the shearing boundaries. The acute angle between the P shears and the shear zone boundary points in the same direction as the sense of shearing motion in the boundary. It has been suggested by Mandl (1988, p. 78) that there is a local compression of the material between two overlapping R shears. This causes the local σ_3-axis to rotate to a closer alignment with the R shears. As a result, new shear fractures develop. Depending on the local orientation of the σ_3-axis, the new shear fractures may appear as P shears, or as low-angle R shears or as shears sub-parallel to the shear zone walls. This last type of synthetic shear fracture is known as *D shear* (Tchalenko 1968) or *Y shear* (Logan *et al.* 1979). In addition, there may be antithetic shears R′ and P′ conjugate with R and P. But these are of relative minor importance. In clay the R, P and D shears may merge with one another to form an anastomosing network which isolates lenses of undeformed materials. These are called shear lenses. If the anastomosing network of shears is concentrated in a narrow zone, as in the case of certain strike-slip faults, the merging shear surfaces may give rise to a through-going master fault parallel to the external shearing motion.

Mechanical paradox of large overthrusts

The large overthrusts such as those found in Scotland, the Alps, the Himalayas, the Scandinavian Caledonides, the Appalachians, the Canadian Rockies and the Peruvian Andes, show horizontal displacements of several kilometres or several tens of kilometres. It was pointed out by Smoluchowski (1909) that movement of such large thrust sheets presents a mechanical paradox. Consider a rectangular block of length b, breadth a and thickness c, resting on a horizontal plane. To move this block by a horizontal force at one of its sides ac, the force required is

$$F = (abc)w\,\mu_s \qquad (18.12)$$

where w is the weight per unit volume of the rock and μ_s is the coefficient of sliding friction at the base of the block. Smoluchowski took the length of the block as 100 miles and μ_s as 0.15. The stress at the side ac is $F/ac = wb\,\mu_s$. This is equal to the weight of a column of rocks of height $b\mu_s$. Smoluchowski found that to withstand the stresses the strength of the rock would be such that it can support a column 15 miles high. The crushing strength of granite, on the other hand, can support a column about 2 miles high (Hubbert & Rubey 1959, p. 122). Smoluchowski concluded that it is impossible to move a slab 100 miles long by a push from the rear.

These calculations were modified by Hubbert & Rubey (1959). By using Coulomb's criterion, they showed that the maximum length of a block which can be pushed from the rear is

$$x_1 = \frac{A}{\rho g\,\tan\varphi} + \frac{B}{2\,\tan\varphi}Z_1 \qquad (18.13)$$

where x_1 is the maximum length, Z_1 is the thickness of the block, $\tan\varphi$ ($= \mu_s$) is the coefficient of sliding friction, ρ is the density of the rock and A and B are given by the following equations:

$$A = 2\tau_0\sqrt{B},$$

$$B = \frac{1 + \sin\varphi}{1 - \sin\varphi}.$$

Hubbert & Rubey took $\tan\varphi = 0.577$, $\tau_0 = 2 \times 10^8$ dynes/cm^2 and $\rho = 2.31$ g/cm^3. With these values the maximum length of the block is found as

$$x_1 = (5.4 + 2.6Z_1)\ \text{km}. \qquad (18.14)$$

The equation shows that a thrust block 1 km thick can have a maximum length of 8 km, while a block 5 km thick can have a maximum length of 18.4 km. These calculations show that, under the conditions assumed, it is impossible to push a thrust block of length 30 km or more over a horizontal surface. Hubbert & Rubey found that the paradox cannot be resolved merely by assuming that the block is pushed down an inclined plane.

To resolve the paradox of overthrust faulting, Hubbert & Rubey (1959) proposed the theory that, in the presence of a high-pore water pressure, there may be a translation of thrust blocks of lengths comparable to those observed in the field.

Rocks often contain fluid-filled pores and cracks. The pressure within an interconnected system of pores, known as pore-fluid pressure, greatly influences the brittle behaviour of a rock. For a fluid-filled rock the behaviour is essentially controlled by the *effective stress*:

$$\sigma'_{11} = \sigma_{11} - p, \ \sigma'_{22} = \sigma_{22} - p, \ \sigma'_{33} = \sigma_{33} - p$$
$$\tau'_{12} = \tau_{12}, \qquad \tau'_{23} = \tau_{23}, \qquad \tau'_{31} = \tau_{31} \qquad (18.15)$$

where p is the pore pressure and σ'_{ij} is the effective stress. The equations show that in a fluid-filled rock the normal stresses are reduced by the amount of pore pressure while the shear stresses remain unchanged. This concept of effective stress, introduced in soil mechanics by Terzaghi in 1923, is of great importance in resolving the paradox of overthrusting. If we consider the effective stresses rather than the actual stresses, Coulomb's criterion has to be replaced by the following equation:

$$|\tau| = \tau_0 + (\sigma - p) \tan\varphi.$$

Evidently, if the pore pressure is large, the fracture can develop at a comparatively low value of shear stress. It is convenient to express the pore pressure in terms of the lithostatic or geostatic pressure P:

$$p = \lambda P$$

where λ may range from 0 to 1. Hubbert & Rubey have shown that the pore water pressure in many oil wells is abnormally high and the value of λ may range between 0.8 and 0.95.

If the pore pressure is taken into account, the maximum length of a horizontal block that can be pushed from the rear is

$$x_1 = \frac{1}{1 - \lambda_1} \left[\frac{A}{\rho g \tan\varphi} + \frac{B + (1 - B)\lambda_1}{2 \tan\varphi} \cdot Z_1 \right]. \qquad (18.16)$$

Taking the values of ρ, g and φ as in the case of eqn. (18.14), we have

$$x_1 = \frac{1}{1 - \lambda_1} [5.4 + (2.6 - 1.73 \lambda_1)Z_1]$$

(Hubbert & Rubey 1959, eqns. 96, 96a), where λ_1 is the value of λ at the base of the block. If Z_1, the thickness of the block, is 5 km and $\lambda_1 = 0.8$, the maximum length of the block will be 57 km. For the same thickness, x_1 is 106 km if $\lambda_1 = 0.9$. Evidently, the moving of a thrust block will be easier if, in the presence of a high fluid pressure, there is a combination of gravitational sliding and a push from the rear.

The model proposed by Hubbert & Rubey is not the only model of overthrusting. There are many regions in which thrusting has taken place without the occurrence of abnormal pore pressures. According to Hubbert & Rubey, the resistance to thrust movement is caused by the sliding friction. Alternatively, there may be a viscous sliding along the thrust surfaces. It is likely that the process of sliding along a thrust surface is a complex one. This problem has been discussed in detail by Elliott (1976a). Elliott has further suggested that a topographic surface slope is an essential factor in the development of large-scale thrusts. Thrusts always move in the direction of surface slope. A detailed analysis of this problem has been given in Elliott (1976b). His analysis suggests that compressive stresses resulting from horizontal pushing from the rear has much less influence on the motion of thrust sheets than the gravitational component down the surface slope.

CHAPTER 19

Joints

19.1. Introduction

Joints are defined as fracture surfaces along or across which the movement is negligibly small. This definition evidently depends on the scale of observation. A joint may not show a displacement in the mesoscopic scale, but may show evidence of displacement in the microscopic scale. A freshly exposed joint appears as a hairline crack. As weathering progresses, its opening may increase. Some joints are filled with minerals. If the mineral fill is very thin (say, a few millimetres in thickness), it is still regarded as a joint when observed in the outcrop scale. However, when the same mineral fill is observed in a thin section, it is described as a vein.

In spite of such a lack of precision in definition, most joints are easily recognized in the field. Joints often occur parallel to one another. An array of parallel joints constitutes a *joint set*. Planar, parallel joints are also described as *systematic joints*. Joints may range from the shortest to many tens of metres in length. On one end of this scale are the *microjoints*, which can be seen only under the microscope. On the other end we have the *master joints*, which can be traced through several metres or tens of metres.

The structures described so far, e.g. folds, faults, cleavages, etc., may not be present in every terrain or in every outcrop of a terrain. Joints, however, are ubiquitous and occur in all kinds of rocks. Yet, till now, the study of joints does not hold an important position in structural geology. Joints form at different stages of the tectonic evolution of an area. They form in sediments which have not been lithified, they develop in buried rocks which are otherwise undeformed, they may form at different stages of formation of a system of folds, and a large number of joints form after the close of the tectonic cycle and during a slow uplift of the rocks. It is difficult, and often impossible, to distinguish among the different generations of joints when they occur together, and therefore it is often difficult to relate the joint pattern with the overall scheme of tectonic evolution of an area.

19.2. Joints in relation to stresses

Joints are brittle fractures which may develop either by tensile failure or by shear failure. If the rock is porous its behaviour will depend upon both the total stress and the fluid pressure or pore pressure. If the total stress and the pore pressure are known, the *effective stress*, i.e the stress that is effective in controlling the behaviour, can be calculated. For isotropic rocks the principal effective stresses are $\sigma_1 - \alpha p$, $\sigma_2 - \alpha p$ and $\sigma_3 - \alpha p$, where p is the pore pressure and α is a constant which is generally taken as 1 (Paterson 1978, p. 72). In the following discussion the stress at brittle failure and all stresses in general are effective stresses, although no special symbols are used for them.

Experiments on deformation of rocks (see reviews by Price 1966, Paterson 1978, Jaeger & Cook 1979) have shown that brittle fractures in isotropic rocks are symmetrically oriented with respect to the effective principal stresses. The fractures are either parallel to the principal compressive stress (σ_3) or occur at an angle of less than 45° with it. When a set of conjugate fractures develop, the axis of principal compressive stress bisects the acute angle between the fractures. The angle between the brittle fractures and the principal stresses is dependent on the absolute value of the stress difference $(\sigma_1 - \sigma_3)$ relative to the tensile strength (T) of the rock.

The relation between shear stress and normal stress at the time of fracturing is often represented by the following equation:

$$\tau^2 = 4T^2 - 4T\sigma \qquad (19.1)$$

(Murrell 1958, Paterson 1978, p. 59), where τ and σ are the shear and normal stresses at the time of failure and T is the tensile strength of the rock. According to the convention followed in this book, σ is considered positive when it is a tensile stress. When the compressive stress is taken as positive, the equation becomes $\tau^2 = 4T^2 + 4T\sigma$. Equation (19.1) represents a parabolic curve (Fig. 19.1) with σ as abscissa and τ as ordinate. It represents the *Mohr envelope*. Brittle fracture develops when the stress condition is such that the Mohr circle touches the Mohr envelope. The assumption of a parabolic Mohr envelope agrees fairly well with the results of experimental rock deformation.

Let the stresses in the Mohr diagram be represented in terms of the tensile strength T. The parabolic Mohr envelope (Fig. 19.1) has then the following characters. Its axis is parallel to the σ-axis of the Mohr diagram. Its vertex is at a point $\sigma = T$, $\tau = 0$. It intersects the τ-axis at two points, $\tau = \pm 2T$. The radius of curvature at the vertex $(T,0)$, as determined from eqn. (19.1), is $2T$.

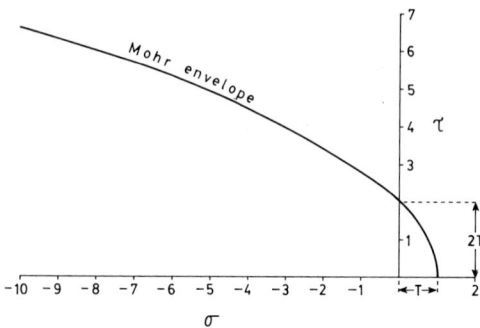

FIG. 19.1. Parabolic Mohr envelope of stress. The scale is in terms of the tensile strength T.

Since the radius of curvature of the Mohr envelope at the vertex is $2T$, the Mohr circle which touches the Mohr envelope at its vertex has a diameter of $4T$ (Fig. 19.2a), with the centre at $(-T,0)$, and with $\sigma_1 = T$ and $\sigma_3 = -3T$. The angle 2θ for this Mohr circle is 0. The joints which develop under this condition are *extension joints* normal to the direction of the principal tensile stress σ_1. Such extension joints also develop as long as the diameter of the Mohr circle $(\sigma_1 - \sigma_3) \leqslant 4T$.

According to the Griffith theory of brittle cracks, if $3\sigma_1 + \sigma_3 < 0$, the criterion of failure becomes

$$(\sigma_1 - \sigma_3)^2 + 8T(\sigma_1 + \sigma_3) = 0 \qquad (19.2)$$

where T is the tensile strength.

Hancock (1985, Hancock *et al.* 1987) has distinguished different types of fractures which form under this condition:

(1) $\sigma_1 \leqslant 0$, $\sigma_3 < 0$. This is a case of either uniaxial or biaxial compression in the $\sigma_1\sigma_3$-plane. If $\sigma_1 = 0$, we find from eqn. (19.2) that $\sigma_3 = -8T$. The Mohr circle (Fig. 19.2b) which touches the Mohr envelope has then a diameter $8T$, with the centre situated at $(-4T,0)$. In general, for all cases in which $(\sigma_1 - \sigma_3) \geqslant 0$, the dihedral angle between the conjugate shear fractures is $2\theta = 60°$. Joints which form under this condition are designated as *shear joints*.

(2) Conjugate joints, with $2\theta < 60°$, have been interpreted as *hybrid joints* (Hancock *et al.* 1987). Among these we may distinguish three cases: (a) The normal stress σ on the joint is zero (Fig. 19.3a). We then find from eqn. (19.1) that $\tau = 2T$. We further find that the slope of the Mohr envelope at this point $(0,2\tau)$ is $d\tau/d\sigma = -1$. Thus the normal to the curve at this point makes an angle of $45°$ with the σ_1-axis. This is also the value of the dihedral angle 2θ. The centre of the Mohr circle is situated at $\sigma = -2T$. From eqn.(5.55) we find that the diameter of the

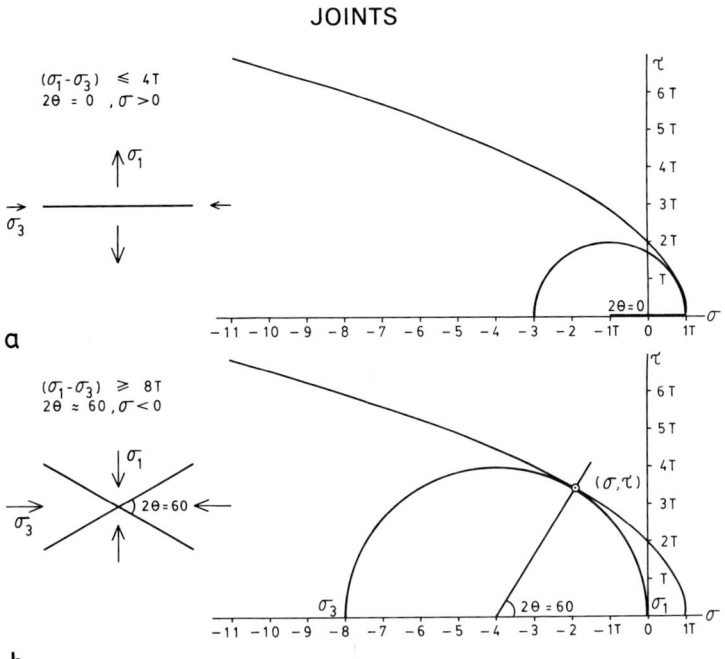

$(\sigma_1 - \sigma_3) \leqslant 4T$
$2\theta = 0$, $\sigma > 0$

$(\sigma_1 - \sigma_3) \geqslant 8T$
$2\theta \approx 60$, $\sigma < 0$

a

b

FIG. 19.2. (a) Mohr circle touching the Mohr envelope at its vertex. The Mohr circle, under the condition of fracture, is for $\sigma_1 = T$, $\sigma_3 = -3T$. Since $\theta = 0$, there is a single tension fracture perpendicular to σ_1. (b) Mohr circle touching the Mohr envelope for $\sigma_1 = 0$, $\sigma_3 = -8T$. Conjugate shear fractures develop at angle of $\theta = 30°$ with the σ_3-axis.

Mohr circle is $(\sigma_1 - \sigma_3) = 5.7T$. (b) If $(\sigma_1 - \sigma_3)$ is less than $5.7T$ but is greater than $4T$, the normal stress σ on the joints will be tensile (Fig. 19.3b). The dihedral angle 2θ will be less than $45°$. These hybrid joints may be called oblique extension fractures or extensional shear fractures (Dennis 1972, pp. 288, 291–295), because there is a dilation across the joints, as well as a shear along them. (c) If $(\sigma_1 - \sigma_3)$ is more than $5.7T$ but is less than $8T$, the 2θ value of the conjugate hybrid joints will be greater than $45°$ and less than $60°$. The normal stress on them will be compressive (Fig. 19.3c).

19.3. Geometrical relation with folds and faults

Joints often show systematic geometric relations with folds. The different geometric relations are sometimes expressed in terms of three mutually perpendicular *tectonic axes a, b* and *c* (Sander 1930), with the *b*-axis parallel to the fold axis and the *c*-axis normal to the bedding. Thus, both *a*- and *b*-axes are parallel to the bedding. The orientation of the *b*-axis remains constant, but the orientations of the *a*- and *c*-axes

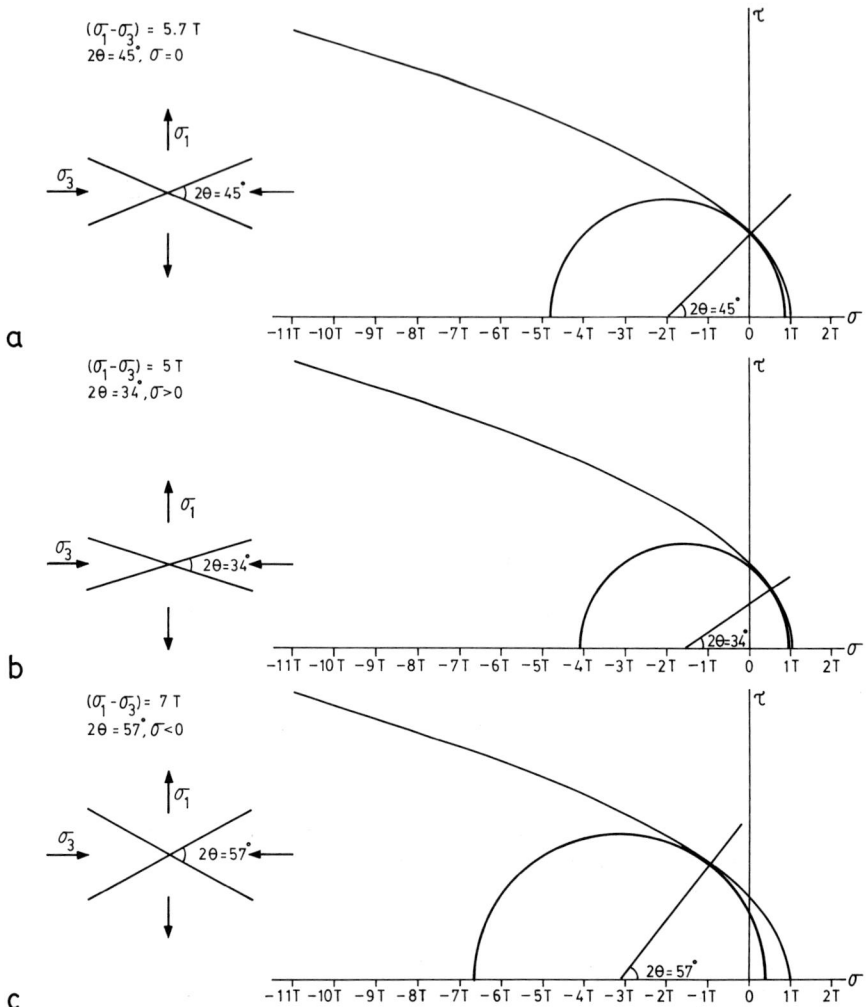

FIG. 19.3. Stress condition for development of conjugate (hybrid) joints for $2\theta < 60°$. Mohr circle touching the Mohr envelope for (a) normal stress $\sigma = 0$, $\sigma_1 - \sigma_3 = 5.7T$, (b) $\sigma > 0$, $\sigma_1 - \sigma_3 = 5T$ and (c) $\sigma < 0$, $\sigma_1 - \sigma_3 = 7T$.

change in different parts of the fold (Fig. 19.4). Joints often develop normal to the fold axis, especially when there is a prominent mineral lineation parallel to it. These are described as *ac-joints* (Fig. 19.5a) or as *cross-joints*. Joints parallel to the axial plane of a fold are described as *bc-joints* or as *longitudinal joints* (Fig. 19.5b). *h0l-joints* (Fig. 19.5c) are conjugate joints intersecting along the fold axis (the *b*-axis) and are symmetrically oriented with respect to the axial plane. The symbol 0 in the middle indicates that these are parallel to the *b*-axis. *hk0-joints*

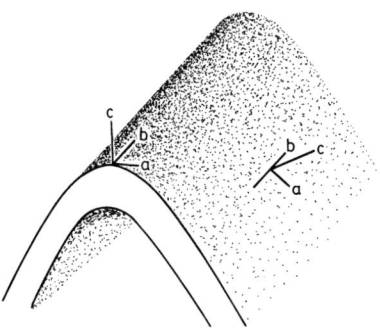

FIG. 19.4. Tectonic axes *a, b, c* over a fold. *b* is everywhere parallel to the fold axis. *a*, perpendicular to *b* and lying on the folded surface, has different orientations at different parts of the fold.

(Fig. 19.5d) are conjugate joints intersecting along a line which is perpendicular to the fold axis and lies parallel to the axial plane. The symbol 0 at the end indicates that these are parallel to the *c*-axis at the hinge. In a similar terminology, 0*kl* indicates that these are parallel to the *a*-axis at the hinge zone. The *radial joints* (Fig. 19.5f), occurring in folded competent layers, are perpendicular to the local orientation of the layer and are parallel to the fold axis.

The *ac*-joints and the radial joints are generally interpreted as extension joints; the *bc*-joints also are probably extension joints, while coeval conjugate joints symmetrically oriented with respect to the fold are likely to be either shear joints or hybrid joints.

Although joints can develop in unconsolidated sediments, they may not survive later compaction and burial. In the same way, if joints do develop at some early stage of deformation, they cannot survive later effects of metamorphism and ductile deformation. Hence, joints in folded rocks must have formed much later than the folding event, when the rocks have passed from the ductile to the brittle field. Why, then, do we get such a systematic relation between the orientations of folds and joints? There is, as yet, no clear answer to this question. Price (1966, p. 130) has suggested that rocks may behave as a Bingham substance which retains residual stress long after the main phase of folding is over. The residual stresses faithfully retain the directions of stresses when the rock had undergone folding. In that case, the orientations of the joints may have a systematic relationship with the orientation of the folds.

Joints of different types may develop during faulting. Among these the *feather joints* (Cloos 1932) or *pinnate joints* are the most important. These are extension joints which develop close to a fault plane. These occur roughly at an angle of 45° with the fault plane. Since feather

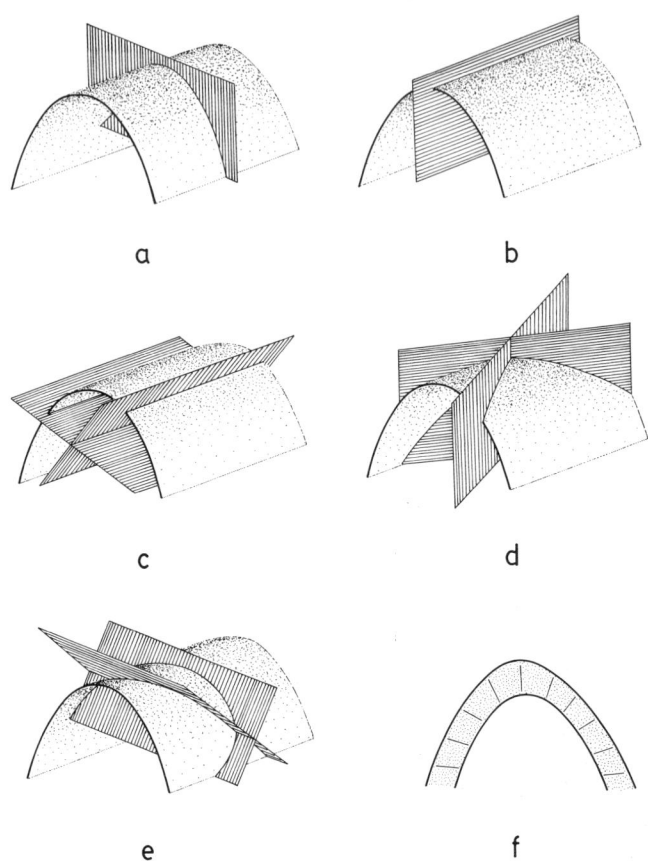

FIG. 19.5. Geometrical relations of joints with folds. (a) *ac*-joint or cross-joint, (b) *bc*-joint or longitudinal joint, (c) conjugate *h0l* joints, (d) conjugate *hk0* joints, (e) *0kl* joints, (f) radial joints.

joints are extension fractures associated with faulting, the sense of movement of the fault can be determined from the joints. The acute angle between the fault plane and the feather joints is sympathetic to the sense of fault movement (Fig. 19.6).

19.4. Surface features

Joint surfaces sometimes have characteristic surface markings (Parker 1942, Hodgson 1961, Roberts 1961, Bahat & Engelder 1984, Engelder 1985). These are generally of two types, *hackle marks* and *rib marks*. Hackle marks are faint ridges on the joint surface. The most common type of hackle mark is the *plume structure*. This is a feather-like marking on the joint surface with a central

JOINTS

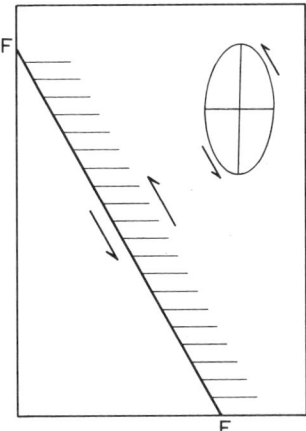

Fɪɢ. 19.6. Feather joints or pinnate joints in the neighbourhood of a fault.

axis from which the rays or barbs branch out into either side. In
some joints, a set of the smaller *en echelon* joints develop at the fringe
of the plume structure. These *fringe joints* or *f-joints* develop at a low
angle to the main joint plane (Fig. 19.7). The plume structures may
be of two types (Bahat & Engelder 1984, Engelder 1985), i.e. the *S-
type plume* (Fig. 19.8a), which runs straight and has an axis parallel
to the bedding, and the *C-type plume* (Fig. 19.8b), with curved pattern which has either a curved axis or shows a fan-like rhythmic
pattern (Fig. 19.8c). Rib marks are circular or concentric ribs on the
joint surface.

Surface marks similar to hackle marks and rib marks also occur on
fracture surfaces of various substances. In the laboratory experiments
the hackle marks generally form during explosive fracture while the rib
marks commonly form on slow-moving fractures (Murgatroyd 1942).
The C-type plumes, with their fan-like rhythmic patterns, develop
during the cyclic propagation of a rupture rather than during a single
explosive fracture (Secor 1969).

There is no general agreement about whether plume structures
develop on extension joints or on shear joints. Parker (1942) showed
that plume structures are common on shear joints while they are rarely
found on extension joints. Roberts (1961) also showed that plume
structures are common on conjugate shear joints while they are absent
on extension joints. Syme-Gash (1971) also suggests plume formation
on shear joints. On the other hand, it has been suggested by several
workers that the plume formation indicates extension failure (Bahat
1979, Engelder 1982, Bahat & Engelder 1984). Plume structures enable
us to determine the direction of propagation of the fractures. The

485

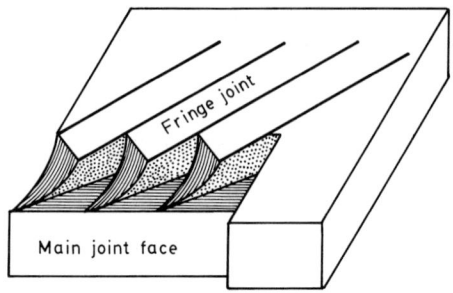

FIG. 19.7. Fringe joints.

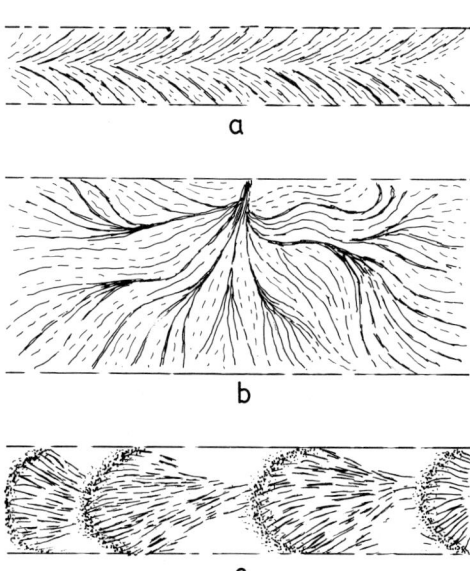

FIG. 19.8. Three different types of plume structure on joint surfaces. (a) S-type, (b) C-type and (c) rhythmic fan-like plumes.

direction of propagation is along the plume axis and opposite to the direction in which the barbs converge.

19.5. Distinction between extension and shear joints

There are several criteria of distinguishing between extension joints and shear joints (see Hancock 1985, pp. 446–448, for a detailed discussion about these criteria):

(1) Joints perpendicular to the fold axis or to a stretching lineation can usually be identified as extension joints, while conjugate joints

symmetrically oriented with respect to folds or lineations can be regarded as shear or hybrid joints.

(2) In certain areas we get only two sets of joints orthogonal to each other. These are generally regarded as extension joints. It has been suggested that these joints developed in different stages in response to a two-dimensional extension, somewhat similar to the mode of development of chocolate tablet boudins (e.g. Bock 1980, Dunne & North 1990, Hancock 1985, Hancock et al. 1987, Ramsay & Huber 1987, p. 664, Stauffer & Gendzill 1987).

The most commonly used criterion for determining the relative ages of joint sets is the abutting relationship. Younger joints abut older joints. The second criterion is the cutting of surface marking. A younger joint cuts the marking of an older joint. By using such relationships, Hancock et al. (1987) and Bahat (1988) found that among orthogonal joints (one set called cross-fold joints and the other set called strike joints), some of the cross-fold joints are younger than the strike joints, while some are older in the same outcrop. This is compatible with what we found in cases of chocolate tablet boudinage (Chapter 17); in a stress field with both σ_1 and σ_3 as tensile, the two sets of extension fractures are not synchronous but one set develops after the other or the two sets develop in alternating stages (cf. Ghosh 1988).

(3) Joints may show displacements in the microscopic scale. If a joint is irregular, the joint walls may be matched by a movement the direction of which may be normal to, at an acute angle to or nearly parallel to the joint walls (Fig. 19.9). This will enable us to decide whether it is an extension joint, an oblique extension joint (hybrid joint) or a shear joint.

(4) Some joints pass strikewise into dilational veins (Hancock 1987). These are evidently extension joints.

(5) As mentioned earlier, feather joints or pinnate joints near faults develop as extension fractures. Similarly, it may be possible to identify conjugate sets of shear joints in the neighbourhood of a fault if their orientations are compatible with the orientations of the principal axes of stress as deduced from the nature of faulting (Turner & Hancock 1990).

19.6. Relationship between joint spacing and bed thickness

Joints are sometimes confined to a particular lithology. This effect has been called joint containment (Engelder 1985). For such joints there is often a clear relation between joint spacing and bed thickness. Harris et al. (1960) showed that the spacing between joints increases

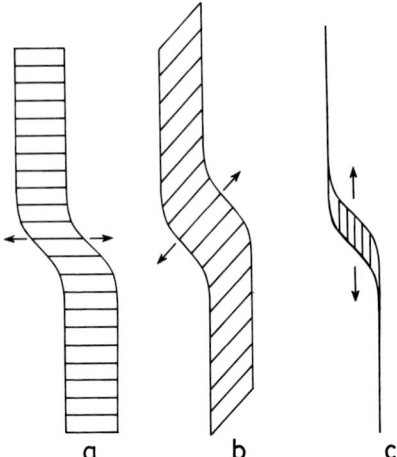

FIG. 19.9. Matching of curved joint walls in different directions of movement in three types of joints: (a) extension joint, (b) hybrid or oblique extension joint, (c) shear joint.

with the bed thickness. Thus, for example, he found in one locality that the average joint spacing is 10 ft in a 10 ft thick dolomite. In a 1 ft thick bed from the same locality the spacing is 1 ft. Similar relations have been observed by several workers (e.g. Bogdanov 1947, Novikova 1947 and Kirollova 1949, referred to by Price 1966; Hobbs 1967, Sowers 1973). Most authors believe that the joint spacing is proportional to the bed thickness. According to Ladeira & Price (1981), however, this relationship is valid for relatively thin competent beds. Very thick massive competent beds may have closely spaced fractures and the fracture spacing is then independent of bed thickness. According to these authors, the spacing between two joints in competent beds is also related to the thickness of adjacent incompetent beds.

19.7. Orientation analysis

Joints in an area have a large range in orientations. This is partly because of the presence of several sets of joints and partly because of the dispersion within each set. The initial objective of the orientation analysis of joints is to make a distinction among the different sets. Although the angle between joint traces on an outcrop face enables us to distinguish between different sets of joints, it should be remembered that differently oriented joints may have sub-parallel traces (Fig. 19.10). Hence, when the plane of a joint is not exposed for direct measurement of both its dip and strike, the attitudes of its traces on any two differently oriented outcrop faces should be measured and the dip and strike should be determined by stereographic projection

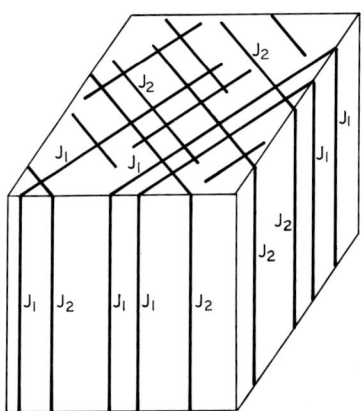

FIG. 19.10. Different sets of joints having parallel traces on some outcrop faces.

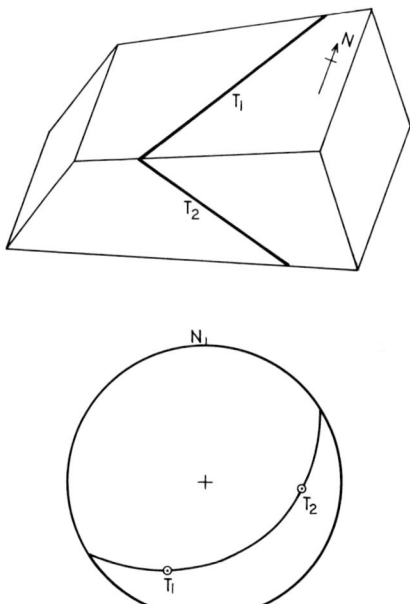

FIG. 19.11. Determination of orientation of a joint by plotting in stereographic projection the attitudes of its traces on two outcrop faces.

(Fig. 19.11). The orientations of joints should no doubt be plotted on a map. This enables us to see whether some joints are prominent in certain localities and are weak or absent elsewhere. We can also see from such a map the gradual change in orientation, if any, and the prominence of different joint sets in different sub-areas. Mapping of the joints also gives us an idea about their relation with the regional folds, faults or boundaries of igneous intrusions.

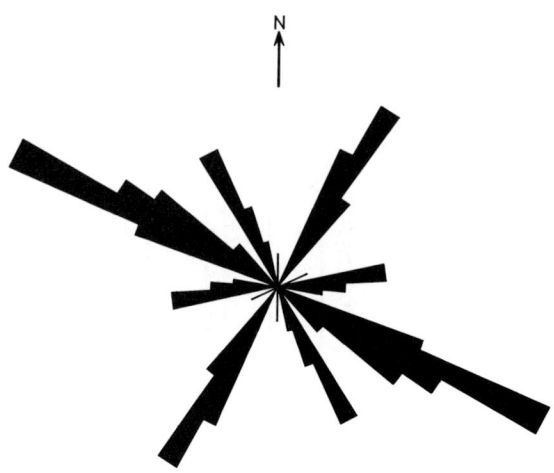

FIG. 19.12. Rose diagram of strike of joints.

The most useful method of handling a large number of orientation data is to plot the joint poles in equal area projection. It is preferable to plot the data separately for different sub-areas and then to combine the data of adjoining sub-areas which have similar patterns of pole distribution. It may not be possible to distinguish all the joint sets from those diagrams. However, the most prominent joint sets can usually be identified from the concentration of the poles. If the area shows uniformly dipping beds or is weakly folded, the pole to the general bedding may be plotted in the joint diagram to see if there is any clear geometrical relation between the orientation of the bedding and that of a joint set. In a folded area we may plot the regional fold axis and the axial plane in the projection diagram to see whether any joint set is perpendicular to the fold axis or to find out whether two sets are symmetrically oriented with respect to the axial plane.

The strike frequency of joints is sometimes represented by a *rose diagram*. To prepare such a diagram the number of joints within a certain interval of orientation, say 5° interval, is determined for a sub-area. The percentage of this number out of the total number of measurements is then calculated. Say, for example, that we have measured the strikes of 225 joints in a sub-area. Out of these, 20 joints have a strike ranging between 030 and 035. The percentage is then 8.9 for joints within this strike range. After calculating the percentage for each 5° interval of the compass directions, a suitable scale is chosen to represent the percentage value by the length of radius of a circle. The rose diagram gives an immediate visual impression of the strike frequency (Fig. 19.12).

JOINTS

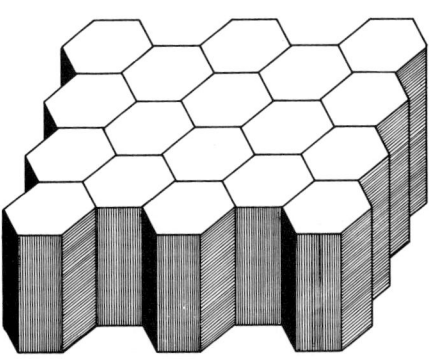

Fig. 19.13. Columnar joints.

The distinction among joint sets cannot be done solely on the basis of a joint map or a projection diagram or a rose diagram. The best method is to identify a joint set in the field itself, from its angular relations with other joints and other structures. In certain cases, different joint sets form in different rocks and have different characteristics, e.g. spacing, straightness, abundance of surface markings, etc. (e.g. Parker 1942, Bahat 1988). A record of such observations will make it easier to distinguish the different joint sets in the projection diagrams.

It is generally believed that, under the action of uniformly oriented principal stresses, we get either a single set or a conjugate set of joints. As mentioned in section 19.2, the orientation of the joints with reference to the directions of principal stresses depends on the absolute value of the stress difference $(\sigma_1 - \sigma_3)$ and the tensile strength (T) of the rock. If the value of $(\sigma_1 - \sigma_3)$ varied with time, from more than $8T$ to less than $4T$ (or vice versa), it may happen that, instead of clearly separable sets, a whole spectrum of joints develops within an angular range of about 60°. Such a continuum of coaxial joints has been described as a *joint spectrum* (Hancock 1986, Hancock *et al.* 1987) and contains extension joints, shear joints and hybrid joints. According to Hancock, partial joint spectra enclosing a maximum angle of 45° are not uncommon.

19.8. Joints in relation to the tectonic cycle

Joints can develop at different stages of the tectonic cycle. It is likely, however, that the majority of joints that are recognized in the field developed after the main tectonic event and during the uplift and unroofing of the rocks (Price 1966). The early formed joints may be involved in later slip, vein filling or development of stylolites. In such cases it may be difficult to recognize their initial mode of origin as joints. It is possible, however, to distinguish between the early and the late joints in weakly folded terrains. Thus, for example, Bahat (1988) has described orthogonal joints from the Lower Eocene chalks near Beer Sheva, Israel. He has shown that, among these, the single-layer joints developed early and during the time of progressive burial of the sediments while the multi-layer joints, having similar orientations, developed after a weak folding and during the uplift and denudation. The multi-layer joints, cutting many layers, have larger openings and irregular spacings which are considerably longer than those of the single-layer joints.

Engelder (1985) has made a detailed analysis of joint development during different stages of the tectonic cycle in a sedimentary basin. From the Catskill Delta in the Appalachian Plateau in the U.S.A., Engelder has shown that extension joints developed during different stages of the tectonic cycle, i.e. burial, lithification, deformation and denudation. He has distinguished four types of joints in this region, i.e. *tectonic joints, hydraulic joints, unloading joints* and *release joints*. The tectonic and hydraulic joints developed at depth during the early stages of the tectonic cycle, while the unloading and release joints formed much later, at shallow depths, during erosion and uplift. Both the tectonic and hydraulic joints formed under the influence of abnormal pore pressures at depth. In the case of tectonic joints the high pore pressures developed in response to tectonic compaction. Tectonic joints can develop even at depths of less than 3 km. On the other hand, the hydraulic joints developed in response to compaction by over-burden loading. These can form only if the depth of burial is larger than 3 km. Both the unloading joints and release joints developed in response to removal of overburden during erosion. The orientation of release joints is controlled by the fabric of the rocks, such as a solution cleavage. In a folded terrain the release joints are generally parallel to the axial planes. Their mode of development is similar to that of the extension fractures which form in the triaxial test specimens when the load is removed. In the Catskill Delta the unloading joints developed after more than half the overburden had been removed. The unloading joints develop in a similar manner as the release joints. However, they are not parallel to a fabric in the rocks and many of them are unrelated to the fold geometry. Their orientation is controlled either by the stress field prevailing at the time of denudation or by the residual stresses of a previous tectonic event.

19.9. Joints in igneous rocks

Primary joints generally develop during the last stage of emplace-ment of an igneous intrusion, when its outer part has already solidified but the partially liquid inner part is still moving. The methods of analysis of such joints and other primary structures have been described in detail by Hans Cloos and co-workers and have been summarized by Balk (1937). The prominent primary lineation in the igneous body is generally described as the flow line. This is considered to be parallel to the local direction of flow of the magma. The primary cross-joints are perpendicular to the flow lines and are extension joints. The diagonal joints are shear fractures which develop at an angle of 45° or more with the flow lines. The longitudinal joints are sub-vertical joints parallel to the flow lines. As noted by Price (1966), most of these fractures are filled with aplite or other vein materials and should not be called joints in the strict sense.

Columnar joints are the most common type of joints in basaltic flows. These form by shrinkage of the lava flow during its cooling. In a horizontal sheet of lava flow the joints appear as vertical fractures approximately at an angle of 120° with one another and dissect the rock into vertical columns with hexagonal cross-sections (Fig. 19.13). This pattern is similar to that of mud cracks which form by shrinkage of a sheet of mud during drying.

CHAPTER 20

Unconformity and Basement–Cover Relation

20.1. Unconformity

An unconformity is a break in the stratigraphic sequence, it records a major time gap and it is caused by erosion or non-deposition. The recognition of an unconformity is of considerable importance in structural geology. It serves as an important marker surface during geological mapping, records vertical movements leading to emergence, may give us the age of folding or of vertical movements and in many cases enables us to distinguish between different tectonic units. The age of an unconformity lies somewhere between that of the youngest rocks of the lower unit and that of the oldest rock of the upper.

In most cases a mappable unconformity represents an erosion surface. A common sequence of geological events is the folding of a sequence of sedimentary rocks, the development of an erosional surface which truncates the folded beds, subsidence and deposition of sediments to cover the surface of erosion. The resulting unconformity is an *angular unconformity* (Fig. 20.1). After its formation the unconformity surface, along with the strata lying above and below it, may also be folded. The upper group of strata is then deformed by a single generation of folds while the lower strata are affected by two periods of folding. The angular unconformity helps us to distinguish between the older and the younger tectonic events. Moreover, because of the presence of the angular unconformity, the geometry of the folds produced by the second deformation will be different in the rocks lying above and below the unconformity. Thus, for example, in the western part of India the older Aravalli Group and the younger Delhi Group of metasediments are separated by an angular unconformity. The earliest deformation in the Delhi Group has produced a series of recumbent folds. Although this deformation has also affected the older rocks of the Aravalli Group, the resulting folds are not recumbent. This difference in the fold geometry is explained by the different initial attitudes of the beds above and below the angular unconformity (Naha *et al.* 1984).

A *nonconformity* is an erosional unconformity above massive plutonic rocks (Fig. 20.2b). A *disconformity* (Fig. 20.2c) is a type of unconformity in which the beds above and below it are parallel (Grabau 1905). In many cases the two units above and below the disconformity are separated by an undulatory erosion surface. This erosion surface is very often a more or less flat surface of peneplanation. Sometimes, however, the erosional surface of unconformity shows a conspicuous relief. In certain cases a sequence of strata may be overlain by a set of parallel strata of much younger age and without a surface of erosion separating the two sets of strata. The unconformity may be difficult to recognize in such cases. Thus, limestone strata may lie disconformably on other strata of limestone without a clastic zone between the two sets (Pettijohn 1975). Such obscure relationships have been described by Dunbar & Rogers (1957) as *paraconformity* (Fig. 20.2d). An unconformity, whatever be its geometrical nature, must represent a major time gap or a major break in the stratigraphic sequence.

The shorter gaps are known as *diastems* (Barrell 1917). For a more detailed description of the different geometrical types of unconformity the reader should consult Dunbar & Rogers (1957).

There are three broad classes of criteria — structural, sedimentary and palaeontologic — by which the unconformities are recognized (Krumbein & Sloss 1963). The most commonly encountered structural criteria are the angular unconformity and the disconformity represented by an undulatory surface which is produced by erosion and emergence and which cuts across the bedding planes of the underlying formation. The truncation of one group of beds by another may result from both faulting and the development of unconformity. An angular unconformity can therefore be established only if the field observations rule out the possibility of occurrence of a fault along the surface of discontinuity. More rarely, an unconformity may be recognized by the abrupt termination of faults in the underlying formation at the surface of discontinuity or the termination of a dyke against it without the evidence of any thermal effects on the overlying strata. Among the sedimentary criteria, the most important is the presence of a basal conglomerate in the unit lying above the surface of discontinuity. A sub-aerial discontinuity is also indicated by the presence of residual chert and buried soil profiles. Submarine disconformities may be represented by zones of glauconite, phosphatized pebbles or by manganiferous zones. There are also several paleontologic criteria, the most important of which are the sudden changes in faunal assemblages and the presence of a significant gap in the evolutionary development of the

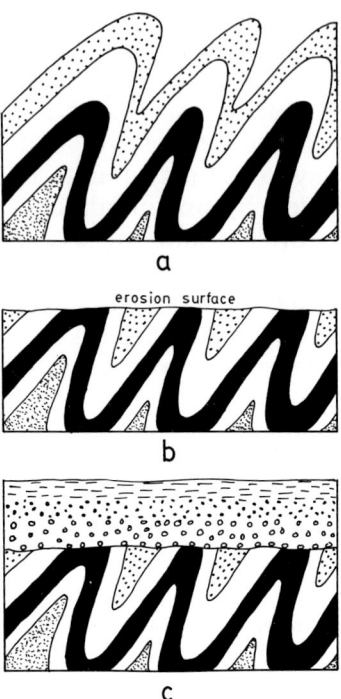

FIG. 20.1. Stages of development of an angular unconformity.

fauna. In addition to these criteria observable at an outcrop or in a vertical section, an unconformity may also be recognized from interpretation of various types of stratigraphic maps (Krumbein & Sloss 1963, chapter 12 and table 12–7). A more detailed discussion about the criteria for the recognition of unconformities is given in Krumbein (1942), Shrock (1948) and Lahee (1961).

An unconformity can be easily recognized when flat-lying, undeformed or weakly deformed sediments of a younger age overlie an older group of schists and gneisses. The older crystalline rocks in such a situation are generally described as a basement while the overlying sedimentary strata are described as a cover. Because of later metamorphism and deformation, it is often extremely difficult to determine the exact nature of the basement–cover relationship in many Precambrian terrains. Such complexities of the basement–cover relationship have been described in the next section.

An unconformity may be of local or of regional extent. There may be several local unconformities at a basin margin which become obscure in the central part of the basin. The unconformities in a basin centre are generally fewer in number and are of larger geographic extent. As

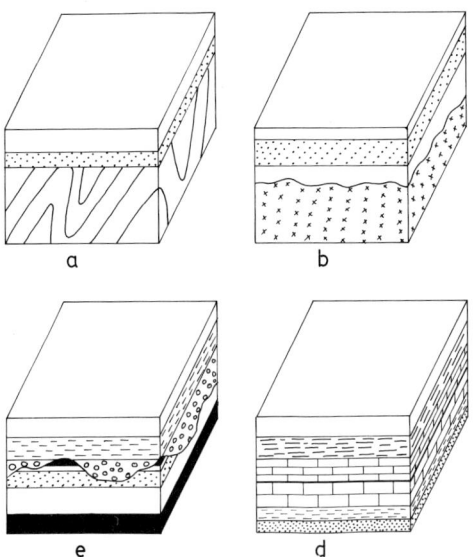

FIG. 20.2. Types of unconformity. (a) Angular conformity, (b) non-conformity, (c) disconformity, (d) paraconformity.

pointed out by Sloss (1963), there is apparently no relationship between the prominence of an unconformity and its areal extent or stratigraphic importance. A prominent angular unconformity may be local in character.

The geographic extent of an unconformity may be very large. Thus, six continent-scale unconformities have been recognized in the stratigraphic sequences of North America (Sloss *et al.* 1949, Sloss 1963). An unconformity may even record a worldwide stratigraphic event. Thus, Suess (1906) suggested that the unconformities associated with Late Cretaceous marine transgression are related with eustatic or worldwide changes in the sea level. Similarly, the basal Cambrian unconformity (Matthews & Cowie 1979) is of virtually global extent. So are the unconformities associated with the major Ordovician transgression (McKerrow 1979, Vail *et al.* 1977) and the post-Cretaceous regression (Hallam 1963). It should be noted that eustatic changes of sea level do not explain all the features associated with the major global-scale unconformities. The angular unconformities must have been associated with some tectonic activities. A more complete discussion on regional unconformities is given in Miall (1984).

20.2. Basement–cover relationship

The basement–cover relationship (Fig. 20.3) is important for stratigraphy, structural geology and also for an overall understanding

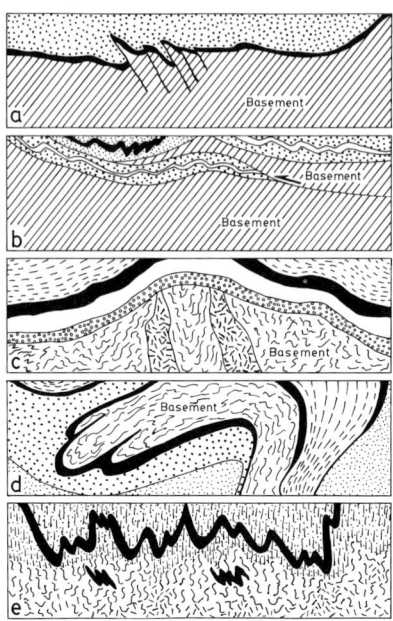

FIG. 20.3. Idealized diagrams showing diverse types of basement–cover relations. (a) Faulted basement with basement wedges. (b) Slice of basement thrust over cover rocks and forming a basement nappe. (c) Partially remobilized basement doming up the cover rocks. (d) Remobilized core of basement interfolded with cover rocks. (e) Remobilized basement. The basement–cover contact is now a migmatite front. The original contact is not recognizable. The gneissic complex lying below may contain the unreconstituted basement in certain places.

of the process of crustal evolution. A discussion on the complex problem is outside the scope of this book. The purpose of this section is to outline the main processes which modify the initial angular unconformity between a basement and its sedimentary cover and to enumerate the problems of identifying the basement in deformed and metamorphic terrains.

When a crystalline basement covered with sediments, and with a nonconformity or an angular unconformity between the two groups of rocks, is deformed in a later orogenic cycle, there may be different degrees of folding of the cover (*plis de couverture* of Argand 1922) while the basement remains more or less passive.

In the Jura Mountains, for example, the cover rocks of Mesozoic age are detached from the basement of crystalline Palaeozoic rocks and are folded independently. This process of detachment, known as *décollement*, usually takes place along extremely incompetent beds of anhydrite or salt-bearing clays (Fig. 12.21e). In spite of the occurrence of the plane of detachment or dislocation, it is not difficult in this case to have an idea of the character of the original interface between the cover and the basement. Indeed, in Jura and elsewhere in Western Europe a thin layer (or *tegument*) of the cover often remains attached to the basement while detachment has taken place on a slightly higher horizon.

The nature of the original interface can also be identified when the basement is weakly deformed to produce large warps on the interface or when it is broken up by a number of faults (Fig. 20.3a) while the cover passively adjusts itself by folding. The interface, however, becomes greatly modified in a type of deformation in which slices of the basement are thrust into the cover as basement wedges or, as in the extreme case of the Eastern Alps, a slice of the basement is dragged over a low-angle thrust (Fig. 20.3b) over rocks which form the sedimentary cover of the Western Alps. The basement along with its sedimentary cover may be deformed together in a ductile manner and may give rise to what has been described by Eskola (1949) as a mantled gneiss dome (Fig. 20.3c). A remarkable ductility of the basement is shown by the Monte Rosa type of nappes in the Western Alps and elsewhere where the basement forms the cores of huge recumbent folds (Fig. 20.3d).

In such strongly deformed terrains as in the Alps, the basement can be identified because of the presence of a good stratigraphic control so that a cover bed of known age can be identified in different regions and sometimes can be traced from a weakly deformed to a strongly deformed terrain. Identification of the basement becomes extremely difficult in many Precambrian terrains where the stratigraphic control is poor and where the cover and the basement have undergone multiple deformations and polyphase metamorphism. Many Precambrian terrains have large expanses of granite gneiss and migmatites and are bordered by schist belts of metasediments and metavolcanics. The problem in these regions is to decide from field evidence whether the gneissic terrain represents a basement or whether the gneisses are younger than the schist belt.

The occurrence of a conglomerate bed along the schist–gneiss contact no doubt strongly suggests that the interface represents an unconformity surface. However, a conglomerate may not be present in all such terrains. While identifying a surface of erosion from the presence of a conglomerate bed in such strongly deformed terrains, one should be careful to distinguish between a true sedimentary conglomerate and an *autoclastic conglomerate*. An autoclastic conglomerate is generally produced by intense deformation and fragmentation of a layer and may look very similar to a true conglomerate.

While establishing the presence of an angular unconformity in such strongly deformed terrains, one should always remember that the angular relation must be between the depositional surfaces of the two

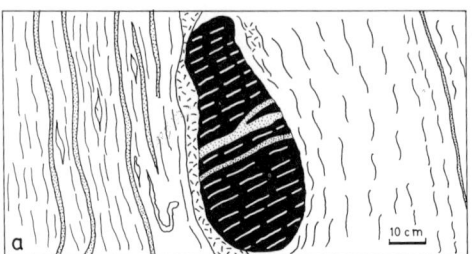

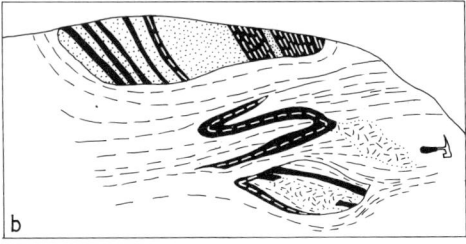

Fig. 20.4. (a) Evidence of remobilized basement with enclave showing evidence of an early deformation. The internal foliation of the enclave is discordant with the outer foliation of the quartzofeldspathic host gneiss. (b) Older enclaves showing diverse relations between internal and external foliations. The enclaves are of metamorphic rocks which are partially migmatized. Both exposures are from Schirmacher Hills, E. Antarctica. Sketched from photographs taken by Sudipta Sengupta.

groups of rocks. Thus, an angular unconformity is not established when the bedding or foliation of the schist belt or the banding and the foliation of the gneisses terminate against the schist–gneiss boundary, because this boundary does not necessarily represent a depositional surface. For the same reason an angular relation between the bedding in the schists and the foliation in gneisses does not indicate an unconformity.

In all such cases one should always make a careful analysis of the structural histories of the schist belt and the gneissic complex. An unconformity is unlikely to be present if the earliest structures of a synkinematic migmatitic complex can also be traced to the bordering schist belt. On the other hand, if we find that the gneissic complex contains a foliation or a set of folds which is earlier than all the structures of the schist belt and if there is no sign of faulting between the two groups of rocks, then it is reasonable to suggest that the schist–gneiss interface represents an unconformity.

The detailed study of some Precambrian terrains has revealed a geological history of great complexity. The nature of the complex relationship between the basement and the cover in such terrains can be understood from the following example of the Precambrians of central and southern Rajasthan in the western part of the Indian shield. Here we have a vast terrain of metamorphosed sediments and lava flows of the Aravalli Group. The Aravalli cycle of orogeny closed around 2000 to 1900 Ma. The Aravalli group of rocks is bordered by terrains of migmatites forming the Banded Gneissic Complex (B.G.C.). The B.G.C. does not have a separate structural entity and all the generations of folds and foliations in it also occur in the Aravalli Group of rocks. The folds of each generation have the same style in both the metasediments and the gneisses. Moreover, it has been demonstrated that migmatization in the B.G.C. is broadly synkinematic with the first deformation affecting the rocks of the Aravalli Group (Naha & Mohanty 1988). These relations apparently indicate that the Aravalli metasediments cannot be younger than the B.G.C. On the other hand, the B.G.C. in southern Rajasthan contains rocks as old as 3500 Ma. Moreover, in certain places a conglomerate containing gneissic pebbles occurs at the contact between the Aravallis and the B.G.C. These relations are in agreement with the conclusion that the B.G.C. represents an ancient basement on which the Aravalli sediments were unconformably deposited. The complexity of the basement–cover relationship arises from these two sets of apparently contradictory findings. All such features are, however, explained if the B.G.C. is assumed to have initially formed an ancient basement which was later remobilized and large parts of which were migmatized during the Aravalli cycle of orogeny. The structural unity of the basement and the cover resulted because, during the remobilization, the basement behaved in a ductile manner and participated in the deformation of the metasedimentary envelope (Naha & Mohanty 1988). Indeed, as Sederholm (1926) and many later workers have shown, each of the vast Precambrian gneissic "basements" did not in general develop in a single event but evolved through different phases of palingenesis. This is clearly indicated by preservation in gneissic terrains of the relics of an earlier basement as small enclaves (e.g. Sutton & Watson 1951 for the Lewisian gneiss in Scotland, Naha et al. 1991 for the Peninsular gneiss in S. India and Sengupta 1988 for the Schirmacher Hills in E. Antarctica). The foliation and axial planes of folds in these enclaves often occur at an angle to the synmigmatitic foliation of the newly formed host gneiss (Fig. 20.4). Parts of such a polycyclic gneissic basement can be younger than the supracrustals while other parts can be older (Fig. 20.3e; Naha et al., 1986, 1991). In the Precambrian gneissic complex of the Schirmacher Hills in E. Antarctica there is a series of such enclaves of an older basement deformed to different degrees during a later event of synkinematic migmatization which produced the host gneiss. In the less-deformed enclaves, the internal foliation of the gneiss is sharply discordant with the external foliation of the newly formed host gneiss (Fig. 20.4a). Other enclaves have been strongly deformed, stretched out in the form of long bands and folded along with the foliation of the host gneiss (Fig. 20.4b). The internal foliation of these enclaves has been brought to parallelism with the external foliation of the host gneiss in most parts, while the discordant relation is preserved only along short segments of the border of the enclave.

The most well-known analysis of the complex history of a gneissic basement is by Sutton & Watson (1951) on the Lewisian basement of the North-west Highlands of Scotland. These authors demonstrated that the Lewisian Gneiss is composed of an older Scourian complex and a younger Laxfordian complex. These two gneiss-forming periods were separated in time by a period of emplacement of a swarm of basic dykes. The older and the younger parts of the Lewisian basement could be separated mainly by the deformation, metamorphism and migmatization of the basic dykes. The dykes cut across the migmatitic foliation of the Scourian gneisses. During the Laxfordian phase the dykes were folded, metamorphosed and migmatized. The Scourian gneiss was also rejuvenated in Laxfordian times. Where this process was intense, the distinctive characters of the earlier gneiss were blotted out.

CHAPTER 21

Ductile Shear Zones

21.1. Types of shear zones

A ductile shear zone is a long narrow zone within which dominantly ductile deformation has caused a localization of large strain. The formation of a ductile shear zone is commonly associated with a drastic reduction of grain size and the development of a well-banded and lineated rock. In the earlier literature ductile shear zones were described simply as shear zones to distinguish them from clean-cut faults. However, the term shear zone, as recently used by Ramsay (1980a), encompasses both clean-cut faults and ductile shear zones. Ramsay classifies the shear zones into three types: (1) brittle shear zone in which tangential (wall-parallel) displacement takes place along brittle fractures and the wall rocks remain unstrained (Fig. 21.1a), (2) brittle–ductile shear zone in which tangential movement along the zone is associated with both ductile deformation and brittle fracture (Fig. 21.1b) and (3) ductile shear zone where the tangential movement is associated with ductile deformation alone (Fig. 21.1c). The most common type of brittle–ductile shear zone is a fault in which the wall rocks show drag effects. Tabular zones with *en echelon* sigmoidal tension gashes (Fig. 21.1d) produced by a bulk tangential movement have also been included by Ramsay (1980a) within the category of brittle–ductile shear zones.

Ductile shear zones may range in scale from the microscopic or grain scale to the scale of a few hundreds of kilometres in length and a few kilometres or a few tens of kilometres in width (e.g. Bak *et al.* 1975)

With respect to deformation of the walls, the ductile shear zones are of two types: (1) discrete shear zones which cut across rocks otherwise undeformed by the particular phase of deformation and (2) shear zones, one or both walls of which have undergone a penetrative deformation during the development of the shear zone itself.

21.2. Shear zone rocks

Ductile shear zones are often characterized by the presence of *mylonite*. This term was coined by Lapworth (1885) to describe the fine-grained rocks of the Moine Thrust Zone at Eriboll in the Scottish Highlands. Lapworth describes the rocks in the following way: "The most intense mechanical metamorphism occurs along the grand dislocation (thrust) planes, where the gneisses and pegmatites resting on those planes are crushed, dragged, and ground out into a finely laminated schist (Mylonite, Gr. mylon, a mill) composed of shattered fragments of the original crystals of the rock set in a cement of secondary quartz, the lamination being defined by minute inosculating lines (fluxion lines) of kaolin or chloritic material and secondary crystals of mica."

The classical idea about the process of development of mylonites has been greatly modified in recent years. It is now realized that among shear zone rocks there are all gradations from those which have yielded by cataclastic flow to those which have been deformed dominantly by crystal-plastic processes. Zeck (1974) has proposed an elaborate classification of shear zone rocks and has introduced such terms as cataclasites, hemiclasites, holoclasites, blastomylonites and myloblastites, depending upon the relative importance of cataclasis and recrystallization. However, most geologists use a simpler classification. For a detailed discussion about the classification of shear zone rocks the reader should consult Higgins (1971), Sibson (1977), Wise *et al.* (1984) and Waters & Campbell (1935).

Faulting in near surface conditions often produces *fault breccia*, *microbreccia* and *gouge*. These are initially incohesive materials but may later become coherent by precipitation of minerals by circulating solutions (e.g. silicified or calcified fault breccia). A fault breccia shows angular fragments of different sizes occurring in a pulverized ground mass. When the angular rock fragments are mostly microscopic the rock is called a microbreccia. Gouge is an extremely fine rock flour which occurs along the fault surface. A fault gouge may sometimes be foliated (Chester *et al.* 1985).

Cohesive shear zone rocks which form at a sufficient depth are generally classified into two broad groups, *cataclasites* and *mylonites*. Cataclasites are those shear zone rocks in which grain refinement has taken place by dominantly cataclastic processes and the deformation has taken place by sliding and rotation of grains. Cataclasites generally do not show a foliation. Under the microscope cataclasites show larger fragments of the original rock set in a crushed fine-grained matrix. If the rock contains 10–50 per cent matrix it may be called a *protocataclasite*. If the matrix is between 50 and 90 per cent, the rock is called a *cataclasite*. In an *ultracataclasite* the matrix is more than 90 per cent.

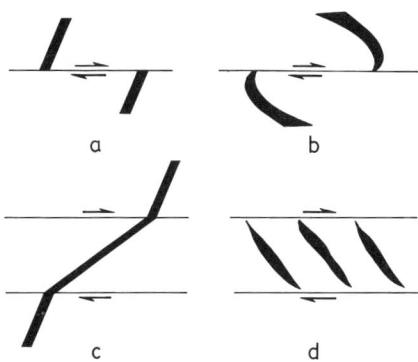

FIG. 21.1. (a) Brittle shear zone, (b) brittle – ductile shear zone, (c) ductile shear zone and (d) tabular zone of *en echelon* veins.

Although some recrystallization may take place during the formation of cataclasites, the dominant process of grain-size reduction is by cataclasis.

A *mylonite* is a foliated and lineated rock which has undergone a drastic reduction in grain size by dominantly crystal-plastic processes. The variety of mylonitic microstructures and the different stages of their evolution are best found when the parent rock is fairly coarse-grained, as in a granite or a quartzite. With the onset of deformation, a mylonite shows a mortar texture with a very fine-grained matrix surrounding large residual grains of the parent rock (Fig. 21.2a). These large grains are called *porphyroclasts* or *residual megacrysts*. In a *protomylonite* there is less than 50 per cent of the matrix. If the porphyroclasts of such rocks are distinctly lenticular, the rock may be called a lenticular protomylonite (Wise *et al.* 1984). For rocks with 50 to 90 per cent of matrix, the term *mylonite* (or *orthomylonite*) is generally used. In an *ultramylonite* the surviving megacrysts should constitute less than 10 per cent of the rock. The process of grain refinement is associated with a very large stretching along the newly formed foliation. The fine-grained product of a large grain is drawn out as a thin sheet and can often be recognized as *tails of porphyroclasts*. This is one of the reasons why mylonites often show a fine colour banding parallel to the schistosity. Because of the extreme stretching, mylonites also acquire a prominent lineation.

Shear zones sometimes contain a dark glassy rock or *pseudotachylite* derived from local melting of the rock by frictional heating. Pseudotachylite may occur either as concordant veins parallel to the shear zone walls or as discordant veins injecting the surrounding rocks. Under the microscope it shows angular rock fragments occurring in a glassy matrix. The matrix may have undergone various degrees of

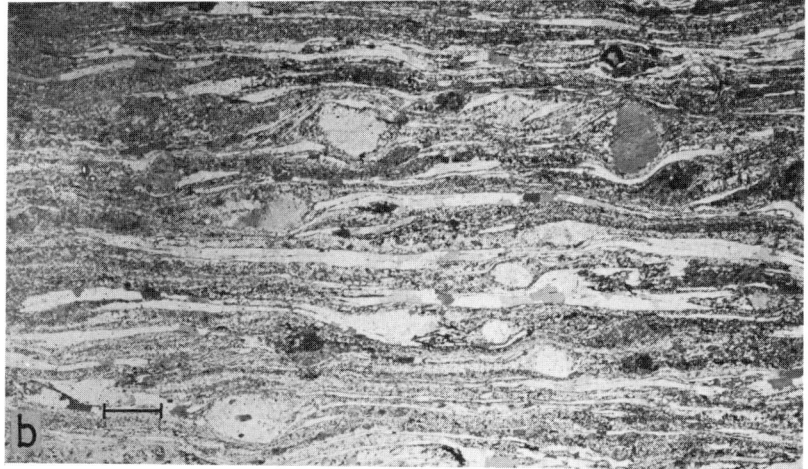

FIG. 21.2. (a) Porphyroclasts of potash feldspar with undulatory extinctions. The mylonitic foliation in the matrix sweeps around the porphyroclasts. (b) Granite mylonite with quartz ribbons, E. Antarctica. Courtesy Sudipta Sengupta. Scale bar 2.5 mm.

recrystallization but, unless it is deformed at a later stage, it does not show a preferred orientation of the recrystallized grains. The original glassy character of the matrix can be often identified by the presence of microlites, spherulites and devitrification structures.

Although the occurrence of pseudotachylite in some shear zones was known long ago (Shand 1916), a detailed study of these rocks has only been started in the last two decades (e.g. Park 1961, Ermanovics *et al.*

DUCTILE SHEAR ZONES

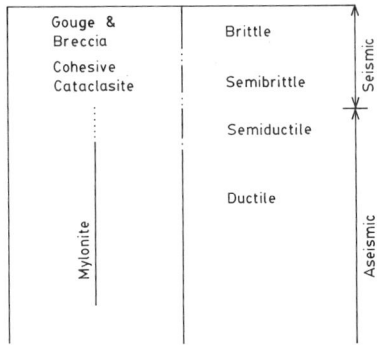

FIG. 21.3. Schematic diagram illustrating the fault zone model of Shimamoto 1989.

1972, Francis 1972, McKenzie & Brune 1972, Sibson 1975, Allen 1979, Masch 1979, Grocott 1981, Passchier 1982, Maddock 1983, Swanson 1988, Magloughlin 1989). The recent interest in pseudotachylite is derived from the fact that it is the only fault rock which unequivocally indicates the occurrence of an ancient earthquake at depth (Sibson 1975). Therefore, when an ancient shear zone is exhumed by erosion the character of pseudotachylite may give us valuable information about the process of earthquake generation at depth. Pseudotachylite is not only associated with cohesive cataclasites, it also occurs sometimes in close association with mylonites (Sibson 1980, Grocott 1981, Passchier 1982, Hobbs *et al.* 1986). Passchier (1982) has shown that, in the Saint Barthelemy Massif of the French Pyrenees, not only has pseudotachylite developed in the mylonite zones, the pseudotachylites themselves have also undergone intense ductile shearing and have been converted to ultramylonites. This close association of pseudotachylite and mylonite is of considerable geological interest, since pseudotachylite is known to develop only along brittle faults and at seismogenic depth, while mylonite was earlier believed to form by slow plastic deformation in aseismic zones. Shimamoto (1989) has suggested that there is a wide depth zone intermediate between the shallow zone of brittle deformation and the deep-lying zone of ductile deformation (Fig. 21.3). Fault gouge and fault breccia form in the shallow part of the brittle zone while cohesive cataclasites form under brittle to semibrittle conditions. According to Shimamoto, while most mylonites develop in the aseismic ductile zone, some of them form in the semiductile regime where deformation occurs by a combination of cataclastic and plastic processes (cf. Hadizadeh *et al.* 1991). From the occurrence of pseudotachylites in this zone, it is concluded that earthquakes may occur in the semi-ductile regime.

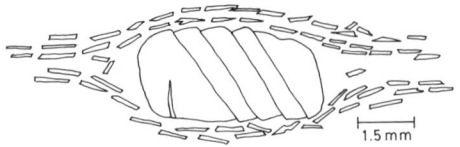

Fɪɢ. 21.4. Porphyroclast of feldspar sliced up by parallel shear fractures. The slices have undergone bookshelf gliding.

21.3. Microstructural characters of mylonites

Recent studies on mylonites by Bell & Etheridge (1973, 1976) and by White and others (White 1973a, b, 1975a, b, 1976, 1977, 1979b, White *et al.* 1980) clearly demonstrate that the dominant process of grain refinement in mylonites is by syntectonic recrystallization and neo-mineralization. The parent grains of the rock first undergo some plastic deformation (Fig. 21.5b). The highly strained peripheral parts of a grain usually form polygonal sub-grains. The sub-grains have very small optical divergence from one another. With progressive misorientation the sub-grains grade into new or recrystallized grains. The rock then acquires a core-and-mantle structure (White 1976) with a strained core of the residual megacryst rimmed by a fine-grained recrystallized mosaic. This process of recrystallization is syntectonic.

Mylonitization is associated with an extreme elongation. The quartz grains in some mylonites are deformed to very long ribbons showing strong undulatory extinctions. Such extremely elongate grains are absent in the recrystallized matrix, although lenticular outlines of the elongated parent grains can often be distinguished by trails of fine opaque minerals surrounding sub-parallel clusters of recrystallized matrix grains in quartzite mylonites.

During mylonitization of polymineralic rocks the different minerals may behave in different ways. In granite mylonites the feldspar grains are more resistant than quartz, and hence the porphyroclasts in granite mylonites are often mostly of feldspars. In some high-temperature granite mylonites the extremely long and thin ribbons of quartz may show little or no optical strain (e.g. White *et al.* 1980, Sengupta 1988). It is likely that these quartz ribbons in high-temperature granite mylonites formed by broadly syntectonic grain growth (cf. Urai & Humphreys 1981), with crystallization outlasting deformation (Fig. 21.2b).

Mylonites may show some cataclastic effects (Fig. 21.5a). In some granite mylonites, we find that feldspar has undergone grain refinement by brittle fracture (Mitra, G. 1978). Porphyroclasts of feldspars in these rocks often show quartz-filled tension fractures. Sometimes the porphyroclasts are sliced up by sub-parallel shear fractures oblique to the foliation while the slices undergo bookshelf gliding (Fig. 21.4).

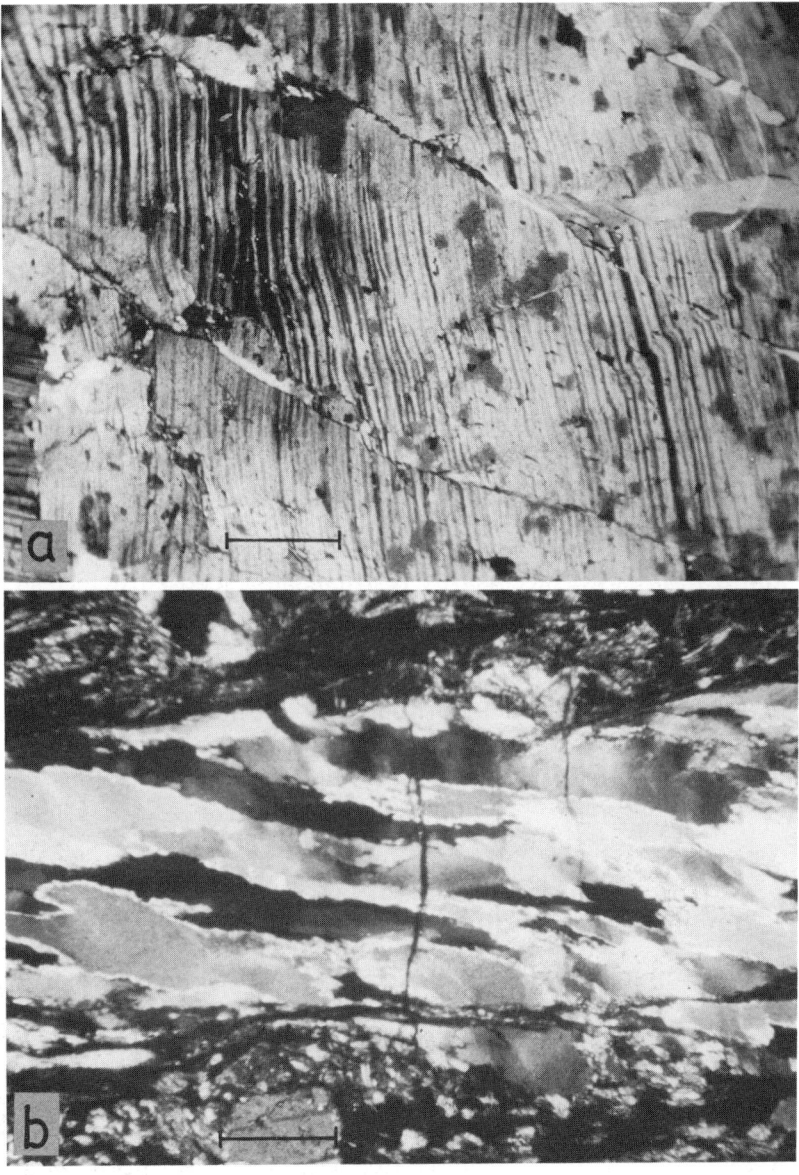

FIG. 21.5. (a) Bent twin lamellae and quartz-filled microfractures. (b) Strong plastic deformation of quartz grains in a concordant microscopic quartz vein in granitic mylonite. Tangarapallium, Andhra Pradesh, India. Scale bar 0.1 mm.

Some of the microstructural characters of mylonites enable us to determine the sense of shearing movement. These characters have been described separately in section 21.7.

21.4. Mylonitization in relation to metamorphism

The assemblage of recrystallized grains of a mylonite reflects the metamorphic conditions under which mylonitization has occurred. Thus, a garnetiferous mica schist mylonitized in the greenschist facies may show a rim of neomineralized chlorite around the core of the parent garnet grain. In some instances the flakes of chlorite in the inner part of the rim are randomly oriented, while those in the outer part show a preferred orientation sub-parallel to the surrounding mylonitic foliation. This type of texture indicates that the chloritization of garnet was broadly syntectonic with reference to the ductile shearing. The relation between metamorphism and ductile shearing was clearly brought out by Teall (1885, 1918) who showed that an unmetamorphosed basic dyke is converted to a hornblende schist when it enters a shear zone.

The majority of mylonites have formed under greenschist facies conditions. Hence, it is rather common to find evidence of retrogression in mylonites derived from higher grade metamorphic rocks. However, mylonitization is not necessarily restricted to greenschist facies conditions; similarly, not all mylonites and phyllonites (phyllite mylonite) have undergone retrogressive metamorphism. It should be noted, nevertheless, that many of the mineral transformations in mylonites, e.g. pyroxene to hornblende, hornblende to biotite or chlorite, kyanite and feldspar to muscovite, and garnet to chlorite, involve hydration. Thus, the process of mylonitization is usually associated with addition of aqueous fluid to the deforming rock system (Beach 1974, 1980).

21.5. Softening processes associated with the development of ductile shear zones

The development of a ductile shear zone involves the concentration or localization of deformation in a long narrow zone. Such a strain localization implies that it is easier to deform the rocks in the particular zone than to deform the rocks outside it. In other words, the rocks in the ductile shear zone are somehow "softened" with respect to the surrounding rocks. Several softening processes have been proposed for the development of mylonite zones (e.g. Poirier 1980, White et al. 1980, 1986).

1. Boullier & Gueguen (1975) and Behrman (1985) have suggested that some mylonites developed by superplastic deformation (see Chapter 7) which involved a change in the deformation mechanism, with a large contribution of grain-boundary sliding. Superplastic deformation is facilitated by a very fine grain size. It has been suggested that as the grain size is reduced by dynamic (syntectonic)

recrystallization, the rock is effectively softened and can undergo superplastic deformation (White *et al.* 1980).

2. A strained grain is difficult to deform. With continual recrystallization the grain becomes strain-free and can be further deformed. However, this softening mechanism is generally associated with the development of a preferred crystallographic orientation. Intracrystalline slip becomes easier as the active slip planes of the grain rotate to favourable orientations. This process of softening is known as *geometric softening* and is thought to have favoured the development of shear bands in some heavily rolled cubic metals (Dillamore *et al.* 1979). Mylonitization often involves the preferential development of phyllosilicate grains sub-parallel to the shear zone walls. This process also facilitates grain-boundary sliding. This is also a type of geometric softening.

3. Mylonitization often involves development of a new mineral assemblage. White & Knipe (1978) concluded that the relevant metamorphic reactions may produce a softening of the rock. This process is known as *reaction softening*. Thus, it is believed that during the breakdown of feldspar to white mica and quartz, the reaction products are weaker than the original grains. Evidently, a phyllonite derived from a granite is likely to have a much smaller strength than that of the parent rock (White *et al.* 1986). However, during the development of a phyllonite from a granite the softening is not only brought about by metamorphic reactions but is also greatly augmented by a drastic reduction in grain size and the development of a very prominent cleavage.

Why do ductile shear zones form at all? In other words, why instead of a uniformly distributed deformation is a large strain localized in a long narrow zone while the surrounding rocks remain less deformed or undeformed? This problem has been discussed in great detail in the metallurgical literature and to a smaller extent in the geological literature (e.g. Anand & Spitzig 1982, Bowden 1970, Bowden & Raha 1970, Cobbold 1977, Chakrabarti & Spretnak 1975, Dillamore *et al.* 1979, Jonas *et al.* 1976, Poirier *et al.* 1979, Poirier 1980, Tvergaard *et al.* 1981). The development of a ductile shear zone (or a *shear band* as it is known in metallurgy) involves its nucleation and growth. In most theories the nucleation can occur only in the presence of heterogeneities. The heterogeneity may be inherent in the material or may be acquired in the course of deformation. In some of the models the heterogeneity is in the form of a strength defect. In others it is assumed that there are local domains of strain concentration. Under certain situations, which depend on the material properties, the nature of deformation and the geometry of the deforming body, the embryonic shear zones propagate and grow into well-defined bands. Cobbold (1977, p. 2519) has suggested that the embryonic shear zones are lens-shaped, the borders of which propagate outward and give rise to a band-shaped zone of ductile shear.

21.6. Shear zone patterns

Ductile shear zones often anastomose around lenses of less-deformed country rock. These lenses may range from the microscopic scale to the scale of a map. Map-scale lenses bordered by shear zones have been reported by Coward (1976), for example, from Rhodesia and by Bell (1978a) from the Woodroffe thrust in Australia. In the meso-scopic scale, the patterns of anastomosing shear zones are best seen where the country rock has not yet been mylonitized. Such mesoscopic-scale shear zones (or shear bands), ranging in width from a few milli-metres to a few centimetres, have been reported by several authors from weakly deformed granite terrains (e.g. Berthé et al. 1979, Simpson 1983, Choukroune & Gapais 1983, Gapais et al. 1987). With progress-ive mylonitization, however, the pods of unmylonitized country rock become fewer in number. A new set of anastomosing shear zones may also form in a rock which has already been mylonitized (Bell 1978a, Ghosh & Sengupta 1987a).

From an analogy with structures produced in sheared clay and granular materials, it seems likely that different types of subsidiary shear zones or shear bands (Tchalenko 1968, 1970, Mandl et al. 1977, Harris & Cobbold 1985) may develop within ductile shear zones. Among the different types of subsidiary shear zones in clay (Fig. 18.44), the R shears or *Riedel shears* form at a low angle to the main shear zone walls, the acute angle pointing against its sense of shear. The P shears or *thrust shears* also develop at a low angle to the main shear zone walls, but the acute angle is sympathetic to the sense of shear of the main shear zone. The D shears or *principal displacement shears* develop parallel to the main shear zone walls and have the same sense of movement as the main shear zone. In addition, there may be R' and P' conjugate with R and P, but these are of minor importance. In clay the R, P and D shears form sequentially and merge with one another to form an anastomos-ing network which isolates lenses of undeformed materials. Depending on the prevalence of R, P and D shears, we may have *Riedel lenses*, *thrust lenses* and *principal displacement lenses* (Tchalenko 1968).

Within the thin shear zones which anastomose through the less-deformed country rock, the mylonitic foliation is generally parallel to the shear zone walls. The foliation in the lenses of the country rock (the shear lenses) occurs at an acute angle to it. This acute angle is always sympathetic to the sense of shear (Fig. 21.6). The geometry of the two foliations can therefore be utilized as a shear sense indicator. The

DUCTILE SHEAR ZONES

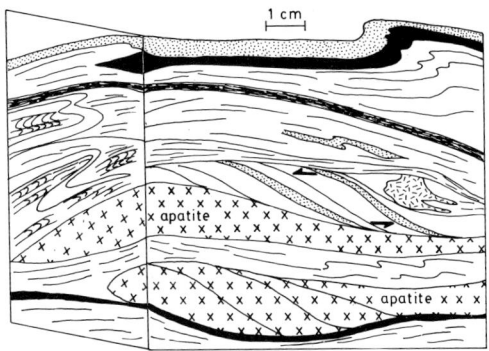

FIG. 21.6. Anastomosing small-scale shear zones separating shear lenses in apatite-bearing granite mylonite. Singhbhum Shear Zone, India.

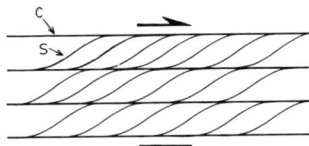

FIG. 21.7. C- and S-surfaces. The acute angle between the two indicates the sense of shear.

foliations in the two domains (in the shear zone and in the shear lens) are not really discordant. The angle between them is gradually reduced towards the borders of the shear lenses. At the contact between the two domains the shear lens foliation swerves to become parallel to the foliation of the shear zone.

The domains of the subsidiary shear zones, with their cleavage parallel to the shear zone walls, may be designated as C zones to distinguish them from the less deformed but foliated S zones which lie in between. These terms were introduced by Berthé et al. (1979) on the basis of their observations from the South Armorican Shear Zone in France. Similar structures had also been described by Ponce de Leon & Choukroune (1980) and Jegouzo (1980). The C zones in the South Armorican Shear Zone are straight and the foliation within them are parallel to the main shear zone (Fig. 21.7). According to Berthé et al., the C-surfaces are shear (cissaillement in French) surfaces while the S-surfaces in the less-deformed S zones are parallel to the local XY-plane of the strain ellipsoid. The acute angle between the C- and S-surfaces points towards the direction of shear displacement. With progressive deformation, the C-surfaces maintain a constant orientation but the S-surfaces rotate so that the angle between the two sets is reduced. As the ultramylonite stage is approached, the C- and S-surfaces become parallel, so that their separate identity is lost.

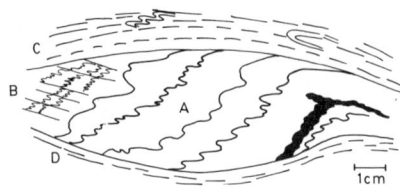

Fɪɢ. 21.8. Axial planes of crenulations in a shear lens making an acute angle with the bounding shear zones. The acute angle between the axial plane or the crenulation cleavage and the shear lens boundary points towards the sense of shear. The bounding shear zones contain remnant crenulation noses, indicating that the shear zone foliation is a transposition structure. Mylonite from Singhbhum Shear Zone, India.

The mere recognition of the C- and S-surfaces is nothing new; it is well known that in ductile shear zones the foliation makes lower and lower angles with the shear zone walls and in zones of intense shear the foliation becomes essentially parallel to the shear zone itself (Ramsay & Graham 1970). However, an important point established by Berthé *et al.* is the simultaneous development of the C- and S-surfaces. The authors showed that the two sets of surfaces are present even in the least deformed rocks, where they occur at an angle of 45° with each other. This proves beyond doubt that the deformation is non-coaxial and the acute angle between C and S points towards the sense of shear.

The shear bands or the subsidiary shear zones are not always planar as in the typical S-C tectonites described above. Nor do the foliations in the shear bands and in the shear lenses always initiate strictly at the same time. We may have three cases: (1) There may be synchronous development of the foliations in the shear bands and in the shear lenses. (2) The two foliations may develop in a single ongoing deformation, with the shear band foliation initiating at a later stage of the deformation. (3) A pre-existing foliation may be transected by the shear bands of a distinctly later deformation (Lister & Snoke 1984). The relations between the two sets of foliation may become rather complex in some of the larger shear zones where, in a single ongoing deformation, the shear bands and the mylonitic foliations have repeatedly developed, one set superimposed on the other. The mylonitic foliation of an earlier stage is then completely replaced by a new mylonitic foliation in the shear bands, while the earlier foliation is only crenulated in the domains in between the shear bands. The axial plane of the crenulations (or a crenulation cleavage forming parallel to it) makes an acute angle with the shear-band foliation and is sympathetic to the sense of shear (Fig. 21.8). This angle is gradually reduced as the high-strain domains of the shear bands are approached (Ghosh & Sengupta 1987a & b, 1990).

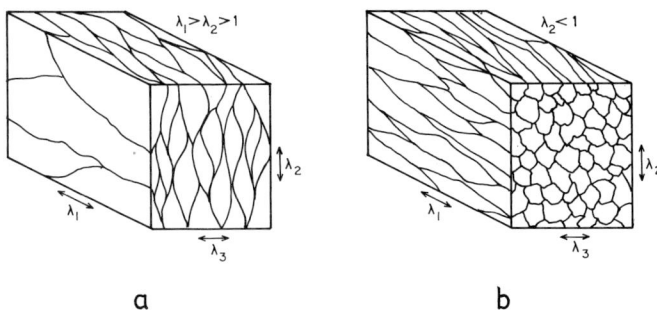

FIG. 21.9. Idealized three-dimensional shapes of shear lenses in (a) flattening type and (b) constriction type of bulk deformation. After Gapais *et al.* 1987.

Although the individual shear zones generally record a non-coaxial deformation, the deformation in the bulk scale, i.e. in the scale of a larger domain containing a number of shear zones, may be either coaxial or non-coaxial. Where the deformation history in the bulk scale is non-coaxial, the pattern of shear zones, and hence the shear lenses enclosed by them, are asymmetric. Symmetric patterns develop when the deformation history in the bulk scale is coaxial (Gapais *et al.* 1987). Gapais *et al.* have suggested that the pattern of shear zones also bears a relation to the general character of the bulk strain ellipsoid. When the bulk strain ellipsoid is of the flattening type, the shear zones curve both in the $\lambda_1 \lambda_3$ and the $\lambda_2 \lambda_3$ principal planes. In a constriction type of strain ellipsoid, the shear zones are likely to be rod-shaped (Fig. 21.9).

Towards the late stage of development of a ductile shear zone, when a prominent mylonitic foliation has already been imprinted on the rocks, a new set of shear bands often form at a low angle (about 30°) to the dominant foliation. The mylonitic foliation is dragged towards the shear bands and, as a result, becomes gently crenulated (Figs 14.18b, 21.10) (Platt 1979, 1984, Platt & Vissers 1980, White *et al.* 1980). This fabric is known as *extensional crenulation cleavage* or *ecc* (Platt 1979) or *extensional shear band* (Marcoux *et al.* 1987). Extensional crenulation cleavage may occur in a single set, or in multiple sets at low angles to one another, or in conjugate sets. The sense of shear along an *ecc* is such that it tends to cause an extension parallel to the dominant mylonitic foliation. Shear bands parallel to *ecc* have been designated as C′ by Ponce de Leon & Choukroune (1980). While the C bands of Berthé *et al.* (1979) initiate at an early stage of development of the mylonitic foliation, record a large strain and are parallel to the main shear zone walls, the C′ bands initiate at a late stage, show a small strain and occur at an angle to the main shear zone.

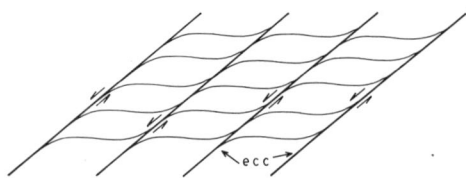

FIG. 21.10. Extensional crenulation cleavage (ecc).

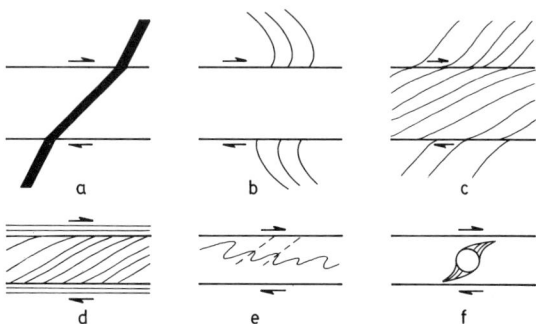

FIG. 21.11. Some shear sense indicators.

21.7. Shear sense indicators

The prominent lineation of the mylonites is a stretching lineation. Within the shear bands and the intensely deformed shear zones it is also the direction of movement. It does not, however, give us the sense of shear. There are several structures, both in the mesoscopic and the microscopic scales, by which the shear sense can be determined (e.g. Simpson & Schmid 1983).

(1) *Rotation of markers*

Where the walls of a mylonite zone can be identified, the sense of shear can be determined by the change in orientation of a dyke or a vein which enters the mylonite zone from the outside. Within the shear zone the dyke or the vein is rotated in accordance with the shear sense in the zone (Fig. 21.11a).

(2) *Rotation of foliation*

In the same type of shear zone, the shear sense can be deduced from the sense of rotation of a pre-existing foliation at the edge of the zone (Fig. 21.11b).

During the formation of a ductile shear zone a foliation may continue to develop outside it. This foliation is also rotated near the edge of the shear zone so that it makes a lower angle with the shear zone boundary (Fig. 21.11c).

(3) *C-S structures*

Where separate shear bands develop within a wide zone of mylonites the shear sense can be determined from the angle between the foliations in the shear bands and in the shear lenses or the angle between the C and S structures (Fig. 21.11d). The acute angle between the two will be sympathetic to the sense of shear.

(4) *Asymmetry of intrafolial folds*

When the foliation in the shear zone is asymmetrically folded, the sense of asymmetry is usually in agreement with the sense of shear in the mylonite zone (Fig. 21.11e). This shear criterion should, however, be used with some caution, because the hinges of the shear zone folds are often rotated in the course of progressive deformation.

(5) *Pressure shadow*

Deformed rocks, including mylonites, sometimes contain large rigid crystals with *pressure shadows* or *pressure fringes*. The pressure shadows are wedge-shaped domains that are somewhat protected from large deformation because of the presence of neighbouring rigid objects. Materials from the neighbouring regions migrate to the pressure shadow zones and crystallize there, the texture and grain size being somewhat different from those of the matrix. Such wedge-shaped pressure shadows of quartz grains growing on crystals of garnet, pyrite, magnetite and porphyroclasts of feldspar are fairly common. The pressure shadow zones are often composed of fibrous quartz. In most cases, the long direction of the wedge-shaped tails or the long axis of the quartz fibres is parallel to the direction of incremental stretching in the matrix. In other words, it tracks the direction of the incremental strain ellipsoid in the matrix. The fibrous pressure fringes enable us very well to determine the incremental strain history of the matrix. The method has been described in detail by Ramsay & Huber (1983, pp. 265–280). For the present discussion it is sufficient to note that the acute angle between the quartz fibres or the pressure shadow tails and the shear direction (parallel to the traces of the C zones, for example) is sympathetic to the sense of shear (Fig. 21.11f).

(6) *δ- and σ-types of porphyroclast tails*

Rigid porphyroblasts or porphyroclasts occurring in a ductile matrix tend to rotate when the bulk deformation is non-coaxial. The foliation in the neighbouring region is disturbed by this rotation. The pattern of the disturbed foliation around the rotating rigid objects may be described in general as *rolling structures* (Van Den Driessche & Brun 1987). The different parameters which control the development of the rolling structures have been described in Chapter 11. We shall be concerned here with certain types of rolling structures which are frequently found in mylonites and which enable us to determine the sense of shear in the mylonite zone.

The surfaces or bands that are distorted by the rigid objects may be a foliation surface, a mylonitic colour banding or *tails of porphyroclasts*. The tail of a porphyroclast is composed of fine recrystallized materials derived by dynamic recrystallization from the periphery of the porphyroclast itself. With increasing deformation, the tails stretch out into the matrix to a greater distance and the major part of a long tail becomes essentially parallel to the foliation in the matrix. Although the tails of porphyroclasts may develop in different compositional types of mylonites, they are commonly observed in the feldspar porphyroclasts of granite mylonites.

To determine the sense of shear, the observation should be made on sections that are parallel to the mylonitic lineation and perpendicular to the dominant foliation. The rigid object may be equiaxed or elongate and with an orthorhombic symmetry. It is also preferable that one of the principal axes of the object is normal to the section plane; this is the axis of rotation of the porphyroclast. On the plane of this section let us take a line passing through the centre of the porphyroclast and parallel to the microscopic shear bands. If such shear bands are not visible in the microscopic scale, we may take a line parallel to the general direction of the foliation and passing through the centre of the porphyroclast. Let this line be designated as the reference line. If a tail is present, a line running along it through the middle of the tail may be designated as the median line. Two types of asymmetric porphyroclast tails (Passchier & Simpson 1986) can be distinguished on the basis of the geometry of the median line with respect to the reference line (Fig. 21.12). In one of these types, designated as the δ-type, the median line crosses the reference line (Fig. 21.12a). In the other type, designated as the σ-type, the median line does not cross the reference line (Fig. 21.12b). In both cases there is a *stair-stepping* of the porphyroclast tails. If the tails near the porphyroclast step up to the right the sense of shear is dextral, and if the tails step up to the left the sense of shear is sinistral.

As discussed in Chapter 11, the sense of rotation of the porphyroclast depends on several factors, i.e. the degree of non-coaxiality of the deformation, the aspect ratio of the porphyroclast and the orientation of the porphyroclast with reference to the shear direction. Depending on these factors we may have three cases: (a) The porphyroclast may continue to rotate throughout the entire history of deformation and in the same sense as that of the shearing (i.e. clockwise rotation in dextral shear and anticlockwise rotation in sinistral shear). Spherical porphyroclasts and those with small aspect ratios will rotate in this manner. (b) An elongate porphyroclast may, under certain circumstances, rotate in the forward direction and then come to attain a stable

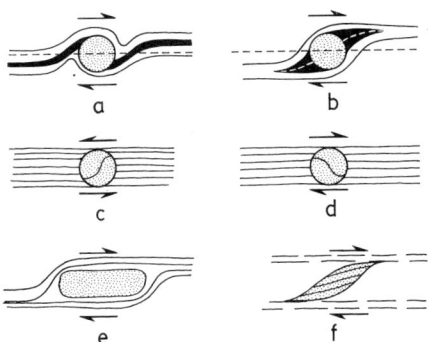

FIG. 21.12. Shear sense indicators.

orientation. (c) An elongate porphyroclast, in rare situations, may rotate backward and then attain a stable orientation. The first case is by far the commonest, and in this situation the sense of rotation of the porphyroclast gives us the sense of shearing. A clockwise rotation distorts the external foliation as well as the tails of porphyroclasts to an S-shaped asymmetric pattern and an anticlockwise rotation distorts them to a Z-shaped pattern. The development of the δ-type of porphyroclast tails can be explained by such rotations. The development of the δ-type of tail is favoured when the porphyroclast is equiaxed or has a small aspect ratio, when the finite strain is large, and when the rate of dynamic recrystallization at the periphery of the porphyroclast is small in comparison with the rate of rotation.

As mentioned earlier, the material of the porphyroclast tail is derived from the porphyroclast itself. With progressive deformation the size of the porphyroclast is reduced while the fine-grained recrystallized material from the periphery is drawn out in the direction of stretching in the matrix. A σ-type of tail develops when the rate of dynamic recrystallization is large, while the rate of rotation of the porphyroclast is small. The finite strain of the rock should also be small. A porphyroclast may undergo a small rotation if its aspect ratio on the $\lambda_1 \lambda_3$-plane is large. Hence the development of a σ-type of tail is favoured when the porphyroclast has an oblong shape (Passchier & Simpson 1986 p. 842). The rotation of the oblong porphyroclast will be further restricted if the nature of bulk deformation is such that the porphyroclast can attain a stable orientation; a σ-type of tail can develop under this condition even when the finite strain is very large (Passchier 1987).

(7) Trails of inclusions in porphyroblasts

It has been shown in Chapter 11 that the sense of rotation of a porphyroblast can be determined from the relation between its trails of inclusions (Si) and the foliation outside it (Se). This criterion can be

519

used to determine the sense of shear only if the Si and the mylonitic foliation of the matrix had developed in the course of a single ongoing deformation. The growth of the porphyroblast might have been completed during an early stage of ductile shearing. Further rotation of the porphyroblast will then cause a discordance between the Si and the Se. Paracrystalline rotation of a porphyroblast may give rise to a snowball structure or an S- or Z-shaped pattern of the inclusion trails. The S pattern will indicate anticlockwise rotation (or sinistral shear) and the Z pattern will indicate clockwise rotation (or dextral shear) (Fig. 21.12c, d).

(8) *Pattern of foliation around unrotated clasts*

An oblong porphyroclast lying parallel to the foliation may not show any evidence of rotation. However, if the deformation is non-coaxial, the foliation around it may be distorted in an asymmetric pattern. The sense of shear can then be determined from the stair-stepping rule (Fig. 21.12e), as in the case of porphyroclast-tail systems.

(9) *Mica fish*

Mylonites sometimes contain (in thin sections parallel to lineation and perpendicular to the foliation) asymmetric lens-shaped coarse grains of mica oriented oblique to the dominant foliation. These have been described as *mica fish* (Lister & Snoke 1984). Sometimes a number of mica fish are linked together by trails of very fine mica. The acute angle between the mica fish and the foliation points towards the direction of shear. Towards each end, the tip of the mica fish bends towards the direction of shear movement. Therefore, the sense of shear can also be determined from the asymmetry of the mica fish (Fig. 21.12f).

(10) *Slip on grain-scale faults*

Megacrysts of some minerals such as feldspar in mylonites may undergo microfaulting under certain circumstances. In some cases a single grain may be divided into several slices by parallel microfaults. The oblique individual fragments slide past each other and rotate at the same time. With reference to the sense of shearing in the matrix, the sense of slip along the microfaults and the sense of rotation of microfaults may be either antithetic or sympathetic. When the sense of shear in a mylonite zone is dextral (Fig. 21.13), a dextral slip on the microfault is considered to be sympathetic and a sinistral slip is regarded as antithetic. Again, in a shear zone with a dextral sense of shear, a clockwise rotation is considered to be sympathetic and an anticlockwise rotation is considered to be antithetic. When the microfaults are oriented at a low angle to the foliation, the slip on them will be sympathetic (Fig. 21.13a, b). Such low-angle microfaults in the grain scale may be safely used to determine the sense of shear in the matrix,

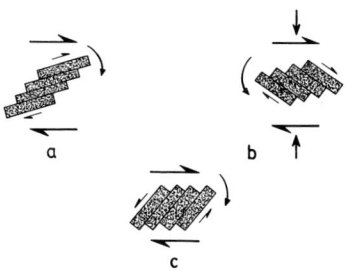

FIG. 21.13. Senses of slip and rotation on grain-scale faults.

although rotation of microfaults may be either sympathetic (Fig. 21.13a) or antithetic (Fig. 21.13b). If the microfaults were initially at a high angle to the foliation, the slip on them is antithetic (Fig. 21.13c). This type of slip is sometimes described as bookshelf sliding or the domino effect, when the acute angle between the foliation and the slip planes is reduced by their rotation in the direction of shear. The shear sense in the matrix may also be deduced from such bookshelf sliding.

(11) *Crystallographic fabric of mylonites*

The crystallographic fabrics of mylonites have been studied for a number of decades; however, there has been a rapid advance in the interpretation of fabric patterns only very recently. Because of the prevalence of quartz-rich mylonites, most of this recent work is on quartz c-axis fabric. To prepare a fabric diagram, the orientations of quartz c-axes are determined from an oriented thin section mounted on a universal stage of the microscope. The c-axes are plotted in the equal area projection. On this pole figure the contours of concentration of the poles are then drawn, each point on a contour representing the percentage of poles lying within 1 per cent area of the equal area net. The detailed procedure of measuring the orientations of the c-axes and of preparing the fabric diagram have been given by Turner & Weiss (1963, pp. 194–205).

The fabric diagram usually shows the foliation as a plane (S) striking left to right and perpendicular to the plane of the diagram, with the lineation occurring at the periphery. The lineation is taken to be parallel to the X-axis of finite strain and the foliation is considered to be parallel to the XY-plane. Thus, the Z-axis occurs at the periphery and at a right angle to S, while the Y-axis is situated at the centre of the diagram (Fig. 21.14). A girdle in a fabric diagram is a diffuse band along which c-axes are distributed. When a girdle shows a branching it is called a crossed girdle. The overall pattern of a contoured pole diagram is sometimes represented in a simplified manner by drawing the *skeletal outline* (Lister & Williams 1979). The skeletal outline

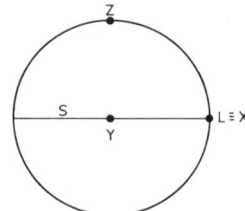

FIG. 21.14. A conventional representation of S and L in petrofabric diagram.

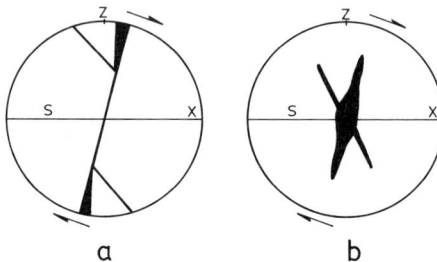

a b

FIG. 21.15. Skeletal outlines in (a) type I and (b) type II crossed girdles.

(Fig. 21.15) emphasizes the gross pattern of the diagram by neglecting the local variations and is drawn as a line through the "peaks" and "ridges" of the contour diagram. Two types of crossed girdles are common in quartz-rich mylonites. A *type I crossed girdle* shows maxima of c-axes near the periphery of the net. The skeletal outline shows a main central segment which bifurcates at each end into two branches or peripheral legs (Fig. 21.15a). A *type II crossed girdle* shows a maximum of c-axes at the centre of the diagram and the skeletal outline shows two lines or girdles intersecting at the centre (Fig. 21.15b).

The c-axis fabric may be symmetrical or asymmetrical. This symmetry is generally defined with reference to the foliation. Thus, a type I crossed girdle is said to be symmetrical or orthorhombic if the central girdle segment is perpendicular to the foliation (Fig. 21.16a); if it is oblique to the foliation, the fabric is regarded as asymmetrical or monoclinic (Fig. 21.16b). Both orthorhombic and monoclinic quartz c-axis fabrics have been described from mylonites, and this raises the problem of the relation between the symmetry of the fabric and the symmetry of the deformation path (e.g. Paterson & Weiss 1961, Christie 1963, Sylvester & Christie 1968, Burg & Laurent 1978, Lister *et al.* 1978, Lister & Hobbs 1980). An orthorhombic fabric in a mylonite may develop by a coaxial phase in a later stage of deformation, i.e. a coaxial deformation following an earlier non-coaxial deformation. A similar deformation history has been postulated for the Moine Thrust belt in Scotland (Christie 1963). Alternatively, at low shear strains an approximately

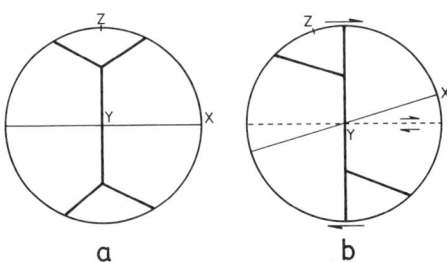

FIG. 21.16. (a) Type I girdle with central segment at a right angle to the XY-plane in progressive pure shear. (b) Type I girdle with central segment orthogonal to the shear plane in progressive simple shear. The sense of shear is sympathetic to the acute angle between the XY-plane and the central segment.

orthorhombic c-axis fabric may develop even in non-coaxial deformations. According to the theoretical prediction by Lister & Hobbs (1980), the c-axis fabrics resulting from coaxial plane strain and progressive simple shear are similar at low strains. The reasons for the occurrence of orthorhombic c-axis fabrics in zones of dominantly non-coaxial deformation are not fully understood as yet.

On the other hand, a number of authors have described asymmetric c-axis fabrics in quartz-rich mylonites, especially in zones of large strain. Where the sense of shear has been independently determined, the asymmetry of the c-axis fabric is clearly related to it. The most strongly marked girdle is deflected in the direction of shear from the Z-axis (or normal to the foliation). This observation is also in agreement with theoretical predictions by Lister & Hobbs (1980) who showed that, in a type I girdle, the central segment should be orthogonal to the shear plane in progressive simple shear (Fig. 21.16b) and it is orthogonal to the XY-plane in progressive pure shear (Fig. 21.16a). Thus, if the central segment of a type I girdle makes an acute angle with the foliation, the sense of shear is sympathetic to it (Fig. 21.16b). It has been suggested by Platt & Behrmann (1986) and by Vissers (1989) that the same relation holds good in other types of plane non-coaxial flow which deviate from progressive simple shear.

There is a large volume of literature on the interpretation of quartz c-axis fabrics of mylonites. In addition to the references given above, the interested reader may consult the following papers and the references given therein: Behrmann & Platt (1982), Bouchez & Pecher (1981), Carreras et al. (1977), Etchecopar (1977), Hara et al. (1973), Law et al. (1984), Mancktelow (1987), Passchier (1983), Schmid & Casey (1986).

21.8. Rotation of shear zone folds

The axes and the axial surfaces of the shear zone folds often show a wide range of attitude. However, in most ductile shear zones the axes of the most dominant set of folds are sub-parallel to the mylonitic lineation. The fold hinges are nearly parallel to the direction of maximum stretching and the axial surfaces are at a low angle to the shear zone boundary. Thus, when a ductile shear zone develops in connection with a thrusting movement, the most dominant folds are often reclined. It has been suggested that shear zone folds initiate with the hinges at low angles to the Y-axis of bulk strain in the shear zone. With progressive deformation, the hinges rotate towards the X-axis (e.g. Bryant & Reed 1969, Sanderson 1973, Escher & Waterson 1974, Mies 1991, Mawer & Williams 1991) and at large strains the hinges become sub-parallel to the X-axis (Fig. 16.8).

Such an idea of rotation of fold hinges through large angles has been confirmed later by the recognition of strongly non-cylindrical and essentially plane folds in shear zones (Figs. 21.17, 21.18, 21.19). In extreme cases the fold hinges show hairpin bends and the folds are shaped somewhat like a flattened test tube. These have been described as *sheath folds*. Sheath-like structures have been described from a number of shear zones such as the Cap de Creus mylonite belt in Spain (Carreras *et al.* 1977), the Seve-Koli Nappe Complex in Sweden (Williams & Zwart 1978), the Kalak Nappe Complex of Finnmark (Rhodes & Gayer 1977), the Ile de Groix blue schist (Quinquis *et al.* 1978) and the South Armorican Shear Zone in France, the shear zones in western Italian Alps (Minnigh 1979), the mylonites of the Grenville front (Dalziel & Bailey 1968) and N.E. Canada (Henderson 1981), Arunta Block and Woodroffe thrust in Central Australia (Goscombe 1991, Bell 1978a) and the Singhbhum Shear Zone in eastern India (Ghosh & Sengupta 1987a, b, 1990). Thus, sheath folds or plane non-cylindrical folds transitional to sheath folds are characteristic of many ductile shear zones.

The development of sheath folds has been explained by accentuation of initial irregularities of the bedding or the foliation surface by a strongly non-coaxial deformation (Ramsay 1980, Cobbold & Quinquis 1980). Although the folds are sometimes localized by the presence of rigid objects like pebbles, boudins and porphyroclasts, such rigid objects are not necessary for the development of sheath folds. In any case, since the initial irregularities are likely to be non-cylindrical, the folds generated from them are likely to have somewhat curved hinges. With progressive non-coaxial deformation, the hinge-line curvatures are accentuated and at large shear strain a sheath-like structure is produced (Fig. 21.20).

The cross-sectional morphology of shear zone folds, including sheath folds, is typical of buckling folds and they are not necessarily concentrated at sites of local irregularities or near islands of less deformed rocks (Platt 1983, Ghosh & Sengupta 1984, 1987a). Thus, although it is well

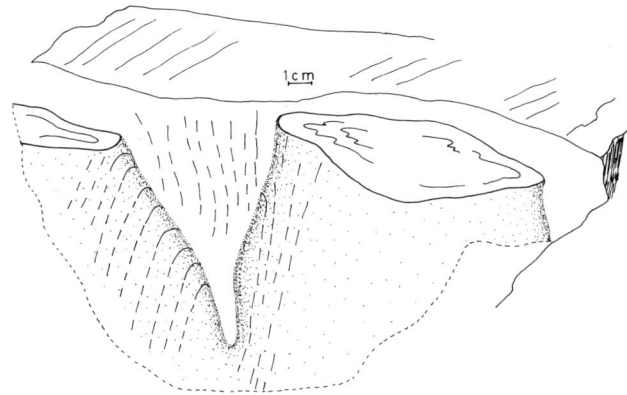

FIG. 21.17. Sheath fold in quartzite mylonite. Near Ramchandra Pahar, Singhbhum Shear Zone. India.

FIG. 21.18. Strongly non-cylindrical plane isoclinal fold (sheath fold) in ductile shear zone from Phulad, Rajasthan, India. The axial surface of the fold is steeply dipping. The striping lineation curves in accordance with the curved hinge line.

established that certain shear zone folds are initiated as non-cylindrical structures which evolve into sheath folds in zones of strong non-coaxial movement, the reason for the development of instabilities which give rise to a regular train of such folds is not quite clear.

525

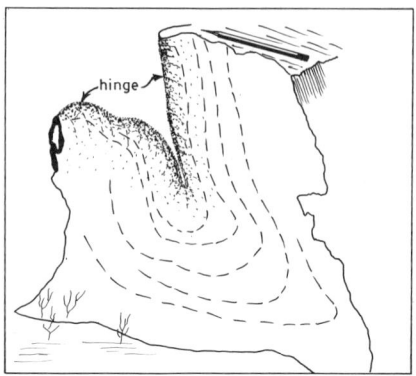

FIG. 21.19. Sketch of the sheath fold shown in Fig. 21.18. Dashed lines represent a hinge-parallel striping lineation.

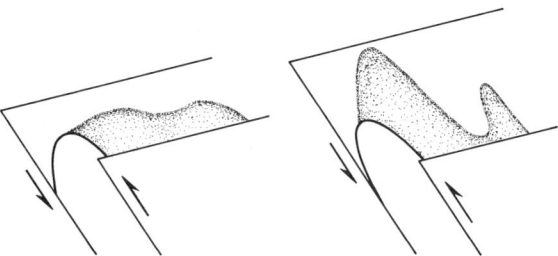

FIG. 21.20. Mechanism of development of sheath folds by accentuation of initial hinge-line curvatures in non-coaxial deformation.

Ductile shear zones often show repeated development of shear zone structures (Fig. 21.21) in the course of a progressive deformation (e.g. Bell 1978a, Platt 1983, Ghosh & Sengupta 1987a, b, Mawer & Williams 1991). Thus, during the development of a major ductile shear zone, successive generations of sheath folds may be initiated and refolded and an axial planar mylonitic foliation may develop repeatedly by transposition of a similar foliation of an earlier phase. Similarly, the characteristic mylonitic lineation sub-parallel to the stretching direction may be a composite structure represented partly by folded lineations of an earlier generation and partly by newly superimposed stretching lineations.

21.9. Nature of deformation in shear zones

The deformation in a ductile shear zone may be considered in different scales. The general character of bulk deformation for a wide shear

FIG. 21.21. Hook-shaped outcrop of coaxial folds in mylonitic foliation of quartz-rich bands in a calcareous matrix. The axial surfaces are steeply dipping and both sets of folds are nearly reclined with axes sub-parallel to a stretching lineation on most segments. The mylonitic foliation in the calcareous matrix has formed by transposition of a mylonitic foliation of an earlier stage. Ductile shear zone in Phulad, Rajasthan, India.

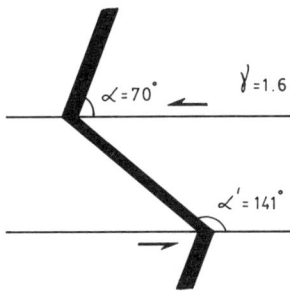

FIG. 21.22. Determination of wall-parallel shear strain in a ductile shear zone from orientation change in a cross-cutting dyke or vein.

zone may not be the same as in the subsidiary shear bands lying within it. The general character of deformation in shear zones has been studied by a number of authors (e.g. Ramsay & Graham 1970, Coward 1976, Ramsay 1980, Gapais *et al.* 1987). Ramsay & Graham considered the development of a tabular zone of ductile shear bounded by undeformed wall rocks. In one model considered by these authors, the shear zone is of infinite extent and there is no variation in strain in any direction

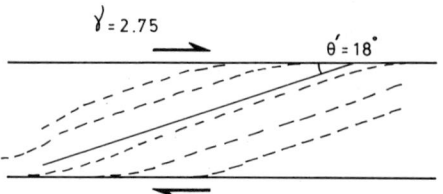

FIG. 21.23. Determination of wall-parallel shear strain from angle between shear zone wall and shear zone cleavage.

parallel to the shear zone walls. Ramsay & Graham showed that, under such conditions, the particle path equation will be

$$x' = x + f(z)z,$$
$$y' = y,$$
$$z' = \frac{1}{1 + \triangle} z$$

where $\triangle$ is the volume change. If $\triangle = 0$, there is no strain along both y- and z-axes and the particle path equation defines a heterogenous simple shear. In this case $f(z)$ denotes the shear strain parallel to the walls. The shear strain (γ) is usually largest in the middle of the shear zone and decreases towards the walls. For such a simple situation Ramsay & Graham showed that the deformation within the shear zone will be a simple shear or a volume change or a combination of the two.

When the constraints mentioned above are not maintained, the deformation within the zone may be of a more complex type. Many shear zones are wavy, thick and thin and are of restricted length. The wall rocks of many shear zones are also deformed. Hence progressive simple shear is not necessarily the only or the most common mode of deformation in zones of ductile shear. From strain measurements in a shear zone in Botswana and in another zone in North Uist, Scotland, Coward (1976) showed that a simple shear movement was accompanied by a pure shear movement across the zones. Coward has also cited another example from a sheared dyke in Sutherland where the angle between the foliation in a lozenge-shaped block and the bounding shear zone is too high to be explained by simple shear. In the central part of the Singbhum Shear Zone in eastern India, synkinematic quartz veins emplaced parallel to the mylonitic foliation often show boudinage and pinch-and-swell structures in two directions, along and across the mylonitic lineation. This feature also indicates a departure from the plane strain model of simple shear (Ghosh & Sengupta 1987b). A deviation from the model of simple shear has also been suggested by Platt & Behrmann (1986) and Vissers (1989) from the Betic Movement Zone in S.E. Spain. From an analysis of the rotation angles of elongate porphyroblasts of garnet with different axial ratios, Vissers has shown that a pure shear component was involved in the non-coaxial flow in this movement zone.

21.10. Shear strain parallel to the shear zone boundary

There are several methods of calculating the wall-parallel shear strain within a ductile shear zone. The different methods have been described in detail by Ramsay (1967, 1980) and Ramsay & Huber (1987). We shall describe here two of the simple cases. The method of determination of wall-parallel shear strain is fairly simple when there are reasons to believe that the deformation within the shear zone has taken place by simple shear. The wall-parallel shear strain can be calculated if, for instance, a dyke or a vein enters the shear zone across the undeformed wall. It is assumed that the plane of observation (Fig. 21.22) is normal to the Y-axis of strain of the simple shear deformation. Let the angle between the shear zone boundary and the dyke or the vein in the undeformed wall rock be α and let this angle within the shear zone be α'. If the deformation is by simple shear, we have by eqn. (8.73)

$$\cot\alpha' = \gamma + \cot\alpha$$

where γ is the wall-parallel shear strain. In Fig. 21.22, $\alpha = 70°$ and $\alpha' = 141°$. This gives a shear strain $\gamma = -1.6$. The shear strain is negative because, according to our convention, the angle between lines parallel to the positive x- and the positive y-axis has increased.

For another simple situatation let us consider the orientation of the shear zone cleavage with reference to the shear zone boundary (Fig. 21.23). Let us consider the hypothetical case in which the cleavage in the central part of the shear zone makes an angle $\theta' = 18°$. The boundary-parallel shear strain can be determined if it is justified to assume that the deformation is by progressive simple shear and that the orientation of the cleavage tracks the XY-plane. The shear strain, as obtained from eqn. (8.69), is

$$\gamma = \frac{2}{\tan 2\theta'}$$

in the middle of the shear zone.

CHAPTER 22

Time Relation Between Crystallization and Deformation

22.1. Introduction

Structures of different stages of evolution are generally recognized by: (1) tracing them from a less deformed area to an area where the deformation is large, (2) distinguishing them by their morphology in different parts of the same area and (3) from the overprinting relation of successive generations of structures. Another method which has been of great importance as a relative dating technique of structures is to determine the time of formation of structures with reference to the time of crystallization of metamorphic minerals. While the primary objective of metamorphic petrologists in this context is to obtain a sequence of crystallization of minerals, the primary objective of a structural geologist is to obtain a time sequence of structures. Such a division of objectives is no doubt artificial. The results obtained from these microstructural studies go far beyond the mere delineation of a time sequence and sometimes into fields where the interests of both structural and metamorphic geologists converge.

Following Read (1949), we may as well enter into this topic through the classical example of the Cowal albites. Porphyroblasts of albite, as well as porphyroblasts of other minerals like garnet and staurolite, often have parallel trails or trends of tiny inclusions of ground-mass minerals or of iron ore, graphite, etc. These *trails* or *trends of inclusions* are relics of an earlier S-surface, bedding or cleavage. The trail of inclusions within the porphyroblast is called *Si* (internal S-surface) while the planar structure in the matrix of the rock is called *Se* (external S-surface). Clough (1897, p. 40) found that the crenulated foliation planes of the albite-gneisses of Cowal, Scotland, can be traced through the albites by the help of the trails of inclusions. Clough concluded that the albite porphyroblasts grew after the deformation. These Cowal albites sometimes show, near their borders, a symmetrical swerving of the inclusion trails. This pattern of *Si* led Bailey (1923) to suggest that, towards a late phase of their crystallization history, the grains of albite rotated while they were still growing. Thus, according to Clough's interpretation the movement was prior to crystallization of albite, while according to Bailey the movement was synchronous with crystallization.

The crystallization of a mineral can be earlier than, synchronous with or later than the deformation. These time relations are described as

(1) *precrystalline deformation, post-tectonic* or *post-kinematic crystallization*,
(2) *paracrystalline deformation, paratectonic, syntectonic* or *synkinematic crystallization*, and
(3) *post-crystalline deformation, pretectonic* or *prekinematic crystallization*.

To emphasize that a certain metamorphic event took place during the late stage of a period of deformation, we may also use the term *late tectonic* or *late kinematic*. Where there is a sequence of crystallization events or a sequence of deformation events, the terms pre-, para- and post-crystalline deformation become meaningful only when we specify the time of crystallization of a certain mineral with reference to a particular deformation event.

When a cold-worked metal is heated up, the deformed grains re-crystallize. This process is described as *static recrystallization* to distinguish it from *dynamic recrystallization* which takes place *during* the deformation of a metal at elevated temperatures. These terms, used also in the geological literature, correspond, respectively, to post-tectonic and syntectonic recrystallizations. Evidently, static recrystallization may also take place when, after hot deformation, the metal is kept at an elevated temperature; such static recrystallization may need an incubation period. On the other hand, when hot deformation comes to an end, but the temperature is still high, recrystallization may continue without an incubation period. This latter process has been described as *metadynamic recrystallization* (Sellars 1978). A similar situation is generally described in the geological literature by saying that crystallization has outlasted deformation (e.g. Read 1949).

22.2. Precrystalline deformation

The porphyroblasts of garnet, chloritoid, albite or other minerals sometimes show a straight *Si* parallel to the more or less straight cleavage of the ground mass; the cleavage does not swerve around the

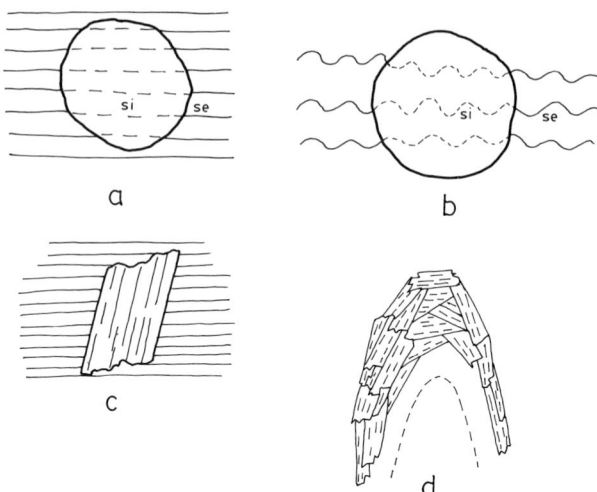

FIG. 22.1. Microstructures indicating precrystalline deformation. (a) Porphyroblast of garnet overgrowing a straight schistosity. (b) Helicitic structure, with a porphyroblast of garnet overgrowing a crenulated schistosity. (c) Underformed transverse mica or chlorite across the schistosity. (d) Polygonal arc of underformed flakes of mica which crystallized after the development of the microfold.

porphyroblasts (Figs. 22.1a, 22.7a). The microstructure clearly indicates that the crystallization of the porphyroblasts was after the development of the cleavage. In a similar way, the porphyroblasts in certain rocks have overgrown the crenulations of the cleavage, the wavy *Si* passing into and having the same form as the crenulations of *Se*. This microstructure, sometimes described as *helicitic structure* (Fig. 22.1b), indicates that the crenulations are precrystalline with reference to porphyroblasts or that the porphyroblasts are post-tectonic with reference to the crenulations. Post-tectonic crystallization is also indicated by the occurrence of large unstrained flakes of mica or chlorite (*encarsioblasts* of Greenly 1919, p. 43) occurring at a high angle to the schistosity (Fig. 22.1c) or by polygonal arcs of unstrained mica flakes (Fig. 22.1d) that retain the form of old microfolds (Knopf & Ingerson 1938).

Certain rocks do not show any evidence of plastic deformation in grain scale in spite of the occurrence of abundant folds in the scale of the outcrop. A good example of this is the Carrara marble of Italy which shows ample evidence of isoclinal folding in the outcrop but no visible traces of deformation in the microfabric (Schmid *et al.* 1985, p. 774). It is likely that, because of the elevated temperature of metamorphism in such cases, an intense post-tectonic recrystallization has overprinted the earlier paracrystalline microfabric.

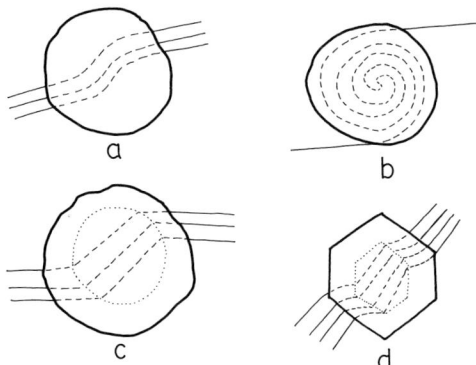

FIG. 22.2. *Si-Se* patterns in paratectonic porphyroblasts. (a) S-shaped trail of inclusions. (b) Spiral trail. (c) and (d) Alternate stages of growth and rotation. In all these cases the *Si*-passes undeviated into the *Se*, indicating that there was no post-crystalline rotation of garnet.

22.3. Paracrystalline deformation

One of the clearest examples of paracrystalline deformation is the occurrence of spiral trails of inclusions (Fig. 22.2b) in *snowball garnets* (Flett 1912, p. 111) or of symmetrical S-shaped trails of inclusion (Fig. 22.2a) in garnet, albite and other minerals. The microstructure develops when a porphyroblast grows and at the same time rotates with respect to the orientation of the cleavage in the rock. Because of the relative rotation, the *Si*, with continually changing orientation, becomes frozen in the successive shells of the growing crystal. The spiral trails of the snowball garnets indicate that the paracrystalline rotation, and hence the deformation of ground mass, was extremely large. To ensure that the *Si* and the *Se* belong to the same generation of the S-surface, it is further required that the *Si*, at the periphery of the porphyroblast, passes undeviated into the *Se*. Paracrystalline rotation is also inferred if, as in Fig. 22.2c and d, the pattern of *Si* develops in alternate episodes of growth and rotation of a porphyroblast (cf. Dixon 1976).

An excellent criterion of paracrystalline deformation, recorded by Zwart (1960a, b) from the metamorphic rocks of the central Pyrenees, is illustrated in Fig. 22.3. In some of the porphyroblasts, as in Fig. 22.3a, the *Si* is either planar or shows weak crenulations in the central part, but towards the border, the crenulations of the *Si* become tighter and merge into similarly crenulated *Se*. The core and the mantle of the crystal must have grown during different stages of the progressive development of the crenulations. Figure 22.3b shows a similar feature, but here the *Si* throughout the porphyroblast is weakly crenulated, while the *Se* shows much stronger crenulations with similar orientation

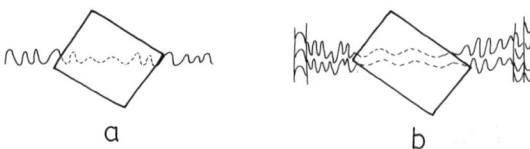

FIG. 22.3. Porphyroblasts which grew during the growth of crenulations. In (a) the core of the porphyroblast overgrew an early stage of crenulation while the rim grew during further tightening of the crenulations. In (b) the porphyroblast grew after an early stage of crenulation but the crenulations continued to grow after the complete crystallization of the porphyroblast.

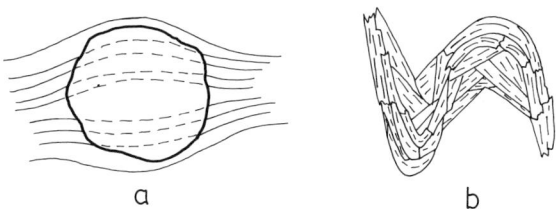

FIG. 22.4. (a) Paratectonic porphyroblast of garnet overgrowing swerving pattern of cleavages. (b) Paracrystalline crenulation with folded mica flakes on outer arcs and polygonal arcs of undeformed flakes at the inner arcs.

of the axial traces. In both these examples the porphyroblasts are syntectonic with reference to the crenulations.

Porphyroblasts in mica schists are generally stiffer than the ground mass. This leads to a bowing out of the schistosity in the neighbourhood of the porphyroblast. If the porphyroblast continues to grow during the deformation, it overgrows the bowed pattern which then becomes frozen in the Si (Zwart 1960a, b, Spry 1963) in the outer part of the porphyroblast (Fig. 22.4a).

The resistant porphyroblasts in mica schists often show pressure shadows (Pabst 1931, Fairbairn 1950). These generally appear as triangular clots of slightly coarser grains of quartz at the opposite edges of a porphyroblast (Fig. 22.5a). Quartz in the pressure shadow zone is therefore syntectonic. In certain cases porphyroblasts of garnet overgrow their tails of quartz in the pressure-shadow zones, indicating that the garnet itself is also syntectonic. In some situations the included pressure-shadow tails of quartz are themselves arranged in a spiral shape within the garnet (Fig. 22.5b). This type of microstructure gives the additional information that the garnet rotated during its growth (Schoneveld 1977). Pressure shadows of another type (pressure fringe of Spry 1969) are composed of fibrous quartz, carbonate, cholorite or micas (Fig. 22.5c). These occur at the edges of resistant grains like

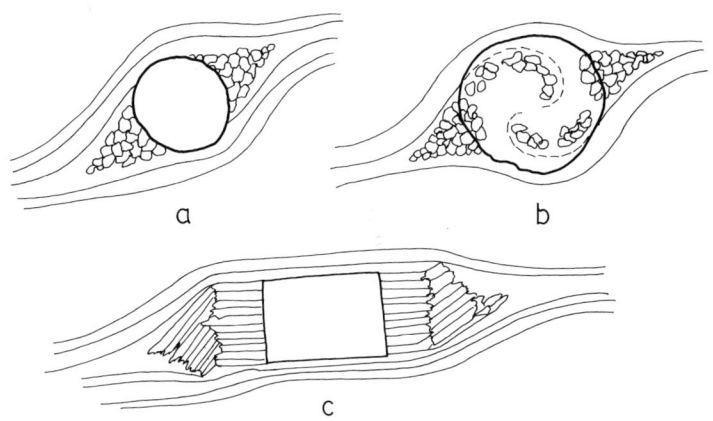

FIG. 22.5. (a) Pressure shadow of quartz at opposite edges of a porphyroblast. The pressure – shadow quartz has grown syntectonically. (b) The porphyroblast of garnet includes a spiral trail of pressure – shadow quartz. Here both the minerals are syntectonic. (c) Fibrous pressure shadow or pressure fringe which grew syntectonically.

pyrite or magnetite and are commonly observed in low-grade metamorphic rocks. These have proved to be of considerable importance, since the syntectonic fibres enable us to estimate the finite and incremental strains as well as the deformation path (e.g. Choukroune 1971, Durney & Ramsay 1973, Ramsay & Huber 1983, Etchecopar & Malavieille 1987).

A microstructure somewhat similar to the pressure shadow is produced by paracrystalline microboudinage (Misch 1969). This process involves extension of an elongate grain either by breaking apart or by necking and recrystallization of the same mineral or crystallization of another mineral in the necked zone (Fig. 22.7b; see pp. 424–425).

Comparison of microstructures of rocks deformed under natural and experimental conditions suggest that deformation of rocks at elevated temperatures is accompanied by recrystallization. Indeed, it is generally believed that regionally metamorphosed rocks have undergone a long period of recrystallization. In certain cases crystallization has outlasted the deformation and static recrystallization has removed the traces of intragranular deformation in the ground mass of the rock. In other situations the rock may show close association of strained and unstrained grains. Thus, crenulations of mica-rich layers (Fig. 22.4b) may show an association of bent mica flakes tracing out a part of the microfold and a polygonal arc of undeformed flakes in another part of the microfold (Knopf & Ingerson 1938, p. 109). The recrystallization of the mica was broadly syntectonic with respect to the crenulations.

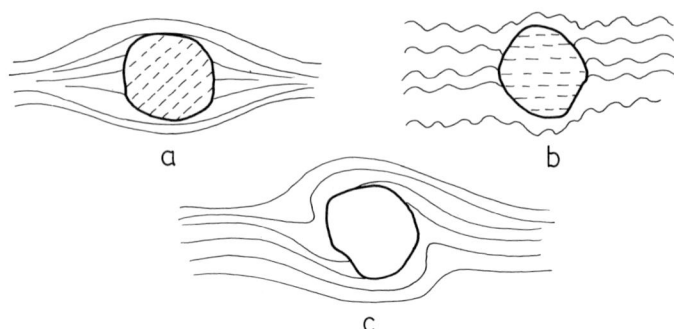

FIG. 22.6. Microstructures indicating post-crystalline deformation. (a) *Si* and *Se* are discordant, indicating post-crystalline rotation of the porphyroblast. (b) Straight *Si* and crenulated *Se* of the same generation, indicating that the crenulation was later than the growth of the porphyroblast. (c) Asymmetric drag pattern of *Se*, indicating anti-clockwise post-crystalline rotation.

22.4. Post-crystalline deformation

Post-crystalline deformation is indicated by the following microstructures.

(1) A straight or S-shaped *Si* discordant with *Se* (Fig. 22.6a). If the *Si* and the *Se* represent the same generation of the S-surface, this microstructure indicates a post-crystalline rotation of the porphyroblast.

(2) A straight *Si* of a porphyroblast continuing into or discordant with a crenulated *Se* (Fig. 22.6b). The crenulation is then post-crystalline with respect to the porphyroblast.

(3) A strongly asymmetric swerving of the cleavage around a porphyroblast (Fig. 22.6c) is indicative of post-crystalline rotation.

(4) Microstructures showing presence of undulatory extinctions, deformation bands, deformation lamellae and Boehm lamellae in quartz, microfolds and kink bands in micas and chlorite, deformation twins, twin lamellae offset by microfaults, separation of grain fragments across tension fractures, grains dissected by parallel shear fractures with slip and rotation of the slices and folding or boudinage of the fine rutile or sillimanite needles occurring as inclusions in quartz grains.

Intense post-crystalline deformation of a metamorphic rock destroys the earlier microstructure and overprints a new microstructure. This is most spectacularly seen when a metamorphic rock is mylonitized. Where the older microstructure is partially preserved, the old grains will show various signs of post-crystalline deformation enumerated in (4). If the mylonitization is at an elevated temperature, there will be a significant amount of recrystallization, giving rise to a fine-grained matrix. With reference to the event of mylonitization, this

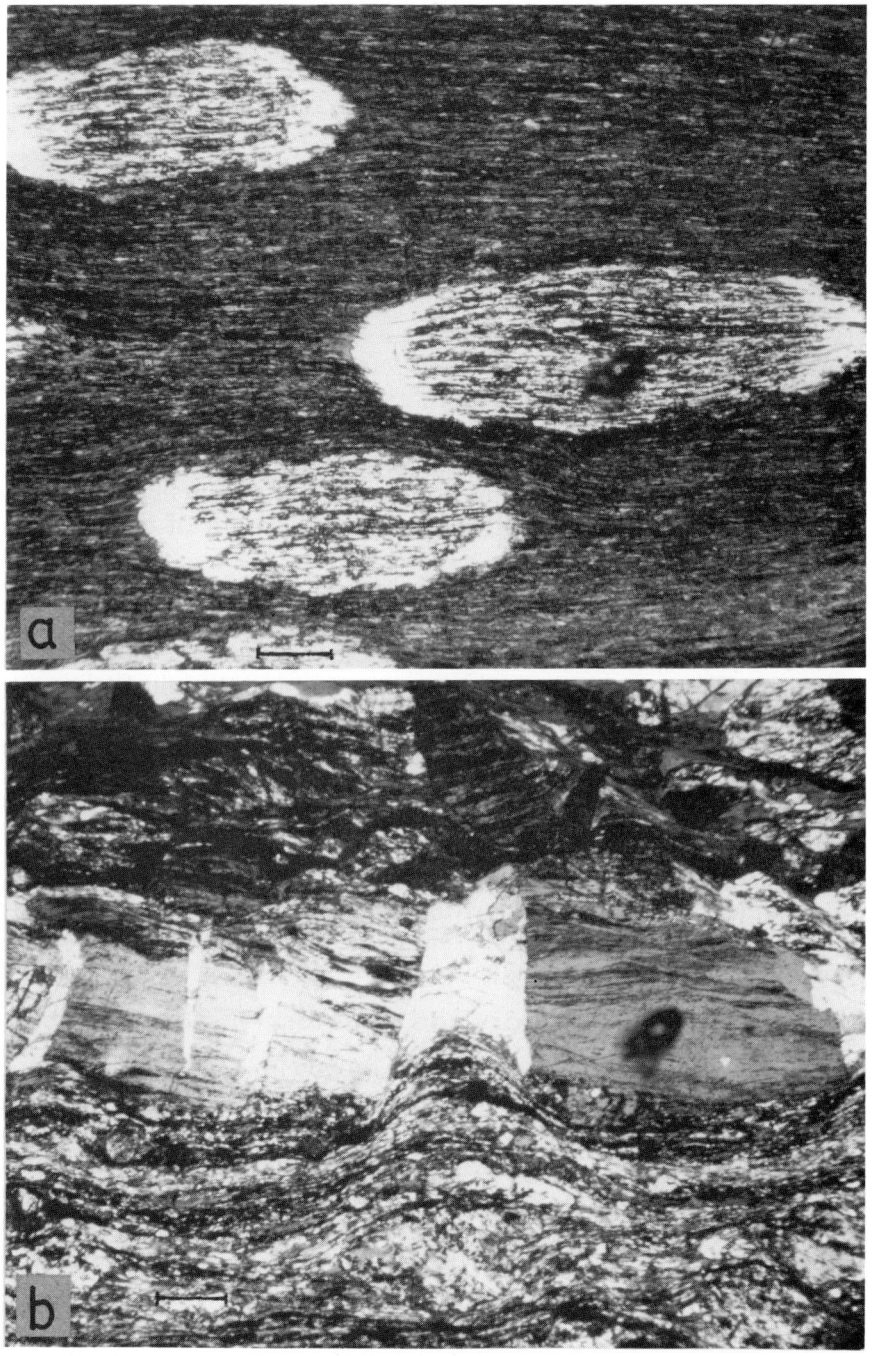

FIG. 22.7. (a) Straight *Si* in albite porphyroblast passing into *Se*, indicating growth of albite after the development of the cleavage. (b) Microboudinage in albite. The separation zone (white) is filled with quartz. Scale bar represents 0.1 mm. Pelona schist, California. Courtesy Carl Jacobson.

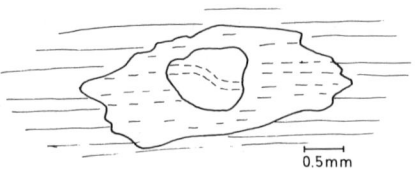

0.5mm

FIG. 22.8. A porphyroblast of garnet included by a larger porphyroblast of staurolite. The garnet shows S-shaped *Si* while staurolite has straight *Si*. After Naha 1965.

crystallization is generally regarded as syntectonic (White 1973a). Evidently, with reference to the same deformation, the older grains are pretectonic. In other words, with reference to the older grains (or the premylonitic metamorphic rock), the mylonitization is post-crystalline. This example shows that the pre-, para- and post-nomenclature is meaningful only if we specify a particular deformation episode and a particular metamorphic event or the crystallization of a particular mineral.

Figure 22.8 illustrates a similar case in which we can relate a single deformation event with the crystallization of two minerals. Here a porphyroblast of garnet is included in a much larger porphyroblast of staurolite. The ground-mass schistosity passes as a straight *Si* through the staurolite, but continues as an S-shaped *Si* through the garnet (Naha 1965, p. 68 and plate 8B). A similar microstructure has also been described by Krige (1916, described by Read 1949, pp. 114–115 or "Granite Controversy", p. 283). The microstructure indicates that the movement on the schistosity is paracrystalline with respect to garnet and precrystalline with reference to staurolite. In other words, garnet is syntectonic and staurolite is post-tectonic with respect to development of schistosity.

22.5. Metamorphic history and structural history

The microstructures of metamorphic rocks with a complex history of deformation enable us to differentiate successive generations of folds and to determine the pressure–temperature conditions at which the different folds formed. The method has been applied in several metamorphic terrains to represent the chronological succession of folding and metamorphism. Thus, Zwart (1960a, b) distinguished seven different stages from the microstructural characters of the metamorphic rocks of the Central Pyrenees. The mica schists in this terrain show three different stages of deformation: (1) folding with E–W axis, (2) folding with N–S axis and (3) folding of the schistosity with NW–SE axis. The different phases of Hercynian metamorphism, as indicated by *Si–Se* relations of porphyroblasts of biotite, andalusite, staurolite, cordierite and garnet, show seven sets of relations: (1) prekinematic (before the E–W folding), (2) first stage of early synkinematic (during the E–W folding), (3) interkinematic I (between the periods of E–W and N–S folding), (4) second stage of early synkinematic (during N–S folding), (5) interkinematic II (between periods of N–S and NW–SE folding), (6) late synkinematic (during NW–SE folding) and (7) post-kinematic (after the last phase of folding).

A particular phase of metamorphism or a particular deformation event or both may not occur simultaneously in all parts of an orogenic belt. Thus, the peak of metamorphism occurred at different times in different parts of the Central Pyrenees. A somewhat different kind of complication may be illustrated by the history of the Finnmarkian nappes of north Norway where the deformation moved diachronously from north-west to south-east and the higher nappes were deformed and were emplaced earlier (Rice 1984). Although each thrust sheet shows structures of three main phases of deformation, their time of formation is not synchronous in all the sheets; in particular, the development of the regional schistosity, related to nappe emplacement, is earlier in the higher thrust sheets. The rocks show two periods of garnet growth Gt_1 and Gt_2. Gt_1 grew earlier in the higher nappes and consequently earlier structural events are preserved in their Si while Gt_2 records successively later structural events higher up in the nappe pile.

The time relation between crystallization of a particular mineral and the events of deformation may appear contradictory in the different thin sections of the same area or sometimes in different parts of the same thin section. An excellent example of such complex relations is given by Jacobson (1983a, b, c, 1984) and Postlewith & Jacobson (1987) from the Pelona schist of southern California. The Si pattern of the albite porphyroblasts with respect to the most dominant variety of isoclinal folds (designated style 1 folds) and their axial plane schistosity is depicted in Fig. 22.9. The relation shown in Fig. 22.9a indicates that albite is post-tectonic with reference to S_1 but pretectonic with reference to its crenulation and complete transposition to S_2. A relation similar to that of Fig. 22.9b indicates that the albite is post-tectonic with respect to an early stage of crenulation of S_1. The rare preservation of Si trails as in Fig. 22.9c further indicates that S_1 itself had developed by transposition of an early schistosity. Albite is later than this early isoclinal folding but is pretectonic with respect to the crenulation of S_1. Jacobson has shown that such complex relations become intelligible when we have the additional information about the chemical character of the minerals and about the regional structural history. Jacobson's studies indicate that the rocks have undergone a single prograde metamorphism and that the successive episodes of folding occur in the course of a continuous process associated with thrusting. The Si-Se

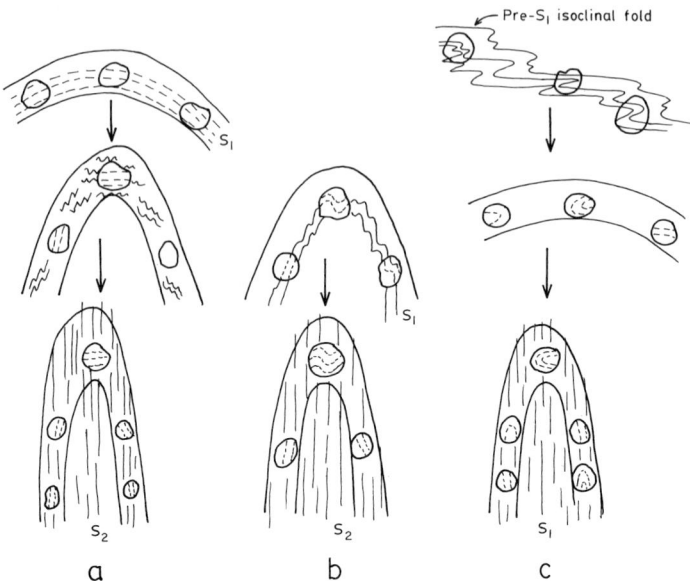

FIG. 22.9. The lowermost figures in (a), (b) and (c) show the present relations between *Si*
and *Se* with respect to albite porphyroblasts in Pelona Schist, California. The upper
figures show the inferred earlier stages of development. Albite formed during a single
prograde metamorphic event. (a) Albite is post-tectonic with reference to early folding
of S_1 and pretectonic with reference to S_2. (b) Albite is post-tectonic with reference to
crenulation of S_1 and is pretectonic with reference to complete transposition of S_1 to S_2.
(c) Albite is post-tectonic with reference to early pre-S_1 isoclinal folds. S_1 had itself
formed by transposition of an earlier cleavage. Based on Jacobson 1983a, b.

relations of the albite porphyroblasts, indicating that albite grew
during different phases of folding, are consistent with such a history of
refolding in a progressive deformation.

The examples cited above show us that a study of the time relation
between crystallization and deformation is of great importance in
working out the structural history of a metamorphic terrain. When
porphyroblasts grow during a progressive deformation, the successive
stages of structural evolution may be frozen in the *Si* of the porphyroblasts. They give us "snapshots" of the developing microstructure (Van
Den Eckhout & Konert 1983, p. 227). In certain instances they may
provide us with the only evidence of an early phase of deformation, the
traces of which are not now decipherable either in the ground mass of
the rock or in the structures of the mesoscopic scale (Fig. 22.10b).
Thus, for example, the khondalites of the Schirmacher Range in E.
Antarctica show a prominent schistosity. This is the earliest recognizable deformational structure as seen in the outcrops or in the ground
mass of the rocks in thin sections. The porphyroblasts of garnet, however, occasionally preserve tight or isoclinal crenulations (Sengupta

1991) of an *Si* marked by trails of sillimanite needles (cf. Vernon 1987). Since such crenulations do not occur in the ground mass, they must have been completely obliterated by transposition, and the *Si* pattern in the garnets gives us the only evidence that the present schistosity had formed by transposition of a still earlier schistosity. Similar structures have also been described by Bell & Rubenach (1983). According to these authors, S-shaped trails of inclusions in garnet porphyroblasts do not necessarily develop by paracrystalline rotation but may form because the garnet overgrows the crenulation of an early cleavage while the *Se* is a new cleavage formed by complete transposition of the earlier cleavage.

22.6. Chemical zoning of garnet

Structural investigations, in conjunction with electron microprobe studies, have recently proved to be useful in determining the time relation between crystallization and deformation. Boulter & Raheim (1974), for example, have described the recrystallization of phengite (celadonitic muscovite) of a distinct composition associated with each of three successive phases of folding. A comparison of the compositions of phengite enabled the authors to assess how closely the different fold phases were related to each other. Again, electron microprobe studies have revealed the presence of chemical zoning in the garnets of prograde, regionally metamorphosed, low and middle amphibolite facies rocks. In nearly all of them there is a decrease in Mn content and an increase in Fe content as we move outward from the centre of the porphyroblasts (Banno 1965, Atherton & Edmunds 1966, Hollister 1966, Harte & Henley 1966, Kretz 1973). The pattern of chemical zoning, taken together with the pattern of *Si* trails, can give us valuable information about the structural history of the rocks (e.g. Spear & Selverstone 1983, Spear *et al.* 1984). For instance, the *Si* pattern in a porphyroblast may show two stages of deformation, e.g. the development of a schistosity and its crenulation. The pattern of concentration profiles of Fe and Mn may then enable us to decide whether the two phases of deformation belong to a continuous process of progressive deformation or are separated by a significant time interval (cf. Rice 1984). The presence of concentric chemical zones has also been used to disprove (Schoneveld 1978, Jamieson & Vernon 1987) the theory (de Wit 1976) that garnets with S-shaped *Si* generally develop by coalescence of veins initiated at multiple nuclei.

An estimate of growth rate of garnet can give us valuable information regarding temperature and pressure changes during metamorphism. It can also give us the rate of deformation if the crystals of garnet contain spiral trails of inclusions. This is one of the most promising techniques

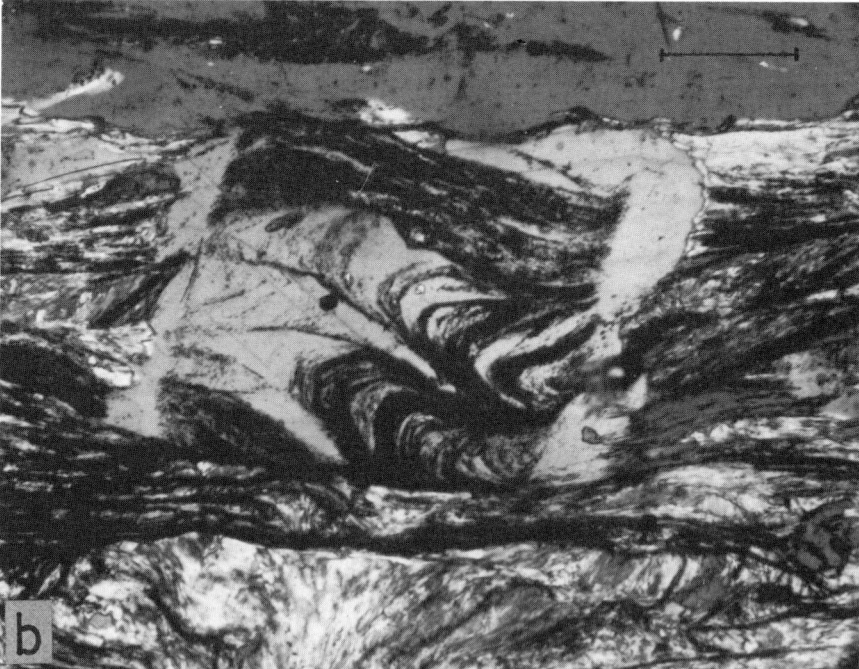

FIG. 22.10. (a) Helcitic structure. The porphyroblast of albite (white) at the centre has overgrown moderately tight crenulations. Photograph by courtesy of C. Jacobson. (b) The rectangular porphyroblast of albite has overgrown a crenulation cleavage. The crenulations in the albite are defined by graphite. The margins are inclusion free. The porphyroblast of albite has undergone slight rotation. Note that the axial surface trace of the crenulated *Si* is slightly curved. The porphyroblast is optically strain free. The crenulations in the matrix have been obliterated by intense deformation after the growth of the porphyroblast. Photograph by courtesy of C. Jacobson.

for the study of rates of petrological processes and of paracrystalline deformation (Christensen *et al.* 1989). The growth rate of single crystals of garnet from south-east Vermont was determined by Christensen *et al.* from measurement of the radial variation of the $^{87}Sr/^{86}Sr$ ratio. This variation records the rate of accumulation of ^{87}Sr by radioactive decay of ^{87}Rb in the rock matrix during the growth of the garnet crystal. The average growth rate was found to be $1.4^{+0.92}_{-0.45}$ millimetres per million years. The average time interval for the growth of the garnet crystals was found to be 10.5 ± 4.2 million years. Electron microprobe traverses of the garnet crystals showed that they are normally zoned in MnO, MgO and FeO. The spiral trails of inclusion in a garnet crystal indicated that it had rotated through an angle of 4 radians. Since, as indicated in Chapter 11, the rotation angle is half the magnitude of simple shear, there is a total shear strain of 8 in 10.5 million years ($\approx 3.3 \times 10^{14}$ seconds). This gives an average rate of shear strain of $2.4^{+1.6}_{-0.7} \times 10^{-14}$ per second. The average rate of rotation is evidently very slow, $1°$ in 45,815 years.

22.7. Time relation between deformation and emplacement of granitic bodies and mafic dykes

Structural geologists working in a granitic terrain have often to determine whether the deformation is earlier than, synchronous with or later than the emplacement of a granitic body. A granitic body may be pretectonic, syntectonic or post-tectonic. There are several criteria for determining the time relation between deformation and the emplacement of granitic, pegmatitic and aplitic bodies (e.g. Read 1949, Misch 1949). It is comparatively easy to identify a post-tectonic granite. Its boundary is discordant with earlier structures such as the cleavage or the axial surface of large-scale or mesoscopic folds. In addition, microscopic examination does not show any evidence of plastic deformation of the grains or modification of the original texture by syn- or post-tectonic recrystallization. Figure 22.11 (a) shows a massive or weakly foliated granite and irregular clots of pegmatite cutting across the foliation of a veined gneiss. The discordant bodies are clearly post-tectonic with reference to the development of the gneissic foliation. Figure 22.11 (b) shows a similar feature. Here the quartzo-feldspathic material has been emplaced to form a nebulite (Sederholm 1926, p. 136), a diffused unfoliated migmatite which cuts across the foliations and the axial surfaces of a banded gneiss. The nebulite is post-tectonic in relation to the development of these structures. On the other hand, a granitic body which has been thoroughly deformed by a later penetrative deformation shows abundant evidence of grain elongation, undulatory

FIG. 22.11. (a) Foliation of a veined gneiss cut across by later granite and diffuse pegmatitic patches. The later granite is post-tectonic in relation to the foliation of the veined gneiss (Sengupta 1991). (b) Development of a nebulite superimposed on the pre-existing structure of a banded gneiss. The nebulite (diffuse white patches) is post-tectonic in relation to the foliation of the banded gneiss. Schirmacher Range, East Antarctica. Sengupta 1988. Photographs by courtesy of Sudipta Sengupta.

extinction of grains, bending of twin lamellae of plagioclase grains (Fig. 21.5a) or evidence of post-tectonic recrystallization of the original grains of the granite.

The majority of the large granite bodies are syntectonic. They are usually gneissose, foliated, migmatitic and contain numerous enclaves

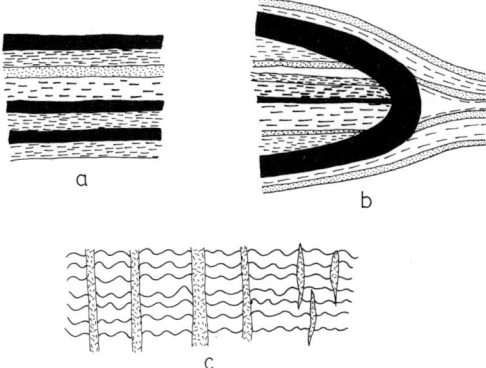

FIG. 22.12. (a) Banded gneiss with parallel bands of foliated amphibolite with concordant bands of granite gneiss of different compositions. (b) Migmatitic banding and foliation axial planar to a fold on a thick band of amphibolite. The migmatization is synkinematic with respect to the folding. (c) Syntectonic pegmatitic veins parallel to axial planes of folds in granite gneiss.

of metamorphic rocks. These gneissic terrains generally show structures of different generations. For such rocks it is usually necessary to relate the time of emplacement with the time of formation of different generations of structures.

There are a variety of field features from which a *synkinematic granite* can be identified. Some of these criteria are described below:

(1) The development of a regular migmatitic banding parallel to a cleavage is generally taken to indicate synkinematic migmatization. The resulting banded gneisses may be of different types. A type commonly encountered in gneissic complexes shows parallel bands of foliated amphibolite alternating with quartzo-feldspathic bands and other bands showing different stages of migmatization (Fig. 22.12a). Another common type is a veined gneiss which develops by *lit-par-lit* emplacement of thin bands of quartzo-feldspathic materials parallel to the schistosity of a mica schist. The remnant of the original metamorphic rock is generally described as palaeosome, while the newly formed quartzo-feldspathic material is called neosome.

In certain cases the synkinematic character of the rock is clearly demonstrated when the coarse banding of an earlier generation is folded and a finer migmatitic banding develops parallel to the axial planes. This later migmatization is evidently synkinematic with reference to that particular generation of folds (Fig. 22.12b). A similar conclusion can be drawn when a set of pegmatitic or aplitic veins cuts across the folded foliation of a granite gneiss and occurs parallel to the axial planes of the folds (Fig. 22.12c). Instead of the thin veins with sharp boundaries as in Fig. 22.12c, there may also be coarse bands of

FIG. 22.13. Pegmatite veins with diffused borders roughly parallel to axial planes of folds in a migmatite. Schirmacher Range, E. Antarctica. From Sengupta 1991.

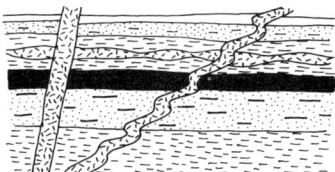

FIG. 22.14. Association of concordant pegmatite vein showing pinch-and-swell structure, folded discordant vein and undeformed discordant vein, indicating broadly syntectonic emplacement of pegmatite.

pegmatite with diffused borders, emplaced roughly parallel to the axial surfaces of the folds (Fig. 22.13). This feature also indicates syntectonic emplacement of the veins.

(2) The broadly synkinematic character of pegmatites and aplites can be established from the association of veins which were emplaced at different times with reference to the development of gneissic foliation. Figure 22.14 shows three types of pegmatite veins: (i) those which are concordant with the foliation and sometimes show pinch-and-swell structures, (ii) folded veins cutting across the foliation and (iii) straight veins cutting across the foliation. The association of these veins indicates that the period of pegmatite emplacement overlapped with the period of development of the gneissic foliation.

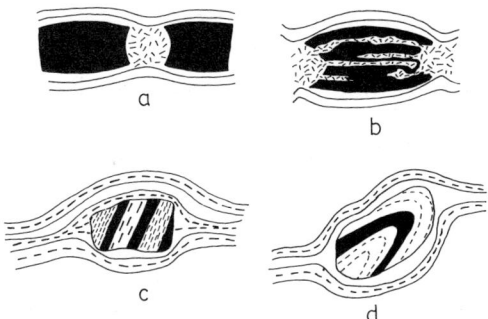

FIG. 22.15. (a) Strictly syntectonic emplacement of pegmatite in separation zone of boudins. (b) Amphibolite boudins partially invaded by diffused pegmatite. (c) Enclave of banded gneiss with foliation discordant with that of the host gneiss. (d) Folded foliation and banding of gneissic enclave discordant with foliation of host gneiss. All four features are from field sketches of migmatitic structures in Jasidih, E. India.

Wegmann & Shaer (1962) have shown that the time relation between the stages of deformation and the emplacement of pegmatitic or aplitic veins can be established with greater precision when, along with the features mentioned above, it is possible to identify different generations of pegmatite and aplite from their geometric relations with different generations of metabasic dykes which occur in the granitic terrain and which show different stages of granitization.

(3) Pegmatites emplaced in a separation zone between boudins (Fig. 22.15a) are strictly syntectonic with reference to the extension leading to boudinage. In certain cases the boudins of amphibolite are themselves migmatized (Fig. 22.15b) by the penetration of a diffused pegmatitic material. With progressive deformation, the migmatized boudins may be further deformed and foliated. Such an interplay of deformation and migmatization during boudinage in granitized terrains has been described in great detail by Wegmann (1932, pp. 487–488).

(4) Migmatitic complexes may contain enclaves of a gneissic rock with the cleavage, banding or the axial surface in the enclaves discordant with the foliation of the host gneiss. The occurrence of the isolated enclaves indicates that an earlier gneiss has been further granitized and the remnants are occurring as enclaves (Fig. 22.15c, d). The gneissic foliation of the host rock sweeps around the resistant enclaves. The enclaves may also have undergone some rotation. If the host gneiss does not show evidence of extensive post-crystalline deformation, we may conclude that its emplacement was broadly syntectonic with respect to the deformation which gave rise to its foliation and which caused a rotation of the enclave. Such features may enable us to distinguish the different stages of granitic activity as well as different generations of structures, an earlier structure preserved in the enclave and a later structure in the host rock.

(5) In certain terrains we find that the granitic activity continued during the development of brittle faults or ductile shear zones in the gneisses. Thus, we may find pegmatite emplaced along a fault or along ductile shear zones. Figure 22.16a shows a pegmatitic body emplaced along a ductile shear zone in granite gneiss. The central part of the pegmatite retains its very coarse-grained original texture. Towards the border of the zone the pegmatite is mylonitized; its new foliation is discordant with the older foliation of the gneiss. Figure 22.16b shows the development of a wider shear zone cutting across the foliation of an augen gneiss. The destruction of the augen structure in the shear zone is a result of both mylonitization and syntectonic migmatization. As a result, both the initial structure and composition have been greatly altered. The occurrence of pegmatite veins cutting across the foliations of both the augen gneiss and of the shear zone rock further indicates that the granitic activity outlasted the period of ductile shearing.

Basic dykes and enclaves are often very useful to reconstruct the tectonothermal history of a gneissic basement (Sederholm 1926, Wegmann 1965). This method was successfully used by Sutton & Watson (1951) in the Lewisian Gneisses of the North-western Highlands of Scotland. The Lewisian basement contains both an older Scourian gneiss and a younger Laxfordian gneiss. A swarm of basic dykes had intruded during the interval between the Scourian and Laxfordian times. During the Laxfordian cycle of orogeny the dykes were folded, metamorphosed and migmatized. The basic dykes thereby provided an index by means of which it was possible to separate two orogenic events of widely different ages.

The different generations of dykes may be identified by their cross-cutting relations (Fig. 22.17c) and also by their discordant relations with different generations of foliations in the host gneiss. Thus, for example, in the Precambrian gneissic basement of the Schirmacher Hills of E. Antarctica, Sudipta Sengupta has distinguished three major groups of events with the help of basic enclaves and dykes (Fig. 22.17) (1) The earliest enclaves occur as isolated bodies with a granulite facies (M_1) assemblage of minerals; the earliest internal foliation (D_1) is sharply truncated by the later D_2-foliation of the enveloping charnockite (Fig. 22.17a). (2) This foliation is cut across elsewhere by metabasic dykes (Fig. 22.17b) which also show a granulite facies assemblage (M_2). The cleavage in the dyke is parallel to the D_2-foliation. These dykes and the second phase of granulite facies metamorphism (M_2) are contemporaneous with some stage of D_2-deformation. (3) The third group of events includes metamorphism of basic bodies in amphibolite facies (M_3) and emplacement of a granite gneiss invading into the charnockite and metabasic rocks. The granitic emplacement is syntectonic with a third phase of deformation D_3. The D_3-foliation of the granite gneiss, in

FIG. 22.16. (a) A ductile shear zone in granite gneiss. The black band below the hammer is amphibolite. The shear zone contains pegmatite, the borders of which are mylonitized. The feature indicates syntectonic emplacement of pegmatite along the shear zone. (b) Ductile shear zone cutting across the foliation of augen gneiss. The shear zone is profusely invaded by a quartzo-feldspathic material. Late pegmatite veins cut across the foliations of both the augen gneiss and the shear zone rocks. Schirmacher Range, E. Antarctica. Photographs by courtesy of Sudipta Sengupta.

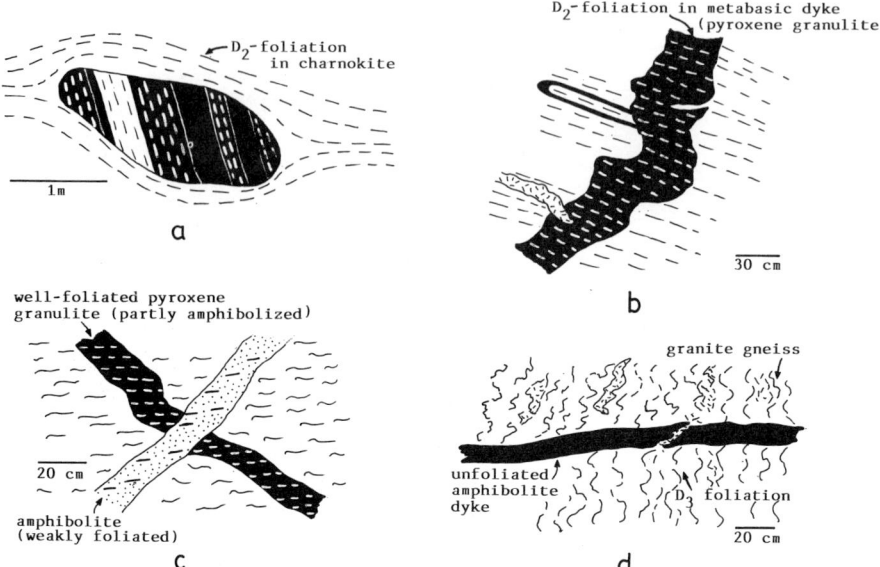

FIG. 22.17. Some relations between different generations of metabasic bodies and their gneissic host in Schirmacher Hills. E. Antarctica. (a) Early metabasic enclave of granulite facies rocks with the earliest foliation (D_1) truncated by D_2-foliation of the host charnockite. (b) Metabasic dyke (pyroxene granulite) cutting across D_2-foliation of charnockite gneiss. The dyke itself is invaded by veins of charnockitic pegmatite. (c) Younger metabasic dyke of weakly foliated amphibolite cross-cutting well-foliated older dyke of pyroxene granulite. Both dykes cut across the foliation of the host gneiss. (d) Unfoliated amphibolite dyke cutting across crenulated D_3-foliation of granite gneiss. The dyke itself is invaded by pegmatites. From such relations the sequence of the following three groups of events have been separated: (1) early granulite facies metamorphism syntectonic with D_1, (2) second granulite facies metamorphism and charnockitization syntectonic with D_2 and (3) amphibolite facies metamorphism and emplacement of granite gneiss partly overlapping with D_3. Sketched from photographs by Sudipta Sengupta.

its turn, is cut across by another generation of basic dykes (Fig. 22.17c, d) with an amphibolite facies assemblage and with a foliation parallel to the D_3-foliation of the gneiss. Some of the amphibolite dykes are themselves partly granitized.

The foliation within or immediately outside a metabasic dyke may have an orientation different from that of the host gneiss (Fig. 22.18). In particular, a dyke oblique to the general orientation of gneissic foliation may show an internal foliation parallel to the dyke wall (Fig. 22.18b), or a narrow zone of dyke-parallel foliation may develop in the gneiss near its contact with the dyke (Fig. 22.18c). Presumably, similar structures may develop either by syntectonic intrusion of a dyke along a shear zone or during deformation of the country rock after the emplacement of the dyke. For the latter case, the difference in foliation

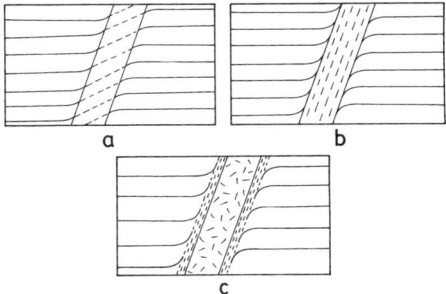

FIG. 22.18. Dissimilar orientations of foliation in the host gneiss and in dykes. In (c) the dyke is unfoliated but a wall-parallel shear zone has developed in the gneiss near the dyke. Schematic representation of three cases described by Sengupta (unpublished work) from the Schirmacher Hills. E. Antarctica. The dyke rock in (a) is an amphibolite and that in (b) is a highly biotitized amphibolite. The dyke in (c) is an unfoliated pyroxene granulite.

orientation (Fig. 22.18) results from a strong competence contrast (Sengupta, unpublished work). As indicated by Sengupta's studies in the Schirmacher Hills of E. Antarctica, this competence contrast is not a fixed entity but changes in the course of metamorphism and migmatization. A zone of high shear strain may develop in the host gneiss at the contact of an extremely competent dyke. On the other hand, if the dyke rock becomes incompetent with respect to the host, its passive rotation may be accompanied by a large wall-parallel shear strain within the dyke itself. Whether a mafic dyke intruded syntectonically along a shear zone or whether the shear zone was localized along a pre-existing dyke may sometimes be decided from the time relation between the metamorphic and deformation histories. Thus, for example, the dyke shown in Fig. 22.18c (a schematic representation of an example described by Sengupta) is an unfoliated two-pyroxene granulite containing a few discrete, microscopic, wall-parallel shear zones along which the minerals have undergone a retrogression to amphibolite facies. The dyke must have been intruded before the close of granulite facies metamorphism, whereas the wall-parallel shear zones could only have developed later when the rocks were under amphibolite facies condition. The possibility of syntectonic intrusion of the dyke along a shear zone can, therefore, be ruled out in this case.

References

Allen, A. R. 1979. Mechanism of frictional fusion in fault zones. *J. Struct. Geol.* **1**, 231–243.

Alvarez, W., Engelder, T. & Lowrie, W. 1976. Formation of spaced cleavage and folds in brittle limestone by dissolution. *Geology* **4**, 698–701.

Alvarez, W., Engelder, T. & Geiser, P. A. 1978. Classification of solution cleavage in pelagic limestones. *Geology* **6**, 263–266.

Amenzade, Yu. A. 1979. *Theory of elasticity*. Mir Publishers, Moscow, 284pp.

Anand, L. & Spitzig, W. A. 1982. Shear-band orientations in plane strain. *Acta Metall.* **30**, 553–561.

Anderson, E. M. 1951. *The dynamics of faulting and dyke formation with application to Britain*, 2nd edition. Oliver & Boyd, 206pp.

Anderson, R. E. 1971. Thin-skin distension in Tertiary rocks of southeastern Nevada. *Bull. Geol. Soc. Am.* **82**, 43–58.

Anderson, T. B. 1964. Kink bands and related geological structures. *Nature* **202**, 272–274.

Anderson, T. B. 1968. The geometry of natural orthorhombic kink bands. In: A. J. Baer & D. K. Norris (eds.), *Kink bands and brittle deformation. Geol. Surv. Can.*, Paper 68–52, pp. 200–226.

Anthony, M. & Wickham, J. 1978. Finite-element simulation of asymmetric folding. *Tectonophysics* **47**, 1–14.

Argand, E, 1911. Les nappes de recouvrement des Alpes penines et leurs prolongements structuraux. *Matér. Carte. Géol. Suisse* n.s. **31** (61), 26pp.

Argand, E. 1912. Sur la segmentation tectonique des Alpes occidentales. *Bull. Soc. Vaud. Sci. Nat.* 5th ser. **48** (176), 345–356.

Argand, E. 1916. Sur l'arc des Alpes occidentales. *Eclogae Geol. Helv.* **14**, 145–191.

Argand, E. 1922. La tectonique de l'Asie. *Int. Geol. Congr., 13th Session, Brussels*, pp. 171–372.

Armstrong. R. L. 1972. Low-angle (denudational) faults, hinterland of the Sevier orogenic belt, eastern Nevada and western Utah. *Bull. Geol. Soc. Am.* **83**, 1729–1754.

Ashby, M. F. 1972. A first report on deformation-mechanism maps. *Acta Metall.* **20**, 887–897.

Ashby, M. F. & Verrall, R. A. 1973. Diffusion accommodated flow and superplasticity. *Acta Metall.* **21**, 149–163.

Atherton, M. P. & Edmunds, W. M. 1966. An electron microprobe study of some zoned garnets from metamorphic rocks. *Earth Planetary Sci. Letters* **1**, 185–193.

Aubert, D. 1945. Le Jura et la tectonique d'écoulement. *Bull. Lab. Géol. Mineral. Géophys. Musée Géol. Univ. Lausanne* 83, 19pp.

Aubouin, J. 1965. *Geosynclines*. Elsevier Publishing Co., Amsterdam, 335pp.

Auden, J. B. 1937. The structure of the Himalaya in Garhwal. *Rec. Geol. Surv. India* **71**, 407–433.

Aydin, A. & Nur, A. 1982. Evolution of pull-apart basins and their scale independence. *Tectonics* **1**, 91–105.

Badgley, P. C. 1959. *Structural methods for the exploration geologist*. Harper & Brothers, New York, 280pp.

Badgley, P. C. 1965. *Structural and tectonic principles*. Harper & Row, 521pp.

Baëta, R. D. & Ashbee, K. H. G. 1967. Plastic deformation and fracture of quartz at atmospheric pressure. *Phil. Mag.* **15**, 931–938.

Baëta, R. D. & Ashbee, K. H. G. 1970. Mechanical deformation of quartz. I. Constant strain-rate compression experiments. II. Stress relaxation and thermal activation parameters. *Phil. Mag.* **22**, 601–635.

Bahat, D. 1979. Theoretical considerations on mechanical parameters of joint surfaces based on studies on ceramics. *Geol. Mag.* **116**, 81–92.

REFERENCES

Bahat, D. 1988. Early single-layer and late multi-layer joints in the Lower Eocene chalks near Beer Sheva, Israel. *Annales Tectonicae* **2**, 3–11.

Bahat, D. & Engelder, T. 1984. Surface morphology on cross-fold joints of the Appalachian Plateau, New York and Pennsylvania. *Tectonophysics* **104**, 299–373.

Bailey, E. B. 1923. Metamorphism of South-West Highlands. *Geol. Mag.* **60**, 317–331.

Bailey, E. B. 1935. *Tectonic essays, mainly alpine.* Oxford University Press, London, 200pp.

Bailey, E. B. & McCallien, W. J. 1937. Perthshire tectonics: Schichallion to Glen Lyon. *Trans. Roy. Soc. Edin.* **59**, 79–117.

Bailey, E. B., Weir, J. & McCallien, W. J. 1939. *Introduction to geology.* Macmillan, London.

Baird, A. K. 1962. Superposed deformations in the Central Sierra Nevada foothills east of the Mother Lode. *Bull. Univ. Calif. Dep. Geol.* 42, 69pp.

Bak, J., Korstgard, J. & Sorensen, K. A. 1975. Major shear zone within the Nagssugtoquidian of West Greenland. *Tectonophysics* **27**, 191–201.

Balderman, M. A. 1974. The effect of strain rate and temperature on the yield point of hydrolytically weakened quartz. *J. Geophys. Res.* **79**, 1647–1652.

Balk, R. 1937. Structural behavior of igneous rocks. *Mem. Geol. Soc. Am.* 5, 177pp.

Bally, A. W., Gordy, P. L. & Stewart, G. A. 1966. Structure, seismic data, and orogenic evolution of southern Canadian Rocky Mountains. *Bull. Can. Petrol. Geol.* **14**, 337–381.

Balsley, J. R. 1941. Deformation of marble under tension. *Trans. Am. Geophys. Union* **22**, 519–525.

Banks, C. J. & Warburton, J. 1986. "Passive roof" duplex geometry in the frontal structures of the Kirthar and Sulaiman mountain belts, Pakistan. *J. Struct. Geol.* **8**, 229–238.

Banno, S. 1965. Notes on rock forming minerals (34). Zonal structure of pyralspite garnet in Sambagwa schists in Bessi area, Sikoku. *J. Geol. Soc. Japan* **71**, 185–188.

Barr, D. 1987. Structure/stratigraphic models for extensional basins of half graben type. *J. Struct. Geol.* **9**, 491–500.

Barrel, J. 1917. Rhythms and measurement of geologic time. *Bull. Geol. Soc. Am.* **28**, 745–904.

Bartlett, W. L., Friedman, M. J. & Logan, J. M. 1981. Experimental folding and fracturing of rocks under confining pressure. Part IX. Wrench faults in limestone layers. *Tectonophysics* **79**, 255–257.

Bayly, M. B. 1964. A theory of similar folding in viscous materials. *Am. J. Sci.* **262**, 753–766.

Bayly, M. B. 1970. Viscosity and anisotropy estimates from measurements on chevron folds. *Tectonophysics* **9**, 459–474.

Bayly, M. B. 1971. Similar folds, buckling, and great-circle patterns. *J. Geol.* **74**, 110–118.

Bayly, M. B. 1974. Energy calculation concerning the roundness of folds. *Tectonophysics* **24**, 291–316.

Bayly, M. B., Borradaile, G. J. & Powell, C. McA. (eds.) 1977. *Atlas of rock cleavage*, Provisional edition. Univ. Tasmania, Hobart, 100 plates, 26pp.

Beach, A. 1974. Amphibolitization of Scourian granulites. *Scott. J. Geol.* **10**, 35–43.

Beach, A. 1979. Pressure solution as a metamorphic process in deformed terrigenous sedimentary rocks. *Lithos* **12**, 51–58.

Beach, A. 1980. Retrogressive metamorphic processes in shear zones with special reference to the Lewisian complex. *J. Struct. Geol.* **2**, 257–263.

Beach, A. 1982. Deformation mechanisms in some cover thrust sheets from the external French Alps. *J. Struct. Geol.* **4**, 137–149.

Becke, F. 1903. Über Minerallestand und Struktur der Kristallischen Schiefer. *Rep. Int. Geol. Congr. Vienna* II, 533–570. Published 1904.

Becker, G. F. 1896. Schistosity and slaty cleavage. *J. Geol.* **4**, 429–448.

Becker, G. F. 1904. Experiments on schistosity and slaty cleavage. *Bull. U.S.G.S.* **241**, 1–34.

Behrmann, J. H. 1985. Crystal plasticity and superplasticity in quartzite: a natural example. *Tectonophysics* **115**, 101–129.

Behrmann, J. H. & Platt, J. P. 1982. Sense of nappe emplacement from quartz-c-axis fabrics: an example from Betic Cordilleras (Spain). *Earth Planet Sci. Lett.* **59**, 208–215.

Behrmann, J. H. & Mainprice, D. 1987. Deformation mechanisms in a high-temperature quartz-feldspar mylonite: evidence for superplastic flow in the lower continental crust. *Tectonophysics* **140**, 297–305.

Bell, A. 1985. Strain paths during slaty cleavage formation — the role of volume loss. *J. Struct. Geol.* **7**, 563–568.

Bell, A. M. 1981. Vergence: an evaluation. *J. Struct. Geol.* **3**, 197–202.

Bell, T. H. 1978a. Progressive deformation and reorientation of fold axes in a ductile mylonite zone: the Woodroffe thrust. *Tectonophysics* **44**, 285–320.

Bell, T. H. 1978b. The development of slaty cleavage across the Nacakra arc of the Adelaide geosyncline. *Tectonophysics* **51**, 171–201.

Bell, T. H. 1985. Deformation partitioning and porphyroblast rotation in metamorphic rocks: a radical reinterpretation. *Metamorphic Geol.* **3**, 109–118.

Bell, T. H. & Etheridge, M. A. 1973. Microstructure of mylonites and their descriptive terminology. *Lithos* **6**, 337–348.

Bell, T. H. & Etheridge, M. A. 1976. The deformation and recrystallization of quartz in a mylonite zone, Central Australia. *Tectonophysics* **32**, 235–267.

Bell, T. H. & Rubenach, M. J. 1980. Crenulation cleavage development — evidence for progressive bulk inhomogeneous shortening from "millipede" microstructures in the Robertson Rivers Metamorphics. *Tectonophysics* **68**, T9–T15.

Bell, T. H. & Rubenach, M. J. 1983. Sequential porphyroblast growth and crenulation cleavage development. *Tectonophysics* **92**, 171–194.

Beloussov, V. V. 1962. *Basic problems in geotectonics*. McGraw-Hill, New York, 809pp.

Berthé, D. & Brun, J. P. 1980. Evolution of folds during progressive shear in South Armorican Shear Zone, France. *J. Struct. Geol.* **2**, 127–133.

Berthé, D., Choukroune, P. & Jegouzo, P. 1979. Orthogneiss, mylonite and non-coaxial deformation in granites: the example from South Armorican Shear Zone. *J. Struct. Geol.* **1**, 31–42.

Berthelsen, A. 1957. The structural evolution of an ultra- and polymetamorphic gneissic complex, west Greenland. *Geol. Rdsch.* **50**, 474–499.

Berthelsen, A. 1960. Structural studies in the Pre-cambrian of Western Greenland. II. Geology of Tovqussap Nunâ. *Medd. Grønland.* **123**, 1–226.

Bertrand, M. 1899. La grande nappe de recouvrement de la Basse-Provence. *Bull. Serv. Carte Géol. France* **10**, 297–462.

Billings, M. P. 1954. *Structural geology*, 2nd edition. Prentice Hall, Englewood Cliffs, New Jersey, 514pp.

Biot, M. A. 1957. Folding instability of a layered viscoelastic medium under compression. *Proc. Roy. Soc. Lond. Ser. A* **242**, 444–454.

Biot, M. A. 1961. Theory of folding of stratified viscoelastic media and its implications in tectonics and orogenesis. *Bull. Geol. Soc. Am.* **72**, 1595–1620.

Biot, M. A. 1964. Theory of internal buckling of a confined multilayered structure. *Bull. Geol. Soc. Am.* **75**, 563–568.

Biot, M. A. 1965a. *Mechanics of incremental deformations*. John Wiley & Sons, Inc., New York, 504pp.

Biot, M. A. 1965b. Further development of the theory of internal buckling of multilayers. *Bull. Geol. Soc. Am.* **75**, 833–840.

Biot, M. A. 1965c. Theory of similar folding of the first and second kind. *Bull. Geol. Soc. Am.* **76**, 251–258.

Biot, M. A. 1965d. Theory of viscous buckling and gravity instability of multilayers with large deformation. *Bull. Geol. Soc. Am.* **76**, 371–378.

Biot, M. A., Odé, H. & Roever, W. L. 1961. Experimental verification of the theory of folding of stratified viscoelastic media. *Bull. Geol. Soc. Am.* **72**, 1621–1631.

Blacic, J. D. 1972. Effect of water on the experimental deformation of olivine In: H. C. Heard, I. Y. Borg, N. L. Carter & C. B. Raleigh (eds.), *Flow and fracture of rocks. Geophys. Monogr. Am. Geophys. Union* **16**, 109–115.

Blacic, J. D. 1975. Plastic deformation mechanisms in quartz: the effect of water. *Tectonophysics* **27**, 271–294.

Blacic, J. D. & Christie, J. M. 1984. Plasticity and hydrolytic weakening of quartz single crystals. *J. Geophys. Res.* **89**, 4223–4239.

Blatt, H., Middleton, G. & Murray, R. 1980. *Origin of sedimentary rocks*. Prentice Hall, Englewood Cliffs, New Jersey, 782pp.

Blenkinsop, T. G. & Rutter, E. H. 1986. Cataclastic deformation of quartzite in Moine thrust zone. *J. Struct. Geol.* **8**, 669–681.

Bock, H. 1980. Das fundamentale kluft system. *Z. Deut. Geol. Ges.* **131**, 627–650.

Bonney, T. G. 1886. Anniversary address of the President. *Proc. Geol. Soc. Lond.* **42**, 38–115.

REFERENCES

Borradaile, G. J. 1978. Transected folds: a study illustrated with examples from Canada and Scotland. *Bull. Geol. Soc. Am.* **89**, 481–493.

Borradaile, G. J., Bayly, M. B. & Powell, C. McA. (eds.) 1982. *Atlas of deformation of rock fabrics.* Springer, New York.

Bosworth, W. 1985. Discussion on the structural evolution of extensional basin margins. *J. Geol. Soc. Lond.* **142**, 939–942.

Bouchez, J. L. 1977. Plastic deformation of quartzites at low temperature in an area of natural strain gradient. *Tectonophysics* **39**, 25–50.

Bouchez, J. L. & Pecher, A. 1976. Plasticité du quartz et sens de cisaillement dans les quartzites du Grand Chevauchement Central Himalayen. *Bull. Soc. Géol. France* **6**, 1375–1383.

Bouchez, J. L. & Pecher, A. 1981. The Himalayan Main Central Thrust pile and its quartz-rich tectonites in central Nepal. *Tectonophysics* **78**, 23–50.

Boullier, A. M. & Gueguen, Y. 1975. SP-mylonites: origin of some mylonites by superplastic flow. *Contr. Min. Pet.* **50**, 93–105.

Boulter, C. A. & Raheim, A. 1974. Variations in Si^{4+} content of phengites through a three stage deformation sequence. *Contr. Min. Pet.* **48**, 57–71.

Bowden, P. B. 1970. A criterion for inhomogeneous plastic deformation. *Phil. Mag.* **22**, 455–462.

Bowden, P. B. & Raha, S. 1970. The formation of microshear bands in polystyrene. *Phil. Mag.* **22**, 463–482.

Boyer, S. E. & Elliott, D. 1982. Thrust systems. *Bull. Am. Assoc. Petrol. Geol.* **66**, 1196–1230.

Breddin, H. 1956. Die tectonische Deformation der Fossilien im Rheinischen Schiefergebirge. *Deut. Geol. Ges. Z.* **106**, 227–305.

Breddin, H. 1957. Tectonische Fossil und Gesteinsdeformation im Gebiet von St. Goarhausen. *Decheniana* **110**, 289–350.

Bretherton, F. P. 1962. The motion of rigid particles in a shear flow at low Reynolds number. *J. Fluid Mech.* **14**, 284–304.

Bretz, J. H. 1940. Solution cavities in the Joliet limestone of northeastern Illinois. *J. Geol.* **50**, 675–811.

Bristol, H. M. 1975. Structural geology and oil production in Gallatin County and southernmost White County, Illinois. *Ill. State Geol. Surv. Ill. Petrol.* **105**, 1–20.

Brown, J. S. & Engel, A. E. J. 1956. Revision of the Grenville stratigraphy and structure in the Balmat–Edwards district northwest Adirondacks, New York. *Bull. Geol. Soc. Am.* **67**, 1599–1622.

Bryant, B. & Reed, J. C. 1969. Significance of lineation and minor folds near major thrust faults in the Southern Appalachians and the British and Norwegian Caledonides. *Geol. Mag.* **106**, 412–429.

Bucher, W. H. 1933. *The deformation of the earth's crust.* Princeton Univ. Press, 518pp.

Burg, J. P. & Laurent, Ph. 1978. Strain analysis of a shear zone in a granodiorite. *Tectonophysics* **47**, 15–42.

Butler, R. W. H. 1982. The terminology of structures in thrust belts. *J. Struct. Geol.* **4**, 239–245.

Buxtorf, A. 1907. Geologische Beschreibung des Weissenstein Tunnel und seiner Umgebung. *Beitr. Geol. Karte Schweiz.* **51**, 1–125.

Byerlee, J. D. 1968. Brittle–ductile transition in rocks. *J. Geophys. Res.* **73**, 4741–4750.

Callaway, C. 1883a. The age of the newer gneissic rocks of the northern Highlands. *Quart. J. Geol. Soc. Lond.* **39**, 355–414.

Callaway, C. 1883b. The Highland problem. *Geol. Mag.* n.s. **10**, 139–140.

Campbell, J. W. 1958. En echelon folding. *Econ. Geol.* **53**, 448–472.

Carey, S. W. 1962. Folding. *J. Alberta Soc. Petrol. Geol.* **10**, 95–144.

Carreras, J., Estrada, A. & White, S. 1977. The effect of folding on the *c*-axis fabric of a quartz mylonite. *Tectonophysics* **39**, 3–24.

Carter, N. L. & Avé Lallemant, H. G. 1970. High temperature flow of dunite and peridotite. *Bull. Geol. Soc. Am.* **81**, 2181–2202.

Carter, N. L., Christie, J. M. & Griggs, D. T. 1961. Experimentally produced deformation lamellae and other structures in quartz sand. *J. Geophys. Res.* **66**, 2518–2519.

Carter, N L., Christie, J. M. & Griggs, D. T. 1964. Experimental deformation and recrystallization of quartz. *J. Geol.* **72**, 687–733.

Casey, M., Dietrich, D. & Ramsay, J. G. 1983. Methods for determining deformation history of chocolate tablet boudinage with fibrous crystals. *Tectonophysics* **92**, 211–239.

Chakrabarti, A. K. & Spretnak, J. W. 1975. Instability of plastic flow in the direction of pure shear. *Met. Trans.* **6A**, 733–747.

Chapman, T. J. & Williams, G. D. 1984. Displacement–distance techniques in the analysis of fold-thrust structures and linked fault systems. *J. Geol. Soc. Lond.* **141**, 121–129.

Chapple, W. M. 1968. A mathematical theory of finite amplitude rock folding. *Bull. Geol. Soc. Am.* **79**, 47–68.

Chapple, W. M. 1969. Fold shape and rheology: the folding of an isolated viscous plastic layer. *Tectonophysics* **7**, 97–116.

Chapple, W. M. & Spang, J. H. 1974. Significance of layer-parallel slip during folding of layered sedimentary rocks. *Bull. Geol. Soc. Am.* **85**, 1523–1534.

Chenet, P., Monterdert, L., Gairaud, H. & Roberts, D. 1983. Extension ratio measurements on the Glacia, Portugal and northern Biscay continental margins. In: J. S. Watkins & C. L. Drake (eds.), *Studies in continental margin geology. Am. Assoc. Petrol. Geol. Mem.* **34**, 703–715.

Chester, F. M., Friedman, M. & Logan, J. M. 1985. Foliated cataclasites. *Tectonophysics* **111**, 139–146.

Chopra, P. N. & Paterson, M. S. 1981. The experimental deformation of dunite. *Tectonophysics* **78**, 453–473.

Chopra, P. N. & Paterson, M. S. 1984. The role of water in the deformation of dunite. *J. Geophys. Res.* **89**, 7861–7876.

Choukroune, P. 1971. Contribution a l'étude des mechanismes de la deformation avec schistosité grace aux crystallisation syncinématiques dans les "zones arbitées" ("pressure shadows"). *Bull. Soc. Géol. France* **13**, 257–271.

Choukroune, P. & Gapais, D. 1983. Strain pattern in the Aar granite (central Alps): orthogneiss developed by bulk inhomogeneous shortening. *J. Struct. Geol.* **5**, 411–418.

Christensen, J. N., Rosenfeld, J. L. & De Paolo, D. J. 1989. Rates of tectonometamorphic processes from rubidium and strontium isotopes in garnet. *Science* **244**, 1465–1469.

Christie, J. M. 1963. The Moine thrust zone in the Assynt region, northwest Scotland. *California Univ. Publ. Geol. Sciences* **40**, 345–440.

Christie, J. M. & Ardell, A. J. 1974. Substructures of deformation lamellae in quartz. *Geology* **2**, 405–408.

Christie, J. M., Ardell, A. J. & Balderman, M. A. 1973. The substructure of basal lamellae in experimentally deformed quartz (abstract). *EOS Trans. Am. Geophys. Union* **54**, 453.

Christie, J. M., Griggs, D. T. & Carter, N. L. 1964. Experimental evidence of basal slip in quartz. *J. Geol.* **72**, 734–756.

Christie, J. M. & Ord, A. 1980. Flow stress from microstructures: example and current assessment. *J. Geophys. Res.* **85**, 6253–6262.

Clark, R. H. & McIntyre, D. B. 1951. The use of terms pitch and plunge. *Am. J. Sci.* **249**, 591–599.

Clifford, P. 1960. The geological structures of the Loch Luichart area, Ross-shire. *Quart. J. Geol. Soc. Lond.* **115**, 365–388.

Clifford, P., Fleuty, M. J., Ramsay, J. G., Sutton, J. & Watson, J. 1957. The development of lineation in complex fold systems. *Geol. Mag.* **94**, 1–24.

Clifford, P. M. 1968. Kink band development in the Lake St. Joseph area, North western Ontario. In: A. J. Baer & D. K. Norris (eds.), *Proc. Conference Research in Tectonics. Geol. Surv. Can. Paper* 68–52, pp. 229–242.

Cloos, E. 1932. "Feather joints" as indicators of movements in faults, thrusts, joints and magmatic contacts. *Proc Nat. Acad. Sci.* **18**, 387–395.

Cloos, E. 1946. Lineation, a critical review and annotated bibliography. *Mem. Geol. Soc. Am.* **18**, 122pp.

Cloos, E. 1947a. Boudinage. *Trans. Am. Geophys. Union* **28**, 626–632.

Cloos, E. 1947b. Oolite deformation in South Mountain fold, Maryland. *Bull. Geol. Soc. Am.* **58**, 843–918.

Cloos, E. 1955. Experimental analysis of fracture patterns. *Bull. Geol. Soc. Am.* **66**, 241–256.

Cloos, E. 1968. Experimental analysis of Gulf Coast fracture patterns. *Bull. Am. Assoc. Petrol. Geol.* **52**, 420–454.

Cloos, E. 1971. *Microtectonics along the western edge of the Blue Ridge, Maryland and Virginia.* The Johns Hopkins Press, Baltimore, 234pp.

Cloos, H. 1928. Über antithetischke Bewegungen. *Geol. Rundsch.* **19**, 246–251.

REFERENCES

Cloos, H. 1930. *Künstliche Gebirge. Natur und Museum Senckenbergische Naturf. Ges. Frankfurt*, Part 2, **6**, 258–269.

Cloos, H. 1936. *Einführung in die Geologie, Ein Lehrbuch der inneren Dynamik*. Gebrüder Borntraeger, Berlin, 503pp.

Clough, C. T. 1897. In: W. Gunn *et al.*, *The geology of Cowal including the part of Argyllshire between the Clyde and Loch Fine. Mem. Geol. Surv. Scotland*, pp. 1–111.

Cobbold, P. R. 1975. Fold propagation in a single embedded layer. *Tectonophysics* **27**, 331–351.

Cobbold, P. R. 1977. Description and origin of banded deformation structures. I. Regional strain, local perturbations and deformation bands. II. Rheology and the growth of banded perturbations. *Can. J. Earth Sci.* **14**, 1721–1731, 2510–2523.

Cobbold, P. R., Cosgrove, J. W. and Summers, J. M. 1971. Development of internal structures in deformed anisotropic rocks. *Tectonophysics* **112**, 23–53.

Cobbold, P. R. & Quinquis, H. 1980. Development of sheath folds in shear regimes. *J. Struct. Geol.* **2**, 119–126.

Coe, K. 1959. Boudinage structure in West Cork, Ireland. *Geol. Mag.* **96**, 191–200.

Coleman, J. M. & Garrison, L. E. 1977. Geological aspects of marine slope instability, north-western Gulf of Mexico. *Mar. Geotech.* **2**, 9–44.

Collet, L. W. 1936. *The structure of the Alps*, 2nd edition, Arnold, London, 289pp.

Collins, J. D. & Thompson, D. B. 1982. *Sedimentary structures*. George Allen & Unwin, 194pp.

Collomb, P. & Donzeau, M. 1974. Relations entre kink-bands décamétriques et fractures de socle dans l'Hercynien des Monts D'Ougarte (Sahara occidentale, Algérie). *Tectonophysics* **24**, 213–242.

Cooper, M. A. & Trayner, P. M. 1986. Thrust-surface geometry: implications for thrust-belt evolution and section balancing techniques. *J. Struct. Geol.* **8**, 305–312.

Cooper, M. A., Garton, M. R. & Hossack, J. R. 1983. The origin of Basse Normandie Duplex, Bullonais, France. *J. Struct. Geol.* **5**, 139–152.

Corin, F. 1932. A propos du boudinage en Ardenne. *Bull. Soc. Belge. Géol.* **42**, 101–117.

Cosgrove, J. W. 1976. The formation of crenulation cleavage. *J. Geol. Soc. Lond.* **132**, 155–178.

Cosgrove, J. W. 1980. The tectonic implication of some small structures in the Mona Complex of Holy Isle, North Wales. *J. Struct. Geol.* **2**, 383–396.

Couples, G. 1977. Stress and shear fracture (fault) patterns resulting from a suite of complicated boundary conditions with applications to the Wind River Mountains. *Pure and Appl. Geophys.* **115**, 113–134.

Coward, M. P. 1976. Strain within ductile shear zones. *Tectonophysics* **34**, 181–197.

Coward, M. P., Dietrich, D. & Park, R. G. 1989. *Alpine tectonics*. Geological Society Spec. Pub. 45. Blackwell Scientific Publications, Oxford, 450pp.

Cox, H. L. 1952. The elasticity and strength of paper and other fibrous materials. *Brit. J. Appl. Phys.* **3**, 72–79.

Crowell, J. C. 1974. Origin of late Cenozoic basins in southern California. In: W. R. Dickinson (ed.), *Tectonics and sedimentation*. Special Publication Soc. Econ. Palaeont. Mineral. No. 22, pp. 190–204.

Crowell, J. C. 1975. The Gabriel fault and Ridge Basin, Southern California. *Calif. Div. Mines Geol. Spec. Rep.* **118**, 208–219.

Cummins, W. A. & Shackleton, R. M. 1955. The Ben Lui recumbent syncline (S.W. Highlands). *Geol. Mag.* **92**, 353–363.

Currie, J. B., Patnode, H. W. & Trump, R. P. 1962. Development of folds in the sedimentary strata. *Bull. Geol. Soc. Am.* **73**, 655–674.

Dahlstrom, C. D. A. 1954. Statistical analysis of cylindrical folds. *Bull. Can. Inst. Min. Metall.* **57**, 140–145.

Dahlstrom, C. D. A. 1969. Balanced cross-sections. *Can. J. Earth Sci.* **6**, 743–757.

Dahlstrom, C. D. A. 1970. Structural geology in the eastern margin of the Canadian Rocky Mountains. *Bull. Can. Pet. Geol.* **18**, 332–406.

Dalziel, I. W. D. & Bailey, S. W. 1968. Deformed garnets in mylonite rock from the Grenville front and their tectonic significance. *Am. J. Sci.* **266**, 542–564.

Darwin, C. 1846. *Geological observation in South America*. Smith-Edler.

Daubrée, G. A. 1876. Experiences sur la schistosité des roches et sur les déformations des fossiles, corelatives de ce phénomène. *Compt. Rend. Acad. Sci., Paris* **82**(13), 710 and **82** (15), 798.

Davis, D., Suppe, G. A. & Dahlen, F. A. 1983. Mechanics of fold-and-thrust belts and accretionary wedges. *J. Geophys. Res.* **88**, 1153–1172.

Davis, G. A., Anderson, J. L., Frost, E. G. & Sackleford, T. J. 1980. Regional Miocene detachment faulting and early Tertiary mylonitization, Whippe-Buckskin-Rawhide Mountains, southeastern California and western Arizona. *Mem. Geol. Soc. Am.* **153**, 79–130.

Davison, I. 1986. Listric normal fault profiles: calculation using bed-length balance in fault displacement. *J. Struct. Geol.* **8**, 209–210.

Decker, C. E. 1920. *Studies in minor folds.* Univ. Chicago Press, 89pp.

Dennis, J. G. 1972. *Structural geology.* Ronald Press Co., New York, 532pp.

Dennison, J. M. & Woodward, H. P. 1963. Palinspastic maps of central Appalachians. *Bull. Am. Assoc. Petrol. Geol.* **47**, 666–680.

de Sitter, L. U. 1954. Schistosity and shear in micro- and macro-folds. *Geol. Mijnb.* **16**, 429–439.

de Sitter, L. U. 1958a. Boudins and parasitic folds in relation to cleavage and folding. *Geol. Mijnb.* **20**, 272–286.

de Sitter, L. U. 1958b. *Structural geology*, 2nd edition. McGraw-Hill Book Co., New York, 551pp.

de Sitter, L. U. & Zwart, H. J. 1960. Tectonic development in supra- and infra-structures of a mountain chain. *Int. Geol. Congr., XXI, Copenhagen, Pt. 18*, pp. 248–256.

Dewey, J. F. 1965. Nature and origin of kink bands. *Tectonophysics* **1**, 459–494.

Dewey, J. F. 1966. Kink bands in lower Carboniferous slates of Rush, Co. Dublin. *Geol. Mag.* **103**, 138–142.

de Wit, M. J. 1976. Metamorphic textures and deformation: a new mechanism for the development of syntectonic porphyroblasts and its implications for interpreting time relationships in metamorphic rocks. *Geol. J.* **11**, 71–100.

Dieter, G. E. 1988. *Mechanical metallurgy*, SI metric edition. McGraw-Hill, New York, 751pp.

Dietrich, J. H. 1970. Computer experiments on mechanics of finite amplitude folds. *Can. J. Earth Sci.* **7**, 467–476.

Dietrich, J. H. & Carter, N. L. 1969. Stress history of folding. *Am. J. Sci.* **267**, 129–154.

Dillamore, I. L., Roberts, J. G. & Bush, S. E. 1979. Occurrence of shear bands in heavily rolled cubic metals. *Metal Sci. J.* **13**, 73–77.

Dixon, J. M. 1976. Apparent "double rotation" of porphyroblasts during a single progressive deformation. *Tectonophysics* **34**, 101–115.

Dixon, Joe S. 1982. Regional structural synthesis, Wyoming salient of western overthrust belt. *Bull. Am. Assoc. Petrol. Geol.* **66**, 1560–1580.

Donath, F. A. 1968. Experimental study of kink band development in strongly anisotropic rock. *Geol. Surv. Can.*, Paper 52, pp. 255–287.

Donath, F. A. & Faill, R. T. 1963. Ductile faulting in experimentally deformed rocks. *Trans. Am. Geophys. Union* **44**, 103 (Abstract).

Donath, F. A. & Fruth, L. S. 1971. Dependence of strain rate effects on mechanism and rock type. *J. Geol.* **79**, 347–371.

Donath, F. A. & Parker, R. B. 1964. Folds and folding. *Bull. Geol. Soc. Am.* **75**, 45–62.

Donath, F. A. & Wood, D. S. 1976. Experimental evaluation of the deformation path concept. *Phil. Trans. Roy. Soc. Lond. Ser. A* **283**, 187–201.

Donath, F. A., Faill, R. T. & Tobin, D. G. 1971. Deformational mode fields in experimentally deformed rock. *Bull. Geol. Soc. Am.* **82**, 1441–1462.

Douglas, R. J. W. 1950. Callum Creek, Langford Creek and Gap map areas. *Mem. Geol. Surv. Can.* no. 255, 124pp.

Douglas, R. J. W. 1952. Preliminary map, Waterton, Alberta. *Can. Geol. Surv. Paper 51–22.*

Doukhan, J. -C. & Trepied, L. 1985. Plastic deformation of quartz single crystals. *Bull. Mineral.* **108**, 97–123.

Dunbar, C. O. & Rogers, J. 1957. *Principles of stratigraphy.* John Wiley & Sons, 356pp.

Duncan, A. C. 1985. Transected folds: a re-evaluation, with the examples from the type area at Sulphur Creek, Tasmania. *J. Struct. Geol.* **7**, 409–419.

Dunne, W. M. & North, C. P. 1990. Orthogonal fracture systems at the limits of thrusting: an example from southwestern Wales. *J. Struct. Geol.* **12**, 207–215.

Dunnet, D. 1969. A technique for finite strain analysis using elliptical particles. *Tectonophysics* **7**, 117–136.

REFERENCES

Dunnet, D. & Siddans, A. W. B. 1971. Non-random sedimentary fabrics and their modification by strain. *Tectonophysics* **12**, 307–325.

Durham, W. B. & Goetze, C. 1977a. Plastic flow of oriented single crystals of olivine. 1. Mechanical data. *J. Geophys. Res.* **82**, 5737–5753.

Durham, W. B. & Goetze, C. 1977b. A comparison of creep properties of pure forsterite and iron bearing olivine. *Tectonophysics* **40**, T15–T18.

Durney, D. W. 1972. Solution transfer, an important geological deformation mechanism. *Nature* **235**, 315–317.

Durney, D. W. 1976. Pressure solution and crystallization deformation. *Phil. Trans. Roy. Soc. Lond. Ser. A* **283**, 229–240.

Durney, D. W. 1978. Early theories and hypotheses on pressure-solution-redeposition. *Geology* **6**, 369–372.

Durney, D. W. & Ramsay, J. G. 1973. Incremental strain measured by syntectonic crystal growth. In: K. A. De Jong & R. Scholtan (eds.), *Gravity and tectonics*, pp. 67–96. Wiley, New York.

Edelman, N. 1949. Structural history of the Gulkrona basin, S.W. Finland. *Bull. Comm. Géol. Finl.* **148**, 1–87.

Ekström, T. K. 1975. Pinch and swell structures from a Swedish locality. *Geol. För. Stock. Förh.* **97**, 72–79.

Elliott, D. 1965. Quantitative mapping of directional minor structures. *J. Geol.* **73**, 865–880.

Elliott, D. 1970. Determination of finite strain and initial shape from deformed elliptical objects. *Bull. Geol. Soc. Am.* **81**, 2221–2236.

Elliott, D. 1972. Deformation paths in structural geology. *Bull. Geol. Soc. Am.* **83**, 2621–2638.

Elliott, D. 1973. Diffusion flow laws in metamorphic rocks. *Bull. Geol. Soc. Am.* **84**, 2645–2664.

Elliott, D. 1976a. Energy balance and deformation mechanisms of thrust sheets. *Phil. Trans. Roy. Soc. Lond. Ser. A* **283**, 289–312.

Elliott, D. 1976b. The motion of thrust sheets. *J. Geophys. Res.* **81**, 949–963.

Elliott, D. & Johnson, M. R. W. 1980. Structural evolution in the northern part of N.W. Scotland. *Trans. Roy. Soc. Edinb.* **71**, 69–96.

Elliott, T. 1986. Deltas. In: H. G. Reading (ed.), *Sedimentary environments and facies*, 2nd edition, pp. 113–154. Blackwell Scientific Publications, London.

Engelder, T. 1982. Reply to comments by A. E. Scheidegger on "Is there a genetic relationship between selected regional joints and contemporary stress within the lithosphere of North America?" *Tectonics* **1**, 465–470.

Engelder, T. 1985. Loading paths to joint propagation during a tectonic cycle: an example from the Appalachian Plateau, U.S.A. *J. Struct. Geol.* **7**, 459–476.

Engelder, T. & Marshak, S. 1985. Development of cleavage in limestone of a fold thrust belt in eastern New York. *J. Struct. Geol.* **7**, 345–359.

Ermanovics, I. F., Helmstaedt, H. & Plant, A. G. 1972. An occurrence of Archaean pseudo-tachylite from southeastern Manitoba. *Can. J. Earth Sci.* **9**, 257–265.

Escher, A. & Waterson, J. 1974. Stretching fabrics, folds and crustal shortening. *Tectonophysics* **22**, 223–231.

Eskola, P. 1949. The problem of mantled gneiss domes. *Quart. J. Geol. Soc. Lond.* **104**, 469–476.

Etchecopar, A. 1977. A plane kinematic model of progressive deformation in polystyrene aggregate. *Tectonophysics* **39**, 121–139.

Etchecopar, A. & Malavieille, J. 1987. Computer models of pressure shadows: a method for strain measurement and shear-sense determination. *J. Struct. Geol.* **9**, 667–677.

Etheridge, M. A. & Lee, M. F. 1975. Microstructure of slate from Lady Loretta, Queensland, Australia. *Bull. Geol. Soc. Am.* **86**, 13–22.

Etheridge, M. A., Paterson, M. S. & Hobbs, B. E. 1974. Experimentally produced preferred orientation in synthetic mica aggregates. *Contr. Min. Pet.* **44**, 274–294.

Etheridge, M. A. & Wilkie, J. C. 1981. An assessment of dynamic recrystallization grain size as a palaeopiezometer in quartz-bearing mylonite zones. *Tectonophysics* **78**, 475–508.

Evamy, B. D., Haremboure, J., Kamerling, P., Knapp, W. A., Molloy, F. A. & Rowlands, P. H. 1978. Hydrocarbon habitat of Tertiary Niger delta. *Bull. Am. Assoc. Petrol. Geol.* **62**, 1–39.

Evans, A. M. 1963. Conical folding and oblique structures in Charnwood Forest, Leicester. *Proc. Yorkshire Geol. Soc.* **34**, 67–80.

Fairbairn, H. W. 1949. *Structural petrology of deformed rocks*. Addison-Wesley Publishing Co. Inc., Cambridge, 344pp.

Fairbairn, H. W. 1950. Pressure shadows and relative movements in shear zones. *Trans. Am. Geophys. Union* **31**, 914–916.

Ferguson, C. C. & Lloyd, G. E. 1982. Paleostress and strain estimates from boudinage structure and their bearing on the evolution of a major Variscan fold-thrust complex in southwest England. *Tectonophysics* **88**, 269–289.

Fernandez, A. 1987. Preferred orientation developed by rigid markers in two-dimensional simple shear strain: a theoretical and experimental study. *Tectonophysics* **136**, 151–158.

Finkelbeiner, D. T. II. 1966. *Introduction to matrices and linear transformation*, 2nd edition. W. H. Freeman & Co., 297pp.

Fisher, O. 1884. On cleavage and distortion. *Geol. Mag.* (Decade 3) **1**, 268–276, 396–406.

Fletcher, R. C. 1974. Wavelength selection in the folding of a single layer with power-law rheology. *Am. J. Sci.* **274**, 1029–1043.

Flett, J. S. 1912. In: B. N. Peach, W. Gunn, C. T. Clough, L. W. Hinxman, C. B. Crampton, E. M. Anderson & J. S. Flett (eds.), *The geology of Ben Wyvis, Carn Chuinneag, Inchbae and their surrounding country, including Garve, Evanton, Alness and Kincardine* (explanation on sheet no. 93). *Mem. Geol. Surv. U.K.* 1912, pp. 1–189.

Fleuty, M. J. 1961. The three fold-systems in the metamorphic rocks of Upper Glen Orrin, Ross-shire and Inverness-shire. *Quart. J. Geol. Soc. Lond.* **117**, 447–479.

Fleuty, M. J. 1964. The description of folds. *Proc. Geol. Assoc.* **75**, 461–492.

Flinn, D. 1956. On the deformation of the Funzie conglomerate, Fetlar, Shetland. *J. Geol.* **64**, 480–505.

Flinn, D. 1962. On folding during three-dimensional progressive deformation. *Quart. J. Geol. Soc. Lond.* **118**, 385–433.

Flinn, D. 1978. Construction and computation of three-dimensional progressive deformations. *J. Geol. Soc. Lond.* **135**, 291–305.

Foster, M. E. & Hudleston, P. J. 1986. "Fracture cleavage" in the Duluth complex, northeastern Minnesota. *Bull. Geol. Soc. Am.* **97**, 85–96.

Francis, P. W. 1972. The pseudotachylite problem. *Comments on Earth Science Geophys.* **3**, 35–53.

Freund, R. 1970a. Rotation of strike slip faults in Sistan, southeastern Iran. *J. Geol.* **78**, 188–200.

Freund, R. 1970b. The geometry of faulting in the Galilee. *Israel J. Earth Sci.* **19**, 117–140.

Freund, R. 1974. Kinematics of transform faults. *Tectonophysics* **21**, 93–134.

Frost, H. J. & Ashby, M. F. 1982. *Deformation mechanism maps*. Pergamon Press, Oxford, 166pp.

Fry. N. 1979a. Density distribution techniques and strained length methods for determination of finite strains. *J. Struct. Geol.* **1**, 221–229.

Fry, N. 1979b. Random point distributions and strain measurement in rocks. *Tectonophysics* **60**, 89–105.

Fullagar, P. K. 1980. A description of nucleation of folds and boudins in terms of vorticity. *Tectonophysics* **65**, 39–55.

Fyson, W. K. 1962. Tectonic structures in the Devonian rocks near Plymouth, Devon. *Geol. Mag.* **99**, 208–226.

Fyson, W. K. 1966. Structures in the Lower Palaeozoic Meguma Group, Nova Scotia. *Bull. Geol. Soc. Am.* **77**, 931–944.

Fyson, W. K. 1971. Fold attitudes in metamorphic rocks. *Am. J. Sci.* **270**, 373–382.

Gabrielsen, R. H. & Robinson, C. 1984. Tectonic inhomogeneities of the Kristiansund–Bodo Fault Complex, offshore mid-Norway. In: A. M. Spencer *et al.* (eds.), *Petroleum geology of the north European margin*, pp. 397–496. Graham Trotman Ltd. for the Norwegian Petroleum Society.

Gallup, W. B. 1951. Geology of Turner Valley oil and gas field, Alberta, Canada. *Bull. Am. Assoc. Petrol. Geol.* **35**, 797–821.

Gangi, A. F., Min, K. D. & Logan, J. M. 1977. Experimental folding of rocks under confining pressure. IV. Theoretical analysis of faulted drape folds. *Tectonophysics* **42**, 227–260.

Gansser, A. 1964. *Geology of the Himalayas*. Interscience Publishers, London, 289pp.

Gapais, D., Bale, P., Choukroune, P., Cobbold, P. R., Mahjoub, Y. & Marquer, D. 1987. Bulk kinematics from shear zone patterns: some field examples. *J. Struct. Geol.* **9**, 635–646.

Garfunkel, Z. 1974. Model for the late Cenozoic tectonic history of Mojave Desert, California, and its relation to adjacent regions. *Bull. Geol. Soc. Am.* **85**, 1931–1944.

Garofalo, F. 1965. *Fundamentals of creep and creep rupture in metals*. Macmillan, New York, 258pp.

REFERENCES

Gaudemer, Y. & Tapponnier, P. 1987. Ductile and brittle deformations in the northern Snake Range, Nevada. *J. Struct. Geol.* **9**, 159–180.

Gay, N. C. 1968. Pure shear and simple shear deformation of inhomogenous viscous fluids. I. Theory. *Tectonophysics* **5**, 211–234.

Gay, N. C. & Jaeger, J. C. 1975. Cataclastic deformation of geologic materials in matrices of differing composition. II. Boudinage. *Tectonophysics* **27**, 323–331.

Ghosh, S. K. 1959. Migmatization in relation to regional deformation in the area around Kuilapal, West Bengal. *Quart. J. Geol. Metall. Soc. India* **31**, 171–176.

Ghosh, S. K. 1963. Structural, metamorphic and migmatitic history of the area around Kuilapal, eastern India. *Quart. J. Geol. Min. Metall. Soc. India* **35**, 211–234.

Ghosh, S. K. 1966. Experimental tests of buckling folds in relation to strain ellipsoid in simple shear deformations. *Tectonophysics* **3**, 169–185.

Ghosh, S. K. 1968. Experiments on buckling of multilayers which permit interlayer gliding. *Tectonophysics* **6**, 207–249.

Ghosh, S. K. 1970. A theoretical study of intersecting fold patterns. *Tectonophysics* **9**, 559–569.

Ghosh, S. K. 1974a. Strain distribution in superposed buckling folds and the problem of reorientation of early lineation. *Tectonophysics* **21**, 249–272.

Ghosh, S. K. 1974b. Outcrop patterns of superposed buckling folds developed in multilayered test models. *Geol. Min. Met. Soc. India* **46**, 73–80. Golden Jubilee volume.

Ghosh, S. K. 1975. Distortion of planar structures around rigid spherical bodies. *Tectonophysics* **28**, 185–208.

Ghosh, S. K. 1977. Drag patterns of planar structures around rigid inclusions. In: S. K. Saxena & S. Bhattacharji (eds.), *Energetics of geological processes*, pp. 98–120. Springer-Verlag, New York.

Ghosh, S. K. 1982. The problem of shearing along axial plane foliations. *J. Struct. Geol.* **4**, 63–67.

Ghosh, S. K. 1987. Measure of non-coaxiality. *J. Struct. Geol.* **9**, 111–113.

Ghosh, S. K. 1988. Theory of chocolate tablet boudinage. *J. Struct. Geol.* **10**, 541–553.

Ghosh, S. K. & Chatterjee, A. 1985. Patterns of deformed early lineations over late folds formed by buckling and flattening. *J. Struct. Geol.* **7**, 651–666.

Ghosh, S. K. & Lahiri, S. 1983. Morphology of penecontemporaneous interpenetrative contortions and their modification by diastrophic movements in the Ghatsila-Galudih area, Singhbhum, eastern India. In: S. Sinha-Roy (ed.), *Structure and tectonics of Precambrian rocks of India*, pp. 144–157. Hindustan Publishing Corpn., Delhi.

Ghosh, S. K. & Lahiri, S. 1990. Soft sediment deformation by vertical movements. *Indian J. Earth Sci.* **17**, 23–43.

Ghosh, S. K. & Mukhopadhyay, A. 1985a. Soft-sediment recumbent folding in a slump-generated bed in Jharia basin, eastern India. *J. Geol. Soc. India* **27**, 194–201.

Ghosh, S. K. & Mukhopadhyay, A. 1985b. Tectonic history of the Jharia basin — an intracratonic basin in eastern India. *Quart. J. Geol. Min. & Metall. Soc. India* **57**, 33–58.

Ghosh, S. K. & Ramberg, H. 1968. Buckling experiments on intersecting fold patterns. *Tectonophysics* **5**, 89–105.

Ghosh, S. K. & Ramberg, H. 1976. Reorientation of inclusions by combination of pure shear and simple shear. *Tectonophysics* **34**, 1–70.

Ghosh, S. K. & Ramberg, H. 1978. Reversal of spiral direction of inclusion-trails in paratectonic porphyroblasts. *Tectonophysics* **51**, 83–97.

Ghosh, S. K. & Sengupta, S. 1973. Compression and simple shear of test models with rigid and deformable inclusions. *Tectonophysics* **17**, 133–175.

Ghosh, S. K. & Sengupta, S. 1984. Successive development of plane noncylindrical folds in progressive deformation. *J. Struct. Geol.* **6**, 703–709.

Ghosh, S. K. & Sengupta, S. 1987a. Progressive evolution of structures in a ductile shear zone. *J. Struct. Geol.* **9**, 277–288.

Ghosh, S. K. & Sengupta, S. 1987b. Structural history of the Singhbhum Shear Zone in relation to Northern belt. In: A. K. Saha (ed.), *Geological evolution of Peninsular India — Petrological and structural aspects*, pp. 31–44. Hindustan Publishing Corpn., New Delhi.

Ghosh, S. K. & Sengupta, S. 1990. Singhbhum Shear Zone: structural transition and a kinematic model. *Proc. Indian Acad. Sci. (Earth and Planetary Sciences)*, **99**, 229–247.

Ghosh, S. K., Mandal, N., Khan, D. and Deb, S. 1992. Modes of superposed buckling in single layers controlled by initial tightness of early folds. *J. Struct. Geol.* **14**, 381–394.

Ghosh, S. K., Mandal, N., Sengupta, Sudipta, Deb, S. K. & Khan, D. 1993. Superposed buckling in multilayers. *J. Struct. Geol.* **15**, 95–111.

Gibbs, A. D. 1983. Balanced cross-section constructions from seismic sections in areas of extensional tectonics. *J. Struct. Geol.* **5**, 152–160.

Gibbs, A. D. 1984. Structural evolution of extensional basin margins. *J. Geol. Soc. Lond.* **141**, 609–620.

Gill, J. E. 1971. Continued confusion in the classification of faults. *Bull. Geol. Soc. Am.* **82**, 1389–1392.

Gilmour, P. & McIntyre, D. B. 1954. The geometry of the Ben Lui fold (S.W. Highlands). *Geol. Mag.* **91**, 161–166.

Glangeaud, L. 1944. Les glissements post-tectoniques dans le Jura et leur rôle dans les interprétations structurales. *Compte Rendue* **218**, 466.

Goetze, C. 1975. Sheared lherzolites: from the point of view of rock mechanics. *Geology* **3**, 172–173.

Goguel, J. 1962. *Tectonics*. W. H. Freeman and Co., 384pp.

Goldstein, A. G. 1988. Factors affecting the kinematic interpretations of asymmetric boudinage in shear zones. *J. Struct. Geol.* **10**, 707–715.

Goscombe, B. 1991. Intense non-coaxial shear and the development of mega-scale sheath folds in the Arunta Block, Central Australia. *J. Struct. Geol.* **13**, 299–318.

Grabau, A. W. 1905. Physical character and history of some New York-formations. *Science*, n.s. **22**, 534.

Gray, D. R. 1977. Morphological classification of crenulation cleavage. *J. Geol.* **85**, 229–235.

Gray, D. R. 1979. Geometry of crenulation-folds and their relationship to crenulation cleavage. *J. Struct. Geol.* **1**, 187–205.

Gray, D. R. 1981. Cleavage-fold relationships and their implications for transected folds: an example from Southwest Virginia, U.S.A. *J. Struct. Geol.* **3**, 265–277.

Gray, D. R. & Durney, D. W. 1979a. Investigation on mechanical significance of crenulation cleavage. *Tectonophysics* **58**, 35–79.

Gray, D. R. & Durney, D. W. 1979b. Crenulation cleavage differentiation: implications of solution deposition process. *J. Struct. Geol.* **1**, 73–80.

Greenly, E. 1919. The geology of Anglesey. *Mem. Geol. Surv. Gt. Britain* **1**, 1–388.

Gregg, W. J. 1985. Microscopic deformation mechanisms associated with mica film formation in cleaved psammitic rocks. *J. Struct. Geol.* **7**, 54–56.

Greszczuk, L. B. 1974. Microbuckling of lamina-reinforced composites. *Composite Materials: Testing and Design* (*Third Conference*), ASTM STP 546, pp. 5–29. American Society for Testing and Materials.

Griffith, A. A. 1920. The phenomena of rupture and flow in solids. *Phil. Trans. Roy. Soc. Lond. Ser. A* **221**, 163–198.

Griggs, D. T. 1935. The strain ellipsoid as a theory of rupture. *Am. J. Sci.* **30**, 121–137.

Griggs, D. T. 1936. Deformation of rocks under high confining pressures. *J. Geol.* **44**, 541–577.

Griggs, D. T. 1967. Hydrolytic weakening of quartz and other silicates. *Geophys. J. Roy. Astron. Soc.* **14**, 19–31.

Griggs, D. T. 1974. A model of hydrolytic weakening in quartz. *J. Geophys. Res.* **79**, 1653–1661.

Griggs, D. T. & Blacic, J. D. 1964. The strength of quartz in ductile regime. *Trans. Am. Geophys. Union* **45**, 102–103 (abstract).

Griggs, D. T. & Blacic, J. D. 1965. Quartz–anomalous weakness of synthetic crystals. *Science* **247**, 292–295.

Griggs, D. T. & Handin, J. 1960. Observations on fracture and a hypothesis of earthquakes. In: D. T. Griggs and J. Handin (eds.), *Rock deformation (A Symposium)*. *Mem. Geol. Soc. Am.* **79**, 347–364.

Griggs, D. T., Turner, F. J., Borg, I. & Sosoka, J. 1951. Deformation of Yule marble. IV. Effects at 150°C. *Bull. Geol. Soc. Am.* **62**, 1385–1406.

Griggs, D. T., Turner, F. J., Borg, I. and Sosoka, J. 1953. Deformation of Yule marble. V. Effects at 300°C. *Bull. Geol. Soc. Am.* **64**, 1327–1342.

Griggs, D. T., Turner, F. J. & Heard, H. C. 1960. Deformation of rocks at 500°–800°C. In: D. T. Griggs & J. Handin (eds.), *Rock deformation*. *Mem. Geol. Soc. Am.* **79**, 39–104.

Grocott, J. 1981. Fracture geometry of pseudotachylite generation zones: a study of shear fractures formed during seismic events. *J. Struct. Geol.* **3**, 169–179.

REFERENCES

Groshong, R. H., Jr. 1975a. Slip cleavage caused by pressure solution in a buckle fold. *Geology* 3, 411–413.

Groshong, R. H., Jr. 1975b. Strain, fractures and pressure solution in natural single-layer folds. *Bull. Geol. Soc. Am.* **86**, 1363–1376.

Gunn, W., Clough, C. T. & Hill, J. B. 1897. The geology of Cowal including the part of Argyllshire between the Clyde and Loch Fine. *Mem. Geol. Surv. U.K.* 333pp.

Hadizadeh, J., Babaie, H. A. & Babaie, A. 1991. Development of interlaced mylonites, cataclasites and breccias: example from the Towaliga fault, south central Appalachians. *J. Struct. Geol.* **13**, 63–70.

Hadizadeh, J. & Rutter, E. H. 1983. The low temperature brittle–ductile transition in a quartzite and the occurrence of cataclastic flow in nature. *Geol. Rundsch.* **72**, 493–509.

Hafner, W. 1951. Stress distribution and faulting. *Bull. Geol. Soc. Am.* **62**, 373–398.

Hallam, A. 1963. Major epeirogenic and eustatic changes since the Cretaceous and their possible relationship to crustal structure. *Am. J. Sci.* **261**, 397–423.

Haman, P. J. 1961. *Manual of stereographic projection of geology and related sciences.* West Canadian Research Publications, Calgary, 67pp.

Hamblin, W. K. 1965. Origin of "reverse drag" on the downthrown side of normal faults. *Bull. Geol. Soc. Am.* **76**, 1145–1164.

Hambrey, M. J. & Milnes, A. G. 1975. Boudinage and glacier ice — some examples. *J. Glaciol.* **14**, 383–393.

Hancock, P. L. 1985. Brittle microtectonics: principles and practices. *J. Struct. Geol.* **7**, 437–457.

Hancock, P. L. 1986. Joint spectra. In: I. Nicol & R. W. Nesbit (eds.), *Geology in the real world — the Kingsley Dunham volume*, pp. 155–165. Inst. Mining and Metallurgy, London.

Hancock, P. L., Al-Kahadi, A., Barka, A. A. & Bevan, T. G. 1987. Aspects of analysing brittle structures. *Annales Tectonicae* **1**, 5–19.

Handin, J. 1969. On the Coulomb-Mohr failure criterion. *J. Geophys. Res.* **74**, 5343–5348.

Handin, J. & Fairbairn, H. W. 1955. Experimental deformation of Hasmark dolomite. *Bull. Geol. Soc. Am.* **66**, 1257–1274.

Handin, J. & Hager, R. V. Jr. 1957. Experimental deformation of sedimentary rocks under confining pressures: Tests at room temperatures on dry samples. *Bull. Am. Assoc. Petrol. Geol.* **41**, 1–50.

Handin, J. & Hager, R. V. Jr. 1958. Experimental deformation of sedimentary rocks under confining pressure: tests at high temperature. *Bull. Am. Assoc. Petrol. Geol.* **42**, 2892–2934.

Hanmer, S. K. 1979. The role of discrete heterogeneities and linear fabrics in the formation of crenulations. *J. Struct. Geol.* **1**, 81–91.

Hanmer, S. K. 1982. Vein arrays as kinematic indicators in kinked anisotropic materials. *J. Struct. Geol.* **4**, 151–160.

Hanmer, S. K. 1986. Asymmetrical pull-aparts and foliation fish as kinematic indicators. *J. Struct. Geol.* **8**, 111–122.

Hanna, S. S. & Fry, N. 1979. A comparison of methods of strain determination in rocks from Southwest Dyfed (Pembrokeshire) and adjacent areas. *J. Struct. Geol.* **1**, 155–162.

Hara, I., Takeda, K. & Kimura, T. 1973. Preferred lattice orientation of quartz in shear deformation. *J. Sci. Hiroshima Univ. Ser. C* **7**, 1–11.

Hardin, F. R. & Hardin, C. C. Jr. 1961. Contemporaneous normal faults of Gulf Coast and the relations to flexures. *Bull. Am. Assoc. Petrol. Geol.* **45**, 238–248.

Harding, T. P. & Lowell, J. D. 1979. Structural styles, their plate tectonic habitats and hydrocarbon traps in Petroleum provinces. *Bull. Am. Assoc. Petrol. Geol.* **63**, 1016–1058.

Harker, A. 1885. The cause of slaty cleavage. *Geol. Mag.* **2**, 15–17.

Harker, A. 1886. On slaty cleavage and allied structures. *Rep. Brit. Assoc. Adv. Sci.* 1885, 813–852.

Harker, A. 1889. On the local thickening of dykes and beds by folding. *Geol. Mag.*, n.s. **6**, 69–70.

Harland, W. B. 1971. Tectonic transpression in Caledonian Spitsbergen. *Geol. Mag.* **108**, 27–42.

Harris, J. F., Taylor, G. L. & Walper, J. L. 1960. Relation of deformational fractures in sedimentary rocks to regional and local structures. *Bull. Am. Assoc. Petrol. Geol.* **44**, 12.

Harris, L. B. & Cobbold, P. R. 1985. Development of conjugate shear bands during bulk simple shearing. *J. Struct. Geol.* **7**, 37–44.

Harte, B. & Henley, K. J. 1966. Occurrence of compositionally zoned almanditic garnets in regionally metamorphosed rocks. *Nature* **210**, 689–692.

Haughton, S. 1856. On slaty cleavage and distortion of fossils. *Phil. Mag.* **12**, 409–412.

Heard, H. C. 1960. Transition from brittle fracture to ductile flow in Solenhofen limestone as a function of temperature, confining pressure and interstitial fluid pressure. In: D. T. Griggs & J. Handin (eds.), *Rock deformation. Mem. Geol. Soc. Am.* **79**, 193–226.

Heim, A. 1878. *Untersuchungen über den Mechanismus der Gebirgsbildung.* B. Schwabe, Basel, 246pp.

Heim, A. 1919. Geologie der Schweiz. *Crh. Herm. Tuchnitz, Leipzig.* 2, pt. 1, 476pp.

Heim, Arn, & Gansser, A. 1939. Central Himalaya, Geological observations of the Swiss Expedition, 1936. *Mem. Soc. Helv. Sci. Nat.* **73**(1), 1–245.

Helm, D. G. & Siddans, A. W. B. 1971. Deformation of a lapillar tuff in the English Lake District: Discussion. *Bull. Geol. Soc. Am.* **82**, 523–531.

Henderson, J. R. 1981. Structural analyses of sheath folds in horizontal *X*-axes, northeast Canada. *J. Struct. Geol.* **3**, 203–210.

Heritsch, F. 1929. *The nappe theory in the Alps.* Methuen & Co., London, 228pp.

Higgins, N. W. 1971. Cataclastic rocks. *Prof. Pap. U.S.G.S.* **687**, 1–97.

Hilbert, H. & Cohn-Vossen, S. 1956. *Geometry and the imagination.* Chelsea Publishing Co., New York, 357pp.

Hill, M. L. 1947. Classification of faults. *Bull. Am. Assoc. Petrol. Geol.* **31**, 1664–1673.

Hills, E. S. 1959. Cauldron subsidence, granitic rocks and crustal fracturing in S.E. Australia. *Geol. Rundsch.* **47**, 543–561.

Hills, E. S. 1963. *Elements of structural geology*, 2nd edition. Chapman & Hall, London, 502pp.

Hobbs, B. E. 1968. Recrystallization of single crystals of quartz. *Tectonophysics* **6**, 353–401.

Hobbs, B. E., McLaren, A. C. & Paterson, M. S. 1972. Plasticity of single crystals of synthetic quartz. In: H. C. Heard, I. Y. Borg, N. L. Carter & C. B. Raleigh (eds.), *Flow and fracture of rocks. The Griggs volume. Geophys. Monogr. Am. Geophys. Union* **16**, 29–53.

Hobbs, B. E., Means, W. D. & Williams, P. F. 1976. *An outline of structural geology.* John Wiley & Sons, 571pp.

Hobbs, B. E., Ord, A. & Teyssier, C. 1986. Earthquakes in ductile regime? *Pure and Appl. Geophys.* **124**, 309–326.

Hobbs, D. W. 1967. The formation of tension joints in sedimentary rocks: an explanation. *Geol. Mag.* **104**, 550–556.

Hobson, D. M. 1973. The origin of kink bands near Tintagel, North Cornwall. *Geol. Mag.* **110**, 133–144.

Hodgson, R. A. 1961. Classification of structures on joint surfaces. *Am. J. Sci.* **259**, 493–502.

Hoeppener, R. 1956. Zum Problem der Bruchbildung, Schieferung und Faltung. *Geol. Rundsch.* **45**, 247–283.

Hollister, L. S. 1966. Garnet zoning: an interpretation based on the Rayleigh fractionation model. *Science* **154**, 1647–1650.

Holmquist, P. J. 1930. An interesting ladder vein structure. *Geol. För. Stockh. Förh.* **52**, 357–365.

Holmquist, P. J. 1931. On the relation of boudinage structure. *Geol. För. Stockh. Förh.* **53**, 193–208.

Honea, E. & Johnson, A. M. 1976. Development of sinusoidal and kink folds in multilayers confined by rigid boundaries. *Tectonophysics* **30**, 197–239.

Hossack, J. R. 1979. The use of balanced cross-section in calculation of orogenic contraction: a review. *J. Geol. Soc. Lond.* **136**, 705–711.

Hubbert, M. K. 1951. Mechanical basis for certain familiar geologic structures. *Bull. Geol. Soc. Am.* **62**, 355–372.

Hubbert, M. K. & Rubey, W. W. 1959. Role of fluid pressure in mechanics of overthrust faulting. I. Mechanics of fluid-filled porous solids and its application to overthrust faulting. *Bull. Geol. Soc. Am.* **70**, 115–166.

Hudleston, P. J. 1973a. Fold morphology and some geometrical implications of theories of fold development. *Tectonophysics* **16**, 1–46.

Hudleston, P. J. 1973b. An analysis of "single-layer" folds developed experimentally in viscous media. *Tectonophysics* **16**, 189–214.

Hudleston, P. J. 1973c. The analysis and interpretation of minor folds developed in the Moine rocks of Monar, Scotland. *Tectonophysics* **17**, 89–132.

Hudleston, P. J. 1976. Recumbent folding in the base of the Barnes Ice Cap, Baffin Island, Northwest Territories, Canada. *Bull. Geol. Soc. Am.* **87**, 1684–1692.

REFERENCES

Hudleston, P. J. 1977. Similar folds, recumbent folds and gravity tectonics in ice and rocks. *J. Geol.* **85**, 113–122.

Hudleston, P. J. 1983. Strain patterns in an ice cap and implications for strain variations in shear zones. *J. Struct. Geol.* **5**, 455–463.

Hudleston, P. J. 1986. Extracting information from folds in rocks. *J. Geol. Education* **34**, 237–245.

Hudleston P. J. & Holst, T. B. 1984. Strain analysis and fold shape in a limestone layer and implications for layer rheology. *Tectonophysics* **106**, 321–347.

Hudleston, P. J. & Stephansson, O. 1973. Layer shortening and fold development in the buckling of single layers. *Tectonophysics* **17**, 299–321.

Hunt, C. W. 1957. Planimetric equation. *J. Alberta Petrol. Geol.* **5**, 259–264.

Jackson, J. A. & McKenzie, D. 1983. The geometrical evolution of normal fault systems. *J. Struct. Geol.* **5**, 471–482.

Jackson, J. A., White, N. J., Garfunkel, Z. & Anderson, H. 1988. Relation between normal fault geometry, tilting and vertical motion in extensional terrains: An example from the Southern Gulf of Suez. *J. Struct. Geol.* **10**, 155–170.

Jacobson, C. E. 1983a. Structural geology of the Pelona Schist and Vincent Thrust, San Gabriel Mountains, California. *Bull. Geol. Soc. Am.* **94**, 753–767.

Jacobson, C. E. 1983b. Relationship of deformation and metamorphism of the Pelona Schist to the movement on the Vincent thrust, San Gabriel Mountains, southern California. *Am. J. Sci.* **283**, 587–604.

Jacobson, C. E. 1983c. Complex refolding history of the Pelona, Orocopia and Rand Schist, southern California. *Geology* **11**, 583–586.

Jacobson, C. E. 1984. Petrological evidence for the development of refolded folds during a single deformational event. *J. Struct. Geol.* **6**, 563–570.

Jaeger, J. C. 1964. *Elasticity, fracture and flow.* Methuen, London, 212pp.

Jaeger, J. C. & Cook, N. G. 1979. *Fundamentals of rock mechanics.* Chapman & Hall, 593pp.

Jamieson, R. A. & Vernon, R. H 1987. Timing of porphyroblast growth in the Fleur de Lys Supergroup, Newfoundland. *J. Metam. Geol.* **5**, 273–288.

Jaoul, O., Tullis, J. & Kronenburg, A. 1984. The effect of varying water content on the creep behaviour of Heavitree quartzite. *J. Geophys. Res.* **89**, 4298–4312.

Jeffery, G. B. 1922. The motion of ellipsoidal particles immersed in a viscous fluid. *Proc. Roy. Soc. Lond. Ser. A* **102**, 161–179.

Jegouzo, P. 1980. The South Armorican shear zone. *J. Struct. Geol.* **2**, 39–47.

Johnson, A. M. 1969. Development of folds within Carmel formation, Arches National monument, Utah. *Tectonophysics* **8**, 31–77.

Johnson, A. M. 1977. *Styles of folding.* Elsevier, Amsterdam, 406pp.

Johnson, A. M. & Ellen, S. D. 1974. A theory of concentric, kink, and sinusoidal folding and of monoclinic flexuring in compressible, elastic multilayers. I. Introduction. *Tectonophysics* **21**, 301–339.

Johnson, A. M. & Honea, E. 1975. A theory of concentric, kink and sinusoidal folding and of monoclinic flexuring in compressible, elastic multilayers. II, III. *Tectonophysics* **25**, 261–280; **27**, 1–38.

Johnson, M. R. W. 1956. Conjugate fold systems in the Moine Thrust Zone in the Loch Caron and Coulin Forest areas of eastern Ross. *Geol. Mag.* **93**, 345–350.

Jonas, J. J., Holt, R. H. & Coleman, C. E. 1976. Plastic stability in tension and compression. *Acta Metall.* **24**, 911–918.

Jones, A. G. 1959. Vernon map-area, British Columbia. *Mem. Geol. Surv. Brch. Can.* 296, 186pp.

Jones, O. T. 1937. On the sliding and slumping of submarine sediments in Denbighshire, north Wales, during the Ludlow period. *Quart. J. Geol. Soc. Lond.* **95**, 335–382.

Jordan, P. G. 1991. Development of asymmetric shale pull-aparts in evaporite shear zones. *J. Struct. Geol.* **13**, 399–409.

Jordan, P., Noack, T. & Widmer, T. 1990. The evaporite shear zone of Jura boundary thrust — new evidence from Wisen well (Switzerland). *Eclog. Geol. Helv.* **83**, 525–542.

Julivert, M. & Marcos, A. 1973. Superimposed folding under flexural conditions in the Cantabrian Zone (Hercynian Cordillera, northwest Spain). *Am. J. Sci.* **273**, 353–375.

Kamb, W. B. 1959. Theory of preferred orientation developed by crystallization under stress. *J. Geol.* **65**, 153–170.

Karato, S. 1988. Stress/strain distribution in deformed olivine aggregates. *EOS, Trans. Am. Geophys. Union* **69**, 474.

Karato, S. 1989. Defects and plastic deformation in olivine. In: S. Karato & M. Toriumi (eds.), *Rheology of solids and of the earth*, pp. 176–208. Oxford University Press.

Karato, S., Paterson, M. S. & Fitzgerald, J. D. 1986. Rheology of synthetic olivine aggregates: effect of grain size and of water. *J. Geophys. Res.* **91**, 8151–8176.

Karato, S., Toriumi, M. & Fujii, T. 1982. Dynamic recrystallization and high temperature rheology of olivine. In: S. Akimoto & M. Manghanani (eds.), *High pressure research in Geophysics*, pp. 171–189. Center for Academic Publications, Tokyo.

Karman, T. von. 1911. Festigkeitsversuche unter allseitigem Druck. *Zeitschr. Ver. deutsch. Ingeieure* **55**, 1749–1757.

Kashyap, B. P., Arieli, A. & Mukherjee, A. K. 1985. Microstructural aspects of superplasticity. *J. Mat. Sci.* **20**, 2661–2686.

Kekulawala, K. R. S. S., Paterson, M. S. & Boland, J. N. 1978. Hydrolytic weakening in quartz. *Tectonophysics* **46**, T1–T6.

Kekulawala, K. R. S. S., Paterson, M. S. & Boland, J. N. 1981. An experimental study of the role of water in quartz deformation. In: N. L. Carter, M. Friedman, J. H. Logan & D. W. Stearns (eds.), *Mechanical behaviour of crustal rocks. Geophys. Mongor. Am. Geophys. Union*, **24**, 49–60.

Kelly, A. 1973 *Strong solids*. Clarendon Press, Oxford.

Kennedy, W. Q. 1946. The Great Glen Fault. *Quart. J. Geol. Soc. Lond.* **102**, 41–76.

Kerrich, R. 1978. An historical review and synthesis of research on pressure solution. *Zentbl. Miner. Geol. Palaeont.* **1**, 512–550.

King, B. C. & Rast, N. 1956. Tectonic styles in the Dalradians and Moines of parts of the Central Highlands of Scotland. *Proc. Geol. Assoc.* **66**, 243–269.

Kingma, J. T. 1958. Possible origin of piercement structures, local unconformities and secondary basins in the Eastern Geosyncline, New Zealand. *N.Z. J. Geol. Geophys.* **1**, 269–274.

Kirby, S. H. & McCormick, J. W. 1979. Creep of hydrolytically weakened synthetic quartz crystals oriented to promote {2110} <0001 > slip: a brief summary of work to date. *Bull. Mineral.* **102**, 124–137.

Kirby, S. H. & Kronenberg, A. K. 1984. Hydrolytic weakening of quartz: uptake of molecular water and the role of microfracturing (abstract). *EOS, Trans. Amer. Geophys. Union* **65**, 277.

Knill, J. N. 1960. The tectonic pattern in Dalradian of the Craignish-Kimelfort District, Argyllshire. *Quart. J. Geol. Soc. Lond.* **115**, 339–364.

Knipe, R. J. 1981. The interaction of deformation and metamorphism of slate. *Tectonophysics* **78**, 249–272.

Knipe, R. J. & White, S. H. 1977. Microstructural variation of an axial plane cleavage around a fold — a H.V.E.M. study. *Tectonophysics* **39**, 355–380.

Knopf, E. B. & Ingerson, E. 1938. Structural petrology. *Mem. Geol. Soc. Am.* **6**, 1–270.

Kohlstedt, D. L., Cooper, R. F., Weathers, M. S. & Bird, J. M. 1979. Palaeostress of deformation-induced microstructures: Moine Thrust Zone and Ikertoq shear zone. Analysis of actual fault zones in bedrock, pp. 394–425. Open File Report 99 1239, U.S.G.S. Menlo Park, Calif.

Kohlstedt, D. L. & Ricoult, D. L. 1984. High temperature creep of silicate olivines. In: R. E. Tressler & R. C. Bradt (eds.), *Deformation of ceramic materials, II, Materials science research* vol. 18, pp. 251–280. Plenum Press, New York.

Kohlstedt, D. L. & Weathers, M. S. 1980. Deformation induced microstructures, palaeopiezometers and differential stresses in deeply eroded fault zones. *J. Geophys. Res.* **85**, 6269–6285.

Kranck, E. H. 1953. Interpretation of gneiss structures with special references to Baffin Island. *Proc. Geol. Assoc. Canada* **6**, 59–68.

Kretz, R. 1973. Kinetics of crystallization of garnet at two localities near Yellowknife. *Can. Min.* **12**, 1–20.

Kronenberg, A. K. & Tullis, J. 1984. Flow strengths of quartz aggregates: grain size and pressure effects due to hydrolytic weakening. *J. Geophys. Res.* **89**, 4281–4297.

Kronenberg, A. K., Kirby, S. H., Aines, R. D. & Rossman, G. R. 1986. Solubility and differential uptake of hydrogen in quartz at high water pressure: implications for hydrolytic weakening. *J. Geophys. Res.* **91**, 12723–12744.

Krumbein, W. C. 1942. Criteria for subsurface recognition of unconformities. *Am. Assoc. Petrol. Geol. Bull.* **26**, 36–62.

REFERENCES

Krumbein, W. C. & Sloss, L. L. 1963. *Stratigraphy and sedimentation*, 2nd edition. W. H. Freeman & Co., San Francisco, 660pp.

Kuenen, Ph. H. & Migliorini, C. I. 1950. Turbidity currents as a cause of graded bedding. *J. Geol.* **58**, 91–127.

Kvale, A. 1948. Petrologic and structural studies in the Bergsdalen Quadrangle, Western Norway. 2. Structural geology. *Bergens. Mus. Årb., 1946–47, Naturv. rekke* 1, 255pp.

Ladeira, F. L. & Price, N. J. 1981. Relationship between fracture spacing and bed thickness. *J. Struct. Geol.* **3**, 179–193.

Lahee, F. H. 1952. *Field geology*, 5th edition. McGraw-Hill, New York.

Lahee, F. H. 1961. *Field geology*. McGraw-Hill Book Co.

Lapworth, C. 1885. The highland controversy in British Geology: its causes and consequences. *Nature* Oct. 8, 558–559.

Larsen, P. H. 1988. Relay structure in a lower Permian basement-involved extension system, East Greenland. *J. Struct. Geol.* **10**, 3–8.

Latham, J.-P. 1985a. The influence of nonlinear material properties and resistance to bending on the development of internal structures. *J. Struct. Geol.* **7**, 225–236.

Latham, J.-P. 1985b. A numerical investigation and geological discussion of the relationship between folding, kinking and faulting. *J. Struct. Geol.* **7**, 237–249.

Laubscher, H. 1972. Some overall aspects of Jura dynamics. *Am. J. Sci.* **272**, 293–304.

Laubscher, H. 1977. Fold development in the Jura. *Tectonophysics* **37**, 337–362.

Laubscher, H. 1979. Elements of Jura kinematics and dynamics. *Eclog. Geol. Helv.* **72**, 467–483.

Laubscher, H. 1981. The 3D propagation of décollement in Jura. In: K. R. McClay & N. J. Price (eds.), *Thrust and nappe tectonics*, pp. 311–318. Special Publ. Geol. Soc. London, Blackwell, Oxford.

Law, R. D., Knipe, R. J. & Dayan, H. 1984. Strain path partitioning within thrust sheets: microstructural and petrofabric evidence from the Moine Thrust Zone, Loch Eriboll, northeastern Scotland. *J. Struct. Geol.* **6**, 477–497.

Leith, C. K. 1905. Rock cleavage. *Bull. U.S.G.S.* 239, 216pp.

Leith, C. K. 1923. *Structural geology*. Henry Holt and Co., New York, 390pp.

Lekhnitskii, S. G. 1981. *Theory of elasticity of an anisotropic body*. Mir Publishers, Moscow, 430pp. Translated from revised 1977 Russian edition.

Lensen, G. J. 1958. A method of graben and horst formation. *J. Geol.* **66**, 579–587.

Le Pichon, X., Angelier, J. & Sibuet, J. C. 1983. Subsidence and stretching. In: J. S. Watkins & C. L. Drake (eds.), *Studies in continental margin geology. Mem. Am. Assoc. Petrol. Geol.* **34**, 731–741.

Linker, M. F. & Kirby, S. H. 1981. Anisotropy in the rheology of hydrolytically weakened synthetic quartz crystals. In: N. L. Carter, M. Friedman, J. M. Logan & D. W. Stearns (eds.), *Mechanical behaviour of crustal rocks. Geophys. Mongor. Am. Geophys. Union*, **24**, 29–48.

Linker, M. F., Kirby, S. H., Ord, A. & Christie, J. M. 1984. Effects of compression direction on the plasticity and rheology of hydrolytically weakened synthetic quartz crystals at atmospheric pressure. *J. Geophys. Res.* **89**, 4241–4255.

Lisle, R. J. 1977a. Estimation of tectonic strain ratio from the mean shape of deformed elliptical markers. *Geol. Mijnb.* **56**, 140–144.

Lisle, R. J. 1977b. Clastic grain shape and orientation in relation to cleavage from the Aberystwyth grits, Wales. *Tectonophysics* **39**, 381–395.

Lisle, R. J. 1979. Strain analysis using deformed pebbles: the influence of initial pebble shape. *Tectonophysics* **60**, 263–277.

Lisle, R. J. 1985. *Geological strain analysis. A manual for the R_f/π technique*. Pergamon Press, 99pp.

Lisle, R. J. 1989. A simple construction for shear stress. *J. Struct. Geol.* **11**, 493–495.

Lister, G. S. 1977. Discussion: Crossed girdle c-axis fabrics in quartzites plastically deformed by plane strain and progressive simple shear. *Tectonophysics* **39**, 51–54.

Lister, G. S. & Hobbs, B. E. 1980. The simulation of fabric development during plastic deformation and its application to quartzite: the influence of deformation history. *J. Struct. Geol.* **2**, 355–370.

Lister, G. S., Paterson, M. S. & Hobbs, B. E. 1978. The simulation of fabric development in plastic deformation and its application to quartzite: The model. *Tectonophysics* **45**, 107–158.

Lister, G. S. & Snoke, A. W. 1984. S-C mylonites. *J. Struct. Geol.* **6**, 617–638.

Lister, G. S. & Williams, P. F. 1979. Fabric development in shear zones: Theoretical controls and observed phenomena. *J. Struct. Geol.* **1**, 283–297.

Lloyd, G. E. & Ferguson, C. C. 1981. Boudinage structure — some new interpretations based on elastic-plastic finite element simulations. *J. Struct. Geol.* **3**, 117–128.

Lloyd, G. E., Ferguson, C. C. & Reading, K. 1982. A stress transfer model for the development of extension fracture boudinage. *J. Struct. Geol.* **4**, 355–372.

Logan, J. M., Friedman, M., Higgs, N. G., Dengo, C. & Shimamoto, T. 1979. Experimental study of simulated gouge and their application to studies of natural fault zones. In: *Proc. Conf. VIII. Analysis of actual fault zones in bedrock.* U.S.G.S. open file report 79–1239, pp. 305–343.

Lohest, M. 1909. De l'origine des veines et des géodes des terrains primaires de Belgique. *Ann. Soc. Géol. Belgique* **36** B, 275–282.

Lohest, M., Stanier, X. & Fourmarier, P. 1909. Compte rendu de la session extraordinaire de la Soc. Géol. Belgique. *Ann. Soc. Géol. Belgique* **35**, 351–434.

Lowell, J. D. 1972. Spitsbergen Tertiary orogenic belt and the Spitsbergen fracture zone. *Bull. Geol. Soc. Am.* **83**, 3091–3102.

Lugeon, M. 1901. Les grandes nappes de recouvrement des Alpes du Chablais et de Suisse. *Bull. Soc. Géol. France, 4th ser.* **1**, 723–825.

Lugeon, M. 1902. Sur la coupe géologique de massif du Simplon. *C.R. Acad. Sci., Paris*, 726–727, 134pp.

Lugeon, M. & Gagnebin, E. 1941. Observations et vues nouvelles sur la géologie des Prealpes romande. *Bull. Lab. Géol. Mineral. Géophys. Musée Géol. Univ. Lausanne* 72, 90pp.

Lugeon, M. & Schneegans, D. 1940. Sur le diastrophisme alpin. *Compte Rendu* **210**, 187.

Lutaud, L. 1924. Étude tectonique et morphologique de la Provence cristalline. *Rev. Geograph.* **12**, 271pp.

Mackin, J. H. 1950. The down-structure method of viewing geologic maps. *J. Geol.* **58**, 55–72.

Mackwell, S. J., Kohlstedt, D. L. & Paterson, M. S. 1985. The role of water in the deformation of olivine single crystals. *J. Geophys. Res.* **90**, 11319–11333.

Maddock, R. H. 1983. Melt origin of fault-generated pseudotachylite. *Geology* **11**, 105–108.

Magloughlin, J. F. 1989. The nature and significance of pseudotachylite from the Nason terrain, North Cascade Mountains, Washington. *J. Struct. Geol.* **11**, 907–917.

Mainprice, D. H. & Paterson, M. S. 1984. Experimental studies on the role of water in the plasticity of quartzites. *J. Geophys. Res.* **89**, 4257–4269.

Malavieille, J. 1987. Kinematics of compressional and extensional ductile shearing deformation in a metamorphic core complex of the northeastern Basin and Range. *J. Struct. Geol.* **9**, 541–554.

Mancktelow, N. S. 1987. Quartz textures from the Simplon Fault Zone, southwest Switzerland and north Italy. *Tectonophysics*, **135**, 133–153.

Mandal, N. & Karmakar, S. 1989. Foliation boudinage in homogeneous foliated rocks. *Tectonophysics* **170**, 151–156.

Mandl, G. 1988. *Mechanics of tectonic faulting.* Elsevier, 407pp.

Mandl, G., De Jong, N. J. & Maltha, A. 1977. Shear zones in granular material: an experimental study of their structure and mechanical genesis. *Rock Mechanics* **9**, 95–144.

March, A. 1932. Mathematische Theorie der Regelung nach der Korngestalt bei Affiner Deformation. *Z. Krist.* **81**, 285–297.

Marcoux, J., Brun, J.-P., Burg, J.-P. & Ricou, L. E. 1987. Shear structures in anhydrite at the base of thrust sheets (Antalya, Southern Turkey). *J. Struct. Geol.* **9**, 555–561.

Masch, L. 1979. Deformation and fusion of two fault rocks in relation to their depth of formation: the hyalomylonite of Langtang (Himalaya) and the pseudotachylites of the Silvretta nappe (Eastern Alps). *Proc. Conf. VIII. Analysis of actual fault zones in bedrock.* National Earthquake Hazards Reduction Programme. U.S.G.S., Menlo Park, California, pp. 528–533.

Mason, S. G. & Manley, R. St. J. 1957. Particle motion in sheared suspension. *Proc. Roy. Soc. Lond. Ser. A* **238**, 117–131.

Mathews, P. E., Bond, R. A. B. & van der Berg, J. J. 1974. An algebraic method of strain analysis using elliptical markers. *Tectonophysics* **24**, 31–67.

Matthews, S. C. & Cowie, J. W. 1979. Early Cambrian transgression. *J. Geol. Soc. Lond.* **136**, 137–146.

REFERENCES

Mawer, C. K. & Williams, P. F. 1991. Progressive folding and foliation development in a sheared, coticule-bearing phyllite. *J. Struct. Geol.* **13**, 539–555.

Maxwell, J. C. 1962. Origin of slaty and fracture cleavage in the Delaware gap area, New Jersey and Pennsylvania. *Geol. Soc. Am., Buddington volume*, pp. 283–331.

McClay, K. R. 1981. What is a thrust? What is a nappe? In: K. R. McClay & N. J. Price (eds.), *Thrust and nappe tectonics*, pp. 7–9. Geol. Soc. Lond., Blackwell Scientific Publications, London.

McClintok, F. A. & Walsh, L. B. 1962. Friction on Griffith cracks under pressure. *Proc. Fourth U.S. Nat. Congress of Appl. Mech.*, pp. 1015–1021.

McCormick, J. W. 1977. Transmission electron microscopy of experimentally deformed synthetic quartzites. Unpublished Ph.D. thesis, University of California, Los Angeles.

McCrea, W. H. 1953. *Analytical geometry of three dimensions*. Oliver & Boyd, Edinburgh & London, 144pp.

McIntyre, D. B. 1950a. Note on two lineated tectonites from Strathavon, Banffshire. *Geol. Mag.* **87**, 331–336.

McIntyre, D. B. 1950b. Note on lineation, boudinage, and recumbent folds in the Struan Flags (Moine), near Dalnacardoch, Perthshire. *Geol. Mag.* **87**, 427–432.

McIntyre, D. B. 1951. The tectonics of the area between Grantown and Tomintoul (mid-Strathspey) *Quart. J. Geol. Soc. Lond.* **107**, 1–22.

McIntyre, D. B. & Weiss, L. E. 1956. Construction of block diagram to scale in orthographic projection. *Proc. Geol. Assoc.* **67**, 142–155.

McKenzie, D. P. & Brune, J. N. 1972. Melting on fault planes during large earthquakes. *Geophys. J. Roy. Astron. Soc.* **29**, 65–78.

McKerrow, W. 1979. Ordovician and Silurian changes in sea level. *J. Geol. Soc. Lond.* **136**, 137–146.

Means, W. D. 1975. Natural and experimental microstructures in deformed micaceous sandstones. *Bull. Geol. Soc. Am.* **86**, 1221–1229.

Means, W. D. 1977. Experimental contributions to the study of foliations in rocks: a review of research since 1960. *Tectonophysics* **39**, 329–355.

Means, W. D. 1981. The concept of steady-state foliation. *Tectonophysics* **78**, 179–199.

Means, W. D. 1987. A newly recognized type of slickenside striation. *J. Struct. Geol.* **9**, 585–590.

Means, W. D. 1989a. Synkinematic microscopy of transparent polycrystals. *J. Struct. Geol.* **11**, 163–174.

Means, W. D. 1989b. A construction for shear stress on a generally-oriented plane. *J. Struct. Geol.* **11**, 625–627.

Means, W. D. & Ree, J. H. 1988. Seven types of subgrain boundaries in octachloropropane. *J. Struct. Geol.* **10**, 765–770.

Means, W. D. & Williams, P. F. 1972. Crenulation cleavage and faulting in an artificial salt-mica schist. *J. Geol.* **80**, 569–591.

Means, W. D. & Williams, P. F. 1974. Compositional differentiation in an experimentally deformed salt-mica specimen. *Geology* **2**, 15–16.

Means, W. D., Williams, P. F. & Hobbs, B. E. 1984. Incremental deformation and fabric development in a KCl/mica mixture. *J. Struct. Geol.* **6**, 391–398.

Mercier, J. C. C., Anderson, D. A. & Carter, N. L. 1977. Stress in the lithosphere: inference from the steady state flow of rocks. *Pure Appl. Geophys.* **115**, 199–226.

Mertie, J. B. 1959. Classification, delineation and measurement of non-parallel folds. U.S.G.S. Prof. Paper 314E, pp. 91–124.

Miall, A. D. 1984. *Principles of sedimentary basin analysis*. Springer-Verlag, New York, 490pp.

Michel, P., Appert, G., Lavigne, J., Lefavrais, A., Bonte, A., Lienhardt, G. & Ricour, J. 1953. Le contact Jura-Bresse dans la région de Lons-le-Saunier. *Bull. Soc. Géol. France* **3**, 593–611.

Mies, J. W. 1991. Planar dispersion of folds in ductile shear zones and kinematic interpretation of fold girdles. *J. Struct. Geol.* **13**, 299–318.

Mimran, Y. 1976. Strain determination using a density distribution technique and its application to deformed Upper Cretaceous Dorset chalks. *Tectonophysics* **31**, 175–192.

Minnigh, L. D. 1979. Structural analysis of sheath folds in a meta-chert from the Western Italian Alps. *J. Struct. Geol.* **1**, 275–282.

Misch, P. 1949. Metasomatic granitization of batholithic dimensions. 1 and 3. *Am. J. Sci.* **247**, 209–245, 673–705.

Misch, P. 1969. Paracrystalline microboudinage of zoned grains and other criteria for synkinematic growth of metamorphic minerals. *Am. J. Sci.* **267**, 43–63.

Misch, P. 1970. Paracrystalline microboudinage in a metamorphic reaction sequence. *Bull. Geol. Soc. Am.* **81**, 2483–2486.

Mitra, G. 1978. Ductile deformation zones and mylonites: the mechanical processes involved in the deformation of crystalline basement rocks. *Am. J. Sci.* **278**, 1057–1084.

Mitra, G. 1984. Brittle to ductile transition zone due to large strains along the White Rock thrust, Wind River mountains, Wyoming. *J. Struct. Geol.* **6**, 51–61.

Mitra, G. & Boyer, S. E. 1986. Energy balance and deformation mechanism of duplexes. *J. Struct. Geol.* **8**, 291–304.

Mitra, G., Yonkee, W. A. & Sentry, D. J. 1984. Solution cleavage and its relation to major structures in the Idaho-Utah-Wyoming thrust belt. *Geology* **12**, 354–358.

Mitra, S. 1976. A quantitative study of deformation mechanisms and finite strain in quartzites. *Contr. Min. Pet.* **59**, 203–226.

Mitra, S. 1978. Microscopic deformation and flow laws in quartzite within the South Mountain anticline. *J. Geol.* **86**, 129–152.

Moretti, I. & Colletta, B. 1988. Fault-block tilting: The Gebel Zeit example, Gulf of Suez. *J. Struct. Geol.* **10**, 9–20.

Morgenstern, N. R. & Tchalenko, J. S. 1967. Microscopic structures in kaolin subjected to direct shear. *Geotechnique* **17**, 309–328.

Morris, A. P. 1981. Competing deformation mechanisms and slaty cleavage in deformed quartzose meta-sediments. *J. Geol. Soc. Lond.* **138**, 455–462.

Morrison-Smith, D. J., Paterson, M. S. & Hobbs, B. E. 1976. An electron microscope study of plastic deformation in single crystals of synthetic quartz. *Tectonophysics* **33**, 43–79.

Mukhopadhyay, D. 1972. Discussion: Deformation of slaty, lapillar tuff in the Lake district, England. *Bull. Geol. Soc. Am.* **83**, 547–548.

Mukhopadhyay, D. & Ghosh, K. P. 1980. Deformation of early lineations in the Aravalli rocks near Fatehpur, Udaipur district, southern Rajasthan, India. *Ind. J. Earth Sci.* **7**, 64–75.

Mukhopadhyay, D. & Sengupta, S. 1979. "Eyed folds" in the Precambrian marbles from southeastern Rajasthan, India. *Bull. Geol. Soc. Am.* **90**, 397–404.

Müller, W. H., Schmid, S. M. & Briegel, U. 1981. Deformation experiments on anhydrite rocks of different grain size: rheology and microfabric. *Tectonophysics* **78**, 527–543.

Murgatroyd, J. B. 1942. The significance of surface marks on fractured glass. *J. Soc. Glass Tech.* **26**, 155.

Murphy, F. C. 1985. Non-axial planar cleavage and Caledonian sinistral transpression in eastern Ireland. *Geol. J.* **20**, 257–279.

Murphy, F. X. 1990. The role of pressure solution and intermicrolithon-slip in the development of disjunctive cleavage domains: a study from Helvick Head in the Irish Variscides. *J. Struct. Geol.* **9**, 69–81.

Murrell, S. A. F. 1958. The strength of coal under triaxial compression. In: W. H. Walton (ed.), *Mechanical properties of non-metallic brittle materials*, pp. 123–145. Butterworths, London.

Muskhelishvili, N. I. 1953. *Some basic problems of mathematical theory of elasticity*. Noordhoff, Groningen, 704pp.

Nadai, A. 1963. *Theory of flow and fracture of solids*, Vol. 2. McGraw-Hill, New York, 705pp.

Naha, K. 1956. Structural set-up and movement plan in parts of Dhalbhum, Bihar. *Sci. Culture* **22**, 43–45.

Naha, K. 1959. Steeply plunging recumbent folds. *Geol. Mag.* **96**, 137–140.

Naha, K. 1965. Metamorphism in relation to stratigraphy, structure and movements in parts of East Singhbhum, eastern India. *Quart. J. Geol. Mining Metall. Soc. India* **37**, 41–88.

Naha, K. 1983. Structural-stratigraphic relations of the pre-Delhi rocks of south-central Rajasthan: a summary. In: S. Sinha Roy (ed.), *Structure and tectonics of Precambrian rocks of India*, pp. 50–52. Hindusthan Publishing Corpn. (India), Delhi.

Naha K., Srinivasan, R. & Naqvi, S. M. 1986. Structural unity in the Early Precambrian Dharwar tectonic province, peninsular India. *J. Geol. Mining Met. Soc. India* **58**, 219–243.

Naha, K. & Chatterjee, A. K. 1982. Axial plane folding in the Bababudan hill ranges of Karnataka. *Indian J. Earth Sci.* **9**, 37–43.

REFERENCES

Naha, K. & Halyburton, R. V. 1977. Structural pattern and strain history of superposed fold system in the Precambrian of central Rajasthan, India, I and II. *Precambrian Res.* **4**, 39–84, 85–111.

Naha, K. & Mohanty, S. 1988. Response of basement and cover rocks to multiple deformation: a study from the Precambrian of Rajasthan, western India. *Precambrian Res.* **42**, 77–96.

Naha, K., Chaudhuri, A. K. & Bhattacharya, A. C. 1966. Superposed folding in the older Precambrian rocks around Sangat, Central Rajasthan. *Neues. Jahrb. Geol. Palaeont. Abh.* **126**, 205–231.

Naha, K., Chaudhuri, A. K. & Mukherjee, P. 1973. The hammer-head syncline between Sangat and Kelwa in the Udaipur district. Rajasthan, a structural synthesis. *J. Geol. Soc. India* **14**, 394–407.

Naha, K., Mukhopadhayay, D. K. & Mohanty, R. 1988. Structural evolution of the rocks of the Delhi Group around Khetri, northeastern Rajasthan. *Geol. Soc. Ind. Mem.* **7**, 207–245.

Naha, K., Mukhopadhyay, D. K., Mohanty, R., Mitra, S. K. & Biswal, T. K. 1984. Significance of contrast in the early stages of the structural history of the Delhi and the pre Delhi rock groups in the Proterozoic of Rajasthan, western India. *Tectonophysics* **105**, 193–206.

Naha, K., Srinivasan, R. & Jayaram, S. 1991. Sedimentational, structural and migmatitic history of the Archaean Dharwar tectonic province, southern India. *Proc. Indian Acad. Sci.* (*Earth Planet. Sci.*) **100**, 413–433.

Naha, K., Venkitasubramanyan, C. S. & Singh, R. P. 1969. Upright folding of varying intensity on isoclinal folds of diverse orientation: a study from early Precambrian of western India. *Geol. Rundsch.* **58**, 929–950.

Naylor, M. A., Mandl, G. & Sijpesteijn, C. H. K. 1986. Fault geometries in basement-induced wrench faulting under different initial stress states. *J. Struct. Geol.* **8**, 737–752.

Nettleton, L. L. 1934. Fluid mechanics of salt domes. *Bull. Am. Assoc. Petrol. Geol.* **18**, 1175–1204.

Nevin, C. M. 1931. *Principles of structural geology.* John Wiley & Sons Inc., New York, 303pp.

Nicholson, R. 1963. Eyed folds and interference patterns in the Sokumfjell Marble Group, Northern Norway. *Geol. Mag.* **100**, 59–68.

Nickelsen, R. P. 1972. Attributes of cleavage in some mudstones and limestones of the Valley and Ridge province, Pennsylvania. *Proc Acad. Sci.* **46**, 107–112.

Noakes, L. C. 1957. *The significance of high angle reverse faults* (abstract). Australia and New Zealand. Association for the Advancement of Science.

Nur, A. & Ron, H. 1987. Block rotations, fault domains and crustal deformation. *Annales Tectonicae* **1**, 40–47.

Nur, A., Ron, H. & Scotti, O. 1986. Fault mechanisms and kinematics of block rotations. *Geology* **14**, 746–749.

Ocamb, R. D. 1961. Growth faults of south Louisiana. *Trans. Gulf-Cst. Assoc. Geol. Socs.* **11**, 139–175.

Odé, H. 1960. Faulting as velocity discontinuity, In: D. T. Griggs & J. Handin (eds.), *Rock deformation. Mem. Geol. Soc. Am.* **79**, 293–321.

Odonne, F. & Vialon, P. 1987. Hinge migration as a mechanism of superposed folding. *J. Struct. Geol.* **9**, 835–844.

Oertel, G. 1970. Deformation of a slaty, lappilar tuff in the Lake District, England. *Bull. Geol. Soc. Am.* **81**, 1172–1187.

Oertel, G. 1983. The relationship of strain and preferred orientation of phyllosilicate grains in rocks — a review. *Tectonophysics* **100**, 413–447.

Oertel, G. 1985. Phyllosilicate textures in slates. In: Hans-Rudolf Wenk (ed.), *Preferred orientation in deformed metals and rocks: An introduction to modern texture analysis,* pp. 431–440. Academic Press.

Olesen, N. O. 1978. Distinguishing between inter-kinematic and synkinematic porphyroblastesis. *Geol. Rdsch.* **67**, 278–287.

Ord, A. & Christie, J. M. 1984. Flow stress from microstructures in mylonitic quartzites of the Moine Thrust zone, Assynt area, Scotland. *J. Struct. Geol.* **6**, 639–654.

Ord, A. & Hobbs, B. E. 1986. Experimental control of the water-weakening effect in quartz. In: B. E. Hobbs & H. C. Heard (eds.), *Mineral and rock deformation: Laboratory studies. Geophys. Monogr. Am. Geophys. Union* **36**, 51–72.

Owens, W. H. 1973. Strain modification of angular density distribution. *Tectonophysics* **16**, 249–261.

Pabst, A. 1931. "Pressure shadows" and the measurements of orientation of minerals in rocks. *Am. Miner.* **16**, 55–61.

Park, R. G. 1961. The pseudotachylite of the Gairloch District, Ross-shire, Scotland. *Am. J. Sci.* **259**, 542–550.

Parker, J. M. 1942. Regional systematic jointing in slightly deformed sedimentary rocks. *Bull. Geol. Soc. Am.* **53**, 381–408.

Parrish, D. K. 1973. A nonlinear finite-element fold model. *Am. J. Sci.* **273**, 318–334.

Passchier, C. W. 1982. Pseudotachylite and the development of ultramylonite bands in the St. Barthelemy Massif, French Pyrenees. *J. Struct. Geol.* **4**, 69–79.

Passchier, C. W. 1983. The reliability of asymmetric c-axis fabrics of quartz to determine sense of vorticity. *Tectonophysics* **99**, T9-T18.

Passchier, C. W. 1987. Stable positions of rigid objects in non-coaxial flow — a study in vorticity analysis. *J. Struct. Geol.* **9**, 679–690.

Passchier, C. W. & Simpson, C. 1986. Porphyroclast systems as kinematic indicators. *J. Struct. Geol.* **8**, 831–843.

Paterson, M. S. 1958. Experimental deformation and faulting in Wombeyan marble. *Bull. Geol. Soc. Am.* **69**, 465–476.

Paterson, M. S. 1978. *Experimental rock deformation. The brittle field.* Springer-Verlag, New York, 254pp.

Paterson, M. S. 1987. Problems in the extrapolation of laboratory rheological data. *Tectonophysics* **133**, 33–43.

Paterson, M. S. 1989. The interaction of water with quartz and its influence in dislocation flow — an overview. In: S. Karato & M. Toriumi (eds.), *Rheology of solids and of the earth*, pp. 107–142. Oxford University Press.

Paterson, M. S. & Kekulawala, K. R. S. S. 1979. The role of water in quartz deformation. *Bull. Mineral.* **102**, 92–98.

Paterson, M. S. & Weiss, L. E. 1961. Symmetry concepts in the structural analysis of deformed rocks. *Bull. Geol. Soc. Am.* **72**, 841–882.

Paterson, M. S. & Weiss, L. E. 1966. Experimental deformation and folding in phyllite. *Bull. Geol. Soc. Am.* **77**, 343–374.

Paterson, M. S. & Weiss, L. E. 1968. Folding and boudinage of quartz-rich layers in experimentally deformed phyllite. *Bull. Geol. Soc. Am.* **79**, 795–812.

Peach, B. N., Horne, J., Gunn, W., Clough, C. T., Hinxman, L. W. & Teall, J. J. H. 1907. *The geological structure of the North-West Highlands of Scotland. Mem. Geol. Surv. U.K.* 668pp.

Peach, B. N., Horne, J., Hinxman, L. W., Crampton, C. B., Anderson, E. M. & Carruthers, R. G. 1913. *The geology of Central Ross-shire* (explanation of sheet 82). *Mem. Geol. Surv. U.K.* 1–114.

Peirce, B. O. & Foster, R. M. 1963. *A short table of integrals.* Ginn & Co. New York, 189pp.

Pettijohn, F. J. 1975. *Sedimentary rocks*, 3rd edition. Harper & Row, 628pp.

Phillips, F. C. 1971. *The use of stereographic projection in structural geology.* Edward Arnold, London, 90pp.

Phillips, J. 1844. On certain movements in the parts of stratified rocks. *Rep. Brit. Assoc. Advan. Sci.* 1843, 60–61.

Phillips, W. W. A., Flegg, A. M. & Anderson, T. B. 1980. Strain adjacent to Iapetus Suture in Ireland. In: A. L. Harris, C. H. Holland & B. E. Leake (eds.), *The Caledonides of The British Isles reviewed. Spec. Publ. Geol. Soc. Lond.* **8**, 257–262.

Platt, J. P. 1979. Extensional crenulation cleavage. *J. Struct. Geol.* **1**, 95–96.

Platt, J. P. 1983. Progressive refolding in ductile shear zones. *J. Struct. Geol.* **5**, 619–622.

Platt, J. P. 1984. Secondary cleavages in ductile shear zones. *J. Struct. Geol.* **6**, 439–442.

Platt, J. P. & Behrmann, J. H. 1986. Structures and fabrics in a crustal scale shear zone. Betic Cordillera. S.E. Spain. *J. Struct. Geol.* **8**, 15–33.

Platt, J. P. & Vissers, R. L. M. 1980. Extensional structures in anisotropic rocks. *J. Struct. Geol.* **2**, 397–410.

Plessman, W. von. 1964. Gesteinlösung, ein Hauptfaktor beim Schieferungsprozess. *Geol. Mitt.* **4**, 69–82.

Poirier, J. P. 1980. Shear localization and shear instability in materials in the ductile field. *J. Struct. Geol.* **2**, 135–142.

Poirier, J. P. 1985. *Creep of crystals.* Cambridge University Press, Cambridge, 260pp.

REFERENCES

Poirier, J. P. & Nicolas, A. 1975. Deformation-induced recrystallization by progressive misorientation of subgrain-boundaries, with special reference to mantle peridotites. *J. Geol.* **83**, 707–720.

Poirier, J. P., Bouchez, J. L. & Jonas, J. J. 1979. A dynamic model for aseismic ductile shear zones. *Earth Planet Sci. Lett.* **43**, 441–453.

Ponce de Leon, M. I. & Choukroune, P. 1980. Shear zones in Iberian arc. *J. Struct. Geol.* **2**, 63–68.

Post, R. L. 1977. High temperature creep of Mt. Burnet dunite. *Tectonophysics* **46**, T1–T6.

Postlewith, C. E. & Jacobson, C. E. 1987. Early history and reactivation of the Rand thrust, southern California. *J. Struct. Geol.* **9**, 195–205.

Poumellec, B. & Jaoul, O. 1984. Influence of pO_2 and pH_2O on the high temperature plasticity of olivine. In: R. E. Tressler & R. C. Bradt (eds.), *Deformation of ceramic materials*, II. *Materials Science Research*, Vol. 18, pp. 281–305. Plenum Press, New York.

Powell, C. McA. 1969. Intrusive sandstone dykes in the Siamo slate near Negaunee, Michigan. *Bull. Geol. Soc. Am.* **80**, 2585–2594.

Powell, C. McA. 1974. Timing of slaty cleavage during folding of Precambrian rocks, northwest Tasmania. *Bull. Geol. Soc. Am.* **85**, 1043–1060.

Powell, C. McA. 1979. A morphological classification of rock cleavage. *Tectonophysics* **58**, 21–34.

Powell, C. McA. & Rickard, M. J. 1985. Significance of early foliation at Bermagui, N.S.W., Australia. *J. Struct. Geol.* **7**, 385–400.

Powell, C. McA., Cole, J. P. & Cudahy, T. J. 1985. Megakinking in the Lachlan fold belt, Australia. *J. Struct. Geol.* **7**, 281–300.

Price, N. J. 1959. Mechanics of jointing in rocks. *Geol. Mag.* **96**, 149–167.

Price, N. J. 1966. *Fault and joint development in brittle and semi-brittle rock*. Pergamon Press.

Price, N. J. & Cosgrove, J. 1990. *Analysis of geological structures*. Cambridge University Press, 502pp.

Price, R. A. 1981. The Cordilleran foreland thrust and fold belt in the southern Canadian Rocky Mountains. In: K. R. McClay & N. J. Price (eds.), *Thrust and nappe tectonics*, pp. 427–448. *Geol. Soc. Lond. Spec. Pub.* Blackwells, Oxford.

Pumpelly, R., Wolff, J. E. & Dale, T. N. 1894. Geology of the Green Mountains. III. Mount Greylock: its areal and structural geology. *U.S.G.S. Mon.* 22.

Quinquis, H., Audren, C., Brun, J. P. & Cobbold, P. R. 1978. Intense progressive shear in Ile de Groix blueschists and compatibilty with subduction or obduction. *Nature* **273**, 43–45.

Quirke, T. T. 1923. Boudinage, an unusual structure phenomenon. *Bull. Geol. Soc. Am.* **34**, 649.

Ramberg, H. 1952. *The origin of metamorphic and metasomatic rocks*. Univ. Chicago Press, 317pp.

Ramberg, H. 1955. Natural and experimental boudinage and pinch-and-swell structures. *J. Geol.* **63**, 512–526.

Ramberg, H. 1959. Evolution of ptygmatic folding. *Norsk. Geol. Tiddskr.* **39**, 99–151.

Ramberg, H. 1961. Relationship between concentric longitudinal strain and concentric shear strain during folding of homogeneous sheets of rocks. *Am. J. Sci.* **259**, 382–390.

Ramberg, H. 1963a. Fluid dynamics of viscous buckling applicable to folding of layered rocks. *Bull. Am. Assoc. Petrol. Geol.* **47**, 484–505.

Ramberg, H. 1963b. Strain distribution and geometry of folds. *Bull. Geol. Inst. Univ. Uppsala* **42**, 1–20.

Ramberg, H. 1963c. Evolution of drag folds. *Geol. Mag.* **100**, 97–106.

Ramberg, H. 1963d. Experimental study of gravity tectonics by means of centrifuged models. *Bull. Geol. Inst. Univ. Uppsala* **43**, 1–72.

Ramberg, H. 1964. Selective buckling of composite layers with contrasted rheological properties, a theory for the formation of several orders of folds. *Tectonophysics* **1**, 307–341.

Ramberg, H. 1966. The Scandinavian Caledonides as studied by centrifuged dynamic models. *Bull. Geol. Inst. Univ. Uppsala* **43**, 1–72.

Ramberg, H. 1967. *Gravity, deformation and the earth's crust as studied by centrifuged models*. Academic Press, London, 241pp.

Ramberg, H. 1968. Instability of layered systems in the field of gravity. I, II. *Phys. Earth Planet. Interiors* **1**, 427–447.

Ramberg, H. 1970. Folding of laterally compressed multilayers in the field of gravity. I, II. *Phys. Earth Planet. Interiors* **2**, 203–232; **4**, 83–120.

Ramberg, H. 1975. Particle paths, displacements and progressive strain applicable to rocks. *Tectonophysics* **28**, 1–37.

Ramberg, H. & Ghosh, S. K. 1968. Deformation structures in the Hovin Group schists in the Hommelvik-Hell region (Norway). *Tectonophysics* **6**, 311–330.

Ramberg, H. & Ghosh, S. K. 1977. Rotation and strain of linear and planar structures in three-dimensional progressive deformation. *Tectonophysics* **40**, 309–337.

Ramberg, H. & Stephansson, O. 1964. Compression of floating elastic and viscous plates affected by gravity: a basis for discussing crustal buckling. *Tectonophysics* **1**, 101–120.

Ramberg, H. & Stromgård, K. 1971. Experimental tests of modern buckling theory applied on multilayered media. *Tectonophysics* **11**, 261–272.

Ramsay, A. C. 1881. The geology of North Wales. *Mem. Geol. Surv. U.K.* 611pp.

Ramsay, J. G. 1958a. Superposed folding in Loch Monar, Inverness-shire and Ross-shire. *Quart. J. Geol. Soc. Lond.* **113**, 271–308.

Ramsay, J. G. 1958b. Moine-Lewisian relation at Glenelg, Inverness-shire. *Quart. J. Geol. Soc. Lond.* **113**, 487–523.

Ramsay, J. G. 1960. The deformation of early linear structures in areas of repeated folding. *J. Geol.* **68**, 75–93.

Ramsay, J. G. 1962a. The geometry and mechanics of formation of "similar" type folds. *J. Geol.* **70**, 309–327.

Ramsay, J. G. 1962b. Interference patterns produced by the superposition of folds of similar types. *J. Geol.* **70**, 466–481.

Ramsay, J. G. 1962c. The geometry of conjugate fold systems. *Geol. Mag.* **99**, 516–526.

Ramsay, J. G. 1965. Structural investigations in the Barberton Mountainland, eastern Transvaal. *Trans. Geol. Soc. South Africa* **66**, 353–401.

Ramsay, J. G. 1967. *Folding and fracturing of rocks.* McGraw-Hill, New York, 568pp.

Ramsay, J. G. 1974. Development of chevron folds. *Bull. Geol. Soc. Am.* **85**, 1741–1754.

Ramsay, J. G. 1980a. Shear zone geometry: a review. *J. Struct. Geol.* **2**, 83–99.

Ramsay, J. G. 1980b. The crack-seal mechanism of rock deformation. *Nature* **284**, 135–139.

Ramsay, J. G. 1981. Tectonics of the Helvetic Alps. In: K. R. McClay & N. J. Price (eds.), *Thrust and nappe tectonics*, pp. 293–309. Geological Society of London, Blackwell, Oxford.

Ramsay, J. G. & Graham, R. H. 1970. Strain variations in shear belts. *Can. J. Earth Sci.* **7**, 786–813.

Ramsay, J. G. & Huber, M. I. 1983. *The techniques of modern structural geology. 1. Strain analysis.* Academic Press, London, 307pp.

Ramsay, J. G. & Huber, M. I. 1987. *The techniques of modern structural geology. 2. Folds and fractures.* Academic Press, London, 400pp.

Ramsay, J. G. & Wood, D. S. 1973. The geometric effects of volume change during deformation processes. *Tectonophysics* **13**, 263–277.

Rast, N. 1956. The origin and significance of boudinage. *Geol. Mag.* **93**, 401–408.

Rast, N. 1958. Tectonics of the Schichallion complex (Perthshire). *Trans. Roy. Soc. Edinburgh* **63**, 413–431.

Rast, N. 1964. Morphology and interpretation of folds — a critical essay. *J. Geol.* **4**, 177–188.

Read, H. H. 1949. A contemplation of time in plutonism. *Quart. J. Geol. Soc. Lond.* **105**, 101–156.

Read, H. H. 1957. *Granite controversy.* Thomas Murby & Co., London, 430pp.

Ree, J.-H. 1991. An experimental steady-state foliation. *J. Struct. Geol.* **13**, 1001–1011.

Reiner, M. 1959. The flow of matter. *Scientific American* **201**, 122–138.

Reynolds, D. & Holmes, A. 1954. The superposition of Caledonoid folds on an earlier fold-system in the Dalradians of Malin Head, Co. Donegal. *Geol. Mag.* **91**, 417–444.

Rhodes, S. & Gayer, R. A. 1977. Non-cylindrical folds, linear structures in the X direction and mylonite development during translation of the Caledonian Kalak Nappe Complex of Finnmark. *Geol. Mag.* **114**, 329–341.

Ricard, M. J. 1961. A note on cleavages in crenulated rocks. *Geol. Mag.* **98**, 324–332.

Rice, A. H. N. 1984. Metamorphic and structural diachroneity in the Finnmarkian nappes of north Norway. *J. Metamorphic Geol.* **2**, 219–236.

Rich, J. L. 1934. Mechanics of low-angle overthrust faulting illustrated by Cumberland Thrust Block, Virginia, Kentucky and Tennessee. *Bull. Am. Assoc. Petrol. Geol.* **18**, 1584–1596.

Riedel, W. 1929. Zur Mechanik geologischer Brucherscheinungen. Ein Beitrag zum Problem der "Fiederspatten". *Centralblatt fur Min. Geol. Paleont.*, Pt. B, 354–368.

Rixon, L. K., Bucknell, W. R. & Rickard, M. J. 1983. Megakink folds and related structures in the Upper Devonian Merrimbula Group, South Coast. N.S.W. *J. Geol. Soc. Aust.* **30**, 277–293.

REFERENCES

Roberts, D. 1971. Abnormal cleavage patterns in fold hinge zones from Varanger Peninsula, Northern Norway. *Am. J. Sci.* **271**, 170–180.

Roberts, J. 1961. Feather fracture and the mechanics of rock jointing. *Am. J. Sci.* **259**, 481–492.

Robertson, E. C. 1955. Experimental study of strength of rocks. *Bull. Geol. Soc. Am.* **66**, 1275–1314.

Robin, P. Y. F. 1977. Determination of geologic strain using randomly oriented strain markers of any shape. *Tectonophysics* **42**, T7–T16.

Ron, H., Aydin, A. & Nur, A. 1986. Strike-slip faulting and block rotation in the Lake Mead fault system. *Geology* **14**, 1020–1023.

Ron, H., Freund, R., Garfunkel, Z. & Nur, A. 1984. Block rotation by strike-slip faulting, structural and paleomagnetic evidence. *J. Geophys. Res.* **89**, B7, 6256–6270.

Rosendahl, B. R. & Livingstone, D. A. 1983. Rift lakes of East Africa. New seismic data and implications for future research. *Episodes* **83**, 14–19.

Rosenfeld, J. L. 1970. Rotated garnets in metamorphic rocks. *Geol. Soc. Amer. Spec. Pap.* **129**, 1–105.

Ross, J. V., Ave Lallement, H. G. & Carter, N. L. 1980. Stress dependence of recrystallized grain and subgrain size in olivine. *Tectonophysics* **70**, 39–61.

Roy, A. B. 1973. Nature and evolution of subhorizontal crenulation cleavage in the type Aravalli rocks around Udaipur, Rajasthan. *Proc. Indian Nat. Sci. Acad.* **39A**, 119–131.

Roy, A. B. 1978. Evolution of slaty cleavage in relation to diagenesis and metamorphism: a study from the Hunsruckshiefer. *Bull. Geol. Soc. Am.* **89**, 1775–1789.

Royse, F., Warner, M. A. & Reese, D. L. 1975. Thrust belt structural geometry and related stratigraphic problems, Wyoming-Idaho-northern Utah. In: D. W. Boylard (ed.), *Deep drilling frontiers of the Central Rocky Mountains*, pp. 41–54. Symposium. Rocky Mountain Association of Geologists.

Rutland, R. W. R. 1959. Structural geology of the Glomfjord area, north Norway. *Norsk. Geol. Tiddskr.* **39**, 287–338.

Rutter, E. H. 1974. The influence of temperature, strain rate and interstitial water in the experimental deformation of calcite rocks. *Tectonophysics* **22**, 311–334.

Rutter, E. H. 1976. The kinetics of rock deformation by pressure solution. *Phil. Trans. Roy. Soc. Lond. Ser. A* **283**, 203–219.

Rutter, E. H. 1986. On the nomenclature of mode of failure transitions in rocks. *Tectonophysics* **122**, 381–387.

Sander, B. 1911. Über Zusammenhange Zuischen Teilbewegung und Gefüge in Geisteinen. *Tscherm. Min. Pet. Mitt.* **30**, 381–384.

Sander, B. 1930. *Gefugekunde der Gesteine.* Springer, Berlin, Vienna, 352pp.

Sander, B. 1948. *Einfuhrung in die Gefügekunde der Geologischen Körper*, Pt. 1. Springer, Berlin, Vienna, 215pp.

Sander, B. 1970. *An introduction to the study of fabrics in geological bodies.* Translated by F. C. Phillips and G. Windsor. Pergamon Press, 641pp.

Sanderson, D. J. 1973. The development of fold axes oblique to the regional trend. *Tectonophysics* **16**, 55–70.

Sanderson, D. J. 1974. Patterns of boudinage and apparent stretching lineation developed in folded rocks. *J. Geol.* **82**, 651–661.

Sanderson, D. J. 1976. The superposition of compaction and plane strain. *Tectonophysics* **30**, 35–54.

Sanderson, D. J. & Marchini, W. R. D. 1984. Transpression. *J. Struct. Geol.* **6**, 449–458.

Sanderson, D. J. & Meneilly, A. W. 1981. Analysis of three-dimensional strain modified distributions: andalusite fabrics from granite aureole. *J. Struct. Geol.* **3**, 109–116.

Sanderson, D. J., Andrews, J. R., Phillips, W. E. A. & Hutton, D. H. W. 1980. Deformation studies in the Irish Caledonides. *J. Geol. Soc. Lond.* **137**, 289–302.

Sanford, A. R. 1959. Analytical and experimental studies of simple geologic structures. *Bull. Geol. Soc. Am.* **70**, 19–52.

Schardt, H. 1893. Sur l'origine des Prealpes romandes. *Eclogae Geol. Helv.* **4**, 14.

Schmid, S. M. & Casey, M. 1986. Complete fabric analysis of some commonly observed quartz *c*-axis patterns. In: B. E. Hobbs and H. C. Heard (eds.), *Mineral and rock deformation: Laboratory Studies — The Paterson volume. Am. Geophys. Union Monogr.* **36**, 263–286.

Schmid, S. M., Boland, J. N. & Paterson, M. S. 1977. Superplastic flow in fine grained limestone. *Tectonophysics* **43**, 257–291.

Schmid, S. M., Paterson, M. S. & Boland, J. N. 1980. High temperature flow and dynamic recrystallization in the Carrara marble. *Tectonophysics* **65**, 245–280.

Schmid, S. M., Panozzo, R. & Bauer, S. 1985. Simple shear experiments on calcite rocks: rheology and microfabric. *J. Struct. Geol.* **9**, 747–778.

Schoneveld, C. 1977. A study of some typical inclusion patterns in strongly paracrystalline-rotated garnets. *Tectonophysics* **39**, 453–471.

Schoneveld, C. 1978. Syntectonic growth of garnets: discussion of a new model proposed by M. J. de Wit. *Geol. J.* **13**, Pt. 1, 37–46.

Schryver, K. 1965. On the measurement of the orientation of axial planes of minor folds. *J. Geol.* **74**, 83–84.

Schweitzer, J. & Simpson, C. 1986. Cleavage development in dolomite of the Elbrook Formation, southwest Virginia. *Bull. Geol. Soc. Am.* **97**, 778–786.

Scotford, D. M. 1955. Metamorphism and axial-plane folding in the Poundridge area, New York. *Bull. Geol. Soc. Am.* **65**, 1155–1198.

Secor, D. T. 1969. Mechanics of natural extension fracturing at depths in the earth's crust. *Geol. Surv. Pap. Can.* **68–52**, 3–48.

Sederholm, J. J. 1913. Über ptygmatische Faltungen. *Neues Jb. Min. Geol. Paläont Beilbd.* **36**, 491–512.

Sederholm, J. J. 1926. On migmatites and associated Precambrian rocks in south-western Finland, Pt. II. *Bull. Comm. Géol. Finlande* **77**, 1–143.

Sedgwick, A. 1835. Remarks on structure of large mineral masses, and specially on the chemical changes produced on aggregation of stratified rocks during different periods after their deformation. *Trans. Geol. Soc. Lond. Ser. 2*, **3**, 461–486.

Seidensticker, C. M., Oldow, J. S. & Ave Lallemant, H. G. 1982. Development of bedding-normal boudins in Pilot Mountains, Nevada. *Tectonophysics* **90**, 335–349.

Sellars, C. M. 1978. Recrystallization of metals during hot deformation. *Phil. Trans. Roy. Soc. Lond. Ser. A* **288**, 147–158.

Sengupta, S. 1983. Folding of boudinaged layers. *J. Struct. Geol.* **5**, 197–210.

Sengupta, S. 1985. Boudinage in a milieu of superposed deformation and migmatization. *Indian J. Earth Sci.* **12**, 159–164.

Sengupta, S. 1988. History of successive deformations in relation to metamorphism — migmatitic events in the Schirmacher Hills, Queen Maud Land, East Antarctica. *J. Geol. Soc. Ind.* **32**, 295–319.

Sengupta, S. 1991. Structural and petrological evolution of basement rocks in the Schirmacher Hills, Queen Maud Land, East Antarctica. In: M. R. A. Thomson, J. A. Crame & J. W. Thomson (eds.), *Geological evolution of Antarctica*, pp. 67–72. Cambridge University Press.

Shand, S. J. 1916. The pseudotachylite of Parijs (Orange Free State) and its relation to "trap-shotten gneiss" and "flinty crush rock". *J. Geol. Soc. Lond.* **72**, 198–217.

Shanmugam, G., Moiola, R. J. & Sales, J. K. 1988. Duplex-like structures in submarine fan channels, Ouachita Mountains, Arkansas. *Geology* **16**, 229–232.

Sharpe, D. 1847. On slaty cleavage. *Quart. J. Geol. Soc. Lond.* **3**, 74–105.

Sherwin, J.-A. & Chapple, W. M. 1968. Wavelengths of single layer folds: A comparison between theory and observation. *Am. J. Sci.* **266**, 167–179.

Shimamoto, T. 1989. The origin of S-C mylonites and a new fault zone model. *J. Struct. Geol.* **11**, 51–64.

Shimamoto, T. & Ikeda, Y. 1976. A simple algebraic method for strain estimation from deformed ellipsoidal objects. I. Basic theory. *Tectonophysics* **36**, 315–337.

Shrock, R. R. 1948. *Sequence in layered rocks*. McGraw-Hill Book Co., 507pp.

Sibson, R. H. 1975. Generation of pseudotachylite by ancient seismic faulting. *Geophys. J. Roy. Astr. Soc.* **43**, 774–794.

Sibson, R. H. 1977. Fault rocks and fault mechanisms. *J. Geol. Soc. Lond.* **133**, 191–233.

Sibson, R. H. 1980. Transient discontinuities in ductile shear zones. *J. Struct. Geol.* **2**, 165–171.

Siddans, A. W. B. 1972. Slaty cleavage — a review of research since 1815. *Earth-Sci. Rev.* **8**, 205–232.

Simpson, C. 1983. Strain and shape fabric variations associated with ductile shear zones. *J. Struct. Geol.* **5**, 61–72.

REFERENCES

Simpson, C. & Schmid, S. M. 1983. An evaluation of criteria to deduce the sense of movement in sheared rocks. *Bull. Geol. Soc. Am.* **94**, 1281–1288.

Skjernaa, L. 1975. Experiments on superimposed buckle folding. *Tectonophysics* **27**, 255–270.

Sloss, L. L. 1963. Sequences in cratonic interior of North America. *Bull. Geol. Soc. Am.* **74**, 93–113.

Sloss, L. L., Krumbein, W. C. & Dapples, E. C. 1949. Integrated facies analysis. In: C. R. Longwell (ed.), *Sedimentary facies in geologic history. Geol. Soc. Amer. Mem.* **39**, 91–124.

Smith, R. B. 1975. Unified theory on the onset of folding, boudinage and mullion structure. *Bull. Geol. Soc. Am.* **86**, 1601–1609.

Smith, R. B. 1977. Formation of folds, boudinage and mullions in non-Newtonian materials. *Bull. Geol. Soc. Am.* **88**, 312–320.

Smith, R. B. 1979. The folding of a strongly non-Newtonian layer. *Am. J. Sci.* **279**, 272–287.

Smoluchowski, M. 1909. Über ein gewisses Stabilitätsproblem der Elastizitätslehre und dessen und Beziehung zur Enstehung von Faltengebirgen. *Abhandl. Akad. Wiss. Krakau, Math. Kl.* 1909, 3–20.

Smoluchowski, M. 1910. Verusche über Faltungserscheinungen schwimmender elasticher Platten. *Abhandl. Akad. Wiss. Krakau, Math. Kl.* 1910, 727–734.

Soper, N. J. & Moseley, F. 1978. Structure. In: F. Moseley (ed.), *The geology of Lake Districts. Yorkshire Geol. Soc. Occasional Publication* **3**, 46–67.

Sorby, H. C. 1853. On the origin of slaty cleavage. *New Phil. J. Edinburgh* **55**, 137–148.

Sorby, H. C. 1855. On slaty cleavage as exhibited in the Devonian limestones of Devonshire. *Phil. Mag. Ser. 4*, **11**, 20.

Sorby, H. C. 1856. On the theory of the origin of slaty cleavage. *Phil. Mag. Ser. 4*, **12**, 127–129.

Sorby, H. C. 1858. On some facts connected with slaty cleavage. *Report of the 27th Meeting of the British Association*, pp. 92–93.

Sorby, H. C. 1863. On the direct correlation of mechanical and chemical forces. *Proc. Roy. Soc. Lond.* **12**, 538–550.

Sorby, H. C. 1879a. The anniversary address of the President. *Quart. J. Geol. Soc. Lond.* **35**, 39–95.

Sorby, H. C. 1879b. The structure and origin of limestones. *J. Geol. Soc. Lond. Proc.* **35**, 56–93.

Sorby, H. C. 1908. On the application of quantitative methods in the study of rocks. *Quart. J. Geol. Soc. Lond.* **61**, 171–233.

Sowers, G. M. 1973. Theory of spacing of extension fracture. *Eng. Geol. Case Hist.* **9**, 27–53.

Spear, F. S. & Selverstone, J. 1983. Quantitative P–T paths from zoned minerals: theory and tectonic applications. *Contrib. Mineral. Petrol.* **83**, 348–357.

Spear, F. S., Selverstone, J., Hickmott, D., Crowley, P. & Hodges, K. V. 1984. P–T paths from garnet zoning: a new technique for deciphering tectonic processes in crystalline terraines. *Geology* **12**, 87–90.

Spry, A. 1963. The chronological analysis of crystallization and deformation of some Tasmanian Precambrian rocks. *J. Geol. Soc. Australia* **10**, 193–208.

Spry, A. 1969. *Metamorphic textures.* Pergamon Press, Oxford, 350pp.

Stabler, C. L. 1968. A simplified Fourier analysis of fold shapes. *Tectonophysics* **6**, 343–350.

Stauffer, M. R. 1973. New methods for mapping fold axial surface. *Bull. Geol. Soc. Am.* **84**, 2307–2318.

Stauffer, M. R. & Gendzill, D. J. 1987. Fractures in the northern plains, stream patterns and midcontinent stress field. *Can. J. Earth Sci.* **24**, 1086–1097.

Stearns, D. W. 1969. Fracture as a mechanism of flow in naturally deformed layered rocks. In: A. J. Baer & D. K. Norris (eds.), *Kink bands and brittle deformation. Proc. Research in Tectonics, Geol. Surv. Canada, Paper* 68–52, pp. 79–90.

Stearns, D. W., Couples, G. & Stearns, M. T. 1978. Deformation of non-layered materials that affect structures in layered rock. *Guidebook Ann. Conf. Wyo. Geol. Assoc.* **30**, 213–225.

Stearns, D. W., Couples, G. D., Jamison, W. R. & Morse, J. D. 1981. Understanding faulting in the shallow crust: contributions of selected experimental and theoretical studies. In: N. L. Carter, M. Friedman, J. M. Logan, & D. W. Stearns (eds.), *Mechanical behaviour of crustal rocks. The Handin volume. Geophys. Mon. Am. Geophys. Union* **24**, 215–229.

Stoces, B. & White, C. H. 1935. *Structural geology with special reference to economic deposits.* Macmillan and Co. Ltd., London, 460pp.

Stockwell, C. H. 1950. The use of plunge in the construction of cross-sections of folds. *Proc. Geol. Assoc. Canada* **3**, 97–121.

Stowe, C. W. 1988. Application of Fourier analysis for computer representation of fold profiles. *Tectonophysics* **156**, 303–311.

Stowell, E. Z. & Liu, T. S. 1961. On the mechanical behaviour of fibre-reinforced crystalline materials. *J. Mech. Phys. Solids* **9**, 242–260.

Stringer, P. 1975. Acadian slaty cleavage noncoplanar with fold axial surfaces in the northern Appalachians *Can. J. Earth Sci.* **12**, 949–961.

Stringer, P. & Lajtai, E. Z. 1979. Cleavage in Triassic rocks of southern New Brunswick, Canada. *Can. J. Earth Sci.* **16**, 2165–2180.

Stringer, P. & Treagus, J. E. 1980. Non-axial planar S_1 cleavage in the Hawick rocks of the Galloway area, Southern Uplands, Scotland. *J. Struct. Geol.* **2**, 317–331.

Stromgård, K. E. 1973. Stress distribution during formation of boudinage and pressure shadows. *Tectonophysics* **16**, 215–248.

Suess, E. 1906. *The face of the earth.* Clarendon Press, Oxford.

Suppe, J. 1980. A retrodeformable cross section of northern Taiwan. *Proc. Geol. Soc. China* **23**, 46–55.

Suppe, J. 1983. Geometry and kinematics of fault bend folding. *Am. J. Sci.* **283**, 684–721.

Suppe, J. & Namson J. 1979. Fault bend origin of frontal folds of the Western Taiwan fold-and-thrust belts. *Petrol. Geol. Taiwan* **16**, 1–18.

Sutton, J. 1960. Some cross folds and related structures in Northern Scotland. *Geol. Mijnb.* **39**, 149–162.

Sutton, J. & Watson, J. 1955. The structure and stratigraphic succession of the Moines of Fannich Forest and Strath Bran, Ross-shire. *Quart. J. Geol. Soc. Lond.* **110**, 21–53.

Sutton, J. & Watson, J. V. 1951. The Pre-Torridonian metamorphic history of the Loch Torridon and Scourie areas of north-west Highlands and its bearing on the chronological classification of the Lewisian. *J. Geol. Soc. Lond.* **106**, 241–296.

Sutton, J. & Watson, J. 1956. The Boindie syncline of the Dalradian of the Banffshire coast. *Quart. J. Geol. Soc. Lond.* **112**, 103–130.

Sutton, J. & Watson, J. 1959. Structures in the Caledonides between Loch Duich and Glenelg, North-West Highlands. *Quart. J. Geol. Soc. Lond.* **114**, 231–257.

Swanson, C. O. 1941. Flow cleavage in folded beds. *Bull. Geol. Soc. Am.* **52**, 1245–1263.

Swanson, M. T. 1988. Pseudotachylite-bearing strike-slip duplex structures in the Fort Foster Brittle Zone, S. Maine. *J. Struct. Geol.* **10**, 813–928.

Sylvester, A. G. & Christie, J. M. 1968. The origin of crossed-girdle orientations of optic axes in deformed quartzites. *J. Geol.* **76**, 571–581.

Sylvester, A. G. & Smith, R. R. 1976. Tectonic transpression and basement controlled deformation in the San Andreas fault zone, Salton trough, California. *Bull. Am. Assoc. Petrol. Geol.* **60**, 2081–2102.

Syme-Gash, P. J. 1971. A study of surface features relating to brittle and semi-brittle fractures. *Tectonophysics* **12**, 349–391.

Takeuchi, S. & Argon, A. S. 1976. Review: Steady state creep of single-phase crystalline matter at high temperatures. *J. Mater. Sci.* **11**, 1542–1566.

Talbot, C. J. 1970. The minimum strain ellipsoid using quartz veins. *Tectonophysics* **9**, 47–76.

Tan, B. K. 1973. Determination of strain ellipses from deformed ammonoids. *Tectonophysics* **16**, 89–101.

Taylor, G. I. 1923. The motion of ellipsoidal particles in a viscous fluid. *Proc. Roy. Soc. Lond. Ser. A* **103**, 58–61.

Tchalenko, J. S. 1968. The evolution of kink bands and the development of compression textures in sheared clays. *Tectonophysics* **6**, 159–174.

Tchalenko, J. S. 1970. Similarities between shear zones of different magnitudes. *Bull. Geol. Soc. Am.* **81**, 1625–1640.

Teall, J. J. H. 1885. The metamorphosis of dolerite into hornblende schist. *Quart J. Geol. Soc. Lond.* **41**, 133–145.

Teall, J. J. H. 1918. Dynamic metamorphism, a review, mainly personal. *Proc. Geol. Assoc.* **29**, 1–15.

Tewari, V. K. 1978. *Mechanics of fiber composites.* Wiley Eastern Ltd., New Delhi.

Thakur, V. C. 1981. An overview of thrusts and nappes of Western Himalaya. In: K. R. McClay & N. J. Price (eds.), *Thrust and nappe tectonics*, pp. 381–392. Special Publication Geol. Soc. Lond. No. 9, Blackwells, Oxford.

REFERENCES

Thiessen, R. 1986. Two-dimensional refold interference patterns. *J. Struct. Geol.* **8**, 563–573.

Thiessen, R. & Means, W. D. 1980. Classification of fold interference patterns: a reexamination. *J. Struct. Geol.* **2**, 311–316.

Thompson, G. A. 1960. Problem of Late Cenozoic structures of the Basin Ranges. *Proc. 21st Int. Geol. Congr., Copenhagen* **18**, 62–68.

Thompson, R. I. 1981. The nature and significance of large "blind" thrusts within the northern Rocky Mountains of Canada. In: K. R. McClay & N. J. Price (eds.), *Thrust and nappe tectonics*, pp. 449–462. Special Publication Geol. Soc. Lond. No. 9, Blackwells, Oxford.

Timoshenko, S. 1936. *Theory of elastic stability.* McGraw-Hill, New York, 518pp.

Timoshenko, S. P. & Gere, J. M. 1961. *Theory of elastic stability.* McGraw-Hill, New York, 541pp.

Tobisch, O. T. 1966. Large scale basin and dome pattern resulting from the interference of major folds. *Bull. Geol. Soc. Am.* **77**, 415–419.

Treagus, S. H. 1973. Buckling stability of a viscous single-layer system oblique to the principal compression. *Tectonophysics* **19**, 271–289.

Treagus, S. H. 1981. A theory of stress and strain variations in viscous layers and its geological implications. *Tectonophysics* **72**, 75–103.

Treagus, S. H. 1983. A theory of finite strain variation through contrasting layers, and its bearing on cleavage refraction. *J. Struct. Geol.* **5**, 351–368.

Treagus, J. E. & Treagus, S. H. 1981. Folds and strain ellipsoid. *J. Struct. Geol.* **3**, 1–171.

Troëng, B. 1975. One natural and some experimental pinch and swell structures. *Geol. För. Stock. Förh.* **97**, 383–386.

Truesdell, C. 1954. *The kinematics of vorticity.* Indiana Univ. Pub. Science Series No. 19, Bloomington.

Trumpy, R. 1973. The timing of orogenic events in the Central Alps. In: K. A. DeJong & R. Schloten (eds.), *Gravity and tectonics*, pp. 229–251. Wilby, New York.

Tullis, T. E. 1971. Experimental development of preferred orientation of mica during recrystallization. Ph.D. thesis, University of California, Los Angeles, 262pp.

Tullis, T. E. 1976. Experiments on slaty cleavage and schistosity. *Bull. Geol. Soc. Am.* **87**, 745–753.

Tullis, T. E. & Wood, D. S. 1975. Correlation of finite strain from both reduction bodies and preferred orientation of mica in slate from Wales. *Bull. Geol. Soc. Am.* **86**, 632–638.

Tullis, J. & Yund, R. A. 1977. Experimental deformation of dry Westerly granite. *J. Geophys. Res.* **82**, 5705–5718.

Tullis, J. & Yund, R. A. 1987. Transition from cataclastic flow to dislocation creep of feldspar: mechanisms and microstructures. *Geology* **15**, 606–609.

Tullis, J. & Yund, R. A. 1991. Diffusion creep in feldspar aggregates: experimental evidence. *J. Struct. Geol.* **13**, 987–1000.

Tullis, J., Christie, J. M. & Griggs, D. T. 1973. Microstructures and preferred orientations in experimentally deformed quartzites. *Bull. Geol. Soc. Am.* **84**, 297–314.

Turner, F. J. 1948. Mineralogical and structural evolution of metamorphic rocks. *Mem. Geol. Soc. Am.* **30**, 1–342.

Turner, F. J. & Weiss, L. E. 1963. *Structural analysis of metamorphic tectonites.* McGraw-Hill Book Co., 545pp.

Turner, F. J., Griggs, D. T., Clark, R. H. & Dixon, R. H. 1956. Deformation of Yule marble. VII. *Bull. Geol. Soc. Am.* **67**, 1259–1294.

Turner, J. P. & Hancock, P. L. 1990. Relationships between thrusting and joint systems in the Jaca thrust-top basin, Spanish Pyrenees. *J. Struct. Geol.* **12**, 217–226.

Tvergaard, V., Needleman, A. & Lo, K. K. 1981. Flow localization in the plane strain tensile test. *J. Mech. Phys. Solids* **29**, 115–142.

Twiss, R. J. 1977. Theory and applicability of a recrystallized grain size palaeopiezometer. *Pure Appl. Geophys.* **115**, 227–244.

Tyndall, J. 1856. Comparative review of the cleavage of crystals and slate rocks. *Phil. Mag. Ser. 4*, 12.

Uemura, T. 1965. Tectonic analysis of the boudin structure in the Muro Group, Kii peninsula, southwest Japan. *J. Earth Sci.* **13**, 99–114.

Urai, J. L. & Humphreys, F. J. 1981. The development of shear folds in polycrystalline camphor. *Tectonophysics* **78**, 677–685.

STRUCTURAL GEOLOGY

Vail, P. R., Mitchum, R. M. Jr. & Thompson, S. III. 1977. Seismic stratigraphy and global changes in sea level. Part 4: Global cycles of relative changes of sea level. *Mem. Am. Assoc. Petrol. Geol.* **26**, 83–98.

Van Den Driessche, J. & Brun. J.–P. 1987. Rolling structures at large shear strains. *J. Struct. Geol.* **9**, 691–704.

Van Den Eckhout, B. & Konert, G. 1983. Plagioclase porphyroblast growth and its relation to deformation in the Alhamilla unit (Sierra Alhamilla, Betic Cordilleras, SE Spain). *J. Metamorphic Geol.* **1**, 227–240.

Van Hise, C. R. 1896. Principles of North American Precambrian geology. *Rep. U.S.G.S.* **16**, 571–843.

Van Hise, C. R. & Leith, C. K. 1911. The geology of the Lake Superior region, *U.S. Geol. Surv. Monogr.* **52**, 1–641.

Vernon, R. H. 1987. Fibrous sillimanite related to heterogeneous deformation in K-feldspar-sillimanite metapelites. *J. Metamorphic Geology* **5**, 51–68.

Verrall, P. 1982. Structural interpretation with application to North Sea problems. *JAPEC Course Notes No. 3.*

Vissers, R. L. M. 1989. Asymmetric quartz c-axis fabrics and flow vorticity: a study using rotated garnets. *J. Struct. Geol.* **8**, 15–33.

Voll, G. 1960. New work on petrofabrics. *Liverpool Manchester Geol. J.* **2**, 503–567.

Wallace, H. & Clifford, P. M. 1983. Kink folds at upper Manitoba Lake northeastern Ontario. *Can. J. Earth Sci.* **20**, 1305–1313.

Waters, A. C. & Campbell, C. D. 1935. Mylonites from the San Andreas fault zone. *Am. J. Sci.* **29**, 473–501.

Watkinson, A. J. 1981. Patterns of fold interference: influence of early fold shapes. *J. Struct. Geol.* **3**, 19–23.

Weathers, M. S., Cooper, R. F., Kohlstedt, D. L. & Bird, J. M. 1979. Differential stress determined from deformation-induced micro-structures of the Moine Thrust Zone. *J. Geophys. Res.* **84**, 7495–7509.

Weber, K. J. 1971. Sedimentological aspects of oilfields of Niger delta. *Geol. Mijnb.* **50**, 559–576.

Wegmann, C. E. 1932. Note sur le boudinage. *Bull. Soc. Géol. France Ser. 5*, **ii**, 477–489.

Wegmann, C. E. 1965. Tectonic patterns at different levels. *Geol. Soc. South Africa* 66 (1963), Annex (A. L. du Toit Mem. Lect. No. 8).

Wegmann, C. E. & Shaer, J.-P. 1962. Chronologie et deformations des filons basiques dans les formations precambriennes du sud de 1a Norvege. *Saertrykk av Norsk. Geol. Tiddskrift* **42**, 371–387.

Weiss, L. E. 1955. Fabric analysis of a triclinic tectonite. *Am. J. Sci.* **253**, 225–236.

Weiss, L. E. 1959a. Structural analysis of the Basement System at Turoka, Kenya. *Overseas Geol. Miner. Resour.* **7**, 3–35, 123–153.

Weiss, L. E. 1959b. Geometry of superposed folding. *Bull. Geol. Soc. Am.* **70**, 91–106.

Weiss, L. E. 1968. Flexural-slip folding of foliated model materials In: A. J. Baer & D. K. Norris (eds.), *Kink bands and brittle deformation. Geol. Surv. Canada Paper* 68–52.

Weiss, L. E. & McIntyre, D. B. 1957. Structural geometry of Dalradian rocks at Loch Leven, Scottish Highlands. *J. Geol.* **65**, 575–602.

Weiss, L. E., McIntyre, D. B. & Kursten, M. 1955. Contrasted styles of folding in the rocks of Ord Ban, Mid-Strathspey. *Geol. Mag.* **92**, 21–36.

Wellman, H. W. 1955. New Zealand quaternary tectonics. *Geol. Rundsch.* **43**, 248–257.

Wernicke, B. 1985. Uniform normal-sense simple shear of continental lithosphere. *Can. J. Earth Sci.* **22**, 108–125.

Wernicke, B. & Burchfield, B. C. 1982. Modes of extensional tectonics. *J. Struct. Geol.* **4**, 105–115.

Wettstein, A. 1886. Über die Fisch fauna des Tertiären Glarner Schiefers. *Schweiz. Paläont. Ges. Abh.* **13**, 1–101.

Wheeler, B. 1987. Variable heave models of deformation above listric normal faults: importance of area conservation. *J. Struct. Geol.* **9**, 1047–1049.

White, N. J., Jackson, J. A. & McKenzie, D. P. 1986. The relationship between geometry of normal faults and that of sedimentary layers in their hanging walls. *J. Struct. Geol.* **8**, 897–910.

White, S. 1973a. Syntectonic recrystallization and texture development in quartz. *Nature* **244**, 276–278.

REFERENCES

White, S. 1973b. The dislocation structures responsible for the optical effects in some naturally deformed quartzites. *J. Mat. Sci.* **8**, 490–499.

White, S. 1975a. Effect of polyphase deformation on the defect structures in quartz. *Neues Jb. Miner. Abh.* **123**, 237–252.

White, S. 1975b. Tectonic deformation and recrystallization of oligoclase. *Contr. Min. Pet.* **50**, 287–304.

White, S. 1976. The effects of strain on microstructures, fabrics and deformation mechanisms in quartzites. *Phil. Trans. Roy. Soc. Lond. Ser. A* **283**, 69–86.

White, S. 1977. Geological significance of recovery and recrystallization process in quartz. *Tectonophysics* **39**, 143–170.

White, S. H. 1979a. Difficulties associated with palaeo-stress estimates. *Bull. Mineral.* **102**, 210–215.

White, S. H. 1979b. Grain and sub-grain size variation across a mylonitic zone. *Contr. Min. Pet.* **70**, 193–202.

White, S. H., Bretan, P. G. & Rutter, E. H. 1986. Fault-zone reactivation: kinematics and mechanisms. *Phil. Trans. Roy. Soc. Lond. Ser.A*, **317**, 81–97.

White, S. H., Burrows, S. E., Carreras, J., Shaw, N. D. & Humphreys, F. J. 1980. On mylonites in ductile shear zones. *J. Struct. Geol.* **2**, 175–187.

White, S. H. & Knipe, R. J. 1978. Transformation-and-reaction-enhanced ductility in rocks. *J. Geol. Soc. Lond.* **135**, 513–516.

Whitten, E. H. T. 1966. *Structural geology of folded rocks.* Rand McNally & Co., Chicago, 680pp.

Wickham, J. S. 1972. Structural history of a portion of the Blue Ridge, northern Virginia. *Bull. Geol. Soc. Am.* **83**, 723–760.

Wickham, J. S. 1973. An estimate of strain increments in a naturally deformed carbonate rock. *Am. J. Sci.* **273**, 23–47.

Wilcox, R. E., Harding, T. P. & Steely, D. R. 1973. Basic wrench tectonics. *Bull. Am. Assoc. Petrol. Geol.* **57**, 74–96.

Williams, G. & Vann, I. 1987. The geometry of listric normal faults and deformation in their hanging walls. *J. Struct. Geol.* **9**, 789–795.

Williams, G. D. & Chapman, T. J. 1979. The geometrical classification of non-cylindrical folds. *J. Struct. Geol.* **1**, 181–185.

Williams, G. D. & Chapman, T. J. 1983. Strains developed in the hanging wall of thrusts due to their slip/propagation rate: a dislocation model. *J. Struct. Geol.* **5**, 563–571.

Williams, H. R. 1987. Stick-slip model for kink band formation in shear zones and faults. *Tectonophysics* **140**, 327–331.

Williams, P. F. 1972. Development of metamorphic layering and cleavage in low grade metamorphic rocks at Bermagui, Australia. *Am. J. Sci.* **272**, 1–47.

Williams, P. F. 1976. Relationships between axial-plane foliations and strain. *Tectonophysics* **30**, 181–196.

Williams, P. F. 1977. Foliation: a review and discussion. *Tectonophysics* **39**, 305–328.

Williams, P. F. 1985. Multiply deformed terrains — problems of correlation. *J. Struct. Geol.* **7**, 269–280.

Williams, P. F. & Zwart, H. J. 1978. A model for the development of the Seve-Köli Caledonian Nappe complex. In: S. K. Saxena and S. Bhattacharji (eds.), *Energetics of geological processes*, pp. 169–187. Springer-Verlag.

Willis, B. 1893. Mechanics of Appalachian structure. *U.S. Geological Survey, 13th Annual Report*, Part 2, pp. 211–281.

Willis, B. & Willis, R. 1929. *Geologic structures*, 2nd edition. Mc-Graw-Hill Book Co., New York, 518pp.

Wilson, G. 1946. The relationship of slaty cleavage and kindered structures to tectonics. *Proc. Geol. Assoc.* **57**, 263–300.

Wilson, G. 1951. The tectonics of Tintagel area, North Cornwall. *Quart. J. Geol. Soc. Lond.* **106**, 393–432.

Wilson, G. 1952. Ptygmatic structures and their formation. *Geol. Mag.* **89**, 1–21.

Wilson, G. 1953. Mullion and rodding structures in the Moine Series of Scotland. *Proc. Geol. Assoc.* **64**, 118–151.

Wilson, G. 1961. Tectonic significance of small-scale structures and their importance to geologists in the field. *Ann. Soc. Géol. de Belgique.* **84**, 423–548. Revised as *Introduction to small-scale geological structures*. George Allen & Unwin, 128 pp.

Wilson, J. T. 1965. A new class of faults and their bearing on continental drift. *Nature* **207**, 343–347.

Wiltschko, D. V. & Dorr, J. A. Jr. 1983. Timing of deformation in overthrust belt and foreland in Idaho, Wyoming and Utah. *Bull. Am. Assoc. Petrol. Geol.* **67**, 1304–1322.

Wise, D. V., Dunn, D. E., Engelder, J. T., Geiser, P. A., Hatcher, R. D., Kish, S. A., Odom, S. A. & Schamel, S. 1984. Fault-related rocks: suggestions for terminology. *Geology* **12**, 391–394.

Withjack, H. 1979. An analytical model of continent rift patterns. *Tectonophysics* **59**, 59–81.

Wood, D. S. 1974. Current views of the development of slaty cleavage. Annual review. *Earth and Planet. Sci.* **2**, 369–401.

Wood, D. S. & Oertel, G. 1980. Deformation in the Cambrian slate belt of Wales. *J. Geol.* **88**, 309–326.

Wood, D. S., Oertel, G., Singh, J. & Bennett, H. R. 1976. Strain and anisotropy in rocks. *Phil. Trans. Roy. Soc.* **283**, 27–42.

Woodcock, N. H. & Fischer, M. 1986. Strike-slip duplexes. *J. Struct. Geol.* **8**, 725–735.

Wunderlich, H. G. 1959. Enzeugung engständiger Scherflächen in plastichem Material. *Neues. Jahrb. Geol. Pal. Abh.* **1**, 34–44.

Zeck, H. P. 1974. Cataclastites, hemiclastites, blasto-ditto and myloblastites–cataclastic rocks. *Am. J. Sci.* **274**, 1064–1073.

Zwart, H. J. 1960a. The chronological succession of folding and metamorphism in the central Pyrenees. *Geol. Rundsch.* **50**, 203–218.

Zwart, H. J. 1960b. Relations between folding and metamorphism in the Central Pyrenees. *Geol. En. Mijnb.* **22**, 219–236.

Author Index

Subject Index